Introduction to the Thermodynamics of Materials

Fifth Edition

Introduction to the Thermodynamics of Materials

Fifth Edition

David R. Gaskell
School of Materials Engineering
Purdue University
West Lafayette, IN

Taylor & Francis
Taylor & Francis Group

New York • London

Vice President	Denise Schanck
Editor	Summers Scholl
Editorial Assistant	Alex Engels
Senior Publisher UK	Jackie Harbor
Production Editor	Georgina Lucas
Cover Designer	Andrew Magee
Typesetter	Sunrise Setting Ltd
Printer	Sheridan Books Inc.

10-digit ISBN   1-5916-9043-9
13-digit ISBN   978-1-5916-9043-6

Library of Congress Cataloging-in-Publication Data

Gaskell, David R., 1940-
  Introduction to the thermodynamics of materials / David R. Gaskell. – 5th ed.
      p. cm.
  Includes index.
  ISBN 978-1-59169-043-6
  1. Metallurgy. 2. Thermodynamics. 3. Materials–Thermal properties. I. Title.
  TN673.G33 2008
  620.1'1296–dc22   2008001951

Published in 2008 by Taylor & Francis Group, LLC,
270 Madison Avenue, New York, NY 10016, USA and
2 Park Square, Milton Park, Abingdon, Oxon, OX14 4RN, UK

Printed in the United States of America on acid-free paper

10 9 8 7 6 5 4 3 2 1

For Hugh, Mac and Roydon

Contents

Preface

The current edition of this text has benefited significantly from the critical comments and suggestions made by Professors William Fahrenholtz and Mark Schlesinger of the University of Missouri–Rolla and Professor Stanley Howard of the South Dakota School of Mines and Technology. Professor Fahrenholtz provided several worked examples, which have been included in the text. I am grateful for these contributions.

I again express my extreme gratitude to Dr. Arthur Morris, Professor Emeritus of the Department of Metallurgical Engineering at the University of Missouri–Rolla for providing a CD-ROM to supplement the text. The examples on this disk have been completely revised and expanded and cover a wider range of subjects than those included in the fourth edition. This CD provides a major improvement in the usefulness of this text as a learning tool.

David R. Gaskell
Purdue University

A word about the CD

The CD contains a thermodynamic database, descriptive material, special programs, and 25 worked-out examples showing how to use Microsoft® Excel* for making thermodynamic calculations. The example solutions are completely revised from the 4th Edition to make better use of Excel's calculational tools. New examples were added to cover statistical treatment of thermodynamic data. Each example is keyed to a chapter or section of the text. The CD contents are described by *CD Introduction.pdf*, which you should print and read before trying to use the Excel programs.

The CD also contains three folders: Examples, Database, and Solver. Examples contains an Excel workbook and a descriptive document. Database contains the thermodynamic database program FREED, which is the source of most of the thermodynamic data used in the examples. The User's Guide for FREED should be downloaded from the Thermart web page. Solver contains two programs for enhancing Excel's calculational ability, along with User's Guides for both programs. The database and solver programs are new to the 5th Edition.

Dr. Arthur E. Morris
Thermart Software
www.thermart.net

---

*Microsoft and Excel Windows are either registered trademarks or trademarks of Microsoft Corporation, Inc.

Introduction to the
Thermodynamics of Materials

Fifth Edition

Chapter 1

INTRODUCTION AND DEFINITION OF TERMS

1.1 INTRODUCTION

Thermodynamics is concerned with the behavior of matter, where matter is anything that occupies space, and the matter which is the subject of a thermodynamic analysis is called a *system*. In materials science and engineering the systems to which thermodynamic principles are applied are usually chemical reaction systems. The central aim of applied thermodynamics is the determination of the effect of environment on the state of rest (equilibrium state), of a given system, where environment is generally determined as the pressure exerted on the system and the temperature of the system. The aim of applied thermodynamics is thus the establishment of the relationships which exist between the equilibrium state of existence of a given system and the influences which are brought to bear on the system.

1.2 THE CONCEPT OF STATE

The most important concept in thermodynamics is that of *state*. If it were possible to know the masses, velocities, positions, and all modes of motion of all of the constituent particles in a system, this mass of knowledge would serve to describe the *microscopic state* of the system, which, in turn, would determine all of the properties of the system. In the absence of such detailed knowledge as is required to determine the microscopic state of the system, thermodynamics begins with a consideration of the properties of the system which, when determined, define the *macroscopic state* of the system; i.e., when all of the properties are fixed then the macroscopic state of the system is fixed. It might seem that, in order to uniquely fix the macroscopic, or thermodynamic, state of a system, an enormous amount of information might be required; i.e., all of the properties of the system might have to be known. In fact, it is found that when the values of a small number of properties are fixed then the values of all of the rest are fixed. Indeed, when a simple system such as a given quantity of a substance of fixed composition is being considered, the fixing of the values of two of the properties fixes the values of all of the rest. Thus only two properties are independent, which, consequently, are called the independent variables, and all of the other properties are dependent variables. The thermodynamic state of the simple system is thus uniquely fixed when the values of the two independent variables are fixed.

In the case of the simple system any two properties could be chosen as the independent variables, and the choice is a matter of convenience. Properties most amenable to control are the pressure P and the temperature T of the system. When P and T are fixed, the state of the simple system is fixed, and all of the other properties have unique values corresponding to this state. Consider the volume V of a fixed quantity of a pure gas as a property, the value of which is dependent on the values of P and T. The relationship between the dependent variable V and the independent variables P and T can be expressed as

$$V = V(P,T) \qquad (1.1)$$

The mathematical relationship of V to P and T for a system is called an *equation of state* for that system, and in a three-dimensional diagram, the coordinates of which are volume, temperature, and pressure, the points in P-V-T space which represent the equilibrium states of existence of the system lie on a surface. This is shown in Fig. 1.1 for a fixed quantity of a simple gas. Fixing the values of any two of the three variables fixes the value of the third variable. Consider a process which moves the gas from state 1 to state 2. This process causes the volume of the gas to change by

$$\Delta V = V_2 - V_1$$

This process could proceed along an infinite number of paths on the P-V-T surface, two of which, $1 \rightarrow a \rightarrow 2$ and $1 \rightarrow b \rightarrow 2$, are shown in Figure 1.1. Consider the path $1 \rightarrow a \rightarrow 2$. The change in volume is

$$\Delta V = V_2 - V_1$$
$$= (V_a - V_1) + (V_2 - V_a)$$

where $1 \rightarrow a$ occurs at the constant pressure P_1 and $a \rightarrow 2$ occurs at the constant temperature T_2:

$$(V_a - V_1) = \int_{T_1}^{T_2} \left(\frac{\partial V}{\partial T} \right)_{P_1} dT$$

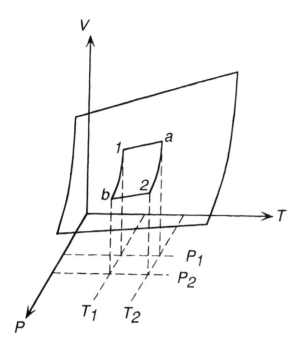

Figure 1.1 The equilibrium states of existence of a fixed quantity of gas in P-V-T space.

and

$$(V_2 - V_a) = \int_{P_1}^{P_2} \left(\frac{\partial V}{\partial P}\right)_{T_2} dP$$

Thus

$$\Delta V = \int_{T_1}^{T_2} \left(\frac{\partial V}{\partial T}\right)_{P_1} dT + \int_{P_1}^{P_2} \left(\frac{\partial V}{\partial P}\right)_{T_2} dP \tag{1.2}$$

Similarly for the path $1 \rightarrow b \rightarrow 2$,

$$(V_b - V_1) = \int_{P_1}^{P_2} \left(\frac{\partial V}{\partial P}\right)_{T_1} dP$$

and

$$(V_2 - V_b) = \int_{T_1}^{T_2} \left(\frac{\partial V}{\partial T}\right)_{P_2} dP$$

and, hence, again

$$\Delta = \int_{P_1}^{P_2} \left(\frac{\partial V}{\partial P}\right)_{T_1} dP + \int_{T_1}^{T_2} \left(\frac{\partial V}{\partial T}\right)_{P_2} dT \tag{1.3}$$

Eqs. (1.2) and (1.3) are identical and are the physical representations of what is obtained when the complete differential of Eq. (1.1), i.e.,

$$dV = \left(\frac{\partial V}{\partial P}\right)_T dP + \left(\frac{\partial V}{\partial T}\right)_P dT \tag{1.4}$$

is integrated between the limits P_2, T_2 and P_1, T_1.

The change in volume caused by moving the state of the gas from state 1 to state 2 depends only on the volume at state 1 and the volume at state 2 and is independent of the path taken by the gas between the states 1 and 2. This is because the volume of the gas is a *state function* and Eq. (1.4) is an exact differential of the volume V.*

1.3 SIMPLE EQUILIBRIUM

In Figure 1.1 the state of existence of the system (or simply the state of the system) lies on the surface in *P-V-T* space; i.e., for any values of temperature and pressure the system is at equilibrium only when it has that unique volume which corresponds to the particular values of temperature and pressure. A particularly simple system is illustrated in Figure 1.2. This is a fixed quantity of gas contained in a cylinder by a movable piston. The system is at rest, i.e., is at equilibrium, when

1. The pressure exerted by the gas on the piston equals the pressure exerted by the piston on the gas, and

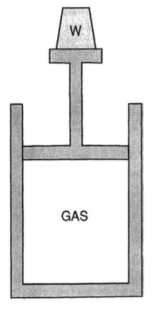

Figure 1.2 A quantity of gas contained in a cylinder by a piston.

*The properties of exact differential equations are discussed in Appendix B.

2. The temperature of the gas is the same as the temperature of the surroundings (provided that heat can be transported through the wall of the cylinder).

The state of the gas is thus fixed, and equilibrium occurs as a result of the establishment of a balance between the tendency of the external influences acting on the system to cause a change in the system and the tendency of the system to resist change. The fixing of the pressure of the gas at P_1 and temperature at T_1 determines the state of the system and hence fixes the volume at the value V_1. If, by suitable decrease in the weight placed on the piston, the pressure exerted on the gas is decreased to P_2, the resulting imbalance between the pressure exerted by the gas and the pressure exerted on the gas causes the piston to move out of the cylinder. This process increases the volume of the gas and hence decreases the pressure which it exerts on the piston until equalization of the pressures is restored. As a result of this process the volume of the gas increases from V_1 to V_2. Thermodynamically, the isothermal change of pressure from P_1 to P_2 changes the state of the system from state 1 (characterized by P_1, T_1), to state 2 (characterized by P_2, T_1), and the volume, as a dependent variable, changes from the value V_1 to V_2.

If the pressure exerted by the piston on the gas is maintained constant at P_2 and the temperature of the surroundings is raised from T_1 to T_2, the consequent temperature gradient across the cylinder wall causes the flow of heat from the surroundings to the gas. The increase in the temperature of the gas at the constant pressure P_2 causes expansion of the gas, which pushes the piston out of the cylinder, and when the gas is uniformly at the temperature T_2 the volume of the gas is V_3. Again, thermodynamically, the changing of the temperature from T_1 to T_2 at the constant pressure P_2 changes the state of the system from state 2 (P_2, T_1) to state 3 (P_2, T_2), and again, the volume as a dependent variable changes from V_2 in the state 2 to V_3 in the state 3. As volume is a state function, the final volume V_3 is independent of the order in which the above steps are carried out.

1.4 THE EQUATION OF STATE OF AN IDEAL GAS

The pressure-volume relationship of a gas at constant temperature was determined experimentally in 1660 by Robert Boyle, who found that, at constant T.

$$P \propto \frac{1}{V}$$

which is known as Boyle's law. Similarly, the volume-temperature relationship of a gas at constant pressure was first determined experimentally by Jacques-Alexandre-Cesar Charles in 1787. This relationship, which is known as Charles' law, is, that at constant pressure

$$V \propto T$$

Thus, in Fig. 1.1, which is drawn for a fixed quantity of gas, sections of the P-V-T surface drawn at constant T produce rectangular hyperbolae which asymptotically approach the P and V axes, and sections of the surface drawn at constant P produce straight lines. These sections are shown in Fig. 1.3a and Fig. 1.3b.

In 1802 Joseph-Luis Gay-Lussac observed that the thermal coefficient of what were called "permanent gases" was a constant. The coefficient of thermal expansion, α, is defined as the fractional increase, with temperature at constant pressure, of the volume of a gas at 0°C; that is

$$\alpha = \frac{1}{V_0}\left(\frac{\partial V}{\partial T}\right)_P$$

where V_0 is the volume of the gas at 0°C. Gay-Lussac obtained a value of 1/267 for α, but more refined experimentation by Regnault in 1847 showed α to have the value 1/273. Later it was found that the accuracy with which Boyle's and Charles' laws describe the behavior of different gases varies from one gas to another and that, generally, gases with lower boiling points obey the laws more closely than do gases with higher boiling points. It was also found that the laws are more closely obeyed by all gases as the pressure of the gas is decreased. It was thus found convenient to invent a hypothetical gas which obeys Boyle's and Charles' laws exactly at all temperatures and pressures. This hypothetical gas is called the *ideal gas,* and it has a value of α of 1/273.15.

The existence of a finite coefficient of thermal expansion sets a limit on the thermal contraction of the ideal gas; that is, as α equals 1/273.15 then the fractional decrease in the volume of the gas, per degree decrease in temperature, is 1/273.15 of the volume at 0°C. Thus, at -273.15°C the volume of the gas is zero, and hence the limit of temperature decrease, -273.15°C, is the absolute zero of temperature. This defines an absolute scale of temperature, called the *ideal gas temperature scale,* which is related to the arbitrary celsius scale by the equation

$$T(\text{degrees absolute}) = T(\text{degrees celsius}) + 273.15$$

combination of Boyle's law

$$P_0V(T,P_0) = PV(T,P)$$

and Charles' law

$$\frac{V(P_0,T_0)}{T_0} = \frac{V(P_0,T)}{T}$$

where

P_0 = standard pressure (1 atm)

T_0 = standard temperature (273.15 degrees absolute)

$V(T,P)$ = volume at temperature T and pressure P

gives

$$\frac{PV}{T} = \frac{P_0V_0}{T_0} = \text{constant} \qquad (1.5)$$

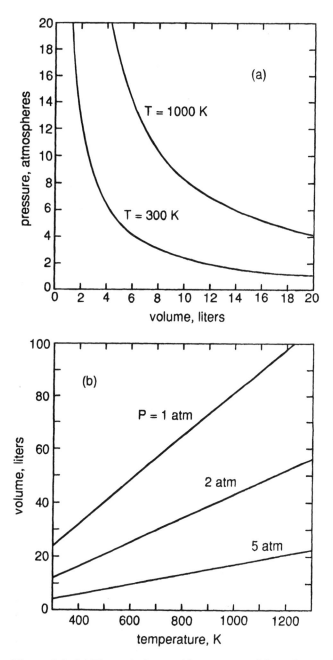

Figure 1.3 (*a*) The variations, with pressure, of the volume of 1 mole of ideal gas at 300 and 1000 K. (*b*) The variations, with temperature, of the volume of 1 mole of ideal gas at 1, 2, and 5 atm.

From Avogadro's hypothesis the volume per gram-mole* of all ideal gases at 0°C and 1 atm pressure (termed *standard temperature and pressure*—STP) is 22.414 liters. Thus the constant in Eq. (1.5) has the value

$$\frac{P_0V_0}{T_0} = \frac{1 \text{ atm} \times 22.414 \text{ liters}}{273.15 \text{ degree} \cdot \text{mole}}$$

$$= 0.082057 \text{ liter} \cdot \text{atm/degree} \cdot \text{mole}$$

This constant is termed R, the *gas constant,* and being applicable to all gases, it is a universal constant. Eq. (1.5) can thus be written as

$$PV = RT \qquad (1.6)$$

which is thus the equation of state for 1 mole of ideal gas. Eq. (1.6) is called the *ideal gas law.* Because of the simple form of its equation of state, the ideal gas is used extensively as a system in thermodynamics discussions.

1.5 THE UNITS OF ENERGY AND WORK

The unit "liter·atmosphere" occurring in the units of R is an energy term. Work is done when a force moves through a distance, and work and energy have the dimensions force × distance. Pressure is force per unit area, and hence work and energy can have the dimensions pressure × area × distance, or pressure × volume. The unit of energy in S.I. is the joule, which is the work done when a force of 1 newton moves a distance of 1 meter. Liter·atmospheres are converted to joules as follows:

$$1 \text{ atm} = 101,325 \text{ newtons/meter}^2$$

Multiplying both sides by liters (10^{-3} m^3) gives

$$1 \text{ liter} \cdot \text{atm} = 101.325 \text{ newton} \cdot \text{meters}$$
$$= 101.325 \text{ joules}$$

and thus

$$R = 0.082057 \text{ liter} \cdot \text{atm/degree} \cdot \text{mole}$$
$$= 8.3144 \text{ joules/degree} \cdot \text{mole}$$

1.6 EXTENSIVE AND INTENSIVE PROPERTIES

Properties (or state variables) are either *extensive* or *intensive.* Extensive properties have values which depend on the size of the system, and the values of intensive properties are independent of the size of the system. Volume is an extensive property, and

---

*A gram-mole (g-mole, or mole) of a substance is the mass of Avogadro's number of molecules of the substance expressed in grams. Thus a g-mole of O_2 has a mass of 32 g, a g-mole of C has a mass of 12 g, and a g-mole of CO_2 has a mass of 44 g.

temperature and pressure are intensive properties. The values of extensive properties, expressed per unit volume or unit mass of the system, have the characteristics of intensive variables; e.g., the volume per unit mass (specific volume) and the volume per mole (the molar volume) are properties whose values are independent of the size of the system. For a system of n moles of an ideal gas, the equation of state is

$$PV' = nRT$$

where V' is the volume of the system. Per mole of the system, the equation of state is

$$PV = RT$$

where V, the molar volume of the gas, equals V'/n.

1.7 PHASE DIAGRAMS AND THERMODYNAMIC COMPONENTS

Of the several ways to graphically represent the equilibrium states of existence of a system, the *constitution* or *phase diagram* is the most popular and convenient. The complexity of a phase diagram is determined primarily by the number of *components* which occur in the system, where components are chemical species of fixed composition. The simplest components are chemical elements and stoichiometric compounds. Systems are primarily categorized by the number of components which they contain, e.g., one-component (unary) systems, two-component (binary) systems, three-component (ternary) systems, four-component (quaternary) systems, etc.

The phase diagram of a one-component system (i.e., a system of fixed composition) is a two-dimensional representation of the dependence of the equilibrium state of existence of the system on the two independent variables. Temperature and pressure are normally chosen as the two independent variables; Fig. 1.4 shows a schematic representation of part of the phase diagram for H_2O. The full lines in

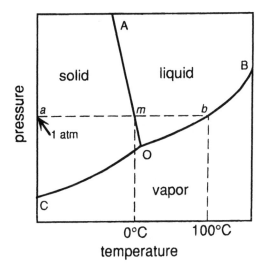

Figure 1.4 Schematic representation of part of the phase diagram for H_2O.

Figure 1.4 divide the diagram into three areas designated solid, liquid, and vapor. If a quantity of pure H_2O is at some temperature and pressure which is represented by a point *within* the area *AOB,* the equilibrium state of the H_2O is a liquid. Similarly, within the areas *COA* and *COB* the equilibrium states are, respectively, solid and vapor. If the state of existence lies on a line, e.g., on the line *AO,* then liquid and solid H_2O coexist in equilibrium with one another, and the equilibrium is said to be two-phase, in contrast to the existence within any of the three areas, which is a one-phase equilibrium. A *phase* is defined as being a finite volume in the physical system within which the properties are uniformly constant, i.e., do not experience any abrupt change in passing from one point in the volume to another. Within any of the one-phase areas in the phase diagram, the system is said to be *homogeneous.* The system is *heterogeneous* when it contains two or more phases, e.g., coexisting ice and liquid water (on the line *AO*) is a heterogeneous system comprising two phases, and the phase boundary between the ice and the liquid water is that very thin region across which the density changes abruptly from the value for homogeneous ice to the higher value for liquid water.

The line *AO* represents the simultaneous variation of *P* and *T* required for maintenance of the equilibrium between solid and liquid H_2O, and thus represents the influence of pressure on the melting temperature of ice. Similarly the lines *CO* and *OB* represent the simultaneous variations of *P* and *T* required, respectively, for the maintenance of the equilibrium between solid and vapor H_2O and between liquid and vapor H_2O. The line *CO* is thus the variation, with temperature, of the saturated vapor pressure of solid ice or, alternatively, the variation, with pressure, of the sublimation temperature of water vapor. The line *OB* is the variation, with temperature, of the saturated vapor pressure of liquid water, or, alternatively, the variation, with pressure, of the dew point of water vapor. The three two-phase equilibrium lines meet at the point *O* (the triple point) which thus represents the unique values of *P* and *T* required for the establishment of the three-phase (solid + liquid + vapor) equilibrium. The path *amb* indicates that if a quantity of ice is heated at a constant pressure of 1 atm, melting occurs at the state *m* (which, by definition, is the normal melting temperature of ice), and boiling occurs at the state *b* (the normal boiling temperature of water).

If the system contains two components, a composition axis must be included and, consequently, the complete diagram is three-dimensional with the coordinates composition, temperature, and pressure. Three-dimensional phase diagrams are discussed in Chapter 14. In most cases, however, it is sufficient to present a binary phase diagram as a constant pressure section of the three-dimensional diagram. The constant pressure chosen is normally 1 atm, and the coordinates are composition and temperature. Figure 1.5, which is a typical simple binary phase diagram, shows the phase relationships occurring in the system Al_2O_3–Cr_2O_3 at 1 atm pressure. This phase diagram shows that, at temperatures below the melting temperature of Al_2O_3 (2050°C), solid Al_2O_3 and solid Cr_2O_3 are completely miscible in all proportions. This occurs because Al_2O_3 and Cr_2O_3 have the same crystal structure and the Al^{3+} and Cr^{3+} ions are of similar size. At temperatures above the melting temperature of Cr_2O_3 (2265°C) liquid Al_2O_3 and liquid Cr_2O_3 are completely miscible in all pro-

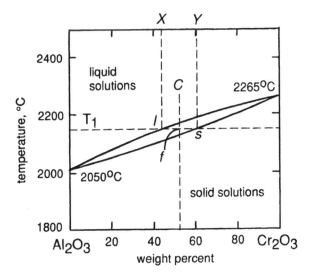

Figure 1.5 The phase diagram for the system Al_2O_3–Cr_2O_3.

portions. The diagram thus contains areas of complete solid solubility and complete liquid solubility, which are separated from one another by a two-phase area in which solid and liquid solutions coexist in equilibrium with one another. For example, at the temperature T_1 a Cr_2O_3–Al_2O_3 system of composition between X and Y exists as a two-phase system comprising a liquid solution of composition l in equilibrium with a solid solution of composition s. The relative proportions of the two phases present depend only on the overall composition of the system in the range $X - Y$ and are determined by the lever rule as follows. For the overall composition C at the temperature T_1 the lever rule states that if a fulcrum is placed at f on the lever ls, then the relative proportions of liquid and solid phases present are such that, placed, respectively, on the ends of the lever at s and l, the lever balances about the fulcrum, i.e., the ratio of liquid to solid present at T_1 is the ratio fs/lf.

Because the only requirement of a component is that it have a fixed composition, the designation of the components of a system is purely arbitrary. In the system Al_2O_3–Cr_2O_3 the obvious choice of the components is Al_2O_3 and Cr_2O_3. However, the most convenient choice is not always as obvious, and the general arbitrariness in selecting the components can be demonstrated by considering the iron-oxygen system, the phase diagram of which is shown in Fig. 1.6. This phase diagram shows the Fe and O form two stoichiometric compounds, Fe_3O_4 (magnetite) and Fe_2O_3 (hematite), and a limited range of solid solution (wustite). Of particular significance is the observation that neither a stoichiometric compound of the formula FeO nor a wustite solid solution in which the Fe/O atomic ratio is unity occurs. In spite of this it is often found convenient to consider the stoichiometric FeO composition as a thermodynamic component of the system. The available choice of the two components of the binary system can be demonstrated by considering the composition X in Fig. 1.6. This composition can equivalently be considered as being in any one of the following systems:

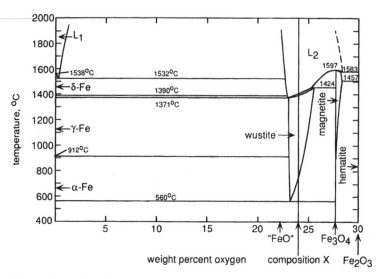

Figure 1.6   The phase diagram for the binary system Fe–O.

1. The system Fe–O (24 weight % O, 76 weight % Fe)
2. The system FeO–Fe$_2$O$_3$ (77.81 weight % FeO, 22.19 weight % Fe$_2$O$_3$)
3. The system FeO–Fe$_3$O$_4$ (67.83 weight % FeO, 32.17 weight % Fe$_3$O$_4$)
4. The system Fe–Fe$_3$O$_4$ (13.18 weight % Fe, 86.82 weight % Fe$_3$O$_4$)
5. The system Fe–Fe$_2$O$_3$ (20.16 weight % Fe, 79.84 weight % Fe$_2$O$_3$)
6. The system FeO–O (97.78 weight % FeO, 2.22 weight % O)

The actual choice of the two components for use in a thermodynamic analysis is thus purely a matter of convenience. The ability of the thermodynamic method to deal with descriptions of the compositions of systems in terms of arbitrarily chosen components, which need not correspond to physical reality, is a distinct advantage. The thermodynamic behavior of highly complex systems, such as metallurgical slags and molten glass, can be completely described in spite of the fact that the ionic constitutions of these systems are not known completely.

1.8 NUMERICAL EXAMPLES

The following examples were provided by Dr. Bill Fahrenholtz, University of Missouri–Rolla.

Example 1

Oxy-acetylene torches can be used to cut or weld metal. Oxygen and acetylene are fed from separate compressed gas cylinders through pressure regulators to the torch, where they are mixed and combusted. To estimate the amount of metal that can be cut or welded per unit volume of gas, the composition of the gases produced by combustion and the heat produced by the combustion reaction must be determined. For this problem:

a) Define a closed system that can be used to calculate the heat produced
b) Define an open system that can be used to calculate the heat produced
c) List at least two intensive variables that are important in the calculation
d) List at least two extensive variables that are important in the calculation

a) The gases can be chosen as being the closed system. With this definition, the initial state of the system could be a mixture of oxygen and acetylene at ambient temperature and pressure. The final state could then be the combustion products at ambient pressure and temperature. The system may do some work against the surroundings (constant pressure, but the volume may change depending on the combustion reaction), and the heat produced by combustion would be transferred out of the system.

b) The torch can be chosen as being the open system. The two gases would flow into the system at ambient temperature and pressure and would be combusted in the system. The combustion products would flow out of the system at ambient temperature and pressure. Although the system would not do any work on its surroundings (the torch has constant volume under steady state operation), heat would be transferred out of the system.

c) Intensive variables do not vary with the size of the system. Temperature, pressure, and density are intensive variables.

d) Extensive variables are functions of the size of the system. Thus volume, mass, and number of moles are extensive variables.

Example 2

Consider one mole of an ideal gas. Using a y-axis of pressure and an x-axis of temperature, plot the variations of pressure with temperature for volumes of 11.2 liters, 22.4 liters, and 44.8 liters. Use ranges of pressure and temperature consistent with Figure 1.3 on page 7.

States of constant volume can be calculated using the ideal gas law ($PV' = nRT$). The table below summarizes the calculations.

| | Pressure (atm) | | |
Temperature (K)	$V = 11.2$ liters	$V = 22.4$ liters	$V = 44.8$ liters
300	2.20	1.10	0.55
400	2.93	1.47	0.73
500	3.66	1.83	0.92
600	4.40	2.20	1.10
700	5.13	2.56	1.28
800	5.86	2.93	1.47
900	6.59	3.30	1.65
1000	7.33	3.66	1.83
1100	8.06	4.03	2.01
1200	8.79	4.40	2.20
1300	9.52	4.76	2.38

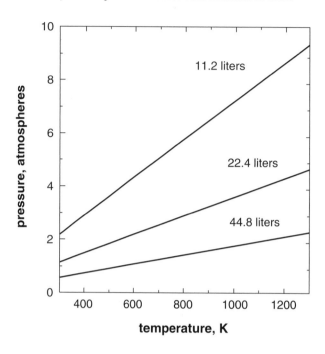

Example 3

A rock with a mass of 10 lb falls 100 feet from a cliff to the floor of a canyon. What is the change in potential energy of the rock?

1 lb = 0.4536 kg $\therefore$ 10 lb = 4.536 kg
1 in = 2.54 cm, 1 m = 100 cm, and 1 ft = 12 in $\therefore$ 100 ft = 30.48 m

$$\Delta PE = mg\Delta h$$
$$= (4.536 \text{ kg})(9.81 \text{ m/s}^2)(-30.48 \text{ m})$$
$$= -1356 \text{ kg} \cdot \text{m}^2 \cdot \text{s}^{-2}$$

1 N = 1 $\frac{\text{kg} \cdot \text{m}}{\text{s}^2}$ and 1 J = 1 N $\cdot$ m

so

$$\Delta PE = -1356 \text{ J}$$

Chapter 2

THE FIRST LAW OF THERMODYNAMICS

2.1 INTRODUCTION

Kinetic energy is conserved in a frictionless system of interacting rigid elastic bodies. A collision between two of these bodies results in a transfer of kinetic energy from one to the other, the work done by the one equals the work done on the other, and the total kinetic energy of the system is unchanged as a result of the collision. If the kinetic system is in the influence of a gravitational field, then the sum of the kinetic and potential energies of the bodies is constant; changes of position of the bodies in the gravitational field, in addition to changes in the velocities of the bodies, do not alter the total dynamic energy of the system. As the result of possible interactions, kinetic energy may be converted to potential energy and vice versa, but the sum of the two remains constant. If, however, friction occurs in the system, then with continuing collision and interaction among the bodies, the total dynamic energy of the system decreases and heat is produced. It is thus reasonable to expect that a relationship exists between the dynamic energy dissipated and the heat produced as a result of the effects of friction.

The establishment of this relationship laid the foundations for the development of the thermodynamic method. As a subject, this has now gone far beyond simple considerations of the interchange of energy from one form to another, e.g., from dynamic energy to thermal energy. The development of thermodynamics from its early beginnings to its present state was achieved as the result of the invention of convenient thermodynamic functions of state. In this chapter the first two of these thermodynamic functions—the internal energy U and the enthalpy H—are introduced.

2.2 THE RELATIONSHIP BETWEEN HEAT AND WORK

The relation between heat and work was first suggested in 1798 by Count Rumford, who, during the boring of cannon at the Munich Arsenal, noticed that the heat produced during the boring was roughly proportional to the work performed during the boring. This suggestion was novel, as hitherto heat had been regarded as being an invisible fluid called caloric which resided between the constituent particles of a substance. In the caloric theory of heat, the temperature of a substance was considered to be determined by the quantity of caloric gas which it contained, and two bodies of differing temperature, when placed in contact with one another, came to an intermediate common temperature as the result of caloric flowing between them. Thermal equilibrium was reached when the pressure of caloric gas in the one body equaled that in the other. Rumford's observation that heat production accompanied the performance of work was accounted for by the caloric theory as being due to the fact that the amount of caloric which could be contained by a body, per unit mass of the body, depended on the mass of the body. Small pieces of metal (the metal turnings produced by the boring) contained less caloric per unit mass than did the original large mass of metal, and thus, in reducing the original large mass to a number of smaller pieces, caloric was evolved as sensible heat. Rumford then demonstrated that when a blunt borer was used (which produced very few metal turnings), the same heat production accompanied the same expenditure of work. The caloric theory "explained" the heat production in this case as being due to the action of air on the metal surfaces during the performance of work.

The caloric theory was finally discredited in 1799 when Humphrey Davy melted two blocks of ice by rubbing them together in a vacuum. In this experiment the latent heat necessary to melt the ice was provided by the mechanical work performed in rubbing the blocks together.

From 1840 onwards the relationship between heat and work was placed on a firm quantitative basis as the result of a series of experiments carried out by James Joule. Joule conducted experiments in which work was performed in a certain quantity of adiabatically* contained water and measured the resultant increase in the temperature of the water. He observed that a direct proportionality existed between the work done and the resultant increase in temperature and that the same proportionality existed no matter what means were employed in the work production. Methods of work production used by Joule included

1. Rotating a paddle wheel immersed in the water
2. An electric motor driving a current through a coil immersed in the water

---

*An adiabatic vessel is one which is constructed in such a way as to prohibit, or at least minimize, the passage of heat through its walls. The most familiar example of an adiabatic vessel is the Dewar flask (known more popularly as a thermos flask). Heat transmission by conduction into or out of this vessel is minimized by using double glass walls separated by an evacuated space, and a rubber or cork stopper, and heat transmission by radiation is minimized by using highly polished mirror surfaces.

3.   Compressing a cylinder of gas immersed in the water
4.   Rubbing together two metal blocks immersed in the water

This proportionality gave rise to the notion of a *mechanical equivalent of heat,* and for the purpose of defining this figure it was necessary to define a unit of heat. This unit is the *calorie* (or 15° calorie), which is the quantity of heat required to increase the temperature of 1 gram of water from 14.5°C to 15.5°C. On the basis of this definition Joule determined the value of the mechanical equivalent of heat to be 0.241 calories per joule. The presently accepted value is 0.2389 calories (15° calories) per joule. Rounding this to 0.239 calories per joule defines the *thermochemical calorie,* which, until the introduction in 1960 of S.I. units, was the traditional energy unit used in thermochemistry.

2.3 INTERNAL ENERGY AND THE FIRST LAW OF THERMODYNAMICS

Joule's experiments resulted in the statement that "the change of a body inside an adiabatic enclosure from a given initial state to a given final state involves the same amount of work by whatever means the process is carried out." The statement is a preliminary formulation of the First Law of Thermodynamics, and in view of this statement, it is necessary to define some function which depends only on the internal state of a body or system. Such a function is U, the internal energy. This function is best introduced by means of comparison with more familiar concepts. When a body of mass m is lifted in a gravitational field from height h_1 to height h_2, the work w done on the body is given by

$$
\begin{aligned}
w &= \text{force} \times \text{distance} \\
&= mg \times (h_2 - h_1) \\
&= mgh_2 - mgh_1 \\
&= \text{potential energy at position } h_2 \text{ minus potential energy at position } h_1
\end{aligned}
$$

As the potential energy of the body of given mass m depends only on the position of the body in the gravitational field, it is seen that the work done on the body is dependent only on its final and initial positions and is independent of the path taken by the body between the two positions, i.e., between the two states. Similarly the application of a force f to a body of mass m causes the body to accelerate according to Newton's Law

$$
f = ma = m \frac{dv}{dt}
$$

where $a = dv/dt$, the acceleration.

The work done on the body is thus obtained by integrating

$$
dw = f \, dl
$$

where l is distance.

$$\therefore dw = m \frac{dv}{dt} dl = m \frac{dl}{dt} dv = mv\, dv$$

Integration gives

$$w = \frac{1}{2} mv_2^2 - \frac{1}{2} mv_1^2$$

= the kinetic energy of the body at velocity v_2 (state 2)

− the kinetic energy of the body at velocity v_1 (state 1)

Thus, again, the work done on the body is the difference between the values of a function of the state of the body and is independent of the path taken by the body between the states.

In the case of work being done on an adiabatically contained body of constant potential and kinetic energy, the pertinent function which describes the state of the body, or the change in the state of the body, is the internal energy U. Thus the work done on, or by, an adiabatically contained body equals the change in the internal energy of the body, i.e., equals the difference between the value of U in the final state and the value of U in the initial state. In describing work, it is conventional to assign a negative value to work done *on* a body and a positive value to work done *by* a body. This convention arises because, when a gas expands, and hence does work against an external pressure, the integral $\int_1^2 P\, dV$, which is the work performed, is a positive quantity. Thus for an adiabatic process in which work w is done on a body, as a result of which its state moves from A to B.

$$w = -(U_B - U_A)$$

If work w is done on the body, then $U_B > U_A$ and if the body itself performs work, then $U_B < U_A$.

In Joule's experiments the change in the state of the adiabatically contained water was measured as an increase in the temperatures of the water. The same increase in temperature, and hence the same change of state, could have been produced by placing the water in thermal contact with a source of heat and allowing heat q to flow into the water. In describing heat changes it is conventional to assign a negative value to heat which flows *out* of a body (an exothermic process) and a positive value to heat which flows *into* a body (an endothermic process). Hence,

$$q = (U_B - U_A)$$

Thus, when heat flows into the body, q is a positive quantity and $U_B > U_A$, whereas if heat flows out of the body, $U_B < U_A$ and q is a negative quantity.

It is now of interest to consider the change in the internal energy of a body which simultaneously performs work and absorbs heat. Consider a body, initially in the state A, which performs work w, absorbs heat q, and, as a consequence, moves to the state B. The absorption of heat q *increases* the internal energy of the body by the amount q, and the performance of work w by the body *decreases* its internal energy by the amount w. Thus the total change in the internal energy of the body, ΔU, is

$$\Delta U = U_B - U_A = q - w \qquad (2.1)$$

This is a statement of the *First Law of Thermodynamics*.

For an infinitesimal change of state, Eq. (2.1) can be written as a differential

$$dU = \delta q - \delta w \qquad (2.2)$$

Notice that the left-hand side of Eq. (2.2) gives the value of the increment in an already existing property of the system, whereas the right-hand side has no corresponding interpretation. As U is a state function, the integration of dU between two states gives a value which is independent of the path taken by the system between the two states. Such is not the case when δq and δw are integrated. The heat and work effects, which involve energy in transit, depend on the path taken between the two states, as a result of which the integrals of δw and δq cannot be evaluated without a knowledge of the path. This is illustrated in Fig. 2.1. In Fig. 2.1 the value of $U_2 - U_1$ is independent of the path taken between state 1 (P_1V_1) and state 2 (P_2V_2). However, the work done by the system, which is given by the integral $\int_1^2 \delta w = \int_1^2 P dV$ and hence is the area under the curve between V_2 and V_1, can vary greatly depending on the path. In Fig. 2.1 the work done in the process $1 \rightarrow 2$ via c is less than that done via b which, in turn, is less than that done via a. From Eq. (2.1) it is seen that the integral of δq must also depend on the path, and in the process $1 \rightarrow 2$ more heat is absorbed by the system via a than is absorbed via b which, again in turn, is greater than the heat absorbed via c. In Eq. (2.2) use of the symbol "d" indicates a differential element of a state function or state property, the integral of which is independent of the path, and use of the symbol "δ" indicates a differential element of some quantity which is not a state function. In Eq. (2.1) note that the algebraic sum of two quantities, neither of which individually is independent of the path, gives a quantity which is independent of the path.

In the case of a cyclic process which returns the system to its initial state, e.g., the process $1 \rightarrow 2 \rightarrow 1$ in Fig. 2.1, the change in U as a result of this process is zero; i.e.,

$$\Delta U = \int_1^2 dU + \int_2^1 dU = (U_2 - U_1) + (U_1 - U_2) = 0$$

The vanishing of a cyclic integral $\oint dU = 0$ is a property of a state function.

In Joule's experiments, where $(U_2 - U_1) = -w$, the process was adiabatic ($q = 0$), and thus the path of the process was specified.

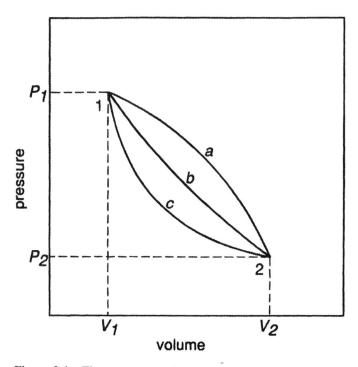

Figure 2.1   Three process paths taken by a fixed quality of gas in moving from the state 1 to the state 2.

As U is a state function, then for a simple system consisting of a given amount of substance of fixed composition, the value of U is fixed once any two properties (the independent variables) are fixed. If temperature and volume are chosen as the independent variables, then

$$U = U(V, T)$$

The complete differential U in terms of the partial derivatives gives

$$dU = \left(\frac{\partial U}{\partial V}\right)_T dV + \left(\frac{\partial U}{\partial T}\right)_V dT$$

As the state of the system is fixed when the two independent variables are fixed, it is of interest to examine those processes which can occur when the value of one of the independent variables is maintained constant and the other is allowed to vary. In this manner we can examine processes in which the volume V is maintained constant (isochore or isometric processes), or the pressure P is maintained constant (isobaric processes), or the temperature T is maintained constant (isothermal processes). We can also examine adiabatic processes in which $q = 0$.

2.4 CONSTANT-VOLUME PROCESSES

If the volume of a system is maintained constant during a process, then the system does no work ($\int PdV = 0$), and from the First Law, Eq. (2.2),

$$dU = \delta q_v \tag{2.3}$$

where the subscript v indicates constant volume. Integration of Eq. (2.3) gives

$$\Delta U = q_v$$

for such a process, which shows that the increase or decrease in the internal energy of the system equals, respectively, the heat absorbed or rejected by the system during the process.

2.5 CONSTANT-PRESSURE PROCESSES AND THE ENTHALPY H

If the pressure is maintained constant during a process which takes the system from state 1 to state 2, then the work done by the system is given as

$$w = \int_1^2 PdV = P \int_1^2 dV = P(V_2 - V_1)$$

and the First Law gives

$$U_2 - U_1 = q_p - P(V_2 - V_1)$$

where the subscript p indicates constant pressure. Rearrangement gives

$$(U_2 + PV_2) - (U_1 + PV_1) = q_p$$

and, as the expression $(U + PV)$ contains only state functions, the expression itself is a state function. This is termed the *enthalpy, H;* i.e.,

$$H = U + PV \tag{2.4}$$

Hence, for a constant-pressure process,

$$H_2 - H_1 = \Delta H = q_p \tag{2.5}$$

Thus the enthalpy change during a constant-pressure process is equal to the heat admitted to or withdrawn from the system during the process.

2.6 HEAT CAPACITY

Before discussing isothermal and adiabatic processes, it is convenient to introduce the concept of heat capacity. The heat capacity, C, of a system is the ratio of the heat added to or withdrawn from the system to the resultant change in the temperature of the system. Thus

$$C = \frac{q}{\Delta T}$$

or if the temperature change is made vanishingly small, then

$$C = \frac{\delta q}{dT}$$

The concept of heat capacity is only used when the addition of heat to or withdrawal of heat from the system produces a temperature change; the concept is not used when a phase change is involved. For example, if the system is a mixture of ice and water at 1 atm pressure and 0°C, then the addition of heat simply melts some of the ice and no change in temperature occurs. In such a case the heat capacity, as defined, would be infinite.

Note that if a system is in a state 1 and the absorption of a certain quantity of heat by the system increases its temperature from T_1 to T_2, then the statement that the final temperature is T_2 is insufficient to determine the final state of the system. This is because the system has two independent variables, and so one other variable, in addition to the temperature, must be specified in order to define the state of the system. This second independent variable could be varied in a specified manner or could be maintained constant during the change. The latter possibility is the more practical, and so the addition of heat to a system to produce a change in temperature is normally considered at constant pressure or at constant volume. In this way the path of the process is specified, and the final state of the system is known.

Thus a heat capacity at constant volume, C_v, and a heat capacity at constant pressure, C_p, are defined as

$$C_v = \left(\frac{\delta q}{dT}\right)_V \quad \text{the heat capacity at constant volume}$$

$$C_p = \left(\frac{\delta q}{dT}\right)_P \quad \text{the heat capacity at constant pressure}$$

Thus, from Eqs. (2.3) and (2.5)

$$C_v = \left(\frac{\delta q}{dT}\right)_V = \left(\frac{dU}{dT}\right)_V \quad \text{or} \quad dU = C_v dT \tag{2.6}$$

$$C_p = \left(\frac{\delta q}{dT}\right)_P = \left(\frac{dH}{dT}\right)_P \quad \text{or} \quad dH = C_p dT \tag{2.7}$$

The heat capacity, being dependent on the size of the system, is an extensive property. However, in normal usage it is more convenient to use the heat capacity per unit quantity of the system. Thus the specific heat of the system is the heat capacity per gram at constant P, and the molar heat capacity is the heat capacity per mole at constant pressure or at constant volume. Thus, for a system containing n moles,

$$nc_p = C_p$$

and

$$nc_v = C_v$$

where c_p and c_v are the molar values.

It is to be expected that, for any substance, c_p will be of greater magnitude than c_v. If it is required that the temperature of a system be increased by a certain amount, then, if the process is carried out at a constant volume, all of the heat added is used solely to raise the temperature of the system. However, if the process is carried out at constant pressure, then, in addition to raising the temperature by the required amount, the heat added is required to provide the work necessary to expand the system at the constant pressure. This work of expansion against the constant pressure per degree of temperature increase is calculated as

$$\frac{PdV}{dT} \text{ or } P\left(\frac{\partial V}{\partial T}\right)_P$$

and hence it might be expected that

$$c_p - c_v = P\left(\frac{\partial V}{\partial T}\right)_P$$

The difference between c_p and c_v is calculated as follows:

$$c_p = \left(\frac{\partial H}{\partial T}\right)_P = \left(\frac{\partial U}{\partial T}\right)_P + P\left(\frac{\partial V}{\partial T}\right)_P$$

and

$$c_v = \left(\frac{\partial U}{\partial T}\right)_V$$

Hence

$$c_p - c_v = \left(\frac{\partial U}{\partial T}\right)_P + P\left(\frac{\partial V}{\partial T}\right)_P - \left(\frac{\partial U}{\partial T}\right)_V$$

but

$$dU = \left(\frac{\partial U}{\partial V}\right)_T dV + \left(\frac{\partial U}{\partial T}\right)_V dT$$

and therefore

$$\left(\frac{\partial U}{\partial T}\right)_P = \left(\frac{\partial U}{\partial V}\right)_T\left(\frac{\partial V}{\partial T}\right)_P + \left(\frac{\partial U}{\partial T}\right)_V$$

Hence,

$$c_p - c_v = \left(\frac{\partial U}{\partial V}\right)_T \left(\frac{\partial V}{\partial T}\right)_P + \left(\frac{\partial U}{\partial T}\right)_V + P\left(\frac{\partial V}{\partial T}\right)_P - \left(\frac{\partial U}{\partial T}\right)_V$$

$$= \left(\frac{\partial V}{\partial T}\right)_P \left[P + \left(\frac{\partial U}{\partial V}\right)_T\right] \tag{2.8}$$

The two expressions for $c_p - c_v$ differ by the term $(\partial V/\partial T)_P \times (\partial U/\partial V)_T$, and in an attempt to evaluate the term $(\partial U/\partial V)_T$ for gases, Joule performed an experiment which involved filling a copper vessel with a gas at some pressure and connecting this vessel via a stopcock to a similar but evacuated vessel. The two-vessel system was immersed in a quantity of adiabatically contained water and the stopcock was opened, thus allowing free expansion of the gas into the evacuated vessel. After this expansion, Joule could not detect any change in the temperature of the system. As the system was adiabatically contained and no work was performed, then from the First Law,

$$\Delta U = 0$$

and hence

$$dU = \left(\frac{\partial U}{\partial V}\right)_T dV + \left(\frac{\partial U}{\partial T}\right)_V dT = 0$$

Thus as $dT = 0$ (experimentally determined) and dV is not 0 then the term $(\partial U/\partial V)_T$ must be zero. Joule thus concluded that the internal energy of a gas is a function only of temperature and is independent of the volume (and hence pressure). Consequently, for a gas

$$c_p - c_v = P\left(\frac{\partial V}{\partial T}\right)_P$$

However, in a more critical experiment performed by Joule and Thomson, in which an adiabatically contained gas of molar volume V_1 at the pressure P_1 was throttled through a porous diaphragm to the pressure P_2 and the molar volume V_2, a change in the temperature of the gas was observed, which showed that, for real gases, $(\partial U/\partial V)_T \neq 0$.

Nevertheless, if

$$\left(\frac{\partial U}{\partial V}\right)_T = 0$$

then, from Eq. (2.8),

$$c_p - c_v = P\left(\frac{\partial V}{\partial T}\right)_P$$

and as, for one mole of ideal gas, $PV = RT$, then

$$c_p - c_v = \frac{R}{P} \times P = R$$

The reason for Joule's not observing a temperature rise in the original experiment was that the heat capacity of the copper vessels and the water was considerably greater than the heat capacity of the gas, and thus the small heat changes which actually occurred in the gas were absorbed in the copper vessels and the water. This decreased the actual temperature change to below the limits of the then-available means of temperature measurement.

In Eq. (2.8) the term

$$P\left(\frac{\partial V}{\partial T}\right)_P$$

represents the work done by the system per degree rise in temperature in expanding against the constant external pressure P acting on the system. The other term in Eq. (2.8), namely,

$$\left(\frac{\partial U}{\partial V}\right)_T\left(\frac{\partial V}{\partial T}\right)_P$$

represents the work done per degree rise in temperature in expanding against the internal cohesive forces acting between the constituent particles of the substance. As will be seen in Chap. 8, an ideal gas is a gas consisting of noninteracting particles, and hence the atoms of an ideal gas can be separated from one another without the expenditure of work. Thus for an ideal gas the above term, and so the term

$$\left(\frac{\partial U}{\partial V}\right)_T$$

are zero.

In real gases the internal pressure contribution is very much smaller in magnitude than the external pressure contribution; but in liquids and solids, in which the interatomic forces are considerable, the work done in expanding the system against the external pressure is insignificant in comparison with the work done against the internal pressure. Thus for liquids and solids the term

$$\left(\frac{\partial U}{\partial V}\right)_T$$

is very large.

2.7 REVERSIBLE ADIABATIC PROCESSES

During a reversible process during which the state of the gas is changed, the state of the gas never leaves the equilibrium surface shown in Fig. 1.1. Consequently, during a reversible process, the gas passes through a continuum of equilibrium states, and the work w is given by the integral $\int_1^2 PdV$ only if the process is conducted reversibly.

In an adiabatic process $q = 0$, and thus, from the First Law, $dU = -\delta w$. Consider a system comprising one mole of an ideal gas. From Eq. (2.6)

$$dU = c_v \, dT$$

and, for a reversible adiabatic process

$$\delta w = P dV$$

Thus

$$c_v \, dT = -P dV$$

As the system is one mole of ideal gas, then $P = RT/V$ and hence

$$c_v \, dT = -\frac{RTdV}{V}$$

Integrating between states 1 and 2 gives

$$c_v \ln\left(\frac{T_2}{T_1}\right) = R \ln\left(\frac{V_1}{V_2}\right)$$

or

$$\left(\frac{T_2}{T_1}\right)^{c_v} = \left(\frac{V_1}{V_2}\right)^{R}$$

or

$$\left(\frac{T_2}{T_1}\right) = \left(\frac{V_1}{V_2}\right)^{R/c_v}$$

For an ideal gas it has been shown that $c_p - c_v = R$. Thus $c_p/c_v - 1 = R/c_v$; and if $c_p/c_v = \gamma$, then $R/c_v = \gamma - 1$, and hence

$$\frac{T_2}{T_1} = \left(\frac{V_1}{V_2}\right)^{\gamma-1}$$

From the ideal gas law,

$$\frac{T_2}{T_1} = \frac{P_2 V_2}{P_1 V_1} = \left(\frac{V_1}{V_2}\right)^{\gamma-1}$$

Thus

$$\frac{P_2}{P_1} = \left(\frac{V_1}{V_2}\right)^{\gamma}$$

and hence

$$P_2 V_2^{\gamma} = P_1 V_1^{\gamma} = PV^{\gamma} = \text{constant} \tag{2.9}$$

This is the relationship between the pressure and the volume of an ideal gas undergoing a reversible adiabatic process.

2.8 REVERSIBLE ISOTHERMAL PRESSURE OR VOLUME CHANGES OF AN IDEAL GAS

From the First Law

$$dU = \delta q - \delta w$$

and as $dT = 0$ (isothermal process), then $dU = 0$. Therefore $\delta w = \delta q = PdV = RTdV/V$ per mole of gas.

Integrating between the states 1 and 2 gives

$$w = q = RT \ln\left(\frac{V_2}{V_1}\right) = RT \ln\left(\frac{P_1}{P_2}\right) \tag{2.10}$$

Thus, for an ideal gas, an isothermal process is one of constant internal energy during which the work done by the system equals the heat absorbed by the system, both of which are given by Eq. (2.10).

A reversible isothermal process and a reversible adiabatic process are shown on a P-V diagram in Fig. 2.2 in which it is seen that, for a given decrease in pressure, the

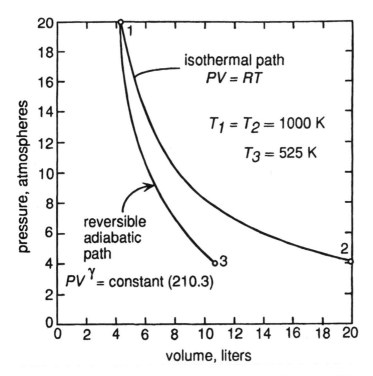

Figure 2.2 Comparison of the process path taken by a reversible isothermal expansion of an ideal gas with the process path taken by a reversible adiabatic expansion of an ideal gas between an initial pressure of 20 atm and a final pressure of 4 atm.

work done by the reversible isothermal process (which is equal to the area under the curve) exceeds that done by the reversible adiabatic process. This difference is due to the fact that during the isothermal process heat is absorbed by the system in order to maintain the temperature constant, whereas during the adiabatic process no heat is admitted to the system. During the isothermal expansion the internal energy of the gas remains constant, and during the adiabatic expansion the internal energy decreases by an amount equal to the work done.

For the reversible isothermal path shown in Fig. 2.2, Eq. (2.10) gives the work done by the gas as

$$w = RT \ln\left(\frac{P_1}{P_2}\right) = 8.3144 * 1000 * \ln\left(\frac{20}{4}\right) = 13.38 \text{ kJ}$$

For the reversible adiabatic path, the area under the curve is obtained as follows. As $PV^\gamma = \text{constant} = 210.3$ and $\gamma = c_p/c_v = (5/2 * R)/(3/2 * R) = 5/3$ then

$$w = \int P dV = 210.3 * \int dV/V^{5/3}$$

The volume in the state 3 is obtained from Eq. (2.9) as $V_3 = (210.4/4)^{3/5} = 10.78$ liters and the volume in the state 1 is $V_1 = (210.3/20)^{3/5} = 4.10$ liters. Thus

$$w = 210.3 * \left(-\frac{3}{2}\right) * (V_3^{-2/3} - V_1^{-2/3})$$

$$= 210.3 * (-1.5) * (10.78^{-2/3} - 4.10^{-2/3})$$

$$= 210.3 * (-1.5) * (0.205 - 0.390) = 58.5 \text{ l.atm}$$

Alternatively, as $q = 0$, $w = -\Delta U = c_v(T_3 - T_1) = -1.5 * 8.3144 * (525 - 1000) = 5924$ J

$$= 5924 * \frac{0.082057}{8.3144} = 58.5 \text{ l.atm}$$

2.9 SUMMARY

1. The establishment of the relationship between the work done on or by a system and the heat entering or leaving the system was facilitated by the introduction of the thermodynamic function U, the internal energy. U is a function of state, and thus the difference between the values of U in two states depends only on the states and is independent of the process path taken by the system in moving between the states. The relationship between the internal energy change, the work done, and the heat absorbed per mole by a system of fixed composition in moving from one state to another is given as $\Delta U = q - w$, or, for an increment of this process, $dU = \delta q - \delta w$. This relationship is called the First Law of Thermodynamics.

2. The integrals of δq and δw can only be obtained if the process path taken by the system in moving from one state to another is known. Process paths which are

convenient for consideration include

a. Constant-volume processes in which $\int \delta w = \int P dV = 0$
b. Constant-pressure processes in which $\int \delta w = P \int dV = P\Delta V$
c. Constant-temperature processes
d. Adiabatic processes in which $q = 0$

3. For a constant-volume process, as $w = 0$, then $\Delta U = q_v$. The definition of the constant-volume molar heat capacity as $c_v = (\delta q/dT)_V = (\partial U/\partial T)_V$ (which is an experimentally measurable quantity) facilitates determination of the change in U resulting from a constant-volume process as $\Delta U = \int_1^2 c_v dT$.

4. Consideration of constant-pressure processes is facilitated by the introduction of the thermodynamic function H, the enthalpy, defined as $H = U + PV$. As the expression for H contains only functions of state, then H is a function of state, and thus the difference between the values of H in two states depends only on the states and is independent of the path taken by the system in moving between them. For a constant-pressure process, $\Delta H = \Delta U + P\Delta V = (q_p - P\Delta V) + P\Delta V = q_p$. The definition of the constant-pressure molar heat capacity as $c_p = (\Delta q/dT)_P = (\partial H/\partial T)_P$ (which is an experimentally measurable quantity) facilitates determination of the change in H as the result of a constant-pressure process as $dH = \int_1^2 c_p dT$.

5. For an ideal gas, the internal energy U is a function only of temperature, and $c_p - c_v = R$.

6. The process path of an ideal gas undergoing a reversible adiabatic change of state is described by $PV^\gamma = $ constant, where $\gamma = c_p/c_v$. During an adiabatic expansion, as $q = 0$, the decrease in the internal energy of the system equals the work done by the system.

7. As the internal energy of an ideal gas is a function only of temperature, the internal energy of an ideal gas remains constant during an isothermal change of state. Thus the heat which enters or leaves the gas as a result of the isothermal process equals the work done by or on the gas, with both quantities being given by

$$w = q = RT \ln\left(\frac{V_2}{V_1}\right) = RT \ln\left(\frac{P_1}{P_2}\right)$$

8. Only the differences in the values of U and H between two states, i.e., the values of ΔU and ΔH, can be measured. The absolute values of U and H in any given state cannot be determined.

2.10 NUMERICAL EXAMPLES

Ten liters of a monatomic ideal gas at 25°C and 10 atm pressure are expanded to a final pressure of 1 atm. The molar heat capacity of the gas at constant volume, c_v, is $3/2 R$ and is independent of temperature. Calculate the work done, the heat absorbed, and the change in U and in H for the gas if the process is carried out (1) isothermally and reversibly, and (2) adiabatically and reversibly. Having determined the final state of the gas after the reversible adiabatic expansion, verify that the change in U for the process is independent of the path taken between the initial and final states by considering the process to be carried out as

i.   An isothermal process followed by a constant-volume process
ii. A constant-volume process followed by an isothermal process
iii. An isothermal process followed by a constant-pressure process
iv. A constant-volume process followed by a constant-pressure process
v.   A constant-pressure process followed by a constant-volume process

The size of the system must first be calculated. From consideration of the initial state of the system (the point a in Fig. 2.3)

$$n = \text{the number of moles} = \frac{P_a V_a}{RT_a} = \frac{10 \times 10}{0.08206 \times 298} = 4.09$$

(a) The isothermal reversible expansion. The state of the gas moves from a to b along the 298 degrees isotherm. As, along any isotherm, the product PV is constant,

$$V_b = \frac{P_a V_a}{P_b} = \frac{10 \times 10}{1} = 100 \text{ liters}$$

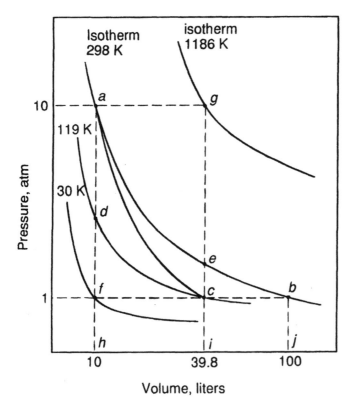

Figure 2.3 The five process paths considered in the numerical example.

For an ideal gas undergoing an isothermal process, $\Delta U = 0$ and hence, from the First Law,

$$q = w = \int_a^b PdV = nRT \int_a^b \frac{dV}{V} = 4.09 \times 8.3144 \times 298 \times \ln \frac{100}{10} \text{ joules}$$

$$= 23.3 \text{ kilojoules}$$

Thus in passing from the state a to the state b along the 298 degree isotherm, the system performs 23.3 kilojoules of work and absorbs 23.3 kilojoules of heat from the constant-temperature surroundings.

As, for an ideal gas, H is a function only of temperature, then $\Delta H_{(a \to b)} = 0$; that is,

$$\Delta H_{(a \to b)} = \Delta U_{(a \to b)} + (P_b V_b - P_a V_a) = (P_b V_b - P_a V_a)$$

$$= nRT_b - nRT_a = nR(T_b - T_a) = 0$$

(b) The reversible adiabatic expansion. If the adiabatic expansion is carried out reversibly, then during the process the state of the system is, at all time, given by $PV^\gamma = \text{constant}$, and the final state is the point c in the diagram. The volume V_c is obtained from $P_a V_a^\gamma = P_c V_c^\gamma$ as

$$V_c = \left(\frac{10 \times 10^{5/3}}{1} \right)^{3/5} = 39.8 \text{ liters}$$

and

$$T_c = \frac{P_c V_c}{nR} = \frac{1 \times 39.8}{4.09 \times 0.08206} = 119 \text{ degrees}$$

The point c thus lies on the 119 degree isotherm. As the process is adiabatic, $q = 0$ and hence

$$\Delta U_{(a \to c)} = -w = \int_a^c nc_v dT = nc_v (T_c - T_a)$$

$$= 4.09 \times 1.5 \times 8.3144 \times (119 - 298) \text{ joules}$$

$$= -9.13 \text{ kilojoules}$$

The work done by the system as a result of the process equals the decrease in the internal energy of the system $= 9.13$ kilojoules.

(i) An isothermal process followed by a constant-volume process (the path $a \to e \to c$; that is, an isothermal change from a to e, followed by a constant-volume change from e to c).

$$\Delta U_{(a \to e)} = 0 \text{ as this is an isothermal change of state}$$

$$\Delta U_{(e \to c)} = q_v \quad (\Delta V = 0 \text{ and hence } w = 0)$$

$$= \int_e^c nc_v dT$$

and as the state e lies on the 298 degree isotherm then

$$\Delta U_{(e \to c)} = 4.09 \times 1.5 \times 8.3144 \times (119 - 298) \text{ joules} = -9.13 \text{ kilojoules}$$

Thus

$$\Delta U_{(a \to c)} = \Delta U_{(a \to e)} + \Delta U_{(e \to c)} = -9.13 \text{ kilojoules}$$

(ii) A constant-volume process followed by an isothermal process (the path $a \to d \to c$; that is, a constant-volume change from a to d, followed by an isothermal change from d to c).

$$\Delta U_{(a \to d)} = q_v (\Delta V = 0 \text{ and hence } w = 0)$$

$$= \int_a^d nc_v dT \text{ and as the state } d \text{ lies on the 119 degree isotherm then}$$

$$\Delta U_{(a \to d)} = 4.09 \times 1.5 \times 8.3144 \times (119 - 298) \text{ joules} = -9.13 \text{ kilojoules}$$

$$\Delta U_{(d \to c)} = 0 \text{ as this is an isothermal process and hence}$$

$$\Delta U_{(a \to c)} = \Delta U_{(a \to d)} + \Delta U_{(d \to c)} = -9.13 \text{ kilojoules}$$

(iii) An isothermal process followed by a constant-pressure process (the path $a \to b \to c$; that is, an isothermal change from a to b, followed by a constant-pressure change from b to c).

$$\Delta U_{(a \to b)} = 0 \text{ as this process is isothermal}$$

$$\Delta U_{(b \to c)} = q_p - w \text{ and as } P_b = P_c \text{ then } w = P_b(V_c - V_b)$$

$$= \int_b^c nc_p dT - P_b(V_c - V_b)$$

As $c_v = 1.5 \ R$ and $c_p - c_v = R$, then $c_p = 2.5 \ R$; and as 1 liter atm equals 101.3 joules,

$$\Delta U_{(b \to c)} = [4.09 \times 2.5 \times 8.3144 \times (119 - 298)]$$

$$- [1 \times (39.8 - 100) \times 101.3] \text{ joules}$$

$$= -15.2 + 6.1 = -9.1 \text{ kilojoules}$$

Thus

$$\Delta U_{(a \to c)} = U_{(a \to b)} + U_{(b \to c)} = -9.1 \text{ kilojoules}$$

(iv) A constant-volume process followed by a constant-pressure process (the path $a \to f \to c$; that is, a constant-volume change from a to f, followed by a constant-pressure change from f to c).

$$\Delta U_{(a \to f)} = q_v \ (V_a = V_f \text{ and hence } w = 0)$$

$$= \int_a^f nc_v dT$$

From the ideal gas law

$$T_f = \frac{P_f V_f}{nR} = \frac{1 \times 10}{4.09 \times 0.08206} = 30 \text{ degrees}$$

i.e., the state f lies on the 30 degrees isotherm. Thus

$$\Delta U_{(a \to f)} = 4.09 \times 1.5 \times 8.3144 \times (30 - 298) \text{ joules} = -13.67 \text{ kilojoules}$$

$$\Delta U_{(f \to c)} = q_p - w = \int_f^c nc_p dT - P_f(V_c - V_f)$$

$$= [4.09 \times 2.5 \times 8.3144 \times (119 - 30)] - [1 \times (39.8 - 10) \times 101.3]$$

$$\text{joules} = +7.57 - 3.02 \text{ kilojoules}$$

Thus

$$\Delta U_{(a \to c)} = \Delta U_{(a \to f)} + \Delta U_{(f \to c)} = 13.67 + 7.57 - 3.02$$

$$= -9.12 \text{ kilojoules}$$

(v) A constant-pressure process followed by a constant-volume process (the path $a \to g \to c$; i.e., a constant-pressure step from a to g, followed by a constant-volume step from g to c).

$$\Delta U_{(a \to g)} = q_p - w$$

From the ideal gas law

$$T_g = \frac{P_g V_g}{nR} = \frac{10 \times 39.8}{4.09 \times 0.08206} = 1186 \text{ degrees}$$

and hence the state g lies on the 1186 degrees isotherm. Thus

$$\Delta U_{(a \to g)} = [4.09 \times 2.5 \times 8.3144 \times (1186 - 298)] \text{ joules}$$

$$- [10 \times (39.8 - 10) \times 101.3] \text{ joules}$$

$$= 75.5 - 30.2 \text{ kilojoules}$$

$$\Delta U_{(g \to c)} = q_v = 4.09 \times 1.5 \times 8.3144 \times (119 - 1186) \text{ joules}$$

$$= -54.4 \text{ kilojoules}$$

Thus

$$\Delta U_{(a \to c)} = \Delta U_{(a \to g)} + \Delta U_{(g \to c)} = 75.5 - 30.2 - 54.4$$

$$= -9.1 \text{ kilojoules}$$

The value of $\Delta U_{(a \to c)}$ is thus seen to be independent of the path taken by the process between the states a and c.

The change in enthalpy from a to c. The enthalpy change is most simply calculated from consideration of a path which involves an isothermal portion over which

$\Delta H = 0$ and an isobaric portion over which $\Delta H = q_p = \int nc_p dT$. For example, consider the path $a \rightarrow b \rightarrow c$.

$$\Delta H_{(a \rightarrow b)} = 0$$

$$\Delta H_{(b \rightarrow c)} = q_p = nc_p(T_c - T_b)$$

$$= 4.09 \times 2.5 \times 8.3144 \times (119 - 298) \text{ joules}$$

$$= -15.2 \text{ kilojoules}$$

and hence

$$\Delta H_{(a \rightarrow c)} = -15.2 \text{ kilojoules}$$

or alternatively

$$\Delta H_{(a \rightarrow c)} = \Delta U_{(a \rightarrow c)} + (P_c V_c - P_a V_a)$$

$$= -9.12 \text{ kilojoules} + [(1 \times 39.8 - 10 \times 10) \times 101.3)] \text{ joules}$$

$$= -9.12 - 6.10 = -15.2 \text{ kilojoules}$$

in each of the paths (i) to (v) the heat and work effects differ, although in each case the difference $q - w$ equals -9.12 kilojoules. In the case of the reversible adiabatic path, $q = 0$ and hence $w = +9.12$ kilojoules. If the processes (i) to (v) are carried out reversibly, then

For path (i) $q = -9.12 +$ the area $aeih$
For path (ii) $q = -9.12 +$ the area $dcih$
For path (iii) $q = -9.12 +$ the area $abjh -$ the area $cbji$
For path (iv) $q = -9.12 +$ the area $fcih$
For path (v) $q = -9.12 +$ the area $agih$

PROBLEMS

2.1 An ideal gas at 300 K has a volume of 15 liters at a pressure of 15 atm. Calculate (1) the final volume of the system, (2) the work done by the system, (3) the heat entering or leaving the system, (4) the change in the internal energy, and (5) the change in the enthalpy when the gas undergoes

a. A reversible isothermal expansion to a pressure of 10 atm
b. A reversible adiabatic expansion to a pressure of 10 atm

The constant volume molar heat capacity of the gas, c_v, has the value 1.5 R.

2.2 One mole of a monatomic ideal gas, in the initial state $T = 273$ K, $P = 1$ atm, is subjected to the following three processes, each of which is conducted reversibly:

a. A doubling of its volume at constant pressure,
b. Then a doubling of its pressure at constant volume,

c. Then a return to the initial state along the path $P = 6.643 \times 10^{-4}V^2 + 0.6667$.

Calculate the heat and work effects which occur during each of the three processes.

2.3 The initial state of a quantity of monatomic ideal gas is $P = 1$ atm, $V = 1$ liter, and $T = 373$ K. The gas is isothermally expanded to a volume of 2 liters and is then cooled at constant pressure to the volume V. This volume is such that a reversible adiabatic compression to a pressure of 1 atm returns the system to its initial state. All of the changes of state are conducted reversibly. Calculate the value of V and the total work done on or by the gas.

2.4 Two moles of a monatomic ideal gas are contained at a pressure of 1 atm and a temperature of 300 K. 34,166 joules of heat are transferred to the gas, as a result of which the gas expands and does 1216 joules of work against its surroundings. The process is reversible. Calculate the final temperature of the gas.

2.5 One mole of N_2 gas is contained at 273 K and a pressure of 1 atm. The addition of 3000 joules of heat to the gas at constant pressure causes 832 joules of work to be done during the expansion. Calculate (a) the final state of the gas, (b) the values of ΔU and ΔH for the change of state, and (c) the values of c_v and c_p for N_2. Assume that nitrogen behaves as an ideal gas, and that the above change of state is conducted reversibly.

2.6 Ten moles of ideal gas, in the initial state $P_1 = 10$ atm, $T_1 = 300$ K, are taken round the following cycle:

a. A reversible change of state along a straight line path on the P-V diagram to the state $P = 1$ atm, $T = 300$ K,
b. A reversible isobaric compression to $V = 24.6$ liters, and
c. A reversible constant volume process to $P = 10$ atm.

How much work is done on or by the system during the cycle? Is this work done on the system or by the system?

2.7 One mole of an ideal gas at 25°C and 1 atm undergoes the following reversibly conducted cycle:

a. An isothermal expansion to 0.5 atm, followed by
b. An isobaric expansion to 100°C, followed by
c. An isothermal compression to 1 atm, followed by
d. An isobaric compression to 25°C.

The system then undergoes the following reversible cyclic process.

a. An isobaric expansion to 100°C, followed by
b. A decrease in pressure at constant volume to the pressure P atm, followed by
c. An isobaric compression at P atm to 24.5 liters, followed by
d. An increase in pressure at constant volume to 1 atm.

Calculate the value of P which makes the work done on the gas during the first cycle equal to the work done by the gas during the second cycle.

2.8 Two moles of an ideal gas, in an initial state $P = 10$ atm, $V = 5$ liters, are taken reversibly in a clockwise direction around a circular path give by $(V - 10)^2 + (P - 10)^2 = 25$. Calculate the amount of work done by the gas as a result of the process, and calculate the maximum and minimum temperatures attained by the gas during the cycle.

Chapter 3

THE SECOND LAW OF THERMODYNAMICS

3.1. INTRODUCTION

In Chap. 2 it was seen that when a system undergoes a change of state, the consequent change in the internal energy of the system, which is dependent only on the initial and final states, is equal to the algebraic sum of the heat and work effects. The question now arises: What magnitudes may the heat and work effects have, and what criteria govern these magnitudes? Two obvious cases occur, namely, the extreme cases in which either $w = 0$ or $q = 0$, in which cases, respectively, $q = \Delta U$ and $w = -\Delta U$. But, if $q \neq 0$ and $w \neq 0$, is there a definite amount of work which the system can do during its change of state? The answers to these questions require an examination of the nature of processes. This examination, which will be made in this chapter, identifies two classes of processes (reversible and irreversible processes) and introduces a state function called the entropy, S. The concept of entropy will be introduced from two different starting points. First, in Secs. 3.2–3.8, entropy will be introduced and discussed as a result of a need for quantification of the degree of irreversibility of a process; and, second, in Secs. 3.10–3.14, it will be seen that, as a result of an examination of the properties of reversibly operated heat engines, there naturally develops a quantity which has all the properties of a state function. This function is found to be the entropy. The examination leads to a statement called the Second Law of Thermodynamics, which, together with the First Law, lays the foundation for the thermodynamic method of describing the behavior of matter.

3.2 SPONTANEOUS OR NATURAL PROCESSES

Left to itself, a system will do one of two things: It will remain in the state in which it happens to be, or it will move, of its own accord, to some other state. That is, if the system is initially in equilibrium with its surroundings, then, left to itself, it will remain in this, its equilibrium, state. On the other hand, if the initial state is not the equilibrium state, the system will spontaneously move toward its equilibrium state. The equilibrium state is a state of rest, and thus, once at equilibrium, a system will only move away from equilibrium if it is acted on by some external agency. Even then, the combined system, comprising the original system and the external agency, is simply moving toward the equilibrium state of the combined system. A process which involves the spontaneous movement of a system from a nonequilibrium state to an equilibrium state is called a *natural* or *spontaneous* process. As such a process cannot be reversed without the application of an external agency (a process which would leave a permanent change in the external agency), such a process is said to be *irreversible.* (The terms natural, spontaneous, and irreversible are synonymous in this context.)

The mixing of gases and the flow of heat down a temperature gradient are common examples of natural processes. If the initial state of a system consisting of two gases, A and B, is that in which gas A is contained in one vessel and gas B is contained in a separate vessel, then, when the vessels are connected to one another, the system spontaneously moves to the equilibrium state in which the two gases are completely mixed, i.e., the composition of the gas mixture is uniform throughout the volume which the gas occupies. If the initial state of a two-body system is that in which one body is at one temperature and the other body is at another temperature, then, when the bodies are placed in thermal contact with one another, the spontaneous process which occurs is the flow of heat from the hotter to the colder body, and the equilibrium state is reached when both bodies attain a common uniform temperature. In both of these examples the reverse process (unmixing of the gases and the flow of heat up a temperature gradient) will never occur spontaneously, and in both examples the simplicity of the system, along with common experience, allows the equilibrium states to be predicted without any knowledge of the criteria for equilibrium. However, in less simple systems, the equilibrium state cannot be predicted from common experience, and the criteria governing equilibrium must be established before calculation of the equilibrium state can be made.

Determination of the equilibrium state is of prime importance in thermodynamics, as knowledge of this state for any materials reaction system will allow determination to be made of the direction in which any reaction will proceed from any starting or initial state. For example, knowledge of the equilibrium state of a chemical reaction system such as

$$A + B = C + D$$

will afford knowledge of whether, from any initial state—which would be some mixture of A, B, C, and D—the reaction will proceed from right to left or from left to right, and, in either case, to what extent before equilibrium is reached.

If a system undergoes a spontaneous process involving the performance of work and the production of heat, then, as the process continues, during which time the system is being brought nearer and nearer to its equilibrium state, the capacity of the system for further spontaneous change decreases. Finally, once equilibrium is reached, the capacity of the system for doing further work is exhausted. In the initial nonequilibrium state of an isolated system (a system of constant internal energy), some of the energy of the system is available for doing useful work, and when the equilibrium state is reached, as a result of the completion of a spontaneous process, none of the energy of the system is available for doing useful work. Thus, as a result of the spontaneous process, the system has become degraded, in that energy, which was available for doing useful work, has been converted to thermal energy (or heat), in which form it is not available for external purposes.

3.3 ENTROPY AND THE QUANTIFICATION OF IRREVERSIBILITY

Two distinct types of spontaneous process are (1) the conversion of work to heat (i.e., the degradation of mechanical energy to thermal energy) and (2) the flow of heat down a temperature gradient. If it is considered that an irreversible process is one in which the energy of the system undergoing the process is degraded, then the possibility that the extent of degradation can differ from one process to another suggests that a quantitative measure of the extent of degradation, or degree of irreversibility, exists. The existence of processes which exhibit differing degrees of irreversibility can be illustrated as follows. Consider the weight-heat reservoir system shown schematically in Fig. 3.1. This system consists of a weight-pulley arrangement which is coupled to a constant-temperature heat reservoir, and the system is at equilibrium when an upward force acting on the weight exactly balances the downward force, W, of the weight. If the upward force is removed, then the equilibrium is upset and the weight spontaneously falls, thus performing work, which is converted, by means of a suitable system of paddle wheels, to heat which enters the constant-temperature heat reservoir. Equilibrium is reattained when the upward force acting on the weight is replaced, and the net effect of this process is that mechanical energy has been converted to thermal energy.

Lewis and Randall* considered the following three processes:

1. The heat reservoir in the weight-heat reservoir system is at the temperature T_2. The weight is allowed to fall, performing work, w, and the heat produced, q, enters the heat reservoir.
2. The heat reservoir at the temperature T_2 is placed in thermal contact with a heat reservoir at a lower temperature T_1, and the same heat q is allowed to flow from the reservoir at T_2 to the reservoir at T_1.
3. The heat reservoir in the weight-heat reservoir system is at the temperature T_1. The weight is allowed to fall, performing work, w, and the heat produced, q enters the reservoir.

---

*G. N. Lewis and M. Randall, "Thermodynamics," revised by K. S. Pitzer and L. Brewer, 3rd ed., p. 78, McGraw-Hill, New York, 1995.

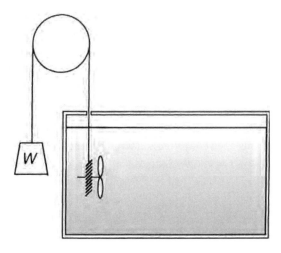

Figure 3.1 A weight-pulley-heat reservoir arrangement in which the work done by the falling weight is degraded to heat, which appears in the heat reservoir.

Each of these process is spontaneous and hence irreversible, and degradation occurs in each of them. However, as process (3) is the sum of processes (1) and (2), the degradation occurring in process (3) must be greater than the degradation occurring in each of the processes (1) and (2). Thus it can be said that process (3) is more irreversible than either process (1) or process (2). Examination of the three processes indicates that both the amount of heat produced, q, and the temperatures between which this heat flows are important in defining a quantitative scale of irreversibility. In the case of comparison between process (1) and process (3), the quantity q/T_2 is smaller than the quantity q/T_1, with is in agreement with the conclusion that process (1) is less irreversible than process (3). The quantity q/T is thus taken as being a measure of the degree of irreversibility of the process, and the value of q/T is called the increase in *entropy*, S, occurring as a result of the process. Thus when the weight-heat reservoir system undergoes a spontaneous process which causes the absorption of heat q at the constant temperature T, the entropy produced by the system, ΔS, is given by

$$\Delta S = \frac{q}{T} \tag{3.1}$$

The increase in entropy, caused by the process, is thus a measure of the degree of irreversibility of the process.

3.4 REVERSIBLE PROCESSES

As the degree of irreversibility of a process is variable, it should be possible for the process to be conducted in such a manner that the degree of irreversibility is minimized. The ultimate of this minimization is a process in which the degree of irreversibility is zero and in which no degradation occurs. This limit, toward which the behavior of real systems can be made to approach, is the *reversible* process. If a process is reversible, then the concept of spontaneity is no longer applicable. It will be recalled that spontaneity occurred as a result of the system moving, of its own accord, from a nonequilibrium state to an equilibrium state. Thus, if the spontaneity is removed, it is apparent that at all times during the process, the system is at equilibrium. Thus a reversible process is one during which the system is never away from equilibrium, and a reversible process which takes the system from the state A to the state B is one in which the process path passes through a continuum of equilibrium states. Such a path is, of course, imaginary, but it is possible to conduct an actual process in such a manner that it is virtually reversible. Such an actual process is one which proceeds under the influence of an infinitesimally small driving force such that, during the process, the system is never more than an infinitesimal distance from equilibrium. If, at any point along the path, the minute external influence is removed, then the process ceases, or, if the direction of the minute external influence is reversed, then the direction of the process is reversed. Natural and reversible processes are illustrated in the following discussion.

3.5 AN ILLUSTRATION OF IRREVERSIBLE AND REVERSIBLE PROCESSES

Consider a system of water and water vapor at the uniform temperature T contained in a cylinder fitted with a frictionless piston, and let the cylinder be placed in thermal contact with a heat reservoir which is also at the constant temperature T. This system is shown in Fig. 3.2.

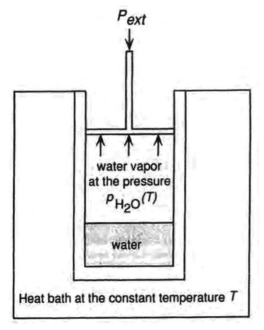

P_{ext}

water vapor at the pressure $P_{H_2O}(T)$

water

Heat bath at the constant temperature T

Figure 3.2 A thermo-stalled piston and cylinder containing water and water vapor.

The water vapor in the cylinder exerts a certain pressure $P_{H_2O}(T)$, which is the saturated vapor pressure of water at the temperature T. The system is exactly at equilibrium when the external pressure acting on the piston, P_{ext}, equals the internal pressure acting on the piston, $P_{H_2O}(T)$, and when the temperature of the water + water vapor in the cylinder equals the temperature T of the constant-temperature heat reservoir. If the external pressure acting on the piston, P_{ext} is suddenly decreased by a finite amount ΔP, then the imbalance in pressure causes the piston to accelerate rapidly out of the cylinder. The consequent rapid expansion of the water vapor decreases the water vapor pressure to a value below its saturation (and hence equilibrium) value, and thus water spontaneously evaporates in an attempt to reestablish equilibrium between the water and its vapor. The spontaneous evaporation, being endothermic, decreases the temperature of the water, and hence sets up a temperature gradient between the heat reservoir and the contents of the cylinder. This, in turn, causes heat to flow spontaneously from the heat reservoir to the cylinder in an attempt to reestablish thermal equilibrium between the two. If, when 1 mole of water has evaporated, the external pressure acting on the piston is instantaneously increased to the original value, P_{ext}, then evaporation of the water vapor ceases, the flow of heat ceases, and complete equilibrium is reestablished. The work done by the system during this process equals $(P_{ext} - \Delta P)V$, where V is the molar volume of water vapor at $P_{H_2O}(T)_2$. If the external pressure acting on the piston is suddenly increased by a finite amount ΔP, then the piston accelerates rapidly into the cylinder. The compression of the water vapor increases its pressure to a value greater than the saturation value, and hence spontaneous condensation occurs, which, being exothermic, increases the temperature in the cylinder to a value higher than T. The temperature gradient between the cylinder and the heat reservoir then causes the spontaneous flow of heat from the cylinder to the reservoir. If, when 1 mole of water vapor has been condensed, the external pressure acting on the piston is instantaneously decreased to its original value, equilibrium is reestablished and the work done on the system equals $(P_{ext} + \Delta P)V$. The permanent change in the external agency caused by the cyclic process is thus $2\Delta PV$.

Consider, again, the evaporation process. If the magnitude of P_{ext} is decreased by an infinitesimal amount δP, the resulting minute imbalance between the pressures acting on the piston causes the cylinder to move slowly out of the piston. The slow expansion of the water vapor decreases its pressure, and when the pressure has fallen by an infinitesimal amount below the saturation value, evaporation of the water begins. The evaporation sets up an infinitesimal temperature gradient between the heat reservoir and the cylinder, down which flows the required latent heat of evaporation of the water. The smaller the value of δP, then the slower the process, the smaller the degree of undersaturation of the water vapor, and the smaller the temperature gradient. The more slowly the process is carried out, then the greater the opportunity afforded to the evaporation and heat flow process to "keep up" with equilibrium. If, after the evaporation of 1 mole of water, the external pressure is instantaneously increased to its original value P_{ext}, then the work done by the system equals $(P_{ext} - \delta P)V$. If the external pressure is then increased by δP, then work $(P_{ext} + \delta P)V$, is done on the system to condense 1 mole of water vapor, and the permanent change in the external agency equals the work done on the system minus the work done by the system during the cyclic process. It is thus seen that the smaller the value of δP, then the more nearly equal are the two work terms, and in the limit that they are equal, no permanent change occurs in the external agency, and hence the cyclic process has been conducted reversibly.

It is thus seen that reversibility is approached when the evaporation or condensation processes are carried out in such a manner that the pressure exerted by the water vapor is never more than infinitesimally different from its saturation value at the temperature T. It can also be seen that, a complete reversibility is approached, the process becomes infinitely slow.

3.6 ENTROPY AND REVERSIBLE HEAT

Consider only the evaporation process. The work done by the system during the evaporation of 1 mole is seen to have its maximum value, $w_{max} = P_{ext}V$, when the process is conducted reversibly. Any irreversible evaporation process performs less work, $w = (P_{ext} - \Delta P)V$. The change in the value of U for the system caused by the evaporation of 1 mole of water is independent of whether the process is conducted reversibly or not, and thus, from the First Law, Eq. (2.1), it is seen that the maximum amount of heat, q_{rev}, is transferred from the heat reservoir to the cylinder when the process is carried out reversibly, where $q_{rev} = \Delta U + w_{max}$. If the process is carried out irreversibly, then less heat q is transferred from the reservoir to the cylinder, where $q = \Delta U + w$. The difference between the work done during the reversible process and that done during the irreversible process, $(w_{max} - w)$, is the mechanical energy which has been degraded to thermal energy (heat) in the cylinder as a result of the irreversible nature of the process. This heat produced by degradation, which is given as $(q_{rev} - q) = (w_{max} - w)$, accounts for the fact that less heat is transferred to the cylinder from the reservoir during the irreversible process than is transferred during the reversible process.

Thus, if the evaporation process is conducted reversibly, then heat q_{rev} leaves the heat reservoir and enters the cylinder at the temperature T. The change in the entropy of the heat reservoir is given by Eq. (3.1) as

$$\Delta S_{\text{heat reservoir}} = -\frac{q_{rev}}{T}$$

where the negative sign corresponds to heat leaving the heat reservoir, and thus the entropy of the reservoir decreases. The change in the entropy of the water and water vapor in the cylinder is

$$\Delta S_{\text{water + vapor}} = \frac{q_{rev}}{T}$$

where the positive sign corresponds to heat entering the cylinder, and thus the entropy of the contents of the cylinder increases. The change in the entropy of the combined water-vapor-heat reservoir system is thus

$$\Delta S_{\text{total}} = \Delta S_{\text{heat reservoir}} + \Delta S_{\text{water + vapor}} = 0$$

This zero change in entropy is due to the fact that the process was carried out reversibly, i.e., no degradation occurred during the process.

If the evaporation is carried out irreversibly then heat q ($q < q_{rev}$) is transferred from the reservoir to the cylinder. The change in the entropy of the heat reservoir is thus

$$\Delta S_{\text{heat reservoir}} = -\frac{q}{T}$$

However, the total heat appearing in the cylinder equals the heat q transferred from the heat reservoir plus the heat which is produced by degradation of work due to the irreversible nature of the process. Thus degraded work, ($w_{max} - w$), equals ($q_{rev} - q$), and thus the change in the entropy of the contents of the cylinder is

$$\Delta S_{\text{water + vapor}} = \frac{q}{T} + \frac{q_{rev} - q}{T}$$

which, it is seen, is simply q_{rev}/T. Thus the change in the entropy of the combined system caused by the irreversible nature of the process is

$$\Delta S_{\text{total}} = \Delta S_{\text{reservoir}} + \Delta S_{\text{water + vapor}}$$
$$= -\frac{q}{T} + \frac{q_{rev}}{T}$$
$$= \frac{q_{rev} - q}{T}$$

As $q_{rev} > q$, this entropy change is positive, and thus entropy has been produced (or created) as a result of the occurrence of an irreversible process. The entropy produced, ($q_{rev} - q$)/T, is termed $\Delta S_{\text{irreversible}}$ (ΔS_{irr}) and is the measure of the degradation which has occurred as a result of the process. Thus, for the evaporation process, irrespective of the degree of irreversibility,

$$\Delta S_{\text{water + vapor}} = \frac{q}{T} + \Delta S_{irr} \tag{3.2}$$

Consideration of the condensation process shows that the work done *on* the system has a minimum value when the condensation is conducted reversibly, and, correspondingly, the heat transferred from the cylinder to the heat reservoir has a minimum value q_{rev}. If the process is conducted irreversibly then a greater amount of work must be performed, and the excess of this work over the minimum required is the work which is degraded to heat in the irreversible process. This extra heat, in turn, is the difference between that leaving the cylinder, q, and the minimum heat q_{rev}.

Thus, for a reversible condensation

$$\Delta S_{\text{water + vapor}} = -\frac{q_{rev}}{T}$$

$$\Delta S_{\text{heat reservoir}} = \frac{q_{rev}}{T}$$

and

$$\Delta S_{total} = 0$$

i.e., entropy is not created.

For an irreversible condensation,

$$\Delta S_{water + vapor} = -\frac{q}{T} + \frac{q - q_{rev}}{T}$$

$$= \frac{\text{heat leaving cylinder}}{T}$$

$$+ \frac{\text{heat produced in cylinder by degradation}}{T}$$

$$\Delta S_{heat \, reservoir} = \frac{q}{T}$$

and

$$\Delta S_{total} = \left(-\frac{q}{T} + \frac{q - q_{rev}}{T} \right) + \frac{q}{T}$$

$$= \frac{q - q_{rev}}{T}$$

And, as $q > q_{rev}$, it is seen that entropy has been created as a result of the irreversible process. The entropy created is ΔS_{irr}, and, thus, again, the change in the entropy of the water and water vapor is given by

$$\Delta S_{water + vapor} = -\frac{q}{T} + \Delta S_{irr} \qquad (3.3)$$

The important feature to be noted from Eqs. (3.2) and (3.3) is that, in going from an initial state to a final state [either the evaporation or condensation of 1 mole of water at the pressure $P_{H_2O}(T)$, and the temperature T], the left-hand sides of Eqs. (3.2) and (3.3) are constants, being equal, respectively, to q_{rev}/T and $-q_{rev}/T$. The difference in entropy between the final and initial states is thus independent of whether the process is conducted reversibly or irreversibly and, being independent of the process path, can be considered as being the difference between the values of a state function. This state function is the entropy, and in going from state A to state B

$$\Delta S = S_B - S_A = \frac{q}{T} + \Delta S_{irr} \qquad (3.4a)$$

$$= \frac{q_{rev}}{T} \qquad (3.4b)$$

Eq (3.4b) indicates that, as the change in entropy can be determined only by measurement of heat flow at the temperature T, then entropy changes can be measured only for reversible processes, in which case the measured heat flow is q_{rev} and $\Delta S_{irr} = 0$.

3.7 THE REVERSIBLE ISOTHERMAL COMPRESSION OF AN IDEAL GAS

Consider the reversible isothermal compression of 1 mole of an ideal gas from the state (V_A,T) to the state (V_B,T). The gas is placed in thermal contact with a heat reservoir at the temperature T, and, by application of a falling weight, the gas is compressed slowly enough that, at all times during its compression, the pressure exerted on the gas is only infinitesimally greater than the instantaneous pressure of the gas, P_{inst}, where $P_{inst} = RT/V_{inst}$. The state of the gas thus lies, at all times, on a section at the constant temperature T of the P-V-T surface (Figs 1.1 and 1.3a), and hence the gas passes through a continuum of equilibrium states in going from the state (V_A,T) to the state (V_B,T). As the gas is never out of equilibrium, i.e., the process is reversible, no degradation occurs, and thus entropy is not created. Entropy is simply transferred from the gas to the heat reservoir, where it is measured as the heat entering divided by the temperature T.* As the compression is conducted isothermally, $\Delta U = 0$ and thus the work done on the gas = the heat withdrawn from the gas, i.e.,

$$w_{max} = q_{rev}$$

where

$$w_{max} = \int_{V_A}^{V_B} PdV = \int_{V_A}^{V_B} \frac{RTdV}{V} = RT \ln \frac{V_B}{V_A}$$

As $V_B < V_A$, w_{max} is a negative quantity, in accordance with the fact that work is done on the gas. The transfer of heat from the gas to the reservoir causes a change in the entropy of the gas

$$\Delta S_{gas} = \frac{q_{rev}}{T} = \frac{w_{max}}{T} = R \ln \frac{V_B}{V_A}$$

---

*The pertinent feature of a constant-temperature heat reservoir is that it experiences only heat effects and neither performs work nor has work performed on it. The "ice calorimeter," which comprises a system of ice and water at 0°C and 1 atm pressure, is an example of a simple constant-temperature heat reservoir. Heat flowing into or out of this calorimeter at 0°C is measured as the change occurring in the ratio of ice to water present as a result of the heat flow, and as the molar volume of ice is larger than that of water, the change in this ratio is measured as a change in the total volume of ice + water in the calorimeter. Strictly speaking, if heat flows out of the calorimeter, thus freezing some of the water, the volume of the system increases, and hence the calorimeter does, in fact, perform work of expansion against the atmospheric pressure. However, the ratio of the work done in expansion to the corresponding heat leaving the system is small enough that the work effects may be neglected, as is illustrated below.

At 0°C and 1 atm, the molar volume of ice is 19.8 cm³, and the molar volume of water is 18 cm³. Thus, the work done against the atmosphere during the freezing of 1 mole of water at 0°C is $1 \times 1.8 \times 10^{-3} \times 101.3 = 0.182$ J. The latent heat of freezing of 1 mole of water is 6 kJ.

On the other hand, the falling weight which performs the work of compression does not experience heat effects, and, as change in entropy is caused by the flow of heat, changes of entropy do not occur then in the falling weight.

which is a also a negative quantity. Consequently, as there is no change in the total entropy of the system during the reversible compression, the change in the entropy of the reservoir is given by

$$\Delta S_{\text{heat reservoir}} = -\Delta S_{\text{gas}} = R \ln \frac{V_A}{V_B}$$

3.8 THE REVERSIBLE ADIABATIC EXPANSION OF AN IDEAL GAS

Consider the reversible adiabatic expansion of 1 mole of an ideal gas from the state (P_A, T_A) to the state (P_B, T_B). For the process to be reversible, the process must be conducted slowly enough that, at all times, the state of the gas lies on its P-V-T surface, and, as has been shown in Chap. 2, this condition, together with the condition that $q = 0$ (adiabatic process), dictates that the process path across the P-V-T surface follows the line $PV^\gamma = $ constant. As the process is reversible, no degradation occurs, and, as the process is adiabatic, no heat flow occurs. The change in the entropy of the gas is thus zero, and consequently, all states of an ideal gas lying on a $PV^\gamma = $ constant line are states of equal entropy (cf. all states of an ideal gas lying on a $PV = RT$ line are states of equal internal energy). A reversible adiabatic process is thus an isentropic process. During a reversible adiabatic expansion the work done by the gas, w_{max}, equals the decrease in the internal energy of the gas.

If the pressure exerted on the gas is suddenly decreased from P_A to P_B, then the state of the gas, which is initially (P_A, T_A), momentarily moves off the P-V-T surface, and, being thus out of equilibrium, the expansion occurs irreversibly and degradation occurs. As the gas is contained adiabatically, the heat produced by the degradation remains in the gas, and thus the final temperature of the gas after an irreversible expansion is higher than the temperature T_B. Thus the final state of a gas after an irreversible adiabatic expansion from P_A to P_B differs from the final state after a reversible expansion from the same initial to the same final pressure. (The irreversible adiabatic expansion does not follow the path $PV^\gamma = $ constant.) The entropy produced in the gas due to the irreversible process is the difference in entropy between the final and initial states, and the final state itself is determined by the degree of irreversibility of the process. That is, for a given decrease in pressure, $(P_A \rightarrow P_B)$, the more irreversible the process, the more heat produced in the gas by degradation, the higher the final temperature and internal energy, and the greater the increase in entropy. Thus, during an irreversible expansion, the work done by the gas still equals the decrease in the internal energy of the gas (as is required by the First Law), but the decrease in U is less than that in the reversible expansion from P_A to P_B, due to the heat appearing in the gas as the result of degradation.

3.9 SUMMARY STATEMENTS

Three points have emerged from the discussion so far:

1. The entropy of a system increases when the system undergoes an irreversible process.
2. Entropy is not created when a system undergoes a reversible process; entropy is simply transferred from one part of the system to another part.
3. Entropy is a state function.

3.10 THE PROPERTIES OF HEAT ENGINES

Traditionally the concept of entropy as a state function is introduced by considering the behavior and properties of heat engines. A heat engine is a device which converts heat into work, and it is interesting to note that the first steam engine, which was built in 1769, was in operation for a considerable number of years before the reverse process, i.e., the conversion of work into heat, was investigated. In the operation of a heat engine, a quantity of heat is withdrawn from a high-temperature heat reservoir, and some of this heat is converted into work, with the remainder being transferred to a low-temperature heat reservoir. The process is shown schematically in Fig. 3.3.

The familiar steam engine is a typical example of a heat engine. In this device, superheated steam is passed from the boiler (the high-temperature heat reservoir) to the cylinders, where it performs work by expanding against the pistons (the engine). As a result of this expansion the temperature of the steam decreases, and at the end of the piston stroke the spent steam is exhausted to the atmosphere (the low-temperature heat reservoir). A flywheel returns the piston to its original position, thus completing the cycle and preparing for the next working stroke.

The efficiency of a heat engine is given by

$$\text{Efficiency} = \frac{\text{work obtained}}{\text{heat input}} = \frac{w}{q_2}$$

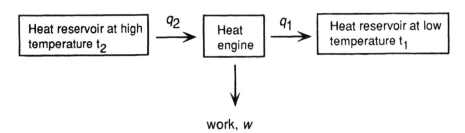

Figure 3.3   Schematic representation of the working of a heat engine.

The factors governing the efficiency of this process were explained in 1824 by Sadi Carnot, who considered the cyclic process illustrated in Fig. 3.4.

In the step $A \rightarrow B$, heat q_2 is reversibly transferred from a heat reservoir at the temperature t_2 to a thermodynamic substance, as a result of which the thermodynamic substance isothermally and reversibly expands from the state A to the state B and performs work w_1 equal to the area *ABba*.

In the step $B \rightarrow C$, the thermodynamic substance undergoes a reversible adiabatic expansion from the state B to the state C, as a result of which its temperature decreases to t_1, and it performs work equal to the area *BCcb*.

In the step $C \rightarrow D$, heat q_1 is isothermally and reversibly transferred from the thermodynamic substance to a heat reservoir at the temperature t_1. Work, w_3, equal to the area *DCcd* is done on the substance.

In the step $D \rightarrow A$, the substance is reversibly and adiabatically compressed during which its temperature increases from t_1 to t_2, and work w_4, equal to the area *ADda,* is done on the substance.

During the cyclic process, which has returned the thermodynamic substance to its initial state, the substance has performed the work $w = w_1 + w_2 - w_3 - w_4$ (equal to the area *ABCD*) and has absorbed heat $q = q_2 - q_1$. For a cyclic process, $\Delta U = 0$, and thus, from the First Law,

$$q = w$$

Thus

$$q_2 - q_1 = w$$

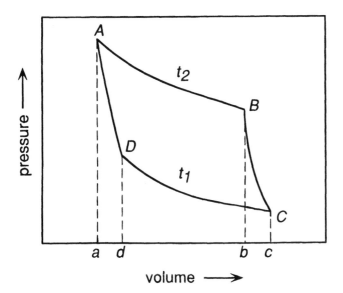

Figure 3.4 The Carnot cycle.

and the efficiency of the cyclic process (which is known as a Carnot cycle) is given by

$$\text{Efficiency} = \frac{w}{q_2} = \frac{q_2 - q_1}{q_2}$$

The consequence of all of the steps in the cyclic process having been conducted reversibly is illustrated in the following discussion. Consider a second engine working with a different substance, again between the temperatures t_1 and t_2, and let this second engine be more efficient than the first one. This greater efficiency could be obtained in either of two ways.

1. q_2 is withdrawn from the heat reservoir at t_2, and more work, w', is obtained from it than was obtained from the first engine, i.e., $w' > w$. Thus, the second engine rejects less heat, q_1' to the cold reservoir at t_1 than does the first engine, i.e., $q' < q$.
2. The same work is obtained by withdrawing less heat q_2' from the heat reservoir at t_2, i.e., $q_2' < q_2$. Thus less heat, q_1', is rejected into the heat reservoir at t_1, i.e., $q_1' < q_1$.

Consider, now, that the second engine is run in the forward direction, and the first engine is run in the reverse direction, i.e., acts as a heat pump. Then, from (1), for the second engine run in the forward direction, $w' = q_2 - q_1'$, for the first engine run in the reverse direction, $-w = -q_2 + q_1$, and the sum of the two processes is

$$(w' - w) = (q_1 - q_1')$$

i.e., an amount of work $(w' - w)$ has been obtained from a quantity of heat $(q_1 - q_1')$ without any other change occurring. Although this conclusion does not contravene the First Law of Thermodynamics, it is contrary to human experience. Such a process corresponds to perpetual motion of the second kind; i.e., heat is converted to work without leaving a change in any other body. (Perpetual motion of the first kind is the creation of energy from nothing.) From (2), for the second engine run in the forward direction, $w = q_2' - q_1'$, for the first engine run in the reverse direction, $-w = -q_2 + q_1$, and the sum of the two processes is

$$(q_2' - q_2) = (q_1 - q_1') = q$$

i.e., an amount of heat q at one temperature has been converted to heat at a higher temperature without any other change occurring. This corresponds to the spontaneous flow of heat up a temperature gradient and is thus more contrary to human experience than is perpetual motion of the second kind.

The above discussion gives rise to a preliminary formulation of the Second Law of Thermodynamics:

1. The principle of Thomsen states that it is impossible, by means of a cyclic process, to take heat from a reservoir and convert it to work without, in the same operation, transferring heat to a cold reservoir.

2. The principle of Clausius states that it is impossible to transfer heat from a cold to a hot reservoir without, in the same process, converting a certain amount of work to heat.

3.11 THE THERMODYNAMIC TEMPERATURE SCALE

The foregoing discussion suggests that all reversible Carnot cycles operating between the same upper and lower temperature must have the same efficiency, namely, the maximum possible. This maximum efficiency is independent of the working substance and is a function only of the working temperatures t_1 and t_2. Thus

$$\text{Efficiency} = \frac{q_2 - q_1}{q_2} = f'(t_1,t_2) = 1 - \frac{q_1}{q_2}$$

or

$$\frac{q_1}{q_2} = f(t_1,t_2)$$

Consider the Carnot cycles shown in Fig. 3.5. The two cycles operating between t_1 and t_2, and between t_2 and t_3, are equivalent to a single cycle operating between t_1 and t_3. Thus

$$\frac{q_1}{q_2} = f(t_1,t_2)$$

$$\frac{q_2}{q_3} = f(t_2,t_3)$$

and

$$\frac{q_1}{q_3} = f(t_1,t_3)$$

so

$$\left(\frac{q_1}{q_3}\right) \times \left(\frac{q_3}{q_2}\right) = \frac{f(t_1,t_3)}{f(t_2,t_3)} = \frac{q_1}{q_2} = f(t_1,t_2)$$

As $f(t_1,t_2)$ is independent of t_3, then $f(t_1,t_3)$ and $f(t_2,t_3)$ must be of the form $f(t_1,t_3) = F(t_1)/F(t_3)$ and $f(t_2,t_3) = F(t_2)/F(t_3)$, i.e., the efficiency function $f(t_1,t_2)$ is the quotient of a function of t_1 alone and t_2 alone.
Thus,

$$\frac{q_1}{q_2} = \frac{F(t_1)}{F(t_2)}$$

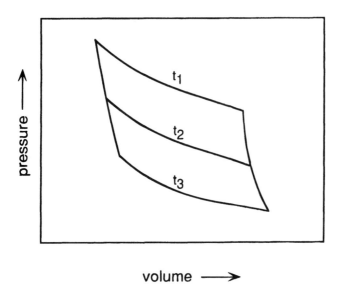

Figure 3.5 Two Carnot cycles.

Kelvin took these functions to have the simplest possible form, namely, T_1 and T_2. Thus

$$\frac{q_1}{q_2} = \frac{T_1}{T_2}$$

in which case the efficiency of a Carnot cycle is

$$\frac{q_2 - q_1}{q_2} = \frac{T_2 - T_1}{T_2} \tag{3.5}$$

This defines an absolute thermodynamic scale of temperature which is independent of the working substance. It is seen that the zero of this temperature scale is that temperature of the cold reservoir which makes the Carnot cycle 100% efficient.

The absolute thermodynamic temperature scale (or Kelvin scale) is identical with the ideal gas temperature scale which was discussed in Chap. 1. This can be demonstrated by considering 1 mole of ideal gas to be the working substance in a Carnot cycle. Referring to Fig. 3.4:

State A to state B. Reversible isothermal expansion at T_2:

$$\Delta U = 0$$

and, from Eq. (2.10)

$$q_2 = w_1 = RT_2 \ln\left(\frac{V_B}{V_A}\right)$$

State B to state C. Reversible adiabatic expansion:

$$q = 0$$

and from Eq. (2.6)

$$w_2 = -\Delta U = -\int_T^{T_1} c_v dT$$

State C to state D. Reversible isothermal compression at T_1:

$$q_1 = w_3 = RT_1 \ln\left(\frac{V_D}{V_C}\right)$$

State D to state A. Reversible adiabatic compression:

$$q = 0$$

$$w_4 = -\int_{T_1}^{T_2} c_v dT$$

Total work done on the gas $= w = w_1 + w_2 + w_3 + w_4$

$$= RT_2 \ln\left(\frac{V_B}{V_A}\right) - \int_{T_2}^{T_1} c_v dT + RT_1 \ln\left(\frac{V_D}{V_C}\right) - \int_{T_1}^{T_2} c_v dT$$

and

The heat absorbed from the hot reservoir $= q_2 = RT_2 \ln\left(\frac{V_B}{V_A}\right)$

It can be shown that

$$\frac{V_B}{V_A} = \frac{V_C}{V_D}$$

and thus

$$w = R(T_2 - T_1) \ln\left(\frac{V_B}{V_A}\right)$$

Thus

$$\text{Efficiency} = \frac{w}{q_2} = \frac{(T_2 - T_1)}{T_2}$$

which is identical with Eq. (3.5).

3.12 THE SECOND LAW OF THERMODYNAMICS

The equation

$$\frac{q_2 - q_1}{q_2} = \frac{T_2 - T_1}{T_2}$$

can be written as

$$\frac{q_2}{T_2} - \frac{q_1}{T_1} = 0 \tag{3.6}$$

Now, any cyclic process can be broken down into a number of Carnot cycles as shown in Fig. 3.6. In going round the cycle ABA in a clockwise direction, the work done by the system equals the area enclosed by the path loop. This loop can be roughly approximated by a number of Carnot cycles as shown, and for the zigzag paths of these cycles, from Eq. (3.6),

$$\sum \frac{q}{T} = 0$$

where the heat entering the system is positive and the heat leaving the system is negative. The zigzag path of the Carnot cycles can be made to coincide with loop ABA by making the Carnot cycles smaller and smaller, and in the limit of coincidence, the summation can be replaced by a cyclic integral, i.e.,

$$\oint \frac{\delta q}{T} = 0$$

The vanishing of the cyclic integral indicates that the integral is a perfect differential of some function of state of the system. This function is called the entropy, S, and is defined as

$$dS = \frac{\delta q}{T} \tag{3.7}$$

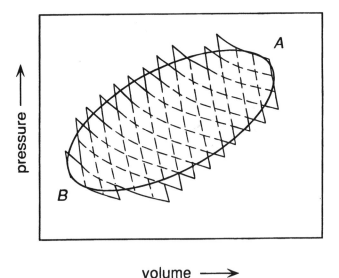

Figure 3.6 A cyclic process broken down into a large number of Carnot cycles.

Thus, for the loop *ABA*

$$\oint dS = 0 = \int_A^B dS + \int_B^A dS = (S_B - S_A) + (S_A - S_B) = 0$$

It is to be emphasized that q in Eq. (3.7) is the reversible heat increment, and thus Eq. (3.7) should be properly written as

$$dS = \frac{\delta q_{rev}}{T} \tag{3.8}$$

(This expression was derived from the consideration of Carnot cycles in which all operations are conducted reversibly.)

The application of Eq. (3.6) to a reversibly operated heat engine, in which q_2 is withdrawn from a constant-temperature source at T_2, work w is performed, and heat q_1 is rejected into a constant-temperature heat sink at T_1, shows that the decrease in the entropy of the heat source, $= q_2/T_2$, equals the increase in the entropy of the heat sink, $= q_1/T_1$, i.e., $\Delta S_{total} = 0$, which is a consequence of the fact that the process is conducted reversibly.

The Second Law of Thermodynamics can thus be stated as

1. The entropy S, defined by the equation $dS = \delta q_{rev}/T$, is a function of state.
2. The entropy of a system in an adiabatic enclosure can never decrease; it increases during an irreversible process and remains constant during a reversible process.

From (2) it is seen that, for an infinitesimal change of state of an adiabatically contained system,

$$\sum dS_i \geq 0 \tag{3.9}$$

i.e., the sum of the incremental changes in entropy of all i parts of the system which are in thermal contact with each other is zero if the infinitesimal change of state is reversible, and is greater than zero if the infinitesimal change of state is irreversible. Eq. (3.9) can be converted to an equality by writing

$$\sum_i dS_i = dS_{irr} \tag{3.10}$$

where dS_{irr} is the entropy created in the given incremental irreversible process.

3.13 MAXIMUM WORK

For a change of state from A to B, the First Law gives

$$U_B - U_A = q - w$$

The First Law gives no indication of the allowed magnitudes of q and w in the given process. It has been seen, however, in the preceding discussion that, although the values of q and w can vary depending on the degree of irreversibility of the path taken between the state A and B, the Second Law sets a definite limit on the maximum amount of work which can be obtained from the system during a given change of state, and, hence, sets a limit on the quantity of heat which the system may absorb. For an infinitesimal change of state, Eq. (3.4a) can be written as

$$dS_{system} = \frac{\delta q}{T} + dS_{irr}$$

and, from the First Law,

$$\delta q = dU_{system} + \delta w$$

Thus

$$dS_{system} = \frac{dU_{system} + \delta w}{T} + dS_{irr}$$

or

$$\delta w = TdS_{system} - dU_{system} - TdS_{irr}$$

or

$$\delta w \leq TdS_{system} - dU_{system} \tag{3.11}$$

If the temperature remains constant throughout the process (and equal to the temperature of the reservoir supplying heat to the system), then integration of Eq. (3.11) from state A to state B gives

$$w \leq T(S_B - S_A) - (U_B - U_A)$$

and as U and S are functions of state, then w cannot be greater than a certain amount w_{max}, the work which is obtained from the system when the process is conducted reversibly, i.e.,

$$w_{max} = T(S_B - S_A) - (U_B - U_A)$$

This work w_{max} corresponds to the absorption of the maximum heat q_{rev}.

As entropy is a state function, then in undergoing any specific change of state from A to B, *the change in the entropy of the system is the same whether the process is conducted reversibly or irreversibly.* The above discussion indicates that *it is the heat effect which is different* in the two cases, i.e., if the process involves the absorption of heat and is conducted reversibly, then the heat absorbed, q_{rev}, is greater than the heat which would have been absorbed if the process had been conducted irreversibly. As has been seen, when 1 mole of an ideal gas is isothermally and reversibly expanded from state A to state B, heat $q = RT \ln (V_B/V_A)$ is reversibly transferred from the heat reservoir to the gas, and the increase in the entropy of the gas, $S_B - S_A$, equals $R \ln (V_B/V_A)$. The entropy of the heat reservoir decreases by an equal amount and entropy is not created, i.e., $\Delta S_{irr} = 0$. However, if the mole of gas is allowed to expand freely from P_A to P_B (as in Joule's experiment discussed in Sec. 2.6) then, as the gas performs no work, no heat is transferred from the reservoir to the gas, and there is no change in the entropy of the reservoir. As entropy is a state function, the value of $S_B - S_A$ is independent of the process path, and hence the entropy created, ΔS_{irr}, equals $S_B - S_A = R \ln (V_B/V_A)$. This entropy is created as a result of the degradation of the work which would have been performed by the gas had the expansion not been carried out against a zero pressure. This degraded work

$$
\begin{aligned}
&= w_{max} - w \\
&= w_{max} - 0 \\
&= w_{max} \\
&= q_{rev}
\end{aligned}
$$

Thus the free expansion represents the limit of irreversibility at which all of the "potential" work is degraded to heat. The degraded heat appearing in the gas accounts for the increase in the entropy of the gas. Thus, for the isothermal expansion of 1 mole of ideal gas from the state A to the state B, the value of ΔS_{irr} can vary between zero and $R \ln (V_B/V_A)$ depending on the degree of irreversibility of the process.

3.14 ENTROPY AND THE CRITERION FOR EQUILIBRIUM

At the beginning of this chapter it was stated that a system, left to itself, would either remain in the state in which it happened to be, or would spontaneously move to some other state, i.e., if the system is initially at equilibrium, then it will remain at equilibrium, and if it is initially in a nonequilibrium state then it will spontaneously move to its equilibrium state. This spontaneous process is, by definition, irreversible, and during the movement of the system from its initial nonequilibrium state to its final equilibrium state the entropy of the system increases. The attainment of the equilibrium state coincides with the entropy reaching a maximum value, and hence entropy can be used as a criterion for determination of the equilibrium state.

In an isolated system of constant internal energy, U, and constant volume, V, equilibrium is attained when the entropy of the system is a maximum, consistent with the fixed values of U and V. Consider the chemical reaction

$$A + B = C + D$$

occurring in an adiabatic enclosure at constant volume. Starting with A and B, the reaction will proceed from left to right as long as the entropy of the system is thereby increased; or, conversely, starting with C and D, the reaction will proceed from right to left, again provided that the entropy of the system is thereby increased. Fig. 3.7 shows a possible variation of entropy with extent of reaction. It is seen that a point is reached along the reaction coordinate at which the entropy of the system has its maximum value. This is the equilibrium state of the system, as further reaction, in either direction, would decrease the entropy and, hence, will not occur spontaneously.

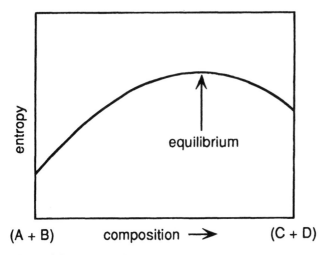

Figure 3.7 Schematic representation of the entropy of a closed system containing $A + B + C + D$ as a function of the extent of the reaction $A + B = C + D$ at constant internal energy and volume.

3.15 THE COMBINED STATEMENT OF THE FIRST AND SECOND LAWS OF THERMODYNAMICS

For an incremental change in the state of a closed system, the First Law of Thermodynamics gives

$$dU = \delta q - \delta w$$

and, if the process occurs reversibly, the Second Law of Thermodynamics gives

$$dS = \frac{\delta q}{T} \quad \text{or} \quad \delta q = TdS$$

and

$$\delta w = PdV$$

Combination of the two laws gives the equation

$$dU = TdS - PdV \tag{3.12}$$

Restrictions on the application of Eq. (3.12) are

1. That the system is closed, i.e., does not exchange matter with its surroundings during the process.
2. That work due to change in volume is the only form of work performed by the system.

Eq. (3.12) relates the dependent variable of the system, U, to the independent variables, S and V, i.e.,

$$U = U(S,V)$$

the total differential of which is

$$dU = \left(\frac{\partial U}{\partial S}\right)_V dS + \left(\frac{\partial U}{\partial V}\right)_S dV \tag{3.13}$$

Comparison of Eq. (3.12) and Eq. (3.13) shows that

> Temperature is defined as $(\partial U/\partial S)_V$
> Pressure is defined as $-(\partial U/\partial V)_S$

The particularly simple form of Eq. (3.12) stems from the fact that, in considering variations in U as the dependent variable, the "natural" choice of independent variables is S and V. Consideration of S as the dependent variable and U and V as the independent variables, i.e.,

$$S = S(U,V)$$

gives

$$dS = \left(\frac{\partial S}{\partial U}\right)_V dU + \left(\frac{\partial S}{\partial V}\right)_U dV \tag{3.14}$$

Rearranging Eq. (3.12) as

$$dS = \frac{dU}{T} + \frac{PdV}{T}$$

and comparing with Eq. (3.14) shows that

$$\left(\frac{\partial S}{\partial V}\right)_U = \frac{P}{T}$$

Equilibrium occurs in a system of constant internal energy and constant volume when the entropy of the system is maximized, and in a system of constant entropy and volume, equilibrium occurs when the internal energy is minimized.

The further development of thermodynamics is a consequence of the fact that S and V are an inconvenient pair of independent variables. In considering a real system, considerable difficulty would be encountered in arranging the state of the system such that, simultaneously, it has the required entropy and occupies the required volume.

3.16 SUMMARY

1. The process paths taken by a system undergoing a change of state can be classified into two types; reversible and irreversible. When the change in the state of the system occurs as the result of the application of a finite driving force, the

process proceeds irreversibly, and the degree of irreversibility of the process increases with increasing magnitude of the driving force. For a process to occur reversibly, the driving force must be infinitesimal, and thus a reversible process proceeds at an infinitesimal rate. The system moves through a continuum of equilibrium states during a reversible process.

2. When a system undergoes a change of state during which it performs work and absorbs heat, the magnitudes of the quantities w and q are maxima (w_{max} and q_{rev}), respectively, when the change of state occurs reversibly. For an irreversible path between the two states, less work is performed by the system, and correspondingly less heat is absorbed.

3. There exists a state function, entropy S, which is defined as

$$dS = \frac{\delta q_{rev}}{T}$$

This difference between the entropy in state B and that in state A is thus

$$\Delta S = S_B - S_A = \int_A^B \frac{\delta q_{rev}}{T}$$

If, in moving between the two states, the temperature of the system remains constant, the increase in the entropy of the system is $\Delta S = q_{rev}/T$, where q_{rev} is the heat absorbed by the system in moving *reversibly* between the two states.

4. If heat q_{rev} is provided by a constant-temperature heat reservoir at the temperature T, the entropy of the reservoir decreases by the amount q_{rev}/T as a result of the system moving from A to B. The entropy of the combined system + heat reservoir is thus unchanged as a result of the reversible process; entropy has simply been transferred from the heat reservoir to the system.

5. If the change in the state of a system from A to B were carried out irreversibly, then less heat q ($q < q_{rev}$) would be withdrawn from the heat reservoir by the system. Thus the magnitude of the decrease in the entropy of the reservoir would be smaller (equal to q/T). However, as entropy is a state function, $S_B - S_A$ is independent of the process path and thus $\Delta S_{system} + \Delta S_{heat\ reservoir} > 0$. Entropy has been created as a result of the occurrence of an irreversible process. The created entropy is termed ΔS_{irr}.

6. In the general case, $S_B - S_A = q/T + \Delta S_{irr}$, and as the degree of irreversibility increases, the heat q withdrawn from the heat reservoir decreases, and the magnitude of ΔS_{irr} increases.

7. The increase in entropy, due to the occurrence of an irreversible process, arises from the degradation of the energy of the system, wherein some of the internal energy, which is potentially available for the doing of useful work, is degraded to heat.

8. A process, occurring in an adiabatically contained system of constant volume (i.e., a system of constant U and V), will proceed irreversibly with a consequent production of entropy until the entropy is maximized. The attainment of maximum entropy is the criterion for equilibrium. Thus the entropy of an adiabatically

contained system can never decrease; it increases as the result of an irreversible process and remains constant at its maximum value during a reversible process.

9. Combination of the First and Second Laws of Thermodynamics gives, for a closed system which does no work other than the work of expansion against a pressure, $dU = TdS - PdV$. U is thus the natural choice of dependent variable for S and V as the independent variables.

3.17 NUMERICAL EXAMPLES

Example 1

Five moles of a monatomic ideal gas are contained adiabatically at 50 atm pressure and 300 K. The pressure is suddenly released to 10 atm, and the gas undergoes an irreversible expansion during which it performs 4000 joules of work. Show that the final temperature of the gas after the irreversible expansion is greater than that which the gas would attain if the expansion from 50 to 10 atm had been conducted reversibly. Calculate the entropy produced as a result of the irreversible expansion. The constant-volume molar heat capacity of the gas, c_v, has the value $1.5R$

In the initial state 1,

$$V_1 = \frac{nRT_1}{P_1} = \frac{5 \times 0.08206 \times 300}{50} = 2.46 \text{ liters}$$

If the adiabatic expansion from 50 to 10 atm is carried out reversibly, then the process path follows $PV^\gamma = $ constant, and in the final state 2,

$$V_2 = \left(\frac{P_1 V_1^\gamma}{P_2}\right)^{1/\gamma} = \left(\frac{50 \times 2.46^{5/3}}{10}\right)^{3/5} = 6.47 \text{ liters}$$

and

$$T_2 = \frac{P_2 V_2}{nR} = \frac{10 \times 6.47}{5 \times 0.08206} = 158 \text{ K}$$

For the irreversible process, which takes the gas from the state 1 to the state 3, as $q = 0$,

$$\Delta U = -w = -4000 = nc_v(T_3 - T_1) = 5 \times 1.5 \times 8.3144 \times (T_3 - 300)$$

and hence $T_3 = 236$ K, which is higher than T_2.

As the irreversible expansion from state 1 to state 3 was conducted adiabatically, no heat entered the system, and hence the difference between the entropy at state 3 and the entropy at state 1 is the entropy created, ΔS_{irr}, as a result of the irreversible process. This difference in entropy can be calculated by considering any reversible path from state 1 to state 3. Consider the reversible path $1 \rightarrow a \rightarrow 3$ shown in Fig. 3.8, which is a reversible decrease in temperature from 300 to 236 K at constant volume followed by a reversible isothermal expansion from V_a to V_3.

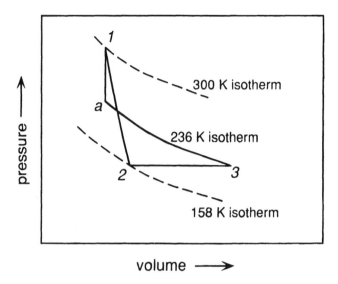

Figure 3.8   The process paths considered in Example 1.

For a reversible constant-volume process

$$\delta q_v = nc_v dT = TdS$$

or

$$dS = \frac{nc_v dT}{T}$$

integration of which, from state 1 to state a, gives

$$S_a - S_1 = nc_v \ln \frac{T_a}{T_1} = 5 \times 1.5 \times 8.3144 \times \ln\frac{234}{300} = -15.0 \text{ J/K}$$

For the reversible isothermal expansion from state a to state 3, as $\Delta U = 0$,

$$q = w = nRT \ln \frac{V_3}{V_a}$$

where

$$V_3 = \frac{nRT_3}{P_3} = \frac{5 \times 0.08206 \times 236}{10} = 9.68 \text{ liters}$$

and thus

$$S_3 - S_a = \frac{q}{T} = nR \ln \frac{V_3}{V_a} = 5 \times 8.3144 \ln \frac{9.68}{2.46} = 57.8 \text{ J/K}$$

The entropy created during the irreversible expansion is thus

$$S_3 - S_1 = -15.0 + 57.0 = 42.0 \text{ joules/degree}$$

Alternatively, the state of the gas could be changed from 1 to 3 along the path $1 \to 2 \to 3$. As the reversible adiabatic expansion from state 1 to state 2 is isentropic,

$$S_3 - S_1 = S_3 - S_2$$

and, for the reversible isobaric expansion from state 2 to state 3,

$$\delta q_p = nc_p dT = TdS$$

or

$$dS = \frac{nc_p dT}{T}$$

integration of which from state 2 to state 3 gives

$$S_3 - S_2 = 5 \times 2.5 \times 8.3144 \times \ln \frac{236}{158} = 42.0 \text{ J/K}$$

which, again, is the entropy created by the irreversible adiabatic expansion of the gas from state 1 to state 3.

Example 2

At a pressure of 1 atm the equilibrium melting temperature of lead is 600 K, and, at this temperature, the latent heat of melting of lead is 4810 J/mole. Calculate the entropy produced when 1 mole of supercooled liquid lead spontaneously freezes at 590 K and 1 atm pressure. The constant-pressure molar heat capacity of liquid lead, as a function of temperature, at 1 atm pressure is given by

$$c_{p(l)} = 32.4 - 3.1 \times 10^{-3} T \text{ J/K}$$

and the corresponding expression for solid lead is

$$c_{p(s)} = 23.56 + 9.75 \times 10^{-3} T \text{ J/K}$$

The entropy produced during the irreversible freezing of the lead equals the difference between the change in the entropy of the lead and the change in the entropy of the constant-temperature heat reservoir (at 590 K) caused by the process.

First calculate the difference between the entropy of 1 mole of solid lead at 590 K and 1 mole of liquid lead at 590 K. Consider the processes illustrated in Fig. 3.9.

1. Step $a \to b$: 1 mole of supercooled liquid lead is heated from 590 to 600 K at 1 atm pressure.

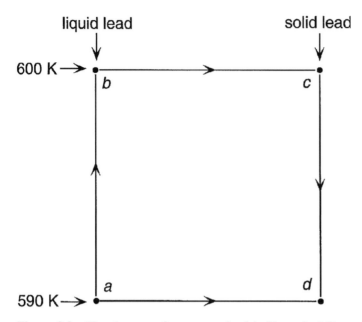

Figure 3.9   The changes of state examined in Numerical Example 2.

2. Step $b \rightarrow c$: 1 mole of liquid lead is solidified reversibly at 600 K (the equilibrium melting or freezing temperature is the only temperature at which the melting or freezing process can be conducted).
3. Step $c \rightarrow d$: 1 mole of solid lead is cooled from 600 to 590 K at 1 atm pressure.

As entropy is a state function,

$$\Delta S_{(a \rightarrow d)} = \Delta S_{(a \rightarrow b)} + \Delta S_{(b \rightarrow c)} + \Delta S_{(c \rightarrow d)}$$

Step $a \rightarrow b$

$$\Delta S_{(a \rightarrow b)} = \int_a^b \frac{\delta q_{rev}}{T} = \int_a^b \frac{\delta q_p}{T} = \int_{590\,K}^{600\,K} \frac{nc_{p,Pb(l)}\,dT}{T}$$

$$= \int_{590\,K}^{600\,K} \left(\frac{32.4}{T} - 3.1 \times 10^{-3} \right) dT$$

$$= 32.4 \ln \frac{600}{590} - 3.1 \times 10^{-3} \times (600 - 590) = +0.514 \text{ J/K}$$

Step b → c

$$\Delta S_{(b \to c)} = \frac{q_{rev}}{T} = \frac{q_p}{T} = \frac{\text{latent heat of freezing}}{\text{equilibrium freezing temperature}}$$

$$= -\frac{4810}{600} = -8.017 \text{ J/K}$$

Step c → d

$$\Delta S_{(c \to d)} = \int_c^d \frac{\delta q_{rev}}{T} = \int_c^d \frac{\delta q_p}{T} = \int_{600 \text{ K}}^{590 \text{ K}} \frac{n c_{p,\text{Pb}(s)} dT}{T}$$

$$= \int_{600 \text{ K}}^{590 \text{ K}} \left(\frac{23.6}{T} + 9.75 \times 10^{-3} \right) dT$$

$$= 23.6 \ln \frac{590}{600} + 9.75 \times 10^{-3}(590 - 600) = -0.494 \text{ J/K}$$

Thus

$$\Delta S_{(a \to d)} = +0.514 - 8.017 - 0.494 = -7.997 \text{ J/K}$$

Consider the heat entering the constant-temperature heat reservoir at 590 K. As the heat is transferred at constant pressure, then $q_p = \Delta H$, where ΔH is the difference between the enthalpies of states d and a. As H is a state function,

$$\Delta H_{(a \to d)} = \Delta H_{(a \to b)} + \Delta H_{(b \to c)} + \Delta H_{(c \to d)}$$

$$\Delta H_{(a \to b)} = \int_a^b n c_{p,\text{Pb}(l)} dT = \int_{590 \text{ K}}^{600 \text{ K}} (32.4 - 3.1 \times 10^{-3}) dT$$

$$= 32.4 \times (600 - 590) - \frac{3.1 \times 10^{-3}}{2} (600^2 - 590^2) = 306 \text{ J}$$

$$\Delta H_{(b \to c)} = -4810 \text{ J}$$

$$\Delta H_{(c \to d)} = \int_c^d n c_{p,\text{Pb}(s)} dT = \int_{600 \text{ K}}^{590 \text{ K}} (23.6 + 9.75 \times 10^{-3}) dT$$

$$= 23.6(590 - 600) + \frac{9.75 \times 10^{-3}}{2} (590^2 - 600^2) = -294 \text{ J}$$

Thus

$$\Delta H_{(a \to d)} = -4799 \text{ J}$$

and so the heat reservoir absorbs 4799 joules of heat at 590 K. Consequently

$$\Delta S_{\text{heat reservoir}} = \frac{4799}{590} = 8.134 \text{ J/K}$$

and thus the entropy created is

$$\Delta S_{irr} = -7.994 + 8.134 = 0.137 \text{ J/K}$$

Examination shows that the lower the temperature of irreversible freezing of the supercooled liquid, the more irreversible the process and the larger the value of ΔS_{irr}.

PROBLEMS

3.1 The initial state of one mole of a monatomic ideal gas is $P = 10$ atm and $T = 300$ K. Calculate the change in the entropy of the gas for (a) an isothermal decrease in the pressure to 5 atm, (b) a reversible adiabatic expansion to a pressure of 5 atm, (c) a constant-volume decrease in the pressure to 5 atm.

3.2 One mole of a monatomic ideal gas is subjected to the following sequence of steps:

a. Starting at 300 K and 10 atm, the gas expands freely into a vacuum to triple its volume.
b. The gas is next heated reversibly to 400 K at constant volume.
c. The gas is reversibly expanded at constant temperature until its volume is again tripled.
d. The gas is finally reversibly cooled to 300 K at constant pressure.

Calculate the values of q and w and the changes in U, H, and S.

3.3 One mole of a monatomic ideal gas undergoes a reversible expansion at constant pressure during which the entropy of the gas increases by 14.41 J/K and the gas absorbs 6236 joules of heat. Calculate the initial and final temperatures of the gas. One mole of a second monatomic ideal gas undergoes a reversible isothermal expansion during which it doubles its volume, performs 1729 joules of work and increases its entropy by 5.763 J/K. Calculate the temperature at which the expansion was conducted.

3.4 Calculate the change in the enthalpy and the change in entropy when 1 mole of SiC is heated from 25°C to 1000°C. The constant pressure molar heat capacity of SiC varies with temperature as

$$c_p = 50.79 + 1.97 \times 10^{-3}T - 4.92 \times 10^6 T^{-2} + 8.20 \times 10^8 T^{-3} \text{ J/mole·K}$$

3.5 One mole of copper at a uniform temperature of 0°C is placed in thermal contact with a second mole of copper which, initially, is at a uniform temperature of 100°C. Calculate the temperature of the 2 mole system, which is contained in an adiabatic enclosure, when thermal equilibrium is attained. Why is the common uniform temperature not exactly 50°C? How much heat is transferred, and how much entropy is produced by the transfer? The constant pressure molar heat capacity of solid copper varies with temperature as

$$c_p = 22.64 + 6.28 \times 10^{-3}T \text{ J/mole·K}$$

3.6 A reversible heat engine, operating in a cycle, withdraws heat from a high-temperature reservoir (the temperature of which consequently decreases), performs work w, and rejects heat into a low-temperature reservoir (the temperature of which consequently increases). The two reservoirs are, initially, at the temperatures T_1 and T_2 and have constant heat capacities C_1 and C_2, respectively. Calculate the final temperature of the system and the maximum amount of work which can be obtained from the engine.

Chapter 4

THE STATISTICAL INTERPRETATION OF ENTROPY

4.1 INTRODUCTION

In Chapter 3 the introduction of entropy as a state function was facilitated by the realization that there exist possible and impossible processes, and by an examination of the heat and work effects occurring during these processes. From the formal statement of the Second Law of Thermodynamics, as developed from Classical Thermodynamics arguments, it is difficult to assign a physical significance or a physical quality to entropy. In this respect entropy differs from internal energy in spite of the fact that, within the scope of Classical Thermodynamics, both properties are simply mathematical functions of the state of a system. The ready acceptance of the First Law of Thermodynamics, after its enunciation, was due to the easily understood physical significance of internal energy, whereas the lack of corresponding understanding of entropy caused the acceptance of the Second Law of Thermodynamics to be slow. From the classical viewpoint the Second Law is only valid as a direct result of the fact that, to date, human ingenuity has failed to invent a perpetual motion machine. Thus, within the scope of Classical Thermodynamics, the Second Law is a "law" only because it has not yet been disproved. The physical interpretation of entropy had to await the development of quantum theory and statistical mechanics.

4.2 ENTROPY AND DISORDER ON AN ATOMIC SCALE

Gibbs described the entropy of a system as being a measure of its "degree of mixed-up-ness" at the atomic or molecular level, i.e., the more mixed up the constituent particles of a system, the larger the value of its entropy. For example, in the crystalline solid state most of the constituent particles are confined to vibrate about their regularly arrayed lattice positions, whereas in the liquid state, confinement of the particles to lattice sites is absent and the particles are relatively free to wander through the communal volume occupied by the liquid. The arrangement of the particles in the crystalline solid state is thus more ordered than that of the liquid state, or, alternatively, is less "mixed up" than that of the liquid state, and, as a consequence, the entropy of the liquid state is greater than that of the solid state. Similarly the atomic disorder in the gaseous state is greater than that in the liquid state, and thus the entropy of the gaseous state is greater than that of the liquid state.

This correlates with phenomena on a macroscopic level; e.g., the transformation of a solid to a liquid at its melting temperature, T_m, requires that the substance absorb a quantity of heat q, called the latent heat of melting. The entropy of the substance being melted is thus increased by the amount q/T_m, and, if the melting process is conducted at constant pressure, from Eq. (2.5), $q = \Delta H$, and thus

$$\Delta S_{melting} = \frac{\Delta H_{melting}}{T_m}$$

The increase in the entropy of the substance which accompanies melting correlates with the corresponding increase in the degree of disorder of its constituent particles. The above correlation does not apply at all temperatures, because, if a supercooled liquid spontaneously freezes, it might appear that a decrease in the degree of disorder causes an increase in entropy (the entropy produced by the irreversible freezing process). The apparent anomaly is removed by considering the influence of the latent heat of freezing released on the entropy and degree of disorder of the heat reservoir which absorbs the heat. If spontaneous freezing of a supercooled liquid occurs, the increase in the degree of order of the freezing system is *less* than the decrease in the degree of order in the heat reservoir, and hence the spontaneous freezing process causes an overall increase in disorder and an overall increase in entropy. If the transformation from liquid to solid occurs at the equilibrium melting temperature of the substance, T_m, then the increase in the degree of disorder of the heat reservoir equals the decrease in the degree of disorder of the substance, and the total degree of disorder of the combined system is unchanged; disorder has simply been transferred from the substance to the heat reservoir. Consequently, the entropy of the combined system is unchanged as a result of the freezing process; entropy has simply been transferred from the substance to the heat reservoir. The equilibrium melting or freezing temperature of a substance can thus be defined as that temperature at which no change in the degree of order of the combined system (substance + heat reservoir) occurs as a result of the phase change. Only at this temperature are the solid and liquid in equilibrium with one another, and hence, only at this temperature can the phase change occur reversibly.

4.3 THE CONCEPT OF MICROSTATE

The development of a quantitative relationship between entropy and "degree of mixed-up-ness" requires quantification of the term "degree of mixed-up-ness," and this can be obtained from a consideration of elementary statistical mechanics. Statistical mechanics arises from the assumption that the equilibrium state of a system is simply the most probable of all of its possible states, and the subject is concerned with determination of, the criteria governing, and the properties of this most probable state.

One of the major developments in physical science which has led to a considerable increase in the understanding of the behavior of matter is the quantum theory. A postulate of the quantum theory is that, if a particle is confined to move within a given fixed volume, then its energy is quantized, i.e., the particle may only have certain discrete allowed values of energy, which are separated by "forbidden energy bands." For any given particle the spacing between the quantized values of energy (the allowed energy levels) decreases as the volume available to the movement of the particle increases, and energy becomes continuous only when no restriction is placed on the position of the particle.

The effect of the quantization of energy can be illustrated by considering a hypothetical system comprising a perfect crystal in which all of the lattice sites are occupied by identical particles. The characteristics of the particles and the crystal structure determine the quantization of the allowed energy levels, in which the lowest energy level, or the ground state, is designated ε_0, and the succeeding levels of increasing energy are designated ε_1, ε_2, ε_3, etc. The crystal contains n particles and has the energy U, and statistical mechanics asks the questions: In how many ways can the n particles be distributed over the available energy levels such that the total energy of the crystal is U, and, of the possible distributions, which is the most probable?

Consider that the crystal contains three identical, and hence indistinguishable, particles which are located on three distinguishable lattice sites A, B, and C. Suppose, for simplicity, that the quantization is such that the energy levels are equally spaced, with the ground level being taken as zero, the first level $\varepsilon_1 = u$, the second level $\varepsilon_2 = 2u$, etc., and let the total energy of the system be $U = 3u$. This system has three different distributions, as shown in Fig. 4.1.

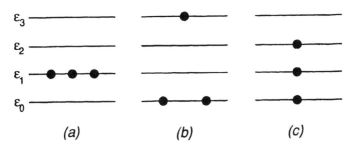

Figure 4.1   The distributions of particles among energy levels in a system of constant energy.

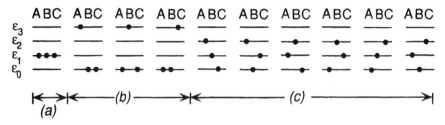

Figure 4.2 Illustration of the complexions or microstates within distributions of particles among energy levels in a system of constant energy.

a. All three particles in level 1
b. One particle in level 3, and the other two particles in level 0
c. One particle in level 2, one particle in level 1, and one particle in level 0

These distributions must now be examined to determine how many distinguishable arrangements they individually contain.

Distribution a. There is only one arrangement of this distribution, as interchange of the particles among the three lattice sites does not produce a different arrangement.

Distribution b. Any of the three distinguishable lattice sites can be occupied by the particle of energy $3u$, and the remaining two sites are each occupied by a particle of zero energy. As interchange of the particles of zero energy does not produce a different arrangement, there are three arrangements in distribution *b*.

Distribution c. Any of the three distinguishable lattice sites can be occupied by the particle of energy, $2u$, either of the two remaining sites can be occupied by the particle of energy, $1u$, and the single remaining site is occupied by the particle of zero energy. The number of distinguishable arrangements in distribution *c* is thus $3 \times 2 \times 1 = 3! = 6$.

These arrangements are shown in Fig. 4.2.

Thus, there are 10 distinguishable ways in which three particles can be placed in the energy levels such that the total energy of the system, U, equals $3u$. These distinguishable arrangements are called *complexions* or *microstates*, and all of the 10 microstates correspond to a single *macrostate*.

4.4 DETERMINATION OF THE MOST PROBABLE MICROSTATE

The concept of macrostate lies within the domain of Classical Thermodynamics, and the macrostate is fixed when the values of the independent variables are fixed. As was shown in Chap. 1, a system of fixed composition has three independent variables—the original two plus a variable which describes the size of the system. In the above example the values of U, V, and n are fixed (constancy of volume being required in order that the quantization of the energy levels be determined), and hence the macrostate of the system is fixed. With respect to the microstates within any macrostate, in view of the absence of any reason to the contrary, it is assumed that each of the microstates is equally probable, and thus the probability of observing the above system in any one of its ten possible microstates is 1/10. However, the ten microstates occur in three distributions, and hence the probability that the system occurs in distribution a is 1/10, the probability that it occurs in distribution b is 3/10, and the probability that it occurs in distribution c is 6/10. Distribution c is thus the "most probable." The physical significance of these probabilities can be viewed in either of two ways: (1) If it were possible to make an instantaneous observation of the system, the probability of observing an arrangement in distribution c is 6/10, or (2) If the system were observed over a finite interval of time, during which the system rapidly changes from one microstate to another, the fraction of this time which the system spends in all of the arrangements in distribution c would be 6/10.

As the total energy of the system and the number of particles which it contains increase, the number of distinguishable arrangements (microstates) increases, and, for fixed values of U, V, and n, these microstates still correspond to a single macrostate. Similarly the number of possible distributions increases, and in real systems, e.g., 1 mole of a system, which contains 6.023×10^{23} particles, the number of arrangements within the most probable distribution is very much larger than the number of arrangements in all of the other distributions. The number of arrangements within a given distribution, Ω, is calculated as follows: If n particles are distributed among the energy levels such that n_0 are in level ε_0, n_1 in level ε_1, n_2 in level $\varepsilon_2 \ldots$, and n_r in the highest level of occupancy ε_r, then the number of arrangements, Ω, is given by

$$\Omega = \frac{n!}{n_0! n_1! n_2! \cdots n_i!} \tag{4.1}$$

$$= \frac{n!}{\displaystyle\prod_{i=0}^{i=r} n_i!}$$

For example, consideration of the system discussed gives

$$\Omega \text{ (distribution } a) = \frac{3!}{3! 0! 0!} = 1$$

$$\Omega \text{ (distribution } b) = \frac{3!}{2! 1! 0!} = 3$$

$$\Omega \text{ (distribution } c) = \frac{3!}{1! 1! 1!} = 6$$

The most probable distribution is obtained by determining the set of numbers $n_0, n_1 \ldots n_r$ which maximizes the value of Ω. When the values of n_i are large, Stirling's approximation can be used (that is $\ln X! = X \ln X - X$). Thus, taking the logarithms of the terms in Eq. (4.1) gives

$$\ln \Omega = n \ln n - n - \sum_{i=0}^{i=r} (n_i \ln n_i - n_i) \qquad (4.2)$$

As the macrostate of the system is determined by the fixed values of U, V, and n, any distribution of the particles among the energy levels must conform with the conditions

$$U = \text{constant} = n_0 \varepsilon_0 + n_1 \varepsilon_1 + n_2 \varepsilon_2 + \cdots + n_r \varepsilon_r$$

$$= \sum_{i=0}^{i=r} n_i \varepsilon_i \qquad (4.3)$$

and

$$n = \text{constant} = n_0 + n_1 + n_2 + \cdots + n_r \qquad (4.4)$$

$$= \sum_{i=0}^{i=r} n_i$$

From Eqs. (4.3) and (4.4), any interchange of particles among the energy levels must conform with the conditions

$$\delta U = \sum_i \varepsilon_i \delta n_i = 0 \qquad (4.5)$$

and

$$\delta n = \sum_i \delta n_i = 0 \qquad (4.6)$$

Also, from Eq. (4.2), any interchange of particles among the energy levels gives

$$\delta \ln \Omega = -\sum \left(\delta n_i \ln n_i + \frac{n_i \delta n_i}{n_i} - \delta n_i \right) \qquad (4.7)$$

$$= -\sum (\delta n_i \ln n_i)$$

If Ω has the maximum possible value then a small rearrangement of particles among the energy levels will not alter the value of Ω or of $\ln \Omega$. Thus, if the set of n_i's is such that Ω has its maximum value then

$$\delta \ln \Omega = -\sum (\delta n_i \ln n_i) = 0 \qquad (4.8)$$

The condition that Ω has its maximum value for the given macrostate is thus that Eqs. (4.5), (4.6), and (4.8) be simultaneously satisfied. The set of values of n_i in the most probable distribution is obtained by the method of undetermined multipliers, in the following manner. Eq. (4.5) is multiplied by the constant β, which has the units of reciprocal energy, to give

$$\sum \beta \varepsilon_i \delta n_i = 0 \tag{4.9}$$

Eq. (4.6) is multiplied by the dimensionless constant α to give

$$\sum \alpha \delta n_i = 0 \tag{4.10}$$

and Eqs. (4.8), (4.9), and (4.10) are added to give

$$\sum_{i=0}^{i=r} (\ln n_i + \alpha + \beta \varepsilon_i) \delta n_i = 0 \tag{4.11}$$

i.e.,

$$(\ln n_0 + \alpha + \beta \varepsilon_0) \delta n_0 + (\ln n_1 + \alpha + \beta \varepsilon_1) \delta n_1$$
$$+ (\ln n_2 + \alpha + \beta \varepsilon_2) \delta n_2 + (\ln n_3 + \alpha + \beta \varepsilon_3) \delta n_3$$
$$+ \cdots + (\ln n_r + \alpha + \beta \varepsilon_r) \delta n_r = 0$$

The solution of Eq. (4.11) requires that each of the bracketed terms be individually equal to zero, i.e.,

$$\ln n_i + \alpha + \beta \varepsilon_i = 0$$

or

$$n_i = e^{-\alpha} e^{-\beta \varepsilon_i} \tag{4.12}$$

Summing over all r energy levels gives

$$\sum_{i=0}^{i=r} n_i = n = e^{-\alpha} \sum_{i=0}^{i=r} e^{-\beta \varepsilon_i}$$

The summation

$$\sum e^{-\beta \varepsilon_i} = e^{-\beta \varepsilon_0} + e^{-\beta \varepsilon_1} + e^{-\beta \varepsilon_2} + \cdots + e^{-\beta \varepsilon_r}$$

which is determined by the magnitude of β and by the quantization of the energy, is called the partition function, P. Thus

$$e^{-\alpha} = \frac{n}{P}$$

and thus

$$n_i = \frac{n e^{-\beta \varepsilon_i}}{P} \tag{4.13}$$

The distribution of particles in the energy levels which maximizes Ω (i.e., the most probable distribution) is thus one in which the occupancy of the levels decreases exponentially with increasing energy, and the shape of this distribution is shown in Fig. 4.3. The actual shape of the exponential curve in Fig. 4.3 (for a given system) is

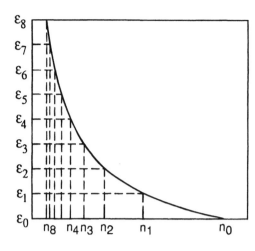

Figure 4.3 Schematic representation of the most probable distribution of particles among the quantized energy levels.

determined by the value of β, and β is inversely proportional to the absolute temperature, being given by

$$\beta = \frac{1}{kT} \tag{4.14}$$

in which k is Boltzmann's constant, an expression of the gas constant per atom or molecule, i.e.,

$$k = \frac{R}{N_O} = \frac{8.3144}{6.0232 \times 10^{23}} = 1.38054 \times 10^{-23} \text{ J/K}$$

where N_O is Avogadro's number.

4.5 THE INFLUENCE OF TEMPERATURE

The nature of the exponential distribution of particles in Fig. 4.3 is determined by the temperature of the system. However, as the macrostate of the system is fixed by fixing the values of U, V, and n, then T, as a dependent variable, is fixed. Eq. (4.14) shows that T increases with decreasing β, and the shape of the exponential distribution changes as shown in Fig. 4.4. An increase in temperature causes the upper energy levels to become relatively more populated, and this corresponds to an increase in the average energy of the particles, i.e., to an increase in the value of U/n, which, for fixed values of V and n, corresponds to an increase in U.

As has been stated, when the number of particles in the system is very large, the number of arrangements within the most probable distribution, Ω_{max}, is the only term which makes a significant contribution to the total number of arrangements, Ω_{total}, which the system may have; that is, Ω_{max} is significantly larger than the sum

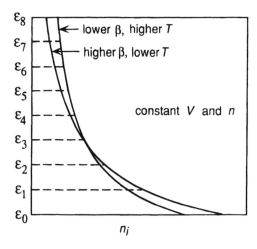

Figure 4.4 The influence of temperature on the most probable distribution of particles among energy levels in a closed system of constant volume.

of all of the other arrangements. Thus, when the number of particles is large, Ω_{total} can be equated with Ω_{max}.

Substituting $\beta = 1/kT$, Eq. (4.2) can be written as

$$\ln \Omega_{\text{total}} = \ln \Omega_{\text{max}} = n \ln n - \sum n_i \ln n_i$$

in which the values of n_i are given by Eq. (4.13). Thus

$$
\begin{aligned}
\ln \Omega_{\text{total}} &= n \ln n - \sum \frac{n}{P} e^{-\varepsilon_i/kT} \ln \left(\frac{n}{P} e^{-\varepsilon_i/kT} \right) \\
&= n \ln n - \frac{n}{P} \sum \left[e^{-\varepsilon_i/KT} \left(\ln n - \ln P - \frac{\varepsilon_i}{kT} \right) \right] \\
&= n \ln n - \frac{n}{P} (\ln n - \ln P) \sum e^{-\varepsilon_i/KT} + \frac{n}{PkT} \sum \varepsilon_i e^{-\varepsilon_i/kT}
\end{aligned}
$$

But

$$U = \sum n_i \varepsilon_i = \sum \frac{n}{P} \varepsilon_i e^{-\varepsilon_i/kT} = \frac{n}{P} \sum \varepsilon_i e^{-\varepsilon_i/kT}$$

Therefore

$$\sum \varepsilon_i e^{-\varepsilon_i/kT} = \frac{UP}{n}$$

and thus

$$\ln \Omega = n \ln P + \frac{U}{kT} \tag{4.15}$$

4.6 THERMAL EQUILIBRIUM AND THE BOLTZMANN EQUATION

Consider now a system of particles in thermal equilibrium with a heat bath and let the state of the combined system (particles + heat bath) be fixed by fixing the values of U, V, and n, where

$$U = U_{\text{particles system}} + U_{\text{heat bath}}$$
$$V = V_{\text{particles system}} + V_{\text{heat bath}}$$

n = number of particles in the system + the heat bath of fixed size

As the particles system and the heat bath are in thermal equilibrium, small exchanges of energy can occur between them, and for such a small exchange at constant U, V, and n, Eq. (4.15) for the particles system gives

$$\delta \ln \Omega = \frac{\delta U}{kT}$$

(P is dependent only on the values of ε_i and T.) As this exchange of energy is carried out at constant total volume, then

$$\delta U = \delta q$$

i.e., the energy exchange occurs as an exchange of heat. Thus

$$\delta \ln \Omega = \frac{\delta q}{kT} \tag{4.16}$$

As the exchange of heat occurs at constant temperature, i.e., occurs reversibly, then, from Chap. 3,

$$\frac{\delta q}{T} = \delta S$$

and thus

$$\delta S = k \delta \ln \Omega$$

As both S and Ω are state functions, the above expression can be written as a differential equation, integration of which gives

$$S = k \ln \Omega \tag{4.17}$$

Eq. (4.17), which is called Boltzmann's equation, is the required quantitative relationship between the entropy of a system and its "degree of mixed-up-ness," in which the latter, given by Ω, is the number of ways in which the energy of the system can be distributed among the particles. The most probable state of the system is that in which Ω has a maximum value, consistent with the fixed values of U, V, and n, and hence the equilibrium state of the system is that in which S is a maximum, consistent with the fixed values of U, V, and n. The Boltzmann equation thus provides a physical quality to entropy.

4.7 HEAT FLOW AND THE PRODUCTION OF ENTROPY

Classical Thermodynamics shows that the transfer of heat from a body at some temperature to a body at a lower temperature is an irreversible process which is accompanied by the production of entropy, and that the reverse process, i.e., the flow of heat up a temperature gradient, is an impossible process. An examination of microstates shows that a microstate in which variations in temperature occur within a system is less probable than a microstate in which the temperature of the system is uniformly constant.

Consider two closed systems, A and B. Let the energy of A be U_A and the number of complexions of A be Ω_A. Similarly let the energy of B be U_B and its number of complexions be Ω_B. When thermal contact is made between A and B, the product $\Omega_A\Omega_B$ will, generally, not have its maximum possible value, and heat will be transferred either from A to B or from B to A. Heat flows from A to B if, thereby, the increase in Ω_B, caused by the increase in U_B, is greater than the decrease in Ω_A, caused by the decrease in U_A. Heat continues to flow from A to B as long as the product $\Omega_A\Omega_B$ continues to increase, and the flow of heat ceases when $\Omega_A\Omega_B$ reaches its maximum value, i.e., when the increase in Ω_B caused by the transfer of an increment of heat from A is exactly compensated by the decrease in Ω_A. The condition for A to be in thermal equilibrium with B is thus that the transfer of a quantity of heat from one body to the other does not cause a change in the value of $\Omega_A\Omega_B$. That is,

$$\delta \ln \Omega_A\Omega_B = 0$$

Consider a rearrangement of the particles in the quantized energy levels in B which causes U_B to increase by a certain amount, and consider a simultaneous rearrangement of the particles in the energy levels of A which causes U_A to decrease by the same amount; i.e., $(U_A + U_B)$ remains constant. If the levels of A are populated in accordance with Eq. (4.16) with $T = T_A$ and if the levels of B are populated in accordance with Eq. (4.16) with $T = T_B$, then

$$\delta \ln \Omega_A = \frac{\delta q_A}{kT_A}$$

and

$$\delta \ln \Omega_B = \frac{\delta q_B}{kT_B}$$

When a quantity of heat is transferred from A to B at total constant energy, then

$$\delta q_A = -\delta q_B$$

Thus

$$\delta \ln \Omega_A\Omega_B = \delta \ln \Omega_A + \delta \ln \Omega_B = \left(\frac{1}{T_A} - \frac{1}{T_B}\right)\frac{\delta q}{k}$$

and hence the condition that $\delta \ln \Omega_A\Omega_B$ be zero is that $T_A = T_B$. The reversible transfer of heat from one body to another thus only occurs when the temperatures of the bodies are equal, as only in such a case does $\Omega_A\Omega_B$, and hence the total entropy of the combined system ($S_A + S_B$), remain constant. An irreversible transfer of heat increases the value of the product $\Omega_A\Omega_B$, and hence entropy is created. From the point of view of microstates, an irreversible process is one which takes the system from a less probable state to the most probable state, and from the point of view of macrostates, an irreversible process takes the system from a nonequilibrium state to the equilibrium state. Thus, what is considered in Classical Thermodynamics to be an impossible process is shown by a consideration of microstates to only be an improbable process.

4.8 CONFIGURATIONAL ENTROPY AND THERMAL ENTROPY

In the preceding discussion entropy was considered in terms of the number of ways in which energy can be distributed among identical particles, and the given example of a mixing process involved the redistribution of thermal energy among the particles of two closed systems when placed in thermal contact. The change in entropy accompanying this redistribution is a change in *thermal entropy*. Entropy can also be considered in terms of the number of ways in which particles themselves can be distributed in space, and this consideration gives rise to the concept of *configurational entropy*.

Consider two crystals at the same temperature and pressure, one containing atoms of the element A and the other containing atoms of the element B. When the two crystals are placed in physical contact with one another the spontaneous process which occurs is the diffusion of A atoms into the crystal of B and the diffusion of B atoms into the crystal of A. As this process is spontaneous, entropy is produced, and intuitively it might be predicted that equilibrium will be reached (i.e., the entropy of the system will reach a maximum value) when the diffusion processes have occurred to the extent that all concentration gradients in the system have been eliminated. This is the mass transport analog of the heat transfer case in which heat flows irreversibly between two bodies until the temperature gradients have been eliminated.

Consider a crystal containing four atoms of A placed in contact with a crystal containing four atoms of B. The initial state of this system, in which all four atoms of A lie to the left of XY and all four of the B atoms lie to the right of XY, is shown in Fig. 4.5. The number of distinguishable ways in which this arrangement can be realized is unity, as interchange among the identical A atoms on the left of XY and/or

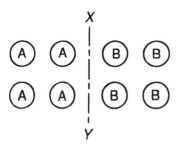

Figure 4.5 Representation of a crystal of A in contact with a crystal of B.

interchange among the identical B atoms on the right of XY does not produce a different configuration. Thus

$$\Omega_{4:0} = 1$$

(in which the notation indicates four atoms of A on the left of XY and none on the right.)

When one A atom is interchanged with one B atom across XY, the B atom can be located on any of four sites, and hence the left side of XY can be realized in four different ways. Similarly the exchanged A atom can be located on any of four sites, and hence the right side can be realized in four different ways. As any of the four former arrangements can be combined with any of the four latter arrangements, the total number of distinguishable configurations of the arrangement 3:1 is $4 \times 4 = 16$, i.e.,

$$\Omega_{3:1} = 16$$

When a second A atom is exchanged with a second B atom across XY, the first B atom on the left of XY can be located in any of four positions, and the second B atom can be located in any of the three remaining positions, giving, thus, $4 \times 3 = 12$ configurations. However, these 12 configurations include those which occur as a result of interchange of the two B atoms themselves, which, being indistinguishable, must be discounted. The number of distinguishable configurations on the left of XY is thus $(4 \times 3)/2! = 6$. Similarly six distinguishable arrangements occur on the right of XY, and hence the total number of distinguishable configurations in the arrangement 2:2 is $6 \times 6 = 36$, i.e.,

$$\Omega_{2:2} = 36$$

When a third A atom is exchanged with a third B atom across XY, the first B atom can be located on any of four sites, the second on any of the three remaining sites and the third on either of the two remaining sites. Factoring out the number of indistinguishable configurations caused by interchange of the three B atoms among themselves gives the number of distinguishable configurations on the left of XY as $(4 \times 3 \times 2)/3! = 4$. Similarly four distinguishable configurations occur on the right of XY, and hence,

$$\Omega_{1:3} = 16$$

Interchange of the final A and B atoms across XY gives

$$\Omega_{0:4} = \left(\frac{4 \times 3 \times 2}{4!} \right) \times \left(\frac{4 \times 3 \times 2}{4!} \right) = 1$$

Thus the total number of spatial configurations available to the system is $1 + 16 + 36 + 16 + 1 = 70$, which is the number of distinguishable ways in which four particles of one kind and four particles of another kind can be arranged on eight sites, i.e.,

$$\text{number or ways} = \frac{8!}{4!4!} = 70$$

If, as before, it is assumed that each of these configurations is equally probable, then the probability of finding the system in the arrangement 4:0 or 0:4 is 1/70, the probability of arrangement 3:1 or 1:3 is 16/70, and the probability of finding the system in the arrangement 2:2 is 36/70. Arrangement 2:2 is thus the most probable and thus corresponds to the equilibrium state, in which the concentration gradients have been eliminated. Again, as

$$S = k \ln \Omega$$

it is seen that maximization of Ω maximizes the entropy. In this case the increase in entropy occurs as a result of the increase in the number of spatial configurations which become available to the system when the crystals of A and B are placed in contact with one another. The increase in the entropy of the system arises from an increase in its configurational entropy, S_{conf}. The mixing process can be expressed as

$$A + B \text{ (unmixed)} \rightarrow A + B \text{ (mixed) at constant } U, V, \text{ and } n$$

i.e.,

$$\text{state (1)} \rightarrow \text{state (2)}$$
$$\Delta S_{conf} = S_{conf(2)} - S_{conf(1)} = k \ln \Omega_{conf(2)} - k \ln \Omega_{conf(1)}$$
$$= k \ln \left(\frac{\Omega_{conf(2)}}{\Omega_{conf(1)}} \right)$$

and, if n_a atoms of A are mixed with n_b atoms of B, then

$$\Omega_{conf(2)} = \frac{(n_a + n_b)!}{n_a! n_b!} \quad \text{and} \quad \Omega_{conf(1)} = 1$$

Thus

$$\Delta S_{conf} = k \ln \frac{(n_a + n_b)!}{n_a! n_b!} \tag{4.18}$$

The total entropy of a system consists of its thermal entropy, S_{th}, which arises from the number of ways in which the energy of the system can be shared among the particles, and its configurational entropy, S_{conf}, which arises from the number of distinguishable ways in which the particles can fill the space available to them. Thus

$$S_{total} = S_{th} + S_{conf}$$
$$= k \ln \Omega_{th} + k \ln \Omega_{conf}$$
$$= k \ln \Omega_{th} \Omega_{conf}$$

The number of spatial configurations available to two closed systems placed in thermal contact, or to two open chemically identical systems placed in thermal contact, is unity. Thus in the case of heat flow down a temperature gradient between two such systems, as only Ω_{th} changes, the increase in the entropy arising from the heat transfer which takes the system from state 1 to state 2 is

$$\Delta S_{total} = k \ln \frac{\Omega_{th(2)}\Omega_{conf(2)}}{\Omega_{th(1)}\Omega_{conf(1)}} = k \ln \frac{\Omega_{th(2)}}{\Omega_{th(1)}} = \Delta S_{th}$$

Similarly in the mixing of particles of A with particles of B, ΔS_{total} only equals ΔS_{conf} if the mixing process does not cause a redistribution of the particles among the energy levels, i.e., if $\Omega_{th(1)} = \Omega_{th(2)}$. This condition corresponds to "ideal mixing" of the particles and requires that the quantization of energy be the same in crystals A and B. Ideal mixing is the exception rather than the rule, and, generally, when two or more components are mixed at constant U, V, and n, $\Omega_{th(2)}$ does not have the same value as $\Omega_{th(1)}$; thus completely random mixing of the particles does not occur. In such cases either clustering of like particles (indicating difficulty in mixing) or ordering (indicating a tendency toward compound formation) occurs. In all cases, however, the equilibrium state of the system is that which, at constant U, V, and n, maximizes the product $\Omega_{th}\Omega_{conf}$.

4.9 SUMMARY

1. A single macrostate of a system, which is determined when the independent variables of the system are fixed, contains a very large number of microstates, each of which is characterized by the manner in which the thermal energy of the system is distributed among the particles and the manner in which the particles are distributed in the space available to them.

2. Although the occurrence of a system in any one of its microstates is equally probable, greatly differing numbers of microstates occur in differing distributions. The distribution which contains the largest number of microstates is the most probable distribution, and in real systems the number of microstates in the most probable distribution is significantly larger than the sum of all of the other microstates occurring in all of the other distributions. This most probable distribution is the equilibrium thermodynamic state of the system.

3. The relationship between the number of microstates available to the system, Ω, and the entropy of the system is given by Boltzmann's equation as $S = k \ln \Omega$, in which k is Boltzmann's constant. Thus, if a situation arises which allows an increase in the number of microstates available to the system, then spontaneous redistribution of the energy among the particles (or particles over the available space) occurs until the newly available most probable distribution occurs. The Boltzmann equation shows that an increase in the number of microstates made available to the system causes an increase in the entropy of the system.

4. The total entropy of a system is the sum of the thermal entropy, S_{th}, and the configurational entropy, S_{conf}. The former arises from the number of ways in which the thermal energy available to the system can be shared among the constituent

particles, Ω_{th}, and the latter arises from the number of ways in which the particles can be distributed over the space available to them, Ω_{conf}. As any of the thermal distributions can be combined with any of the configurational distributions, the total number of microstates available to the system is the product $\Omega_{th}\Omega_{conf}$, and hence, from the logarithmic form of Boltzmann's equation, the total entropy of the system is the sum of S_{th} and S_{conf}.

4.10 NUMERICAL EXAMPLES

Example 1

Spectroscopic observation of molecular N_2 in an electrical discharge shows that the relative numbers of molecules in excited vibrational states with energies given by

$$\varepsilon_i = \left(i + \frac{1}{2}\right)hv \tag{4.19}$$

are

i	0	1	2	3
$\dfrac{n_i}{n}$	1.00	0.250	0.062	0.016

Show that the gas is in thermodynamic equilibrium with respect to the distribution of vibrational energy, and calculate the temperature of the gas. In Eq. (4.19), i is an integer which has values in the range zero to infinity, h is Planck's constant of action ($= 6.6252 \times 10^{-34}$ joules·s), and the vibration frequency, v, is $7.00 \times 10^{13}\,s^{-1}$.

From Eqs. (4.13), (4.14), and (4.19),

$$\frac{n_i}{n} = \frac{\exp\left[-\left(i + \frac{1}{2}\right)\dfrac{hv}{kT}\right]}{P}$$

Observation shows that

$$\frac{n_i}{n_0} = 0.250$$

Thus,

$$\frac{n_i}{n_0} = \exp\left(-\frac{3}{2}\frac{hv}{kT}\right)\exp\left(\frac{1}{2}\frac{hv}{kT}\right) = \exp\left(-\frac{hv}{kT}\right) = 0.250$$

which gives

$$\frac{hv}{kT} = 1.386$$

Then, from Eq. (4.13),

$$P\frac{n_0}{n} = \exp\left(-\frac{1}{2}\frac{hv}{kT}\right) = 0.5$$

$$P\frac{n_1}{n} = \exp\left(-\frac{3}{2}\frac{hv}{kT}\right) = 0.125$$

$$P\frac{n_2}{n} = \exp\left(-\frac{5}{2}\frac{hv}{kT}\right) = 0.031$$

and

$$P\frac{n_3}{n} = \exp\left(-\frac{7}{2}\frac{hv}{kT}\right) = 0.008$$

Normalizing gives

$$\frac{n_0}{n} = 0.5/0.5 = 1.0$$

$$\frac{n_1}{n} = 0.125/0.5 = 0.25$$

$$\frac{n_2}{n} = 0.031/0.5 = 0.062$$

$$\frac{n_3}{n} = 0.008/0.5 = 0.016$$

which shows that the gas is in equilibrium with respect to the distribution of vibrational energy. The temperature of the gas is obtained as

$$T = \frac{hv}{1.386k} = \frac{(6.6252 \times 10^{-34})(7.00 \times 10^{13})}{1.386 \times 1.38054 \times 10^{-23}} = 2420 \text{ K}$$

Example 2

The isotopic composition of lead in atomic percent is

atomic weight	atomic percent
204	1.5
206	23.6
207	22.6
208	52.3

Calculate the molar configurational entropy of lead. The configurational entropy is obtained from Boltzmann's equation

$$S = k \ln \Omega \qquad (4.17)$$

where

$$\Omega = \frac{(N_O)!}{(0.015N_O)!(0.236N_O)!(0.226N_O)!(0.523N_O)!}$$

Stirling's theorem gives

$$
\begin{aligned}
\ln \Omega &= N_O \ln N_O - 0.015N_O \ln 0.015N_O - 0.236N_O \ln 0.236N_O \\
&\quad - 0.226N_O \ln 0.226N_O - 0.523N_O \ln 0.523N_O \\
&= 3.298 \times 10^{25} - 4.567 \times 10^{23} - 7.578 \times 10^{24} - 7.251 \times 10^{24} \\
&\quad - 1.704 \times 10^{25} \\
&= 6.498 \times 10^{23}
\end{aligned}
$$

Therefore, the molar configurational entropy is

$$S = k \ln \Omega = (1.38054 \times 10^{-23}) \times (6.498 \times 10^{23}) = 8.97 \text{ J/K}$$

PROBLEMS

4.1 A rigid container is divided into two compartments of equal volume by a partition. One compartment contains 1 mole of ideal gas A at 1 atm, and the other contains 1 mole of ideal gas B at 1 atm. Calculate the increase in entropy which occurs when the partition between the two compartments is removed. If the first compartment had contained 2 moles of ideal gas A, what would have been the increase in entropy when the partition was removed? Calculate the corresponding increases in entropy in each of the above two situations if both compartments had contained ideal gas A.

4.2 Show that, when n atoms of A and n atoms of B form a randomly mixed solution, the fraction of the total number of distinguishable complexions which occur in the most probable distribution decreases with increasing value of n.

4.3 Assuming that a silver-gold alloy is a random mixture of gold and silver atoms, calculate the increase in entropy when 10 g of gold are mixed with 20 g of silver to form a homogeneous alloy. The gram atomic weights of Au and Ag are, respectively, 198 and 107.9.

4.4 On the assumption that copper-nickel alloys are random mixtures of copper and nickel atoms, calculate the mass of copper which, when mixed with 100 g of nickel, causes an increase in entropy of 15 J/K. The gram atomic weights of Cu and Ni are, respectively, 63.55 and 58.69.

Chapter 5

AUXILIARY FUNCTIONS*

5.1 INTRODUCTION

The main power of the thermodynamic method stems from its provision of criteria for equilibrium in materials systems and its determination of the influence, on an equilibrium state, of changes in the external influences acting on the system. The practical usefulness of this power is, however, determined by the practicality of the equations of state for the system, i.e., the relationships which can be established among the thermodynamic properties of the system.

Combination of the First and Second Laws of Thermodynamics leads to the derivation of Eq. (3.12)

$$dU = TdS - PdV$$

This equation of state gives the relationship between the dependent variable U and the independent variables S and V for a closed system of fixed composition which is undergoing a process involving a change of volume against the external pressure as the only form of work performed on, or by, the system. Combination of the First and Second Law also provides the criteria for equilibrium that

1. In a system of constant internal energy and constant volume, the entropy has its maximum value, and
2. In a system of constant entropy and constant volume, the internal energy has its minimum value.

The further development of thermodynamics beyond Eq. (3.12) arises, in part, from the fact that, from a practical point of view, S and V are an inconvenient choice of independent variables. Although the volume of a system can be measured with relative ease and, in principle, can be controlled, entropy can be neither simply measured nor simply controlled. It is thus desirable to develop a simple expression, similar in form to Eq. (3.12), which contains a more convenient choice of independent variables and which can accommodate changes in the composition of the system. From a practical point of view the most convenient pair of independent variables would be temperature and pressure, as these variables are most easily measured and controlled. The derivation of an equation of state of the simple form of Eq. (3.12), but using P and T as the independent variables, and a criterion for equilibrium in

---

*The derivation of the auxiliary functions is presented in Appendix C.

a constant-pressure constant-temperature system are thus desirable. Alternatively, from the theoretician's point of view the most convenient choice of independent variables would be V and T, as constant-volume constant-temperature systems are most easily examined by the methods of statistical mechanics. This arises because fixing the volume of a closed system fixes the quantization of its energy levels, and thus the Boltzmann factor, $\exp(-\varepsilon_i/kT)$, and the partition function, both of which appear in Eq. (4.13), have constant values in constant-volume constant-temperature systems. The derivation of an equation of state using T and V as the independent variables and the establishment of a criterion for equilibrium in a system of fixed volume and fixed temperature are thus desirable.

Eq. (3.12) cannot be applied to systems which undergo changes in composition caused by chemical reactions or to systems which perform work other than work of expansion against an external pressure (so-called P-V work). As systems which experience changes in composition, such as the transfer of an impurity element from a metal to a refining slag or the precipitation of a second phase in a ceramic oxide, are of prime importance to the materials engineer, composition variables must be included in any equation of state and in any criterion for equilibrium. Also any equation of state must be capable of accommodating forms of work other than P-V work, such as the electrical work performed by a galvanic cell.

Thus, although Eq. (3.12) lays the foundation of thermodynamics, it is necessary to develop auxiliary thermodynamic functions, which, as dependent variables, are related in simple form to more convenient choices of independent variables. Also, with this increase in the number of thermodynamic functions, it is necessary to establish the relationships which exist among them. It is often found that some required thermodynamics expression which, itself, is not amenable to experimental measurement is related in a simple manner to some measurable quantity. Examples of this have been presented in Chap. 3, where it was found that $(\partial U/\partial S)_V = T$, $-(\partial U/\partial V)_S = P$, and $(\partial S/\partial V)_U = P/T$.

In this chapter the thermodynamic functions A (the Helmholtz free energy), G (the Gibbs free energy), and μ_i (the chemical potential of the species i) are introduced, and their properties and interrelationships are examined. The functions A and G are defined as

$$A = U - TS \qquad (5.1)$$

and

$$G = U + PV - TS$$
$$= H - TS \qquad (5.2)$$

5.2 THE ENTHALPY H

As has been seen in Chap. 2, for a closed system undergoing a change of state at constant pressure, P, from the state 1 to the state 2, the First Law gives

$$U_2 - U_1 = q_p - P(V_2 - V_1)$$

which, on rearrangement, gives

$$(U_2 + PV_2) - (U_1 + PV_1) = q_p$$

or

$$\Delta H = H_2 - H_1 = q_p$$

Thus, when a closed system undergoes a change of state at constant pressure (during which only P-V work is done), the change in the enthalpy of the system equals the heat entering or leaving the system. The properties of H are examined in detail in Chap. 6.

5.3 THE HELMHOLTZ FREE ENERGY A

For a system undergoing a change of state from state 1 to state 2, Eq. (5.1) gives

$$(A_2 - A_1) = (U_2 - U_1) - (T_2 S_2 - T_1 S_1)$$

and, if the system is closed,

$$(U_2 - U_1) = q - w$$

in which case

$$(A_2 - A_1) = q - w - (T_2 S_2 - T_1 S_1)$$

If the process is isothermal, that is, $T_2 = T_1 = T$, the temperature of the heat reservoir which supplies or withdraws heat during the process, then, from the Second Law, Eq. (3.4a),

$$q \leq T(S_2 - S_1)$$

and hence

$$(A_2 - A_1) \leq -w$$

Comparison with Eq. (3.11) shows that the equality can be written as

$$(A_2 - A_1) + T\Delta S_{irr} = -w \qquad (5.3)$$

and thus, during a reversible isothermal process, for which ΔS_{irr} is zero, the amount of work done by the system w_{max} is equal to the decrease in the value of the Helmholtz free energy. Furthermore, for an isothermal process conducted at constant volume, which, necessarily, does not perform P-V work, Eq. (5.3) gives

$$(A_2 - A_1) + T\Delta S_{irr} = 0 \qquad (5.4)$$

or, for an increment of such a process,

$$dA + TdS_{irr} = 0$$

As dS_{irr} is always positive during a spontaneous process it is thus seen that A decreases during a spontaneous process, and as $dS_{irr} = 0$ is a criterion for a reversible process, equilibrium requires that

$$dA = 0 \tag{5.5}$$

Thus in a closed system held at constant T and V, the Helmholtz free energy can only decrease or remain constant, and equilibrium is attained in such a system when A achieves its minimum value. The Helmholtz free energy thus provides a criterion for equilibrium in a system at constant temperature and constant volume.

This criterion can be illustrated by examination of the following system. Consider n atoms of some element occurring in both a solid crystalline phase and a vapor phase contained in a constant-volume vessel, which, in turn, is immersed in a constant-temperature heat reservoir. The problem involves determining the equilibrium distribution of the n atoms between the solid phase and the vapor phase. At constant volume and constant temperature this distribution must be that which gives the Helmholtz free energy its minimum value. From Eq. (5.1)

$$A = U - TS$$

which shows that low values of A are obtained with low values of U and high values of S. The two extreme states of existence which are available to the system are

1. That in which all of the atoms exist in the solid crystalline phase and none occurs in the vapor phase, and
2. That in which all of the atoms exist in the vapor phase and none occurs in the solid phase.

Consider the system occurring in the first of these two states. The equilibrium distance between the centers of neighboring atoms in a solid phase is that at which there is a balance between the attractive and repulsive forces operating between the atoms, and thus, if an atom is to be removed from the surface of a solid and placed in a gas phase, work must be done against the attractive forces operating between the atom and its neighbors. For the separation of the atom to be conducted isothermally, the energy required for the separation must be supplied as heat which flows from the heat reservoir. This flow of heat into the system increases both the internal energy and the entropy of the constant-volume system. A system comprising 1 atom in the vapor phase and $n - 1$ atoms in the solid phase is more random than the system comprising n atoms in the solid phase, and hence the entropy of the former is greater than that of the latter.

As more atoms are removed from the solid and placed in the vapor phase, heat continues to flow from the heat reservoir, and the internal energy and the entropy of the system continue to increase. Eventually when all n atoms occur in the vapor phase, the internal energy and the entropy of the constant-volume constant-temperature system have their maximum values. In contrast, the state in which all of the n atoms occur in the solid is that in which the internal energy and the entropy have their minimum values. The variations of the internal energy and the entropy of the system with the number of atoms occurring in the vapor phase, n_v, are shown, respectively, in Figs. 5.1a and b. As the transfer of an atom from the solid to the vapor causes a fixed increment in the internal energy of the system, the internal energy increases linearly with n_v as shown in Fig. 5.1a. In contrast, as the magnitude of the increase in the disorder in the system, caused by the transfer of an atom from the solid to the vapor, decreases with increasing number of atoms in the vapor phase, the entropy of the system, shown in Fig. 5.1b, is not a linear function of n_v; the rate of increase of S decreases with increasing n_v. The variation of A, which is obtained as the sum of U and $-TS$, with n_v is shown in Fig. 5.2. This figure shows that A has a minimum value at a unique value of n_v, designated as $n_{v(eq,T)}$. This state is the compromise between minimization of U and maximization of S, and in this state the solid exerts its equilibrium vapor pressure at the temperature T. If the vapor behaves ideally, then the vapor pressure, which is called the saturated vapor pressure, is given by

$$p = \frac{n_{v(eq,T)}kT}{(V - V_S)}$$

in which V is the volume of the containing vessel, V_S is the volume of the solid phase present, and k is Boltzmann's constant. The saturated vapor pressure is proportional to the concentration of atoms in the vapor phase, $n_{v(eq,T)}/(V - V_S)$, and is thus not dependent on the volume in the system available to the vapor phase. As the magnitude of the entropy contribution, $-TS$, increases with increasing temperature and the internal energy contribution is independent of temperature, the contribution of the entropy term to A becomes increasingly predominant as the temperature is increased, and the compromise between U and $-TS$ which minimizes A occurs at larger values of n_v (or larger concentrations of atoms in the vapor phase). This is illustrated in Fig. 5.3 which is drawn for the temperatures T_1 and T_2, where $T_1 < T_2$.

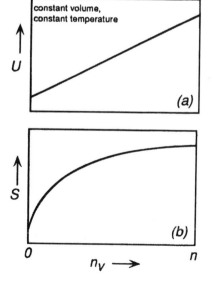

Figure 5.1 The variations of (a) internal energy, U, and (b) entropy, S, with the number of atoms in the vapor phase of a closed solid-vapor system at constant temperature and constant volume.

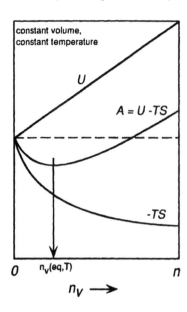

constant volume, constant temperature

U

$A = U - TS$

$-TS$

$0 \quad n_{v(eq,T)}$

n

$n_v \longrightarrow$

Figure 5.2 Illustration of the criterion for equilibrium in a closed solid-vapor system at constant temperature and constant volume.

An increase in the temperature from T_1 to T_2 increases the saturated vapor pressure of the solid from

$$p_{(at\,T_1)} = \frac{n_{v(eq,T_1)}kT_1}{(V - V_{S(at\,T_1)})}$$

to

$$p_{(at\,T_2)} = \frac{n_{v(eq,T_2)}kT_2}{(V - V_{S(at\,T_2)})}$$

In general the saturated vapor pressures of solids and liquids increase exponentially with increasing temperature. For example, the saturated vapor pressure of solid CO_2, "dry ice," varies with temperature as

$$\ln p\,(atm) = \frac{-3116}{T} + 16.01$$

The saturated vapor pressure of dry ice thus increases from 0.01 atm at 151.2 K, to 0.1 atm at 170.2 K, to 1 atm at 194.6 K.

For the constant-volume system the maximum temperature at which both solid and vapor phases occur is that temperature at which minimization of A occurs at $n = n_v$. Above this temperature the entropy contribution overwhelms the internal energy contribution, and hence all n atoms occur in the vapor phase. At such temperatures the pressure in the vessel is less than the saturated value, and the solid phase could only be made to reappear by either decreasing the volume of the system or increasing the number of atoms in the system, both of which increase the concentration of atoms in the vapor phase. Conversely, the concentration of atoms in the vapor phase decreases with decreasing temperature, and, in the limit $T \to 0$ K,

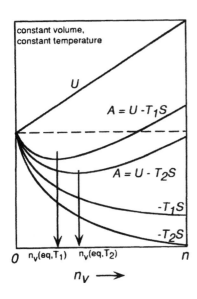

constant volume,
constant temperature

U

$A = U - T_1 S$

$A = U - T_2 S$

$-T_1 S$

$-T_2 S$

0 $n_v(\text{eq}, T_1)$ $n_v(\text{eq}, T_2)$ n

$n_v \longrightarrow$

Figure 5.3 The influence of temperature on the equilibrium state of a closed solid-vapor system of constant volume.

the entropy contribution to A vanishes and minimization of A coincides with minimization of U. Consequently, all of the n atoms occur in the solid phase.

If the constant-temperature heat reservoir containing the constant-volume system is, itself, of constant volume and is adiabatically contained, then the combined system is one of constant U and constant V. Thus the occurrence of the equilibrium concentration of atoms in the vapor phase coincides with the combined system having a maximum entropy. This is illustrated as follows. The entropy of the combined system is equal to the sum of the entropy of the heat reservoir and the entropy of the constant-volume particles system contained within it. The flow of heat from the reservoir to the particles system decreases the entropy of the former and increases the entropy of the latter. However, if less than the equilibrium number of atoms occurs in the vapor phase, then spontaneous evaporation of the solid occurs until the saturated vapor pressure is reached. During this process heat q flows spontaneously from the heat reservoir to the particles system, the entropy of the reservoir decreases by the amount q/T, the entropy of the particles system increases by the amount $q/T + \Delta S_{\text{irr}}$, and the increase in the entropy of the combined system is ΔS_{irr}. From Eq. (5.4) the corresponding decrease in the Helmholtz free energy is

$$\Delta A = -T\Delta S_{\text{irr}}$$

and hence minimization of A in the constant T, constant V particles system corresponds to maximization of S in the constant U, constant V combined system.

Classical Thermodynamics is concerned only with equilibrium states, and hence only that single state at the given values of V and T at which A has its minimum value is of interest from the point of view of thermodynamic consideration. In contrast, from the viewpoint of statistical mechanics all values of n_v in the range $0 \le n_v \le n$ are possible at the given values of V and T, although the probability that n_v deviates from the equilibrium value of $n_{v(eq,T)}$ is exceedingly small. This probability is small enough that, in practical terms, it corresponds with the thermodynamic statement that spontaneous deviation of a system from its equilibrium state is impossible.

5.4 THE GIBBS FREE ENERGY G

For a system undergoing a change of state from 1 to 2, Eq. (5.2) gives

$$(G_2 - G_1) = (H_2 - H_1) - (T_2 S_2 - T_1 S_1)$$
$$= (U_2 - U_1) + (P_2 V_2 - P_1 V_1) - (T_2 S_2 - T_1 S_1)$$

For a closed system, the First Law gives

$$(U_2 - U_1) = q - w$$

and thus

$$(G_2 - G_1) = q - w + (P_2 V_2 - P_1 V_1) - (T_2 S_2 - T_1 S_1) \tag{5.6}$$

If the process is carried out such that $T_1 = T_2 = T$, the constant temperature of the heat reservoir which supplies or withdraws heat from the system, and also if $P_1 = P_2 = P$, the constant pressure at which the system undergoes a change in volume, then

$$(G_2 - G_1) = q - w + P(V_2 - V_1) - T(S_2 - S_1)$$

In the expression for the First Law the work w is the *total* work done on or by the system during the process. Thus if the system performs chemical or electrical work in addition to the work of expansion against the external pressure, then these work terms are included in w. Thus w can be expressed as

$$w = w' + P(V_2 - V_1)$$

in which $P(V_2 - V_1)$ is the *P-V* work done by the change in volume at the constant pressure P and w' is the sum of all of the non–*P-V* forms of work done. Substituting into Eq. (5.6) gives

$$(G_2 - G_1) = q - w' - T(S_2 - S_1)$$

and again, as

$$q \le T(S_2 - S_1)$$

then

$$w' \le -(G_2 - G_1) \tag{5.7}$$

Again, the equality can be written as

$$-w' = (G_2 - G_1) + T\Delta S_{irr}$$

Thus, for an isothermal, isobaric process, during which no form of work other than P-V work is performed, i.e., $w' = 0$

$$(G_2 - G_1) + T\Delta S_{irr} = 0 \tag{5.8}$$

Such a process can only occur spontaneously (with a consequent increase in entropy) if the Gibbs free energy decreases. As $dS_{irr} = 0$ is a criterion for thermodynamic equilibrium, then an increment of an isothermal isobaric process occurring at equilibrium requires that

$$dG = 0 \tag{5.9}$$

Thus, for a system undergoing a process at constant T and constant P, the Gibbs free energy can only decrease or remain constant, and the attainment of equilibrium in the system coincides with the system having the minimum value of G consistent with the values of P and T. This criterion of equilibrium, which is of considerable practical use, will be used extensively in the subsequent chapters.

5.5 SUMMARY OF THE EQUATIONS FOR A CLOSED SYSTEM

Eq. (3.12) gives

$$dU = TdS - PdV$$

Now,

$$H = U + PV \text{ and thus } dH = TdS + VdP \tag{5.10}$$

$$A = U - TS \text{ and thus } dA = -SdT - PdV \tag{5.11}$$

$$G = H - TS \text{ and thus } dG = -SdT + VdP \tag{5.12}$$

5.6 THE VARIATION OF THE COMPOSITION AND SIZE OF THE SYSTEM

The discussion thus far has been restricted to consideration of closed systems of fixed size and fixed composition, and in such cases it was found that the system has two independent variables which, when fixed, uniquely, fixed the state of the system. However, if the size and composition can vary during a process, then the specification of only two variables is no longer sufficient to fix the state of the system. For example, it has been shown that, for a process conducted at constant temperature and constant pressure, equilibrium is attained when G attains its minimum value. If the composition of the system is variable, in that the numbers of moles of the various species present can vary as the consequence of a chemical reaction occurring in the system, then minimization of G at constant P and T occurs when the system has a unique composition. For example, if the system contained the gaseous species CO, CO_2, H_2, and H_2O, then at constant T and P minimization of

G would occur when the reaction equilibrium $CO + H_2O = CO_2 + H_2$ was established. Similarly, as G is an extensive property, i.e., is dependent on the size of the system, it is necessary that the number of moles within the system be specified.

The Gibbs free energy, G, is thus a function of T, P, and the numbers of moles of all of the species present in the system, i.e.,

$$G = G(T, P, n_i, n_j, n_k, \ldots) \qquad (5.13)$$

in which $n_i, n_j, n_k, \ldots$ are the numbers of moles of the species $i, j, k, \ldots$ present in the system and the state of the system is only fixed when all of the independent variables are fixed. Differentiation of Eq. (5.13) gives

$$dG = \left(\frac{\partial G}{\partial T}\right)_{P, n_i, n_j, \ldots} dT + \left(\frac{\partial G}{\partial P}\right)_{T, n_i, n_j, \ldots} dP + \left(\frac{\partial G}{\partial n_i}\right)_{T, P, n_j, n_k, \ldots} dn_i \qquad (5.14)$$

$$+ \left(\frac{\partial G}{\partial n_j}\right)_{T, P, n_i, n_k, \ldots} dn_j + \text{etc.}$$

If the numbers of moles of all of the individual species remain constant during the process, Eq. (5.14) simplifies to Eq. (5.12), i.e.,

$$dG = -SdT + VdP$$

from which it is seen that

$$\left(\frac{\partial G}{\partial T}\right)_{P, n_i, n_j, \ldots} = -S$$

and

$$\left(\frac{\partial G}{\partial P}\right)_{T, n_i, n_j, \ldots} = V$$

Substitution into Eq. (5.14) gives

$$dG = -SdT + VdP + \sum_{i=1}^{i=k} \left(\frac{\partial G}{\partial n_i}\right)_{T, P, n_j, \ldots} dn_i \qquad (5.15)$$

where

$$\sum_{i=1}^{i=k} \left(\frac{\partial G}{\partial n_i}\right)_{T, P, n_j, \ldots} dn_i$$

is the sum of k terms (one for each of the k species) each of which is determined by partial differentiation of G with respect to the number of moles of the ith species at constant T, P, and n_j, where n_j represents the numbers of moles of every species other than the ith species.

5.7 THE CHEMICAL POTENTIAL

The term $(\partial G/\partial n_i)_{T,P,n_j,\ldots}$ is called *the chemical potential of the species i* and is designated as μ_i, i.e.,

$$\left(\frac{\partial G}{\partial n_i}\right)_{T,P,n_j,\ldots} = \mu_i \tag{5.16}$$

The chemical potential of i in a homogeneous phase is thus the rate of increase of G with n_i when the species i is added to the system at constant temperature, pressure, and numbers of moles of all of the other species. Alternatively, if the system is large enough that the addition of 1 mole of i, at constant temperature and pressure, does not measurably change the composition of the system, then μ_i is the increase in the Gibbs free energy of the system caused by the addition.

Eq. (5.15) can thus be written as

$$dG = -SdT + VdP + \sum_{1}^{k}\mu_i dn_i \tag{5.17}$$

in which from G is expressed as a function of T, P, and composition. Eq. (5.17) can thus be applied to open systems which exchange matter as well as heat with their surroundings, and to closed systems which undergo changes in composition caused by chemical reactions.

Similarly, Eqs. (3.12), (5.10), and (5.11) can be made applicable to open systems by including the terms describing the dependences on composition of, respectively, U, H, and A:

$$dU = TdS - PdV + \sum_{1}^{k}\left(\frac{\partial U}{\partial n_i}\right)_{S,V,n_j,\ldots} dn_i \tag{5.18}$$

$$dH = TdS - VdP + \sum_{1}^{k}\left(\frac{\partial H}{\partial n_i}\right)_{S,P,n_j,\ldots} dn_i \tag{5.19}$$

$$dA = -SdT + PdV + \sum_{1}^{k}\left(\frac{\partial A}{\partial n_i}\right)_{T,V,n_j,\ldots} dn_i \tag{5.20}$$

Inspection of Eqs. (5.16), (5.18), (5.19), and (5.20) shows that

$$\left(\frac{\partial G}{\partial n_i}\right)_{T,P,n_j} = \mu_i = \left(\frac{\partial U}{\partial n_i}\right)_{S,V,n_j} = \left(\frac{\partial H}{\partial n_i}\right)_{S,P,n_j} = \left(\frac{\partial A}{\partial n_i}\right)_{T,V,n_j} \tag{5.21}$$

and, hence, the complete set of equations is

$$dU = TdS - PdV + \sum \mu_i dn_i \tag{5.22}$$

$$dH = TdS + VdP + \sum \mu_i dn_i \tag{5.23}$$

$$dA = -SdT - PdV + \sum \mu_i dn_i \tag{5.24}$$

$$dG = -SdT + VdP + \sum \mu_i dn_i \tag{5.25}$$

U is thus the "characteristic function" of the independent variables S, V, and composition, H is the characteristic function of the independent variables S, P, and composition, A is the characteristic function of the independent variables T, V, and composition, and G is the characteristic function of the independent variables T, P, and composition. Although all four of the above equations are basic in nature, Eq. (5.25) is called the *fundamental equation* because of its practical usefulness.

The First Law gives

$$dU = \delta q - \delta w$$

which, on comparison with Eq. (5.22), indicates that, for a closed system undergoing a process involving a reversible change in composition (e.g., a reversible chemical reaction),

$$\delta q = TdS$$

and

$$\delta w = PdV + \sum \mu_i dn_i$$

The term $\sum \mu_i dn_i$ is thus the chemical work done by the system which was denoted as w' in Eq. (5.8), and the total work w is the sum of the P-V work and the chemical work.

5.8 THERMODYNAMIC RELATIONS

The following relationships are obtained from Eqs. (5.22)–(5.25).

$$T = \left(\frac{\partial U}{\partial S} \right)_{V,\text{comp}} = \left(\frac{\partial H}{\partial S} \right)_{P,\text{comp}} \tag{5.26}$$

$$P = -\left(\frac{\partial U}{\partial V} \right)_{S,\text{comp}} = -\left(\frac{\partial A}{\partial V} \right)_{T,\text{comp}} \tag{5.27}$$

$$V = \left(\frac{\partial H}{\partial P} \right)_{S,\text{comp}} = \left(\frac{\partial G}{\partial P} \right)_{T,\text{comp}} \tag{5.28}$$

$$S = -\left(\frac{\partial A}{\partial T} \right)_{V,\text{comp}} = -\left(\frac{\partial G}{\partial T} \right)_{P,\text{comp}} \tag{5.29}$$

5.9 MAXWELL'S EQUATIONS

If Z is a state function and x and y are chosen as the independent variables in a closed system of fixed composition then

$$Z = Z(x, y)$$

differentiation of which gives

$$dZ = \left(\frac{\partial Z}{\partial x} \right)_y dx + \left(\frac{\partial Z}{\partial y} \right)_x dy$$

If the partial derivative $(\partial Z/\partial x)_y$ is itself a function of x and y, being given by $(\partial Z/\partial x)_y = L(x, y)$, and similarly the partial derivative $(\partial Z/\partial y)_x = M(x, y)$ then

$$dZ = L\,dx + M\,dy$$

Thus

$$\left[\frac{\partial}{\partial y}\left(\frac{\partial Z}{\partial x}\right)_y\right]_x = \left(\frac{\partial L}{\partial y}\right)_x$$

and

$$\left[\frac{\partial}{\partial x}\left(\frac{\partial Z}{\partial y}\right)_x\right]_y = \left(\frac{\partial M}{\partial x}\right)_y$$

But, as Z is a state function, the change in Z is independent of the order of differentiation, i.e.,

$$\left[\frac{\partial}{\partial y}\left(\frac{\partial Z}{\partial x}\right)_y\right]_x = \left[\frac{\partial}{\partial x}\left(\frac{\partial Z}{\partial y}\right)_x\right]_y = \frac{\partial^2 Z}{\partial x \partial y}$$

and hence

$$\left(\frac{\partial L}{\partial y}\right)_x = \left(\frac{\partial M}{\partial x}\right)_y \tag{5.30}$$

Application of Eq. (5.30) to Eqs. (3.12) and (5.10)–(5.12) gives a set of relationships which are known as Maxwell's equations. These are

$$\left(\frac{\partial T}{\partial V}\right)_S = -\left(\frac{\partial P}{\partial S}\right)_V \tag{5.31}$$

$$\left(\frac{\partial T}{\partial P}\right)_S = \left(\frac{\partial V}{\partial S}\right)_P \tag{5.32}$$

$$\left(\frac{\partial S}{\partial V}\right)_T = \left(\frac{\partial P}{\partial T}\right)_V \tag{5.33}$$

$$\left(\frac{\partial S}{\partial P}\right)_T = -\left(\frac{\partial V}{\partial T}\right)_P \tag{5.34}$$

Similarly, equations can be obtained by considering variations in composition. The value of the above equations lies in the fact that they contain many experimentally measurable quantities. Consider the dependence of the entropy of an ideal gas on the independent variables T and V, i.e.,

$$S = S(T, V)$$

differentiation of which gives

$$dS = \left(\frac{\partial S}{\partial T}\right)_V dT + \left(\frac{\partial S}{\partial V}\right)_T dV \tag{i}$$

From combination of the definition of the constant volume heat capacity, Eq. (2.6), and Eq. (3.8) applied to a reversible process conducted at constant volume,

$$TdS = \delta q_v = dU = nc_v dT$$

the first partial derivative on the right-hand side of Eq. (i) is obtained as

$$\left(\frac{\partial S}{\partial T}\right)_V = \frac{nc_v}{T}$$

and the second partial derivative in Eq. (i) is obtained from Maxwell's equation (5.33). Thus, Eq. (i) can be written as

$$dS = \frac{nc_v}{T}dT + \left(\frac{\partial P}{\partial T}\right)_V dV \tag{ii}$$

From the ideal gas law,

$$\left(\frac{\partial P}{\partial T}\right)_V = \frac{nR}{V} \tag{iii}$$

and thus Eq. (ii) can be written as

$$dS = \frac{nc_v}{T}dT + \frac{nR}{V}dV \tag{iv}$$

integration of which between the states 1 and 2 gives

$$S_2 - S_1 = nc_v \ln\left(\frac{T_2}{T_1}\right) + nR\ln\left(\frac{V_2}{V_1}\right) \tag{v}$$

Eq. (v) could have been used for the solution of Ex. 1 presented in Chap. 3, and, for a reversible isentropic process, Eq. (v) collapses to Eq. (2.9).

A similar example of the use of Maxwell's equations is as follows. For a closed system of fixed composition Eq. (3.12) gives

$$dU = TdS - PdV$$

Thus

$$\left(\frac{\partial U}{\partial V}\right)_T = T\left(\frac{\partial S}{\partial V}\right)_T - P$$

Use of Maxwell's equation (5.33) allows this to be written as

$$\left(\frac{\partial U}{\partial V}\right)_T = T\left(\frac{\partial P}{\partial T}\right)_V - P \tag{vi}$$

which is an equation of state relating the internal energy, U, of a closed system of fixed composition to the measurable quantities T, V, and P. If the system is 1 mole of ideal gas substitution of Eq. (iii) into Eq. (vi) gives $(\partial U/\partial V)_T = 0$, which shows that the internal energy of an ideal gas is independent of the volume of the gas.

Similarly, for a closed system of fixed composition, Eq. (5.10) gives $dH = TdS + VdP$ in which case

$$\left(\frac{\partial H}{\partial P}\right)_T = T\left(\frac{\partial S}{\partial P}\right)_T + V$$

Substituting Maxwell's equation (5.34) gives

$$\left(\frac{\partial H}{\partial P}\right)_T = -\left(\frac{\partial V}{\partial P}\right)_T + V$$

which is an equation of state which gives the dependence of enthalpy on T, P, and V. Again, as the system is an ideal gas, this equation of state shows that the enthalpy of an ideal gas is independent of its pressure.

5.10 THE UPSTAIRS-DOWNSTAIRS-INSIDE-OUT FORMULA

Given three state functions x, y, and z and a closed system of fixed composition, then

$$x = x(y, z)$$

or

$$dx = \left(\frac{\partial x}{\partial y}\right)_z dy + \left(\frac{\partial x}{\partial z}\right)_y dz$$

For an incremental change of state at constant x,

$$\left(\frac{\partial x}{\partial y}\right)_z dy = -\left(\frac{\partial x}{\partial z}\right)_y dz$$

or

$$\left(\frac{\partial x}{\partial y}\right)_z \left(\frac{\partial y}{\partial z}\right)_x = -\left(\frac{\partial x}{\partial z}\right)_y$$

This can be written as

$$\left(\frac{\partial x}{\partial y}\right)_z \left(\frac{\partial y}{\partial z}\right)_x \left(\frac{\partial z}{\partial x}\right)_y = -1 \qquad (5.35)$$

Eq. (5.35) can be used with any three state functions and is called the upstairs-downstairs-inside-out formula because each of the state functions appears once in the numerator, once in the denominator, and once outside the bracket.

5.11 THE GIBBS-HELMHOLTZ EQUATION

Eq. (5.2) gives

$$G = H - TS$$

and Eq. (5.12) gives

$$\left(\frac{\partial G}{\partial T}\right)_P = -S$$

Therefore, at constant pressure and composition,

$$G = H + T\left(\frac{dG}{dT}\right)$$

or

$$GdT = HdT + TdG$$

Dividing throughout by T^2 and rearranging gives

$$\frac{TdG - GdT}{T^2} = -\frac{HdT}{T^2}$$

which, on comparison with the identity $d(x/y) = (y\,dx - x\,dy)/y^2$, shows that

$$\frac{d(G/T)}{dT} = -\frac{H}{T^2} \tag{5.36}$$

Eq. (5.36) is known as the Gibbs-Helmholtz equation and is applicable to a closed system of fixed composition undergoing processes at constant composition. For an isobaric change of state of a closed system of fixed composition, Eq. (5.36) gives the relation of the change in G to the change in H as

$$\frac{d(\Delta G/T)}{dT} = -\frac{\Delta H}{T^2} \tag{5.36a}$$

This equation is of particular use in experimental thermodynamics, as it allows ΔH, the heat of a reaction, to be obtained from a measurement of the variation of ΔG, the free energy change for the reaction, with temperature, or, conversely, it allows ΔG to be obtained from a measurement of ΔH. The usefulness of this equation will be developed in Chap. 10.

The corresponding relationship between A and U is obtained as follows: Eq. (5.1) gives

$$A = U - TS$$

$$= U + T\left(\frac{\partial A}{\partial T}\right)_V$$

and manipulation similar to the above gives

$$\frac{d(A/T)}{dT} = -\frac{U}{T^2} \tag{5.37}$$

This equation is only applicable to closed systems of fixed composition under-going processes at constant volume and again, for a change of state under these conditions,

$$\frac{d(\Delta A/T)}{dT} = -\frac{\Delta U}{T^2} \tag{5.37a}$$

5.12 SUMMARY

1. The Helmholtz free energy, A, is given by $A = U - TS$.
2. The Gibbs free energy, G, is given by $G = H - TS$.
3. For a change of state at constant pressure, $\Delta H = q_p$.
4. In a closed system held at constant T and V, A can only decrease or remain constant. Equilibrium is attained when A achieves its minimum value.
5. During an isothermal, isobaric process during which no form of work other than P-V work is performed, i.e., $w' = 0$, G can only decrease or remain constant. Equilibrium is attained when G reaches its minimum value.
6. Internal energy varies with S, V, and composition as

$$dU = TdS - PdV + \sum \mu_i dn_i$$

Enthalpy varies with S, P, and composition as

$$dH = TdS + VdP + \sum \mu_i dn_i$$

Helmholtz free energy varies with T, V, and composition as

$$dA = -SdT - PdV + \sum \mu_i dn_i$$

and Gibbs free energy varies with T, P, and composition as

$$dG = -SdT + VdP + \sum \mu_i dn_i$$

7. Maxwell's equations are

$$\left(\frac{\partial T}{\partial V}\right)_S = -\left(\frac{\partial P}{\partial S}\right)_V, \qquad \left(\frac{\partial T}{\partial P}\right)_S = \left(\frac{\partial V}{\partial S}\right)_P$$

$$\left(\frac{\partial S}{\partial V}\right)_T = \left(\frac{\partial P}{\partial T}\right)_V, \qquad \left(\frac{\partial S}{\partial P}\right)_T = -\left(\frac{\partial V}{\partial T}\right)_P$$

8. With x, y, and z as state functions, the upstairs-downstairs-inside-out formula is

$$\left(\frac{\partial x}{\partial y}\right)_z \left(\frac{\partial y}{\partial z}\right)_x \left(\frac{\partial z}{\partial x}\right)_y = -1$$

9. The Gibbs-Helmholtz equations are

$$\left[\frac{\partial\left(\frac{G}{T}\right)}{\partial T}\right]_P = -\frac{H}{T^2}$$

and

$$\left[\frac{\partial\left(\frac{A}{T}\right)}{\partial T}\right]_V = -\frac{U}{T^2}$$

5.13 EXAMPLE OF THE USE OF THE THERMODYNAMIC RELATIONS

Eq. (2.8) gives the relationship between c_p and c_v as

$$c_p - c_v = \left(\frac{\partial V}{\partial T}\right)_P\left[P + \left(\frac{\partial U}{\partial V}\right)_T\right] \tag{2.8}$$

The use of the thermodynamic relations allows the difference between c_p and c_v to be expressed in terms of experimentally measurable quantities. Eq. (5.27) gives

$$P = -\left(\frac{\partial A}{\partial V}\right)_T$$

Thus

$$c_p - c_v = \left(\frac{\partial V}{\partial T}\right)_P\left[\left(\frac{\partial U}{\partial V}\right)_T - \left(\frac{\partial A}{\partial V}\right)_T\right]$$

By definition

$$A = U - TS$$

Thus

$$c_p - c_v = \left(\frac{\partial V}{\partial T}\right)_P\left[\left(\frac{\partial U}{\partial V}\right)_T - \left(\frac{\partial U}{\partial V}\right)_T + T\left(\frac{\partial S}{\partial V}\right)_T\right]$$

$$= \left(\frac{\partial V}{\partial T}\right)_P\left[T\left(\frac{\partial S}{\partial V}\right)_T\right]$$

Maxwell's equation (5.33) gives

$$\left(\frac{\partial S}{\partial V}\right)_T = \left(\frac{\partial P}{\partial T}\right)_V$$

and the upstairs-downstairs-inside-out formula, using P, V, and T, gives

$$\left(\frac{\partial P}{\partial T}\right)_V = -\left(\frac{\partial P}{\partial V}\right)_T\left(\frac{\partial V}{\partial T}\right)_P$$

Thus

$$c_p - c_v = -T \left(\frac{\partial V}{\partial T} \right)_P \left(\frac{\partial P}{\partial V} \right)_T \left(\frac{\partial V}{\partial T} \right)_P$$

In Chap. 1, the isobaric thermal expansivity was defined as

$$\alpha = \frac{1}{V} \left(\frac{\partial V}{\partial T} \right)_P$$

Similarly, the isothermal compressibility of a substance, or system, is defined as

$$\beta = -\frac{1}{V} \left(\frac{\partial V}{\partial P} \right)_T$$

which is the fractional decrease in the volume of the system for unit increase in pressure at constant temperature. The negative sign is used to make β a positive number. Thus

$$c_p - c_v = \frac{VT\alpha^2}{\beta} \tag{5.38}$$

and the right-hand side of this equation contains only experimentally measurable quantities.

At 20°C aluminum has the following properties:

$c_p = 24.36$ J/mole·K
$\alpha = 7.05 \times 10^{-5}$ K^{-1}
$\beta = 1.20 \times 10^{-6}$ atm^{-1}
density, $\rho = 2.70$ g/cm^3

The atomic weight of aluminum is 26.98, and thus, at 20°C, the molar volume of aluminum, V, is

$$V = \frac{26.98}{2.70 \times 1000} = 0.010 \text{ liters/mole}$$

and thus the difference between c_p and c_v is

$$\frac{0.010 \times 293 \times (7.05 \times 10^{-5})^2}{1.20 \times 10^{-6}} = 0.0121 \text{ liter·atm/mole·K}$$

$$0.0121 \times \frac{8.3144}{0.08206} = 1.23 \text{ J/mole·K}$$

The constant volume molar heat capacity of aluminum at 20°C is thus

$$24.36 - 1.23 = 23.13 \text{ J/mole·K}$$

5.14 NUMERICAL EXAMPLE

The following example was provided by Dr. Bill Fahrenholtz, University of Missouri–Rolla.

Example 1

The adiabatic thermoelastic effect describes the change in temperature with pressure for a brittle solid when it is loaded rapidly (i.e., the rate of loading is much more rapid that the rate of heat transfer). This effect has been used to measure the stresses that develop around defects in composite materials using cyclic loading and a high speed thermal imaging camera. Using Maxwell's relations, derive an expression for the adiabatic thermoelastic effect. Estimate the change in temperature for a piece of alumina that is loaded to 500 MPa.

Data:
Initial temperature 298 K
$\alpha = 2.2 \times 10^{-5}/K$
$c_P = 80$ J/mole·K

The problem requires calculation of the variation in temperature with pressure at constant entropy, i.e.

$$\left(\frac{\partial T}{\partial P}\right)_S$$

Use of the variables T, P and S in Eq. (5.35) gives

$$\left(\frac{\partial T}{\partial P}\right)_S \left(\frac{\partial P}{\partial S}\right)_T \left(\frac{\partial S}{\partial T}\right)_P = -1 \qquad (i)$$

The definition of the molar constant pressure heat capacity

$$c_p = T\left(\frac{\partial S}{\partial T}\right)_P \quad \text{gives}$$

$$\left(\frac{\partial S}{\partial T}\right)_P = \frac{c_p}{T} \qquad (ii)$$

The inverse of Maxwell's equation, Eq. (5.34), gives

$$\left(\frac{\partial P}{\partial S}\right)_T = -\left(\frac{\partial T}{\partial V}\right)_P \qquad (iii)$$

and the isobaric thermal expansivity is defined as

$$\alpha = \frac{1}{V}\left(\frac{\partial V}{\partial T}\right)_P \quad \text{or} \quad \left(\frac{\partial V}{\partial T}\right)_P = \alpha V \qquad (iv)$$

Substituting Eqs. (ii)–(iv) into Eq. (i) and rearranging gives

$$\left(\frac{\partial T}{\partial P}\right)_S \left(\frac{\partial P}{\partial S}\right)_T \left(\frac{c_p}{T}\right) = 1$$

$$\left(\frac{\partial T}{\partial P}\right)_S \left(-\frac{1}{V\alpha}\right) \left(\frac{c_p}{T}\right) = -1$$

or

$$\left(\frac{\partial T}{\partial P}\right)_S = \frac{TV\alpha}{c_p}$$

It is reasonable to assume that Al_2O_3 is an incompressible solid and that its heat capacity does not vary significantly over small ranges of temperature and pressure. Thus the variables can be separated for integration

$$\left(\frac{\partial T}{\partial P}\right)_S = \frac{TV\alpha}{c_p}$$

$$\frac{dT}{T} = \frac{V\alpha dP}{c_p} \quad \text{at constant } S \text{ (an adiabatic process)}$$

$$\ln\frac{T_2}{T_1} = \left(\frac{V\alpha}{c_p}\right)\Delta P$$

$$\ln\left(\frac{T_2}{298}\right) = \left[\frac{(2.56 \times 10^{-5} \text{ m}^3/\text{mole})(2.2 \times 10^{-5} \text{ K}^{-1})}{80 \text{ J/mole·K}}\right](500 \times 10^6 \text{ N/m}^2)$$

$$= 0.00352$$

$$\therefore \ T_2 = 299 \text{ K and } \Delta T = 299 - 298 \text{ K } = 1.0 \text{ K}$$

PROBLEMS

5.1 Show that

$$\left(\frac{\partial S}{\partial V}\right)_P = \frac{c_p}{TV\alpha}$$

5.2 Show that

$$\left(\frac{\partial S}{\partial P}\right)_V = \frac{c_p\beta}{T\alpha} - V\alpha$$

5.3 Show that

$$\left(\frac{\partial A}{\partial P}\right)_V = -\frac{S\beta}{\alpha}$$

5.4 Show that

$$\left(\frac{\partial A}{\partial V}\right)_P = -\left(\frac{S}{V\alpha} + P\right)$$

5.5 Show that

$$\left(\frac{\partial H}{\partial S}\right)_V = T\left(1 + \frac{V\alpha}{c_v\beta}\right)$$

5.6 Show that

$$\left(\frac{\partial H}{\partial V}\right)_S = -\frac{c_p}{c_v\beta}$$

5.7 Show that

$$\left(\frac{\partial c_p}{\partial P}\right)_T = -TV\left(\alpha^2 + \frac{d\alpha}{dT}\right)$$

5.8 Show that

$$\left(\frac{\partial T}{\partial P}\right)_S = \frac{T\alpha V}{c_p}$$

5.9 Show that

$$\left(\frac{\partial P}{\partial V}\right)_S = -\frac{c_p}{c_v V\beta}$$

5.10 Show that

$$\left(\frac{\partial^2 G}{\partial P^2}\right)_T = -\frac{1}{\left(\dfrac{\partial^2 A}{\partial V^2}\right)_T}$$

5.11 Joule and Thomson showed experimentally that when a steady stream of non-ideal gas is passed through a thermally insulated tube, in which is inserted a throttle valve, the temperature of the gas changes and the state of the gas is changed from P_1, T_1 to P_2, T_2. Show that this process is isenthalpic. The change in T is described in terms of the Joule-Thomson coefficient, μ_{J-T}, as

$$\mu_{J-T} = \left(\frac{\partial T}{\partial P}\right)_H$$

Show that

$$\mu_{J-T} = -\frac{V}{c_p}(1 - \alpha T)$$

and show that the Joule-Thomson coefficient for an ideal gas is zero.

5.12 Determine the values of ΔU, ΔH, ΔS, ΔA, and ΔG for the following processes. (In (c), (d), and (e) show that an absolute value of the entropy is required.)

a. The four processes in Prob. 4.1.
b. One mole of an ideal gas at the pressure P and the temperature T expands into a vacuum to double its volume.
c. The adiabatic expansion of mole of an ideal gas from P_1, T_1 to P_2, T_2.
d. The expansion of mole of an ideal gas at constant pressure from V_1, T_1 to V_2, T_2.
e. The expansion of mole of an ideal gas at constant volume from P_1, T_1 to P_2, T_2.

Chapter 6

HEAT CAPACITY, ENTHALPY, ENTROPY, AND THE THIRD LAW OF THERMODYNAMICS

6.1 INTRODUCTION

Eqs. (2.6) and (2.7) defined two heat capacities, the heat capacity at constant volume

$$C_v = \left(\frac{\partial U}{\partial T} \right)_V \tag{2.6}$$

and the heat capacity at constant pressure

$$C_p = \left(\frac{\partial H}{\partial T} \right)_P \tag{2.7}$$

At this point it is convenient to introduce a notation which allows distinctions to be made between the values of extensive properties per mole of a system and the values of the extensive properties for the entire system. If E is an extensive property, then E' will be used to denote the value of the property for the entire system containing n moles, and E will be used to denote the value of the property per mole of the system. Thus E is the molar value of the property and, for a system containing n moles,

$$E' = nE$$

Eqs. (2.6) and (2.7) can thus be written equivalently as

$$dU' = C_v dT = n c_v dT \quad \text{or} \quad dU = c_v dT \tag{2.6a}$$

$$dH' = C_p dT = n c_p dT \quad \text{or} \quad dH = c_p dT \tag{2.7a}$$

in which c_p and c_v are, respectively, the constant-pressure, and constant-volume molar heat capacities. Integration of Eq. (2.7a) between the states (T_2, P) and (T_1, P) gives the difference between the molar enthalpies of the two states as

$$\Delta H = H(T_2, P) - H(T_1, P) = \int_{T_1}^{T_2} c_p dT \tag{6.1}$$

from which it is seen that a knowledge of the variation of c_p with temperature is required for determination of the temperature dependence of enthalpy, and, as will be seen, for the determination of the temperature dependence of entropy. Similarly, integration of Eq. (2.6a) between T_2 and T_1 at constant volume shows that knowledge of the variation of c_v with temperature is required for determination of the temperature dependence of internal energy.

6.2 THEORETICAL CALCULATION OF THE HEAT CAPACITY

As a result of experimental measurement, Dulong and Petit introduced an empirical rule in 1819 which states that the molar heat capacities of all solid elements have the value $3R(= 24.9$ J/K$)$, and, in 1865, Kopp introduced a rule which states that, at ordinary temperatures, the molar heat capacity of a solid chemical compound is approximately equal to the sum of molar heat capacities of its constituent chemical elements. Although the molar heat capacities of most elements at room temperature have values which are very close to $3R$, subsequent experimental measurement shows that heat capacity usually increases with increasing temperature and can have values significantly lower than $3R$ at low temperatures. Fig. 6.1 shows that, although lead and copper closely obey Dulong and Petit's rule at room temperature, the constant-volume heat capacities of silicon and diamond are significantly less than $3R$. Fig. 6.1 also shows the significant decrease in the heat capacities at low temperatures.

Calculation of the heat capacity of a solid element, as a function of temperature, was one of the early triumphs of the quantum theory. The first such calculation was made in 1907 by Einstein, who considered the properties of a crystal containing n atoms, each of which behaves as a harmonic oscillator vibrating independently about its lattice point. As the behavior of each oscillator is not influenced by the behavior of its neighbors, each oscillator vibrates with a single fixed frequency v, and a system of such oscillators is called an Einstein crystal.

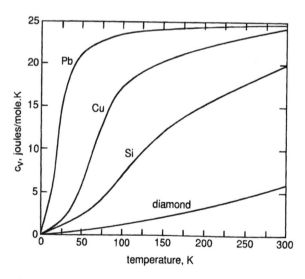

Figure 6.1 The constant-volume molar heat capacities of Pb, Cu, Si, and diamond as functions of temperature.

Quantum theory gives the energy of the ith energy level of a harmonic oscillator as

$$\varepsilon_i = \left(i + \frac{1}{2}\right)hv \tag{6.2}$$

in which i is an integer which has values in the range zero to infinity, and h is Planck's constant of action. As each oscillator has three degrees of freedom, i.e., can vibrate in the x, y, and z directions, the energy, U', of the Einstein crystal (which can be considered to be a system of $3n$ linear harmonic oscillators) is given as

$$U' = 3\sum n_i \varepsilon_i \tag{6.3}$$

where, as before, n_i is the number of atoms in the ith energy level. Substituting Eqs. (6.2) and (4.13) into Eq. (6.3) gives

$$U' = 3\sum \left(i + \frac{1}{2}\right)hv\left(\frac{ne^{-hv(i+\frac{1}{2})/kT}}{\sum e^{-hv(i+\frac{1}{2})/kT}}\right)$$

$$= 3nhv\left(\frac{\sum ie^{-hv(i+\frac{1}{2})/kT}}{\sum e^{-hv(i+\frac{1}{2})/kT}} + \frac{\frac{1}{2}\sum e^{-hv(i+\frac{1}{2})/kT}}{\sum e^{-hv(i+\frac{1}{2})/kT}}\right)$$

$$= 3nhv\left(\frac{\sum ie^{-hvi/kT}}{\sum e^{-hvi/kT}} + \frac{1}{2}\right)$$

$$= \frac{3}{2}nhv\left(1 + \frac{2\sum ie^{-hvi/kT}}{\sum e^{-hvi/kT}}\right)$$

Taking

$$\sum ie^{-hvi/kT} = \sum ix^i$$

where $x = e^{-hv/kT}$, gives

$$x(1 + 2x + 3x^2 + \cdots) = \frac{x}{(1-x)^2}$$

and

$$\sum e^{-hvi/kT} = \sum x^i = 1 + x + x^2 + \cdots = \frac{1}{(1-x)}$$

in which case

$$U' = \frac{3}{2}nhv\left(1 + \frac{2x}{1+x}\right)$$

$$= \frac{3}{2}nhv\left(1 + \frac{2e^{-hv/kT}}{1 - e^{-hv/kT}}\right)$$

$$= \frac{3}{2}nhv + \frac{3nhv}{(e^{hv/kT} - 1)} \tag{6.4}$$

Eq. (6.4) gives the variation of the energy of the system with temperature, and differentiation of Eq. (6.4) with respect to temperature at constant volume gives, by definition, the constant-volume heat capacity C_v. Maintaining a constant volume causes constant quantization of the energy levels. Thus

$$C_v = \left(\frac{\partial U'}{\partial T}\right)_V = 3nh\nu(e^{h\nu/kT} - 1)^{-2}\frac{h\nu}{kT^2}e^{h\nu/kT}$$

$$= 3nk\left(\frac{h\nu}{kT}\right)^2\frac{e^{h\nu/kT}}{(e^{h\nu/kT} - 1)^2}$$

Defining $h\nu/k = \theta_E$, where θ_E is the Einstein characteristic temperature, and taking n equal to Avogadro's number, gives the constant-volume molar heat capacity of the crystal as

$$c_v = 3R\left(\frac{\theta_E}{T}\right)^2\frac{e^{\theta_E/T}}{(e^{\theta_E/T} - 1)^2} \tag{6.5}$$

The variation of c_v with T/θ_E is shown in Fig. 6.2, which shows that as T/θ_E (and hence T) increases, $c_v \to 3R$ in agreement with Dulong and Petit's law, and as $T \to 0$, $c_v \to 0$, which is in agreement with experimental observation. The actual value of θ_E for any element and its vibration frequency, ν, are obtained by curve-fitting Eq. (6.5) to experimentally measured heat capacity data. Such curve-fitting, which is shown in Fig. 6.2, shows that although the Einstein equation adequately represents actual heat capacities at higher temperatures, the theoretical values approach zero more rapidly than do the actual values. As T/θ_E decreases from 0.02 to 0.01 the theoretical molar heat capacity decreases from 1.2×10^{-17} to 9.3×10^{-39} J/K. This discrepancy is caused by the fact that the oscillators do not vibrate with a single frequency.

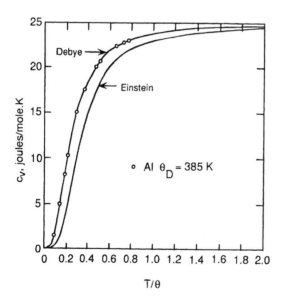

Figure 6.2 Comparison among the Debye heat capacity, the Einstein heat capacity, and the actual heat capacity of aluminum.

The next step in the theory was made in 1912 by Debye, who assumed that the range of frequencies of vibration available to the oscillators is the same as that available to the elastic vibrations in a continuous solid. The lower limit of these vibrations is determined by the interatomic distances in the solid, i.e., if the wavelength is equal to the interatomic distance then neighboring atoms would be in the same phase of vibration and, hence, vibration of one atom with respect to another would not occur. Theoretically, the shortest allowable wavelength is twice the interatomic distance, in which case neighboring atoms vibrate in opposition to one another. Taking this minimum wavelength, λ_{min}, to be in the order of 5×10^{-8} cm, and the wave velocity, v, in the solid to be 5×10^5 cm/sec, gives the maximum frequency of vibration of an oscillator to be in the order of

$$v_{max} = \frac{v}{\lambda_{min}} = \frac{5 \times 10^5}{5 \times 10^{-8}} = 10^{13} \text{ sec}^{-1}$$

Debye assumed that the frequency distribution is one in which the number of vibrations per unit volume per unit frequency range increases parabolically with increasing frequency in the allowed range $0 \leq v \leq v_{max}$, and, by integrating Einstein's equation over this range of frequencies, he obtained the heat capacity of the solid as

$$C_v = \frac{9nh^3}{k^2\theta_D^3} \int_0^{v_D} v^2 \left(\frac{hv}{kT}\right)^2 \frac{e^{hv/kT}}{(1 - e^{hv/kT})^2} dv$$

which, with $x = hv/kT$, gives

$$C_v = 9R\left(\frac{T}{\theta_D}\right)^3 \int_0^{\theta_D/T} \frac{x^4 e^{-x}}{(1 - e^{-x})^2} dx \tag{6.6}$$

where v_D (the Debye frequency) $= v_{max}$ and $\theta_D = hv_D/k$ is the characteristic Debye temperature of the solid.

Eq. (6.6) is compared with Einstein's equation in Fig. 6.2, which shows that Debye's equation gives an excellent fit to the experimental data at lower temperatures. Fig. 6.3 shows the curve-fitting of Debye's equation to the measured heat capacities of Pb, Ag, Al, and diamond. The curves are nearly identical except for a horizontal displacement and the relative horizontal displacement is a measure of θ_D. When plotted as c_v versus $\log T/\theta_D$, all of the datum points in Fig. 6.3 fall on a single line.

The value of the integral in Eq. (6.6) from zero to infinity is 25.98, and thus, for very low temperatures, Eq. (6.6) becomes

$$c_v = 9R \times 25.98 \left(\frac{T}{\theta_D}\right)^3 = 1943 \left(\frac{T}{\theta_D}\right)^3 \tag{6.7}$$

which is called the Debye T^3 law for low-temperature heat capacities.

Debye's theory does not consider the contribution made to the heat capacity by the uptake of energy by electrons, and, since $c_v = (\partial U/\partial T)_v$, it follows that a contribution to the heat capacity will be made in any range of temperature in which

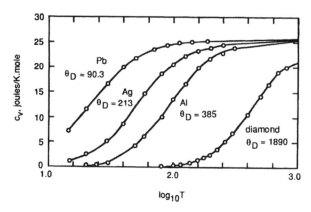

Figure 6.3 The constant-volume molar heat capacities of several solid elements. The curves are the Debye equation with the indicated values of θ_D.

the energy of the electrons changes with temperature. The electron gas theory of metals predicts that the electronic contribution to the heat capacity is proportional to the absolute temperature, and thus the electronic contribution becomes large in absolute value at elevated temperatures. Thus, at high temperatures, where the lattice contribution approaches the Dulong and Petit value, the molar heat capacity should vary with temperature as

$$c_v = 24.94 + bT \text{ J/K·mole}$$

in which bT is the electronic contribution. Theoretical calculation of the value of the coefficient b is made difficult by a lack of knowledge of the number of electrons per atom present in the electron gas. Also, the theoretical approach to heat capacities does not consider the contribution made by the anharmonicity of the lattice vibrations at elevated temperatures.

As a consequence of the various uncertainties in the theoretical calculation of heat capacities, it is normal practice to measure the variation of the constant-pressure molar heat capacity with temperature and express the relationship analytically.

6.3 THE EMPIRICAL REPRESENTATION OF HEAT CAPACITIES

The experimentally measured variation of the constant-pressure molar heat capacity of a material with temperature is normally fitted to an expression of the form

$$c_p = a + bT + cT^{-2}$$

and it should be noted that the analytical expression is only applicable in that stated temperature range over which the values of the heat capacity were measured. For example ZrO_2 exists as monoclinic α-ZrO_2 from room temperature to 1478 K and as tetragonal β-ZrO_2 in the range of temperature 1478–2670 K and each polymorph has its own equation giving the variation of its heat capacity with temperature.

$$\alpha\text{-}ZrO_2 \quad c_p = 69.62 + 7.53 \times 10^{-3}T - 14.06 \times 10^5 T^{-2} \text{ J/K·mole}$$

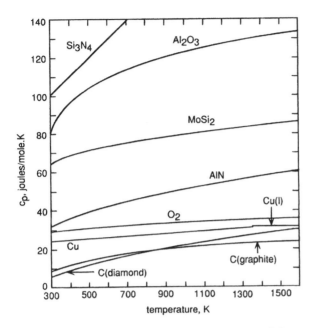

Figure 6.4 The variations, with temperature, of the constant-pressure heat capacities of several elements and compounds.

over the temperature range 298–1478 K, and

$$\beta\text{-ZrO}_2 \quad c_p = 74.48 \text{ J/K·mole}$$

from 1478 to 2670 K. In fitting the analytical expression to the measured heat capacities all of a, b, and c have non-zero values in the expression for α-ZrO$_2$, whereas the molar heat capacity of β-ZrO$_2$ is independent of temperature, in which case b and c are zero in the analytical expression. The variations, with temperature, of c_p for several elements and compounds which do not undergo phase transitions in the solid state are shown in Fig. 6.4, and the variations for some elements which exhibit allotropy and compounds which exhibit polymorphism are shown in Fig. 6.5. The data for α-ZrO$_2$ and β-ZrO$_2$ are included in Fig. 6.5.

6.4 ENTHALPY AS A FUNCTION OF TEMPERATURE AND COMPOSITION

For a closed system of fixed composition undergoing a change in temperature from T_1 to T_2 at the constant pressure P, integration of Eq. (2.7) gives Eq. (6.1):

$$\Delta H = H(T_2, P) - H(T_1, P) = \int_{T_1}^{T_2} c_p \, dT \tag{6.1}$$

ΔH is thus the area under a plot of c_p vs. T between the limits T_1 and T_2, and, from Eq. (2.7), $\Delta H = q_p$, which is simply the amount of heat required to increase the temperature of 1 mole of the system from T_1 to T_2 at the constant pressure P.

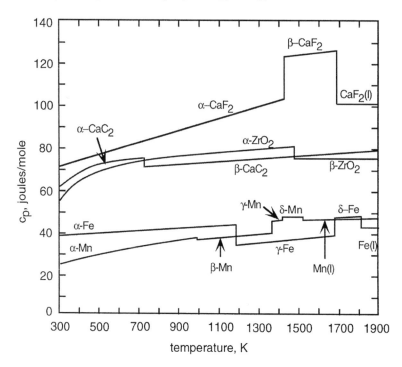

Figure 6.5 The variations, with temperature, of the constant-pressure molar heat capacities of some elements which exhibit allotropy and some compounds which exhibit polymorphism.

When a system undergoes a chemical reaction or a phase transformation at constant temperature and pressure, e.g., the reaction $A + B = AB$, ΔH is the difference between the enthalpy of the products of the reaction (the state 2), and the enthalpy of the reactants (the state 1), i.e.,

$$\Delta H(T, P) = H_{AB}(T, P) - [H_A(T, P) + H_B(T, P)] \tag{6.8}$$

and Eq. (6.8) is a statement of Hess's law. If ΔH is a positive quantity the reaction causes the system to absorb heat from its thermostatting heat bath, and the reaction is thus *endothermic*. Conversely, if ΔH is a negative quantity the reaction occurs with an evolution of heat and is thus an *exothermic* process. This convention is the same as that used with the First Law for the sign of q, the heat entering or leaving the system.

Changes in enthalpy caused by changes in temperature and/or composition can be graphically represented on an enthalpy-temperature diagram. Consider the change of state

$$A_{(s)} \rightarrow A_{(l)}$$

i.e., the melting of pure A. ΔH_{T_1} for this process is the difference between the molar enthalpies of liquid and solid A at the temperature T_1:

$$\Delta H_{T_1} = H_{A(l)}(T_1) - H_{A(s)}(T_1)$$

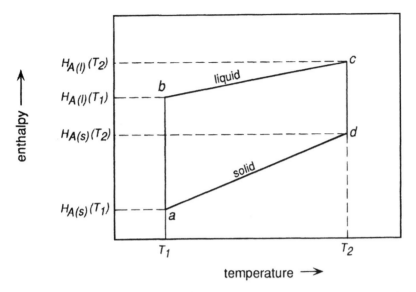

Figure 6.6 The variation, with temperature, of the molar enthalpies of the solid and liquid phases of a substance.

This change in enthalpy is represented by line ba in Fig. 6.6. For the change of phase occurring at the temperature T_2,

$$\Delta H_{T_2} = H_{A(l)}(T_2) - H_{A(s)}(T_2)$$

which is represented in Fig. 6.6 by the line of cd. As H is a state function, then

$$\Delta H(a \to b) = \Delta H(a \to d) + \Delta H(d \to c) + \Delta H(c \to b) \qquad \text{(i)}$$

where $\Delta H(a \to d)$ is the heat required to increase the temperature of one mole of solid A from T_1 to T_2 at constant pressure.

$$= \int_{T_1}^{T_2} c_{P_A(s)} dT$$

in which $c_{P_A(s)}$ is the molar heat capacity of solid A

$$\Delta H(d \to c) = \Delta H_{T_2}$$

$\Delta H(c \to b)$ is the heat which is evolved by 1 mole of liquid A when its temperature is decreased from T_2 to T_1 (or the negative of the amount of heat required to raise the temperature of a mole of liquid A from T_1 to T_2)

$$\int_{T_2}^{T_1} c_{P_A(l)} dT = - \int_{T_1}^{T_2} c_{P_A(l)} dT$$

in which $c_{P_{A(l)}}$ is the molar heat capacity of liquid A. Substitution of the individual expressions into Eq. (i) gives

$$\Delta H(A_{(s)} \rightarrow A_{(l)} \text{ at } T_1) = \Delta H(A_{(s)} \rightarrow A_{(l)} \text{ at } T_2) + \int_{T_1}^{T_2} c_{P_{A(s)}} dT - \int_{T_1}^{T_2} c_{P_{A(l)}} dT$$

or

$$\Delta H_{T_2} = \Delta H_{T_1} + \int_{T_1}^{T_2} \Delta c_p dT \tag{6.9}$$

where

$$\Delta c_p = c_{P_{A(l)}} - c_{P_{A(s)}}$$

Thus if the heat of the reaction is known at one temperature and the constant-pressure heat capacities of the products and the reactants are known (along with their dependencies on temperature), then the heat of the reaction at any other temperature can be calculated. It is to be noted that if $\Delta c_p = 0$, then $\Delta H_{T_2} = \Delta H_{T_1}$, i.e., the heat of the reaction, ΔH, is independent of temperature. In Fig. 6.6 the slope of the line bc, which is $(\partial H/\partial T)_p$, is c_p, for the liquid A, and bc is a straight line only if c_p is independent of temperature.

As H does not have an absolute value (only changes in H can be measured), it is convenient to introduce a convention which will allow the comparison of different enthalpy-temperature diagrams. This convention assigns the value of zero to the enthalpy of *elements in their stable states at 298 K* (25°C). Thus the enthalpy of a compound at 298 K is simply the heat of formation of the compound from its elements at 298 K. For example, for the oxidation

$$M_{(solid)} + \frac{1}{2} O_{2(gas)} = MO_{(solid)} \text{ at } 298 \text{ K}$$

$$\Delta H_{298} = H_{MO(s)298} - H_{M(s)298} - \frac{1}{2} H_{O_2(g)298}$$

and, as $H_{M,298}$ and $H_{O_2,298}$, are by convention, equal to zero, then

$$\Delta H_{298} = H_{MO,298}$$

The variation of heats of chemical reaction (or heats of formation) with temperature at constant pressure can be represented on an enthalpy-temperature diagram such as Fig. 6.7, which is drawn for the oxidation

$$Pb + \frac{1}{2} O_2 = PbO$$

The pertinent thermochemical data for this system are listed in Table 6.1. In Fig. 6.7a, a represents the enthalpy of $\frac{1}{2}$ mole of oxygen gas and 1 mole of $Pb_{(s)}$ at 298 K ($= 0$ by convention); ab represents the variation of $H_{Pb(s)}$ with temperature in the range $298 \leq T \leq 600$, where $H_{Pb(s)}$, T is given by $\int_{298}^{T} c_{p,Pb(s)} dT$; ac represents variation of $H_{\frac{1}{2}O_2(g)}$ with temperature in the range $298 \leq T \leq 3000$ K, where

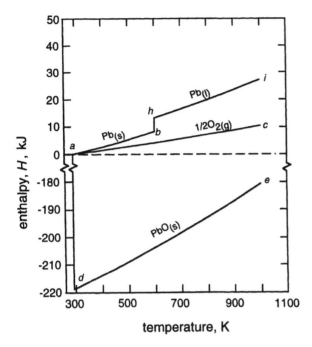

Figure 6.7 (a) The variation, with temperature, of the enthalpies of $Pb_{(s)}$, $Pb_{(l)}$, $\frac{1}{2}O_{2(g)}$, and $PbO_{(s)}$.

$H_{\frac{1}{2}O_{2(g)}}$ is given by $\frac{1}{2}\int_{298}^{T} c_{p,O_2(g)}dT$; $\Delta H_{PbO(s),298\,K} = -219,000$ J; and de represents the variation of $H_{PbO(s)}$ with temperature in the range $298\,K \leq T \leq 1159\,K$ where

$$H_{PbO(s),T} = -219,000 + \int_{298}^{T} c_{p,PbO(s)}dT \text{ J}$$

In Fig. 6.7b, a represents the enthalpy of $\frac{1}{2}$ mole of $O_{2(g)}$ and 1 mole of $Pb_{(s)}$ at 298 K; f represents the enthalpy of $\frac{1}{2}$ mole of $O_{2(g)}$ and 1 mole of $Pb_{(s)}$ at the temperature T; and g represents the enthalpy of 1 mole of $PbO_{(s)}$ at the temperature T.

Thus

$$\Delta H_{PbO,298\,K} = \Delta H(a \rightarrow f) + \Delta H(f \rightarrow g) + \Delta H(g \rightarrow d)$$
$$= \int_{298}^{T}\left(\frac{1}{2}c_{p,O_2(g)} + c_{p,Pb(s)}\right)dT + \Delta H_{PbO,T} + \int_{T}^{298} c_{p,PbO(s)}dT$$

and thus

$$\Delta H_T = \Delta H_{298} + \int_{298}^{T} \Delta c_p dT$$

where

$$\Delta c_p = c_{p,PbO(s)} - c_{p,Pb(s)} - \frac{1}{2}c_{p,O_2(g)}$$

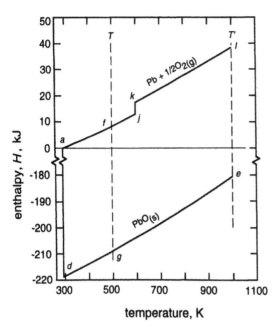

Figure 6.7 (b) The variation, with temperature, of the enthalpies of $(Pb + \frac{1}{2}O_2)$ and PbO.

Table 6.1 Thermochemical data for Pb, PbO, and O_2.

$H_{PbO(298)} = -219,000$ J/K
$c_{p,Pb(s)} = 23.6 + 9.75 \times 10^{-3}T$ J/K from 298 K to $T_{m,Pb}$
$c_{p,Pb(l)} = 32.4 - 3.1 \times 10^{-3}T$ J/K from $T_{m,Pb}$ to 1200 K
$c_{p,PbO(s)} = 37.9 + 26.8 \times 10^{-3}T$ J/K from 298 K to $T_{m,PbO}$
$c_{p,O_2(g)} = 29.96 + 4.18 \times 10^{-3}T - 1.67 \times 10^5 T^{-2}$ J/K from 298 K to 3000 K
$\Delta H_{m,Pb} = 4810$ J at $T_{m,Pb} = 600$ K
$T_{m,PbO} = 1159$ K

From the data in Table 6.1,

$$\Delta c_p = -0.7 + 14.96 \times 10^{-3}T + 0.85 \times 10^5 T^{-2}$$

and, thus, in the range of temperature from 298 to 600 K $(T_{m,Pb})$,

$$\Delta H_T = -219,000 + \int_{298}^{T} (-0.7 + 14.96 \times 10^{-3}T + 0.85 \times 10^5 T^{-2})dT$$

$$= -219,000 - 0.7(T - 298) + 7.48 \times 10^{-3}(T^2 - 298^2)$$

$$-0.85 \times 10^5 \left(\frac{1}{T} - \frac{1}{298} \right)$$

With $T = 500$ K, this gives $\Delta H_{500\,K} = -217,800$ J, as can be seen in Figs. 6.7b and 6.8. If a phase change occurs in one or more of the reactants or products,

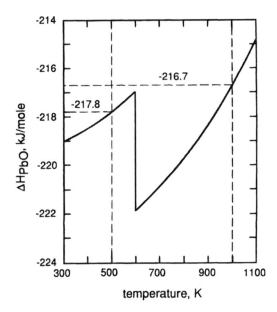

Figure 6.8 The variation, with temperature, of the enthalpy change for the reaction $Pb + \frac{1}{2}O_2 = PbO$.

between the two temperatures at which the reaction is being considered, then the latent heats of the phase changes must be considered. In Fig. 6.7a, h represents the enthalpy of 1 mole of $Pb_{(l)}$ at the melting temperature of 600 K, given as

$$H_{Pb(l),600\,K} = \int_{298}^{600} c_{p,Pb(s)}\,dT + \Delta H_{m,Pb}$$

hb is the latent heat of melting of Pb at the melting temperature of 600 K ($= 4810$ J); and hi represents the variation of the enthalpy of 1 mole of $Pb_{(l)}$ with temperature in the range 600 to 1200 K:

$$H_{Pb(l),T} = \int_{298}^{600} c_{p,Pb(s)}\,dT + \Delta H_{m,Pb} + \int_{600}^{T} c_{p,Pb(l)}\,dT$$

In Fig. 6.7b, $ajkl$ represents the variation of the enthalpy of 1 mole of Pb and $\frac{1}{2}$ mole of $O_{2(g)}$, and hence $\Delta H_{T'}$ is calculated from the cycle

$$\Delta H_{298\,K} = \Delta H(a \rightarrow d) = \Delta H(a \rightarrow j) + \Delta H(j \rightarrow k) + \Delta H(k \rightarrow l)$$
$$+ \Delta H(l \rightarrow e) + \Delta H(e \rightarrow g) + \Delta H(g \rightarrow d)$$

where

$$\Delta H(a \rightarrow j) = \int_{298}^{T_{m,Pb}} \left(c_{p,Pb(s)} + \frac{1}{2}c_{pO_2(g)} \right) dT$$

$$\Delta H(j \rightarrow k) = \text{the latent heat of melting of Pb at } T_{m,Pb} = 4810 \text{ J}$$

$$\Delta H(k \to l) = \int_{T_{m,\text{Pb}}}^{T} \left(c_{p,\text{Pb}(l)} + \frac{1}{2} c_{p,O_2(g)} \right) dT$$

$$\Delta H(l \to e) = \Delta H_T$$

$$\Delta H(e \to g) = \int_{T}^{T_{m,\text{Pb}}} c_{p,\text{PbO}(s)} dT$$

$$\Delta H(g \to d) = \int_{T_{m,\text{Pb}}}^{298} c_{p,\text{PbO}(s)} dT$$

Thus

$$\Delta H_T = \Delta H_{298\,\text{K}} + \int_{298}^{T_{m,\text{Pb}}} \left(c_{p,\text{PbO}(s)} - c_{p,\text{Pb}(s)} - \frac{1}{2} c_{p,O_2(g)} \right) dT$$

$$- \Delta H_{m,\text{Pb}} + \int_{T_{m,\text{Pb}}}^{T} \left(c_{p,\text{PbO}(s)} - c_{p,\text{Pb}(l)} - \frac{1}{2} c_{p,O_2(g)} \right) dT$$

$$= -219,000 + \int_{298}^{600} (-0.7 + 14.96 \times 10^{-3} T + 0.85 \times 10^5 T^{-2}) dT$$

$$- 4810 + \int_{600}^{T} (-9.3 + 27.8 \times 10^{-3} T + 0.85 \times 10^5 T^{-2}) dT$$

This gives $\Delta H_{1000} = -216,700$ joules at $T' = 1000$ K, as is seen in Figs. 6.7b and 6.8. Fig. 6.8 shows the variation of $\Delta H_{\text{PbO},T}$ with temperature in the range 298–1100 K.

If the temperature of interest is higher than the melting temperatures of both the metal and its oxide, then both latent heats of melting must be considered. For example, with reference to Fig. 6.9, which is drawn for the general oxidation,

$$M + \frac{1}{2} O_2 = MO$$

$$\Delta H_T = \Delta H_{298} + \int_{298}^{T_{m,M}} \left(c_{p,\text{MO}(s)} - c_{p,\text{M}(s)} - \frac{1}{2} c_{p,O_2(g)} \right) dT$$

$$- \Delta H_{m,M} + \int_{T_{m,M}}^{T_{m,\text{MO}}} \left(c_{p,\text{MO}(s)} - c_{p,\text{M}(l)} - \frac{1}{2} c_{p,O_2(g)} \right) dT$$

$$+ \Delta H_{m,\text{MO}} + \int_{T_{m,\text{MO}}}^{T} \left(c_{p,\text{MO}(l)} - c_{p,\text{M}(l)} - \frac{1}{2} c_{p,O_2(g)} \right) dT$$

When phase transformations of the reactants or products have to be considered, care must be taken with the signs of the changes in enthalpy. The signs can be obtained from a consideration of Le Chatelier's principle, which states that "when a system, which is at equilibrium, is subjected to an external influence, the system moves in that direction which nullifies the effects of the external influence." Thus if the system contains a low-temperature phase in equilibrium with a high-temperature phase at the equilibrium phase transition temperature, such as a solid coexisting with a liquid at the equilibrium melting temperature, then introduction of heat to

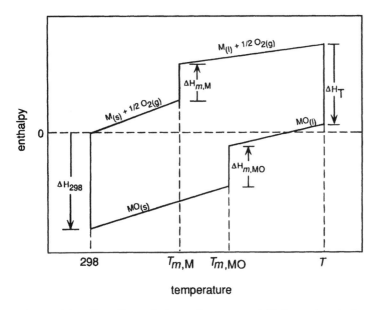

Figure 6.9 The effect of phase changes on ΔH for a chemical reaction.

the system (the external influence) would be expected to increase the temperature of the system (the effect). However, the system undergoes an endothermic change, which absorbs the heat introduced at constant temperature, and hence nullifies the effect of the external influence. The endothermic process is the melting of some of the solid. A phase change from a low- to a high-temperature phase is always endothermic, and hence ΔH for the change is always a positive quantity. Thus ΔH_m, the molar latent heat of melting, which is the difference between the enthalpy of a mole of liquid and the enthalpy of a mole of solid, is always positive. The general Eq. (6.9) can be obtained as follows:

$$\text{For the state 1} \quad \left(\frac{\partial H_1}{\partial T}\right)_P = c_{p(1)}$$

$$\text{For the state 2} \quad \left(\frac{\partial H_2}{\partial T}\right)_P = c_{p(2)}$$

Subtraction gives

$$\left(\frac{\partial H_2}{\partial T}\right)_P - \left(\frac{\partial H_1}{\partial T}\right)_P = c_{p(2)} - c_{p(1)}$$

$$\left[\frac{\partial (H_2 - H_1)}{\partial T}\right]_P = \Delta c_p$$

or

$$\left(\frac{\partial \Delta H}{\partial T}\right)_P = \Delta c_p \tag{6.10}$$

and integrating from state 1 to state 2 gives

$$\Delta H_{T_2} - \Delta H_{T_1} = \int_{T_1}^{T_2} \Delta c_p dT \qquad (6.11)$$

Equations (6.10) and (6.11) are expressions of *Kirchhoff's Law.*

6.5 THE DEPENDENCE OF ENTROPY ON TEMPERATURE AND THE THIRD LAW OF THERMODYNAMICS

For a closed system undergoing a reversible process, the Second Law gives

$$dS = \frac{\delta q}{T} \qquad (3.8)$$

If the process is conducted at constant pressure, then

$$dS = \left(\frac{\delta q}{T} \right)_P = \left(\frac{dH}{T} \right)_P = c_p \frac{dT}{T}$$

and thus if the temperature of a closed system of fixed composition is increased from T_1 to T_2 at constant pressure, the increase in the entropy per mole of the system, ΔS, is given by

$$\Delta S = S(T_2, P) - S(T_1, P) = \int_{T_1}^{T_2} \frac{c_p}{T} dT \qquad (6.12)$$

This change of entropy is obtained as the area under a plot of c_p/T vs. T between the limits T_2 and T_1, or, equivalently, as the area under a plot of c_p vs. $\ln T$ between the limits $\ln T_2$ and $\ln T_1$. Generally, S_T, the molar entropy of the system at any temperature T is given by

$$S_T = S_0 + \int_0^T \frac{c_p}{T} dT \qquad (6.13)$$

where S_0 is the molar entropy of the system at 0 K. Consideration of the value of S_0 leads to the statement of what is commonly called the Third Law of Thermodynamics. In 1906 Nernst postulated that, for chemical reactions between pure solids or pure liquids, the terms

$$\left(\frac{\partial \Delta G}{\partial T} \right)_P \quad \text{and} \quad \left(\frac{\partial \Delta H}{\partial T} \right)_P$$

approach zero as temperature approaches absolute zero. For any change in the state of a system, e.g., a chemical reaction at the constant temperature T, Eq. (5.2) gives

$$\Delta G_T = \Delta H_T - T \Delta S_T$$

and thus Nernst's postulate is that ΔG for the reaction varies with temperature as shown in Fig. 6.10. The slope of the line in Fig. 6.10, at any temperature is equal

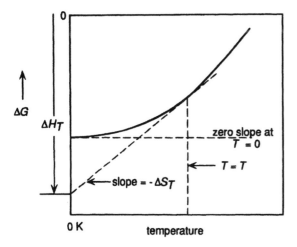

Figure 6.10 The variation of the change in the Gibbs free energy for a reaction with temperature as the temperature approaches absolute zero.

to $-\Delta S_T$, and the intercept, with the ΔG axis at $T = 0$, of the tangent to the line at any temperature is equal to ΔH_T, the change in the enthalpy at the temperature T. As the temperature approaches zero, the slope of the line approaches zero and the variation of the tangential intercept with temperature approaches zero, the consequences of which are that, as $T \to 0$, then $\Delta S \to 0$ and $\Delta c_p \to 0$. This can be seen by differentiating Eq. (5.2) with respect to T at constant P:

$$\left(\frac{\partial \Delta G}{\partial T}\right)_P = \left(\frac{\partial \Delta H}{\partial T}\right)_P - T\left(\frac{\partial \Delta S}{\partial T}\right)_P - \Delta S$$

From Eq. (5.12)

$$\left(\frac{\partial \Delta G}{\partial T}\right)_P = -\Delta S$$

and thus

$$\left(\frac{\partial \Delta H}{\partial T}\right)_P = T\left(\frac{\partial \Delta S}{\partial T}\right)_P = \Delta c_p$$

Thus if $(\partial \Delta G / \partial T)_P$ and $(\partial \Delta H / \partial T)_P$ approach zero as $T \to 0$, the values of ΔS and Δc_p approach zero as $T \to 0$ (provided that $(\partial \Delta S / \partial T)_p$ is not infinite at $T = 0$).

Nernst's heat theorem states that "for all reactions involving substances in the condensed state, ΔS is zero at the absolute zero of temperature." Thus, for the general reaction

$$A + B = AB$$

$\Delta S = S_{AB} - S_A - S_B = 0$ at $T = 0$, and if S_A and S_B are assigned the value of zero at 0 K, then the compound AB also has zero entropy at 0 K.

The incompleteness of Nernst's theorem was pointed out by Planck, who stated that "the entropy of any homogeneous substance, which is in complete internal equilibrium, may be taken to be zero at 0 K." The requirement that the substance be in complete internal equilibrium can be illustrated as follows:

1. Glasses are noncrystalline solids which can be regarded as being supercooled liquids in which the disordered atomic arrangements occurring in the liquid state have been "frozen into" the solid state. Substances which form glasses usually have complex atomic, ionic, or molecular structures in the liquid state, and the structures would require extensive atomic reorganization in order to assume the periodic structure characteristic of the crystalline state. In the absence of the ability of the glass-forming substance to undergo the necessary atomic rearrangement at a unique freezing temperature, the liquid on cooling simply becomes more and more viscous and eventually forms a solid glass. If the solid glass were to crystallize, its enthalpy, internal energy, and entropy would decrease, and the decreases in the enthalpy and entropy would be, respectively, the latent heat and entropy of crystallization at the temperature at which devitrification occurred. At temperatures lower than its equilibrium freezing temperature the glassy state is metastable with respect to the crystalline state, and a glass, not being in internal equilibrium, has an entropy at 0 K which is greater than zero by an amount which is dependent on the degree of atomic disorder in the glass.

2. Solutions are mixtures of atoms, ions, or molecules, and a contribution is made to their entropies by the fact that they are mixtures [see Eq. (4.18)]. This contribution is called the entropy of mixing and is determined by the randomness with which the particles are mixed in the solution. The atomic randomness of a mixture determines its degree of order, e.g., in a mixture containing 50 atomic percent of A and 50 atomic percent of B, complete ordering occurs when every atom of A is coordinated only by B atoms and vice versa, and complete randomness occurs when, on average, 50 percent of the neighbors of every atom are A atoms and 50 percent are B atoms. Respectively, the degrees of order in these two extreme configurations are unity and zero. The equilibrium degree of order is temperature-dependent and increases with decreasing temperature. However, the maintenance of the equilibrium degree of order is dependent on the abilities of the particles to change their positions in the solution, and, with ever-decreasing temperature, as atomic mobility decreases exponentially with decreasing temperature, the maintenance of internal equilibrium becomes increasingly difficult. Consequently a nonequilibrium degree of order can be frozen into the solid solution, in which case the entropy will not decrease to zero at 0 K.

3. Even chemically pure elements are mixtures of isotopes, and because of the chemical similarity between isotopes it is to be expected that completely random mixing of the isotopes occurs. Thus an entropy of mixing occurs, and consequently, the entropy does not decrease to zero at 0 K. For example, solid chlorine at 0 K is a solid solution of $Cl^{35} - Cl^{35}$, $Cl^{35} - Cl^{37}$, and $Cl^{37} - Cl^{37}$ molecules. However, as this entropy of mixing is present in any other substance which contains the element, it is customary to ignore it.

4. At any finite temperature a pure crystalline solid contains an equilibrium number of vacant lattice sites, which, because of their random positioning in the crystal, give rise to an entropy of mixing which is exactly the same as the entropy of mixing in a chemical solution. Both the equilibrium number of vacancies and the diffusivity of the atoms in the crystal decrease exponentially with decreasing temperature, and as the vacancies "disappear" by diffusing to the free surface of the crystal, nonequilibrium concentrations of vacancies can be frozen into the crystal at low temperatures, causing a non-zero entropy at 0 K. Random crystallographic orientation of molecules in the crystalline state can also give rise to a non-zero entropy at 0 K. Such is the case with solid CO, in which a structure such as

CO	CO	OC	CO	OC	CO	CO
OC	CO	CO	OC	CO	OC	
OC	CO	CO	OC	OC	CO	CO

can occur. The entropy would have its maximum value if equal numbers of molecules were oriented in opposite directions and random mixing of the two orientations occurred. From Eq. (4.18) the molar configurational entropy of mixing would be

$$\Delta S_{conf} = k \ln \frac{(N_O)!}{\left(\frac{1}{2}N_O\right)! \left(\frac{1}{2}N_O\right)!}$$

where N_O is Avogadro's Number, 6.0232×10^{23}. Thus, using Stirling's approximation,

$$\Delta S_{conf} = k \left[N_O \ln N_O - \frac{1}{2} N_O \ln \left(\frac{1}{2}N_O\right) - \frac{1}{2} N_O \ln \left(\frac{1}{2}N_O\right) \right]$$
$$= k \times 4.175 \times 10^{23}$$
$$= 1.38054 \times 10^{-23} \times 4.175 \times 10^{23}$$
$$= 5.76 \text{ J/mole·K}$$

Comparison of this value with the measured value of 4.2 J/mole·K indicates that the actual molecular orientations are not fully random.

In view of the above considerations, the statement of the Third Law of Thermodynamics requires the inclusion of the qualification that the homogeneous phase be in complete internal equilibrium.

6.6 EXPERIMENTAL VERIFICATION OF THE THIRD LAW

The Third Law can be verified by considering a phase transition in an element such as

$$\alpha \rightarrow \beta$$

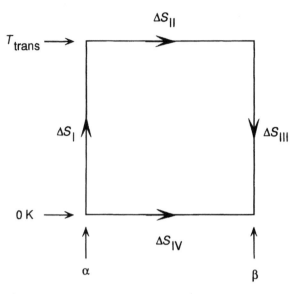

Figure 6.11 The cycle used for the experimental verification of the Third Law of Thermodynamics.

where α and β are allotropes of the element. In Fig. 6.11 T_{trans} is the temperature, at atmospheric pressure, at which the α and β phases are in equilibrium with one another. For the cycle shown in Fig. 6.11

$$\Delta S_{IV} = \Delta S_I + \Delta S_{II} + \Delta S_{III}$$

For the Third Law to be obeyed, $\Delta S_{IV} = 0$, which requires that

$$\Delta S_{II} = -(\Delta S_I + \Delta S_{III})$$

where

$$\Delta S_I = \int_0^{T_{trans}} \frac{c_{p(\alpha)}}{T} dT$$

$$\Delta S_{II} = \frac{\Delta H_{trans}}{T_{trans}}$$

$$\Delta S_{III} = \int_{T_{trans}}^0 \frac{c_{p(\beta)}}{T} dT$$

ΔS_{II}, which is called the experimental entropy change, and $-(\Delta S_I + \Delta S_{III})$, which is called the Third Law entropy change, are equal to one another if the Third Law is obeyed.

The cycle shown in Fig. 6.11 has been examined for the case of sulfur, which has two allotropes; a monoclinic form which is stable above 368.5 K and an orthorhombic form which is stable below 368.5 K, with a molar latent heat of transformation of 400 J/mole at the equilibrium transformation temperature of 368.5 K. As monoclinic sulfur can be supercooled with relative ease, the variations, with temperature, of the heat capacities of both allotropes have been measured experimentally at temperatures below 368.5 K. The measured heat capacities give

$$\Delta S_I = \int_0^{368.5} \frac{c_{p(\text{rhombic})}}{T} dT = 36.86 \text{ J/K}$$

$$\Delta S_{II} = \frac{\Delta H_{\text{trans}}}{T_{\text{trans}}} = \frac{400}{368.5} = 1.09 \text{ J/K}$$

and

$$\Delta S_{III} \int_{368.5}^0 \frac{c_{p(\text{monoclinic})}}{T} dT = -37.8 \text{ J/K}$$

Thus

$$-(\Delta S_I + \Delta S_{III}) = -(36.86 - 37.8) = 0.94 \text{ J/K}$$

and

$$\Delta S_{II} = 1.09 \text{ J/K}$$

As the difference between the experimental and the Third Law entropy changes is less than the experimental error, the equality is taken as being experimental verification of the Third Law.

Assigning a value of zero to S_0 allows the absolute value of the entropy of any material to be determined as

$$S_T = \int_0^T \frac{c_p}{T} dT \text{ J/K}$$

and molar entropies are normally tabulated at 298 K, where

$$S_{298} = \int_0^{298} \frac{c_p}{T} dT \text{ J/K}$$

The variations, with temperature, of the molar entropies of several elements and compounds are shown in Fig. 6.12. With the constant-pressure molar heat capacity of the solid expressed in the form

$$c_{p(s)} = a + bT + cT^{-2}$$

the molar entropy of the solid at the temperature T is obtained as

$$S_T = S_{298} + a \ln\left(\frac{T}{298}\right) + b(T - 298) - \frac{1}{2}c\left(\frac{1}{T^2} - \frac{1}{298^2}\right)$$

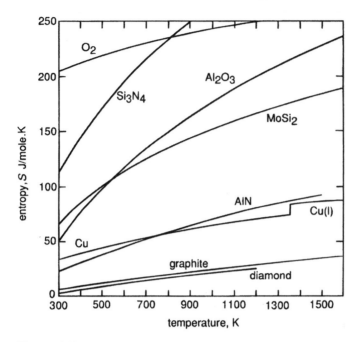

Figure 6.12 The variation, with temperature, of the molar entropies of several elements and compounds.

At temperatures higher than the melting temperature, T_m, the molar entropy of the liquid is obtained as

$$S_T = S_{298} + \int_{298}^{T_m} \frac{c_{p(s)}}{T} dT + \Delta S_m + \int_{T_m}^{T} \frac{c_{p(l)}}{T} dT$$

where the molar entropy of the melting, ΔS_m, is obtained as $\Delta H_m / T_m$.

In 1897 Richards suggested that the entropies of fusion of metals should have the same value, which would require that a plot of ΔH_m vs. T_m be a straight line. Fig. 6.13 is a plot of the molar enthalpies of fusion vs. the melting temperatures of 11 face-centered cubic metals (open circles) and 27 body-centered cubic metals (closed circles) which have melting temperatures below 3000 K. A least-squares analysis of the data for the face-centered cubic metals gives

$$\Delta H_m = 9.61 T_m \pm 0.01 \text{ J/K}$$

and the corresponding equation for the body-centered cubic metals is

$$\Delta H_m = 8.25 T_m \pm 0.19 \text{ J/K}$$

The slopes of the lines indicate that the molar entropy of melting of face-centered cubic (FCC) metals is approximately 9.6 J/K and that of body-centered cubic (BCC) metals is approximately 8.3 J/K, although the scatter in the data for BCC

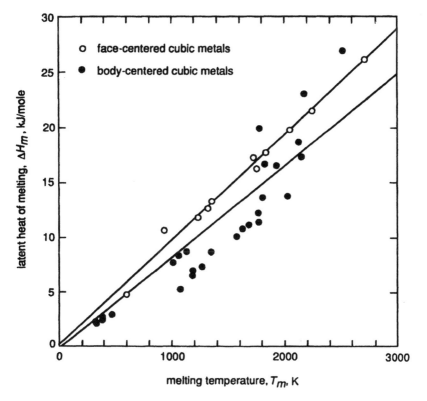

Figure 6.13 An illustration of Richard's rule.

metals is much greater than that for FCC metals. This observation, which is known as Richards's rule, indicates that the difference between the degree of disorder in the liquid structure and that in both the FCC and BCC crystal structures is approximately the same for all FCC and BCC metals. Trouton's rule states that the molar entropy of boiling of a liquid metal is 88 J/K. Fig. 6.14 shows a plot of ΔH_b vs. the boiling temperature, T_b, for 29 liquid metals with boiling temperatures below 4000 K. A least-squares fit of the data, shown as the full line, gives

$$\Delta H_b = 121 T_B - 43 \text{ J/K}$$

However, a least-squares fit of the data for the 13 metals with boiling temperatures below 2100 K, shown as the broken line, gives

$$\Delta H_b = 87 T_b - 0.4 \text{ J/K}$$

which indicates a common molar entropy of boiling of approximately 87 J/K for these metals.

Figure 6.15 is the entropy-temperature diagram for the reaction

$$\text{Pb} + \frac{1}{2} O_2 = \text{PbO}$$

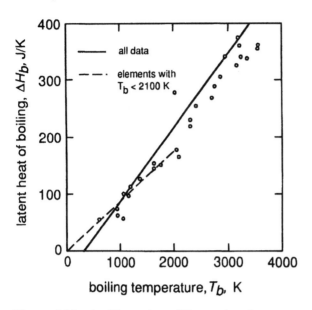

Figure 6.14 An illustration of Trouton's rule.

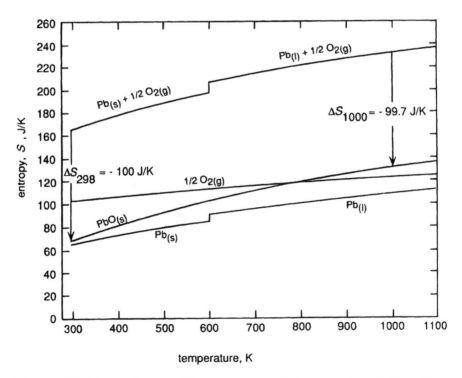

Figure 6.15 The variation, with temperature, of the entropies of $Pb_{(s)}$, $Pb_{(l)}$, $PbO_{(s)}$, $\frac{1}{2}O_{2(g)}$, and the entropy change for the reaction $Pb + \frac{1}{2}O_2 = PbO$.

corresponding to the enthalpy-temperature diagram shown in Fig. 6.7. Because of the similar magnitudes of the molar entropies of the condensed phases Pb and PbO, it is seen that the entropy change for the reaction,

$$\Delta S_T = S_{T,\text{PbO}} - S_{T,\text{Pb}} - \frac{1}{2}S_{T,\text{O}_2}$$

is very nearly equal to $-\frac{1}{2}S_{T,\text{O}_2}$. For example, at 298 K

$$\Delta S_{298} = S_{298,\text{PbO}} - S_{298,\text{Pb}} - \frac{1}{2}S_{298,\text{O}_2}$$

$$= 67.4 - 64.9 - \frac{1}{2} \times 205$$

$$= -100 \text{ J/K}$$

which is similar in magnitude to the decrease in entropy caused by the disappearance of $\frac{1}{2}$ mole of oxygen gas. This approximation is generally valid; i.e., in reactions in which a gas reacts with a condensed phase to produce another condensed phase, the change in the entropy is of similar magnitude to that caused by the disappearance of the gas.

6.7 THE INFLUENCE OF PRESSURE ON ENTHALPY AND ENTROPY

For a closed system of fixed composition undergoing a change of pressure at constant temperature,

$$dH = \left(\frac{\partial H}{\partial P}\right)_T dP$$

Eq. (5.10) gives $dH = TdS + VdP$, and thus

$$\left(\frac{\partial H}{\partial P}\right)_T = T\left(\frac{\partial S}{\partial P}\right)_T + V$$

Maxwell's equation (5.34) gives $(\partial S/\partial P)_T = -(\partial V/\partial T)_P$, in which case

$$\left(\frac{\partial H}{\partial P}\right)_T = -T\left(\frac{\partial V}{\partial T}\right)_P + V$$

The isobaric coefficient of thermal expansion, α, is defined as

$$\alpha = \frac{1}{V}\left(\frac{\partial V}{\partial T}\right)_P$$

and thus

$$\left(\frac{\partial H}{\partial P}\right)_T = -T\alpha V + V = V(1 - \alpha T)$$

The change in molar enthalpy caused by the change in state from (P_1, T) to (P_2, T) is thus

$$\Delta H = H(P_2, T) - H(P_1, T) = \int_{P_1}^{P_2} V(1 - \alpha T) dP \qquad (6.14)$$

For an ideal gas, $\alpha = 1/T$, and thus Eq. (6.14) shows again that the enthalpy of an ideal gas is independent of pressure.

The molar volume and expansivity of Fe are, respectively, 7.1 cm^3 and 0.3×10^{-4} K^{-1}. Thus an increase in the pressure exerted on Fe from 1 to 100 atm at 298 K causes the molar enthalpy to increase by

$$7.1 \times 10^{-3} \times (1 - 0.3 \times 10^{-4} \times 298) \times (100 - 1) = 0.696 \text{ liter·atm}$$
$$= 0.696 \times 101.3 = 71 \text{ J}$$

The same increase in molar enthalpy would be obtained by heating Fe from 298 to 301 K at 1 atm pressure.

For a closed system of fixed composition undergoing a change of pressure at constant temperature,

$$dS = \left(\frac{\partial S}{\partial P} \right)_T dP$$

Maxwell's equation (5.34) is $(\partial S/\partial P)_T = -(\partial V/\partial T)_P$ which, with the definition of α, gives

$$\left(\frac{\partial S}{\partial P} \right)_T = -\alpha V$$

Thus, for the change of state (P_1, T) to (P_2, T),

$$\Delta S = S(P_2, T) - S(P_1, T) = -\int_{P_1}^{P_2} \alpha V dP \qquad (6.15)$$

For an ideal gas, as $\alpha = 1/T$, Eq. (6.15) simplifies to

$$\Delta S = -\int_{P_1}^{P_2} R d \ln P = -R \ln \left(\frac{P_2}{P_1} \right) = R \ln \left(\frac{V_2}{V_1} \right)$$

as was obtained in Sec. 3.7.

An increase in the pressure exerted on Fe from 1 to 100 atm decreases the molar entropy by 0.0022 J/K, which is the same as is obtained by decreasing the temperature by 0.02 degrees from 298 K at 1 atm pressure. It is thus seen that the molar enthalpies and entropies are relatively insensitive to changes in pressure. In the majority of materials applications, in which the range of pressure is 0 to 1 atm, the influence of pressure on the enthalpies and entropies of condensed phases can be ignored.

For a closed system of fixed composition undergoing changes in both pressure and temperature, combination of Eqs. (6.1) and (6.14) gives

$$\Delta H = H(P_2, T_2) - H(P_1, T_1) = \int_{T_1}^{T_2} c_p dT + \int_{P_1}^{P_2} V(1 - \alpha T) dP \qquad (6.16)$$

and combination of Eqs. (6.12) and (6.15) gives

$$\Delta S = S(P_2, T_2) - S(P_1, T_1) = \int_{T_1}^{T_2} \frac{c_p}{T} dT - \int_{P_1}^{P_2} \alpha V dP \qquad (6.17)$$

Just as it was required that the temperature dependence of c_p be known for integration of Eqs. (6.1) and (6.13), strictly, it is required that the pressure dependence of V and α be known for integration of Eqs. (6.14) and (6.15). However, for condensed phases being considered over small ranges of pressure, these pressure dependencies can be ignored.

6.8 SUMMARY

Knowledge of the heat capacities and the entropies of substances and the heats of formation of compounds allows enthalpy and entropy changes to be evaluated for any process, i.e., for phase changes and chemical reactions. As enthalpy does not have an absolute value, it is conventional to assign the value of zero to the enthalpy of all elements in their stable states of existence at 298 K and to consider changes in enthalpy with respect to this reference state. In contrast, entropy has an absolute value, and the entropy of all substances which are in complete internal equilibrium is zero at 0 K.

Although both enthalpy and entropy are dependent on pressure and temperature, the dependence of the enthalpy and entropy of condensed phases is normally small enough to be ignored when the pressure of interest is in the range 0–1 atm.

The determination of ΔH_T and ΔS_T for any change of state at any temperature and pressure allows the all-important change in the Gibbs free energy for the change of state to be calculated as

$$\Delta G_T = \Delta H_T - T\Delta S_T$$

As consideration of the Gibbs free energy in any isothermal, isobaric process provides the criterion for equilibrium, the equilibrium state of a system can be determined from a knowledge of the thermochemical properties of the system.

6.9 NUMERICAL EXAMPLES

Example 1

Uranium can be produced by reacting a uranium-bearing compound with a more reactive metal, e.g., Mg can be used to reduce UF_4 according to the reaction

$$UF_4 + 2Mg \rightarrow 2MgF_2 + U$$

This reaction is exothermic, and the sensible heat released is used to increase the temperature of the reaction products. In order to facilitate a good separation of the U from the MgF_2, it is desirable to produce them as liquids (which are immiscible). If the reactants are placed in an adiabatic container in the molar ratio $Mg/UF_4 = 2.0$ and are allowed to react completely at 298 K, is the quantity of sensible heat released by the reaction sufficient to increase the temperature of the reaction products to 1773 K?

The required thermochemical data are

$$H_{MgF_2,298} = -1,124,200 \text{ J/mole}$$

$$H_{UF_4,298} = -1,919,600 \text{ J/mole}$$

The heat released by the reaction occurring at 298 K is thus

$$\Delta H_{298} = -(2 \times 1,124,200) + 1,919,600 = -328,800 \text{ J}$$

Calculation of the amount of heat required to increase the temperature of the products from 298 to 1773 K requires knowledge of the constant-pressure molar heat capacities and the changes in enthalpy caused by any phase transformations. These data are

$c_{p,U(\alpha)} = 25.10 + 2.38 \times 10^{-3}T + 23.68 \times 10^{-6}T^2$ J/K in the range 298–941 K
$c_{p,U(\beta)} = 42.93$ J/K in the range 941–1049 K
$c_{p,U(\gamma)} = 38.28$ J/K in the range 1049–1408 K
$c_{p,U(l)} = 48.66$ J/K
$c_{p,MgF_2(s)} = 77.11 + 3.89 \times 10^{-3}T - 14.94 \times 10^5 T^{-2}$ in the range 298–1536 K
$c_{p,MgF_2(l)} = 94.56$ J/K

For $U_{(\alpha)} \rightarrow U_{(\beta)}$, $\Delta H_{trans} = 2800$ J at $T_{trans} = 941$ K
For $U_{(\beta)} \rightarrow U_{(\gamma)}$, $\Delta H_{trans} = 4800$ J at $T_{trans} = 1049$ K
For $U_{(\gamma)} \rightarrow U_{(l)}$, $\Delta H_m = 9200$ J at $T_m = 1408$ K
For $MgF_{2,(s)} \rightarrow MgF_{2,(l)}$, $\Delta H_m = 58,600$ J at $T_m = 1536$ K

The heat required to increase the temperature of 1 mole of U from 298 to 1773 K is

$$\Delta H_1 = \int_{298}^{941} c_{p,U(\alpha)}dT + \Delta H_{trans(\alpha \rightarrow \beta)} + \int_{941}^{1049} c_{p,U(\beta)}dT + \Delta H_{trans(\beta \rightarrow \gamma)}$$

$$+ \int_{1049}^{1408} c_{p,U(\gamma)}dT + \Delta H_{U(m)} + \int_{1408}^{1773} c_{p,U(l)}$$

Thus

$$\Delta H_1 = [25.10 \times (941 - 298)] + \left[\frac{2.38 \times 10^{-3}}{2}(941^2 - 298^2)\right]$$

$$+ \left[\frac{23.68 \times 10^{-6}}{3}(941^3 - 298^3)\right] + 2800 + [42.93 \times (1049 - 941)]$$

$$+ 4800 + [38.28 \times (1408 - 1049)] + 9200 + [48.66 \times (1773 - 1408)]$$

$$= 76,395 \text{ J}$$

The heat required to increase the temperature of 2 moles of MgF_2 from 298 to 1773 K is

$$\Delta H_2 = 2\left(\int_{298}^{1536} c_{p,MgF_2(s)}dT + \Delta H_{MgF_2(m)} + \int_{1536}^{1773} c_{p,MgF_2(l)}dT\right)$$

$$= [2 \times 77.11 \times (1536 - 298)] + \left[\frac{2 \times 3.89 \times 10^{-3}}{2}(1536^2 - 298^2)\right]$$

$$+ \left[2 \times 14.94 \times 10^5\left(\frac{1}{1536} - \frac{1}{298}\right)\right]$$

$$+ 2 \times [58{,}600 + 94.56 \times (1773 - 1536)]$$

$$= 353{,}696 \text{ J}$$

The total heat required is thus

$$\Delta H_1 + \Delta H_2 = 76{,}395 + 353{,}696 = 430{,}091 \text{ J}$$

which is

$$430{,}091 - 328{,}800 = 101{,}291 \text{ J}$$

more than is made available by the exothermic reaction. The actual temperature attained by the adiabatically contained reaction products is calculated as follows: Assume that the temperature attained is at least 941 K. The heat required to raise the temperature of 1 mole of $U_{(\alpha)}$ and 2 moles of MgF_2 from 298 to 941 is

$$\Delta H_{(i)} = \int_{298}^{941} (c_{p,U(\alpha)} + 2c_{p,MgF_2(s)})dT$$

$$= [(25.10 + 2 \times 77.11)(941 - 298)]$$

$$+ \left[(2.38 + 2 \times 3.89)\frac{10^{-3}}{2}(941^2 - 298^2)\right]$$

$$+ \left[2 \times 14.94 \times 10^5\left(\frac{1}{941} - \frac{1}{298}\right)\right] + \frac{23.68 \times 10^{-6}}{3}(941^3 - 298^3)$$

$$= 118{,}866 \text{ J}$$

which is less than the 328,800 J released by the exothermic reaction. Assume, now, that the temperature attained is at least 1048 K. The heat required to transform 1 mole of U from α to β at 941 K and heat 1 mole of $U_{(\beta)}$ and 2 moles of MgF_2 from 941 to 1048 K is

$$\Delta H_{(ii)} = \Delta H_{U,(\alpha\to\beta)} + \int_{941}^{1048} (c_{p,U(\beta)} + 2c_{p,MgF_2})dT$$

$$+ 2800 + [(42.93 + 2 \times 77.11)(1048 - 941)]$$

$$+ \left[\frac{2 \times 3.89 \times 10^{-3}}{2}(1048^2 - 941^2)\right]$$

$$+ 2 \times 14.94 \times 10^5\left(\frac{1}{1048} - \frac{1}{941}\right)$$

$$= 24{,}399 \text{ J}$$

Heating to 1048 K consumes $118,866 + 24,399 = 143,265$ J, which leaves $328,800 - 143,265 = 185,535$ J of sensible heat available for further heating. Assume that the temperature attained is at least 1408 K. The heat required to transform 1 mole of U from β to γ at 1048 K and increase the temperature of 1 mole of $U_{(\gamma)}$ and 2 moles of MgF_2 from 1048 K is

$$\Delta H_{(iii)} = \Delta H_{U,(\beta \to \gamma)} + \int_{1048}^{1408} (c_{p,U(\gamma)} + 2c_{p,MgF_2})dT$$

$$+4800 + [(38.28 + 2 \times 77.11)(1408 - 1048)]$$

$$+ \left[\frac{2 \times 3.89 \times 10^{-3}}{2}(1408^2 - 1048^2) \right]$$

$$+2 \times 14.94 \times 10^5 \left(\frac{1}{1408} - \frac{1}{1048} \right)$$

$$= 76,810 \text{ J}$$

To reach 1408 K, requires $143,265 + 76,810 = 220,075$ of the available heat, which leaves 108,725 J for further heating. Assume that the temperature reaches 1536 K. The heat required to melt 1 mole of U at 1408 K and increase the temperature of 1 mole of liquid U and 2 moles of MgF_2 from 1408 to 1536 K is

$$\Delta H_{(iv)} = \Delta H_{U,(m)} + \int_{1408}^{1536} (c_{p,U(l)} + 2c_{p,MgF_2})dT$$

$$= 9200 + [(48.66 + 2 \times 77.11)(1536 - 1408)]$$

$$+ \left[\frac{2 \times 3.89 \times 10^{-3}}{2}(1536^2 - 1408^2) \right]$$

$$+2 \times 14.94 \times 10^5 \left(\frac{1}{1536} - \frac{1}{1408} \right)$$

$$= 36,457 \text{ J}$$

To reach 1536 K thus requires $220,075 + 36,457 = 256,532$ J, which leaves $108,725 - 36,457 = 72,268$ J. The remaining sensible heat is less than the heat of melting of two moles of MgF_2, $(\Delta H_m = 58,600$ J), and thus is used to melt

$$\frac{72,268}{58,600} = 1.23$$

moles of MgF_2 at its melting temperature of 1536 K. The reaction products are thus liquid U, liquid MgF_2, and solid MgF_2 occurring in the ratio 1:1.23:0.77 at 1536 K.

The attainment, by the reaction products, of a final temperature of 1773 K requires that an extra 100,303 J be supplied to the adiabatic reaction container, and this is achieved by preheating the reactants to some temperature before allowing the reaction to occur. The required temperature, T, is obtained from

$$100,303 = \int_{298}^{T} c_{p,reactants} dT$$

Assume that T is less than the melting temperature of Mg, $T_{Mg(m)} = 923$ K. The required thermochemical data are

$$c_{p,Mg} = 21.12 + 11.92 \times 10^{-3}T + 0.15 \times 10^5 T^{-2} \quad \text{in the range 298–923 K}$$

$$c_{p,UF_4} = 107.53 + 29.29 \times 10^{-3}T - 0.25 \times 10^5 T^{-2} \quad \text{in the range 298–1118 K}$$

Thus

$$100,303 = \int_{298}^{T} (2c_{p,Mg} + c_{p,UF_4})dT$$
$$= [(2 \times 21.13 + 107.53)(T - 298)]$$
$$+ \left[(2 \times 11.92 + 29.29)\frac{10^{-3}}{2}(T^2 - 298^2) \right]$$
$$- (2 \times 0.15 - 0.25) \times 10^5 \left(\frac{1}{T} - \frac{1}{298} \right)$$

which has the solution $T = 854$ K, which is less than the melting temperature of Mg. Thus in order to produce liquid U and liquid MgF_2 at 1773 K, the stoichiometric reactants must be preheated to 854 K. The enthalpy-temperature diagram for the process is shown in Fig. 6.16. Taking the relative zero of enthalpy to be $H_{UF_4,298} + H_{Mg,298}$, the line ab represents the influence of the supply of 100,303 joules of heat to 1 mole of UF_4 and 2 moles of Mg, which is to increase the temperature of the system from 298 to 854 K. At the point b the reactants are placed in an adiabatic container and are allowed to react completely, which causes the change in enthalpy

$$\Delta H_{854} = \Delta H_{298} + \int_{298}^{854} (c_{p,U} + 2c_{p,MgF_2} - c_{p,UF_4} + 2c_{p,Mg})dT$$
$$= -328,800 + [29.55(854 - 298)] - \left[\frac{42.97 \times 10^{-3}}{2}(854^2 - 298^2) \right]$$
$$+ 29.93 \times 10^5 \left(\frac{1}{854} - \frac{1}{298} \right) + \left[\frac{23.68 \times 10^{-6}}{3}(854^3 - 298^3) \right]$$
$$= -328,000 \text{ J}$$

The reaction causes the enthalpy to decrease from b to c and the sensible heat produced increases the temperature of the products along cl. The line contains four jogs; ed at 941 K (for the heat of transformation of α-U to β-U), fg at 1049 K (the heat of transformation of β-U to γ-U), hi at 1409 K (the heat of melting of U), and kj at 1536 K which represents the heat of melting of 2 moles of MgF_2. As the reaction has been conducted adiabatically, $H_l = H_b$. In practice the system does not follow the line $b \rightarrow c \rightarrow l$, which would require that all of the heat of the reaction be released isothermally before being made available to increase the temperature of the products. In practice the temperature of the system begins to increase as soon as the reaction begins, but, as enthalpy is a state function, the difference between

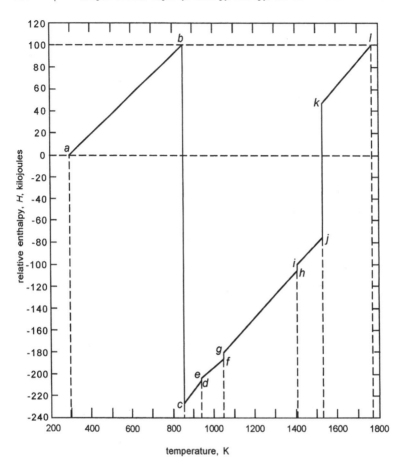

Figure 6.16 The enthalpy-temperature diagram considered in Example 1.

its value in state 1 and its value in state 2 is independent of the process path taken by the system between the states.

Fig. 6.16 shows that the temperature, T, to which the reactants must be raised is lower than the temperature at which α-U transforms to β-U. If this had been assumed, the temperature, T, could have been obtained from computer solution of the equation

The heat required to increase the temperature of the reactants from 298 to T,

(the term A) + The heat of the reaction at the temperature T,

(the term B) + The heat required to increase the temperature of the reaction products from T to 1773 (the term C) = 0

Where the term A is

$$\int_{298}^{T} (c_{p,\mathrm{UF_4}} + 2c_{p,\mathrm{Mg}})\,dT$$

the term B is

$$\Delta H_T = \Delta H_{298} + \int_{298}^{T} \Delta c_p (\text{products} \rightarrow \text{reactants})$$

and the term C is

$$\int_{T}^{941} c_{p,U(\alpha)} dT + \Delta H_{U(\alpha \rightarrow \beta)} + \int_{941}^{1049} c_{p,U(\beta)} dT + \Delta H_{U(\beta \rightarrow \gamma)} + \int_{1049}^{1408} c_{p,U(l)} dT + \Delta H_{U(\gamma \rightarrow l)}$$

$$+ \int_{1408}^{1773} c_{p,U(l)} dT + \int_{T}^{1536} 2 c_{p,MgF_2(S)} dT + 2 \Delta H_{m,MgF_2} + \int_{1536}^{1773} 2 c_{p,MgF_2(l)} dT$$

Example 2

A mixture of Fe_2O_3 and Al, present in the molar ratio 1:2, is placed in an adiabatic container at 298 K, and the Thermit reaction

$$2 Al + Fe_2O_3 \rightarrow 2 Fe + Al_2O_3$$

is allowed to proceed to completion. Calculate the state and the temperature of the reaction products.

From the thermochemical data

$$H_{Al_2O_3,298} = -1,675,700 \text{ J/mole}$$

and

$$H_{Fe_2O_3,298} = -823,400 \text{ J/mole}$$

the heat released by the Thermit reaction at 298 K is calculated as

$$\Delta H_{298} = -1,675,700 + 823,400 = -852,300 \text{ J}$$

and this heat raises the temperature of the reaction products. Assume, first, that the sensible heat raises the temperature of the products to the melting temperature of Fe, 1809 K, in which state the reactants occur as 2 moles of liquid Fe and 1 mole of solid Al_2O_3. The molar heat capacities and molar heats of transformation are

$c_{p,Al_2O_3(s)} = 117.49 + 10.38 \times 10^{-3} T - 37.11 \times 10^5 T^{-2}$ J/K in the range 298–2325 K
$c_{p,Fe(\alpha)} = 37.12 + 6.17 \times 10^{-3} T - 56.92 T^{-0.5}$ J/K in the range 298–1187 K
$c_{p,Fe(\gamma)} = 24.48 + 8.45 \times 10^{-3} T$ in the range 1187–1664 K
$c_{p,Fe(\delta)} = 37.12 + 6.17 \times 10^{-3} T - 56.92 T^{-0.5}$ J/K in the range 1667–1809 K

For $Fe_{(\alpha)} \rightarrow Fe_{(\gamma)}$, $\Delta H_{trans} = 670$ J at 1187 K
For $Fe_{(\gamma)} \rightarrow Fe_{(\delta)}$, $\Delta H_{trans} = 840$ J at 1664 K
For $Fe_{(\delta)} \rightarrow Fe_{(l)}$, $\Delta H_m = 13,770$ J at 1809 K

The heat required to raise the temperature of 1 mole of Al_2O_3 from 298 to 1809 K is

$$\Delta H_1 = [117.48 \times (1809 - 298)] + \left[\frac{10.38}{2} \times 10^{-3}(1809^2 - 298^2) \right]$$

$$+ 37.11 \times 10^5 \left(\frac{1}{1809} - \frac{1}{298} \right)$$

$$= 183,649 \text{ J}$$

and the heat required to raise the temperature of 2 moles of Fe from 298 to 1809 K and melt the 2 moles at 1809 K is

$$\Delta H_2 = [2 \times 37.12 \times (1187 - 298)] + \left[\frac{2 \times 6.17}{2} \times 10^{-3}(1187^2 - 298^2) \right]$$

$$+ \left[\frac{2 \times 56.92}{0.5} (1187^{0.5} - 298^{0.5}) \right] + (2 \times 670)$$

$$+ [2 \times 24.28 \times (1664 - 1187)] + \left[\frac{2 \times 8.45}{2} \times 10^{-3}(1664^2 - 1187^2) \right]$$

$$+ (2 \times 840) + [2 \times 37.12 \times (1809 - 1664)]$$

$$+ \left[\frac{2 \times 6.17}{2} \times 10^{-3}(1809^2 - 1664^2) \right] + \left[\frac{2 \times 56.92}{0.5} (1809^{0.5} - 1664^{0.5}) \right]$$

$$+ (2 \times 13,770)$$

$$= (78,058 + 1340 + 34,654 + 1680 + 14,268 + 27,540)$$

$$= 157,541 \text{ J}$$

The total heat required is thus

$$\Delta H_1 + \Delta H_2 = 183,649 + 157,541 = 341,190 \text{ J}$$

The remaining available sensible heat is $852,300 - 341,190 = 511,110$ J.

Consider that the remaining sensible heat raises the temperature of the system to the melting temperature of Al_2O_3, 2325 K, and melts the mole of Al_2O_3. The heat required to increase the temperature of the mole of Al_2O_3 is

$$\Delta H_3 = [117.49 \times (2325 - 1809)] + \left[\frac{10.38}{2} \times 10^{-3}(2325^2 - 1809^2) \right]$$

$$+ 37.11 \times 10^5 \left[\frac{1}{2325} - \frac{1}{1809} \right]$$

$$= 71,240 \text{ J}$$

and, with $c_{p,Fe(l)} = 41.84$ J/K, the heat required to increase the temperature of the 2 moles of liquid Fe is

$$\Delta H_4 = 2 \times 41.84 \times (2325 - 1809) = 43,178 \text{ J}$$

The molar latent heat of melting of Al_2O_3 at its melting temperature of 2325 K is 107,000 J, and thus the sensible heat consumed is

$$71,240 + 43,178 + 107,000 = 221,418 \text{ J}$$

which still leaves $511,110 - 221,418 = 289,692$ J of sensible heat. Consider that this is sufficient to raise the temperature of the system to the boiling point of Fe, 3343 K. The constant-pressure molar heat capacity of liquid Al_2O_3 is 184.1 J/K, and thus the heat required to increase the temperature of 1 mole of liquid Al_2O_3 and 2 moles of liquid Fe from 2325 to 3343 K is

$$(2 \times 41.84 + 184.1) \times (3343 - 2325) = 272,600 \text{ J}$$

which leaves $289,692 - 272,600 = 17,092$ J. The molar heat of boiling of Fe at its boiling temperature of 3343 K is 340,159 J, and thus the remaining 17,092 J of sensible heat is used to convert

$$\frac{17,092}{340,159} = 0.05$$

moles of liquid iron to iron vapor. The final state of the system is thus 1 mole of liquid Al_2O_3, 1.95 moles of liquid Fe, and 0.05 mole of iron vapor at 3343 K.

Suppose, now, that it is required that the increase in the temperature of the products of the Thermit reaction be limited to 1809 K to produce liquid Fe at its melting temperature. This could be achieved by including Fe in the reactants in an amount sufficient to absorb the excess sensible heat. The sensible heat remaining after the temperature of the mole of Al_2O_3 and the 2 moles of Fe has been increased to 1809 K has been calculated as 511,110 J, and the heat required to raise the temperature of 2 moles of Fe from 298 to 1809 K and melt the Fe has been calculated as $\Delta H_2 = 157,541$ J. The number of moles of Fe which must be added to the reacting mole of Fe_2O_3 and 2 moles of Al_2O_3 is thus

$$\frac{511,110}{0.5 \times 157,541} = 6.49$$

The required final state is thus achieved by starting with Fe, Al, and Fe_2O_3 at 298 K occurring in the ratio 6.49:2:1. The Thermit reaction is used to weld steel in locations which are not amenable to conventional welding equipment.

Example 3

A quantity of supercooled liquid tin is adiabatically contained at 495 K. Calculate the fraction of the tin which spontaneously freezes. Given

$$\Delta H_{m,(Sn)} = 7070 \text{ J at } T_m = 505 \text{ K}$$

$$c_{p,Sn(l)} = 34.7 - 9.2 \times 10^{-3} T \text{ J/K}$$

$$c_{p,Sn(s)} = 18.5 + 26 \times 10^{-3} T \text{ J/K}$$

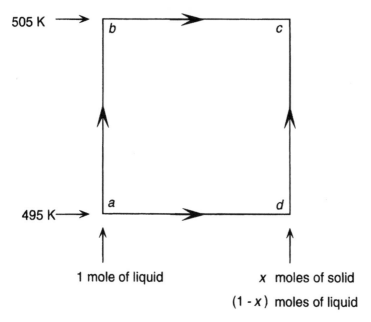

Figure 6.17 Changes in the state of Pb considered in Example 3.

The equilibrium state of the adiabatically contained system is that in which the solid, which has formed spontaneously, and the remaining liquid coexist at 505 K. Thus the fraction of the liquid which freezes is that which releases just enough heat to increase the temperature of the system from 495 to 505 K.

Consider 1 mole of tin and let the molar fraction which freezes be x. In Fig. 6.17 the process is represented by a change of state from a to c, and, as the process is adiabatic, the enthalpy of the system remains constant, i.e.,

$$\Delta H = H_c - H_a = 0$$

Either of two paths can be considered:

Path I, $a \to b \to c$ during which the temperature of the 1 mole of liquid is increased from 495 to 505 K and then x moles freeze. In this case

$$\Delta H_{(a \to b)} = -\Delta H_{(b \to c)}$$

$$\Delta H_{(a \to b)} = \int_{495}^{505} c_{p,Sn(l)} dT = 34.7 \times (505 - 495) - \frac{9.2}{2} \times 10^{-3} (505^2 - 495^2)$$

$$= 301 \text{ J}$$

$$\Delta H_{(b \to c)} = -7070x \text{ J}$$

and thus

$$x = \frac{301}{7070} = 0.0426$$

i.e., 4.26 molar percent of the tin freezes.

Path II, $a \to d \to c$, i.e., the fraction x freezes at 495 K, and then the temperature of the solid and the remaining liquid is increased from 495 to 505 K. In this case

$$\Delta H_{(a \to d)} = -\Delta H_{(d \to c)}$$

$$\Delta H_{(a \to d)} = \text{the heat of freezing } x \text{ moles of tin at 495 K}$$

$$= -x \Delta H_m (495 \text{ K})$$

But

$$\Delta H_m(495 \text{ K}) = \Delta H_m(505 \text{ K}) + \int_{505}^{495} \Delta c_{p(s \to l)} dT$$

$$= 7070 + 16.2(495 - 505) - \frac{35.2}{2} \times 10^{-3}(495^2 - 505^2)$$

$$= 7084 \text{ J}$$

Thus

$$\Delta H_{(a \to d)} = -7084x \text{ J}$$

$$\Delta H_{(d \to c)} = x \int_{495}^{505} c_{p(s)} dT + (1-x) \int_{495}^{505} c_{p(l)} dT$$

$$= x \left[18.5(505 - 495) + \frac{26}{2} \times 10^{-3}(505^2 - 495^2) \right]$$

$$+ (1-x) \left[34.7(505 - 495) - \frac{9.2}{2} \times 10^{-3}(505^2 - 495^2) \right]$$

$$= 301 + 14x$$

Thus

$$-7084x = -14x - 301$$

which gives

$$x = \frac{301}{7070} = 0.0426$$

The actual path which the process follows is intermediate between paths I and II, i.e., the process of freezing and increase in temperature occur simultaneously. The entropy produced by the spontaneous freezing is

$$\Delta S_{(a \to b)} + \Delta S_{(b \to c)} = 34.7 \times \ln\left(\frac{505}{495}\right) - [9.2 \times 10^{-3}(505 - 495)]$$

$$- \left(0.0426 \times \frac{7070}{505} \right)$$

$$= 0.602 - 0.596 = 0.006 \text{ J/K·mole}$$

Example 4

The adiabatic flame temperature is that temperature reached when all of the sensible heat released by the combustion of a fuel is used to raise the temperature of the gaseous products of combustion. Consider the adiabatic flame temperature reached when acetylene, C_2H_2, is combusted at 298 K with (1) the stoichiometric amount of oxygen and (2) the number of moles of air containing the stoichiometric number of moles of oxygen. Air is, by mole or volume percent, 21% O_2 and 79% N_2.

The combustion reaction with stoichiometric oxygen is

$$C_2H_2 + 2.5\,O_2 = 2\,CO_2 + H_2O \tag{i}$$

For C_2H_2, $\Delta H_{298} = +226,700$ J

CO_2 $\Delta H_{298} = -393,500$ J

H_2O $\Delta H_{298} = -241,800$ J

Thus, for the reaction given by Eq. (i)

$$\Delta H(i)_{298} = (-2*393,500) - 241,800 - 226,700 = -1,255,500 \text{ J}$$

The constant pressure molar heat capacities of the products of reaction are,

for H_2O, $c_p, H_2O = 30.00 + 10.71*10^{-3}T + 0.33*10^5 T^2$ J/K·mole, and

for CO_2 $c_p, CO_2 = 44.14 + 9.04*10^{-3}T - 8.54*10^5 T^2$ J/K·mole

The adiabatic flame temperature, T, is then obtained from the requirement

$$\Delta H(i)_{298} + \int_{298}^{T} (2c_p, CO_2 + c_p, H_2O)dT = 0$$

or

$$-1,255,500 + 118.28*(T - 298) + 14.40*10^{-3}*(T^2 - 298^2)$$
$$+ 16.75*10^5(1/T - 1/298) = 0$$

which has the solution $T = 6,236$ K.

For combustion with the stoichiometric amount of air, the reaction is written as

$$C_2H_2 + 2.5\,O_2 + 2.5*(79/21)N_2 = 2\,CO_2 + H_2O + 9.41\,N_2$$

for N_2 $c_p, N_2 = 27.87 + 4.27*10^{-3}T$ J/mole·K

and the adiabatic flame temperature, T is obtained from

$$\Delta H(i)_{298} + \int_{298}^{T} (2c_p, CO_2 + c_p, H_2O + 9.41c_p, N_2)dT = 0$$

or

$$-1,255,500 + 380.1 * (T - 298) + 40.16 * 10^{-3} * (T^2 - 298^2)$$
$$+ 16.75 * 10^5 (1/T - 1/298) = 0$$

as $T = 2,797$ K.

This high adiabatic flame temperature facilitates the use of acetylene for welding metals with high melting temperatures.

PROBLEMS*

6.1 Calculate ΔH_{1600} and ΔS_{1600} for the reaction $Zr_{(\beta)} + O_2 = ZrO_{2(\beta)}$.

6.2 Which of the following two reactions is the more exothermic?

1. $C_{(graphite)} + \dfrac{1}{2}O_{2(g)} \rightarrow CO_{(g)}$ at 1000 K

2. $C_{(diamond)} + \dfrac{1}{2}O_{2(g)} \rightarrow CO_{(g)}$ at 1000 K

6.3 Calculate the change in enthalpy and the change in entropy at 1000 K for the reaction $CaO_{(s)} + TiO_{2(s)} \rightarrow CaTiO_{3(s)}$.

6.4 Copper exists in the state $T = 298$ K, $P = 1$ atm. Calculate the temperature to which the copper must be raised at 1 atm pressure to cause the same increase in molar enthalpy as is caused by increasing its pressure to 1000 atm at 298 K. The molar volume of Cu at 298 K is 7.09 cm^3, and the volumetric coefficient of thermal expansion is 0.501×10^{-4} K^{-1}. These values can be taken as being independent of pressure in the range 1–1000 atm.

6.5 Calculate ΔH_{298} and ΔS_{298} for the following reactions:

a. $2\,TiO + \dfrac{1}{2}O_2 = Ti_2O_3$

b. $3\,Ti_2O_3 + \dfrac{1}{2}O_2 = 2\,Ti_3O_5$

c. $Ti_3O_5 + \dfrac{1}{2}O_2 = 3\,TiO_2$

6.6 An adiabatic vessel contains 1000 g of liquid aluminum at 700°C. Calculate the mass of Cr_2O_3 at room temperature, which, when added to the liquid aluminum (with which it reacts to form Cr and Al_3O_3) raises the temperature of the resulting mixture of Al_2O_3, Cr_2O_3, and Cr to 1600 K.

6.7 Calculate the adiabatic flame temperature attained when methane, at 298 K, is combusted (a) with oxygen in the molar ratio $O_2/CH_4 = 2.0$, and (b) with air in the molar ratio air/$CH_4 = 9.524$. Assume that CO_2 and H_2O are the products of combustion. The adiabatic flame temperature is that temperature reached if all of the heat of the oxidation reaction is used to increase the

*Thermodynamic data required for the solution of the end-of-chapter problems are tabulated in the appendices.

temperature of the products of the reaction. Air is 21 molar percent O_2 and 79 molar percent N_2.

6.8 Calculate the value of ΔG for the reaction

$$Si_3N_4 + 3\,O_2 = 3\,SiO_{2(\alpha-\text{quartz})} + 2\,N_2$$

at 800 K. What percentage error occurs if it is assumed that Δc_p for the reaction is zero?

6.9 Determine the stoichiometric coefficients for the reaction

$$(3\,CaO \cdot Al_2O_3 \cdot 3\,SiO_2) + a(CaO \cdot Al_2O_3 \cdot SiO_2) =$$
$$b(CaO \cdot Al_2O_3 \cdot 2\,SiO_2) + c(2\,CaO \cdot Al_2O_3 \cdot SiO_2)$$

and calculate ΔH_{298}, ΔS_{298}, and ΔG_{298} for the reaction.

6.10 How much heat is required to increase the temperature of 1 kg of cordierite, $2\,MgO \cdot 2\,Al_2O_3 \cdot 5\,SiO_2$, from 298 K to its incongruent melting temperature of 1738 K?

Chapter 7

PHASE EQUILIBRIUM IN A
ONE-COMPONENT SYSTEM

7.1 INTRODUCTION

The intensive thermodynamic properties of a system are temperature, pressure, and the chemical potentials of the various species occurring in the system, and these properties are measures of potentials of one kind or another.

The temperature of a system is a measure of the potential or intensity of heat in the system, and temperature is thus a measure of the tendency for heat to leave the system. For example, if two parts of a system are at different temperatures, a heat-potential gradient exists which produces a driving force for the transport of heat down the gradient from the part at the higher temperature to the part at the lower temperature. Spontaneous heat flow occurs until the thermal gradient has been eliminated, in which state the heat is distributed at uniform intensity throughout the system. Thermal equilibrium is thus established when the heat potential, and hence the temperature, are uniform throughout the system.

The pressure of a system is a measure of its potential for undergoing massive movement by expansion or contraction. If, in a system of fixed volume, the pressure exerted by one phase is greater than that exerted by another phase, then the tendency of the first phase to expand exceeds that of the second phase. The pressure gradient is the driving force for expansion of the first phase, which decreases its pressure and hence its tendency for further expansion, and contraction of the second phase, which increases its pressure and hence its tendency to resist further contraction. Mechanical equilibrium is established when the massive movement of the two phases has occurred to the extent that the pressure gradient has been eliminated, in which state the pressure is uniform throughout the system.

The chemical potential of the species i in a phase is a measure of the tendency of the species i to leave the phase. It is thus a measure of the "chemical pressure" exerted by i in the phase. If the chemical potential of i has different values in different phases of the system, which are at the same temperature and pressure, then, as the escaping tendencies differ, the species i will tend to move from the phases in which it occurs at the higher chemical potential to the phases in which it occurs at the lower chemical potential. A gradient in chemical potential is the driving force for chemical diffusion, and equilibrium is attained when the species i is distributed throughout the various phases in the system such that its chemical potential has the same value in all phases.

In a closed system of fixed composition, e.g., a one-component system, equilibrium, at the temperature T and the pressure, P, occurs when the system exists in that state which has the minimum value of G'. The equilibrium state can thus be determined by means of an examination of dependence of G on pressure and temperature. In the following discussion the system H_2O will be used as an example.

7.2 THE VARIATION OF GIBBS FREE ENERGY WITH TEMPERATURE AT CONSTANT PRESSURE

At a total pressure of 1 atm, ice and water are in equilibrium with one another at $0°C$, and hence, for these values of temperature and pressure, the Gibbs free energy, G', of the system has its minimum value. Any transfer of heat to the system causes some of the ice to melt at $0°C$ and 1 atm pressure, and, provided that some ice remains, the equilibrium between the ice and the water is not disturbed and the value of G' for the system is unchanged. If, by the addition of heat, 1 mole of ice is melted, then for the change of state

$$H_2O_{(solid)} \rightarrow H_2O_{(liquid)} \text{ at 1 atm and 273 K}$$

$$\Delta G = G_{H_2O_{(l)}} - G_{H_2O_{(s)}} = 0$$

Thus, at the state of equilibrium between ice and water,

$$G_{H_2O_{(l)}} = G_{H_2O_{(s)}} \tag{7.1}$$

where $G_{H_2O_{(s)}}$ is the molar Gibbs free energy of H_2O in the solid (ice) phase, and $G_{H_2O_{(l)}}$ is the molar Gibbs free energy of H_2O in the liquid (water) phase. For the system of ice + water containing n moles of H_2O, $n_{H_2O_{(s)}}$ of which are in the ice phase and $n_{H_2O_{(l)}}$ of which are in the water phase, the Gibbs free energy of the system, G', is

$$G' = n_{H_2O_{(s)}} G_{H_2O_{(s)}} + n_{H_2O_{(l)}} G_{H_2O_{(l)}} \tag{7.2}$$

and from Eq. (7.1) it is seen that, at $0°C$ and 1 atm pressure, the value of G' is independent of the proportions of the ice phase and the water phase present.

The equality of the molar Gibbs free energies of H_2O in the solid and liquid phase at $0°C$ and 1 atm corresponds with the fact that, for equilibrium to occur, the escaping tendency of H_2O from the liquid phase must equal the escaping tendency of

H_2O from the solid phase. Hence it is to be expected that a relationship exists between the molar Gibbs free energy and the chemical potential of a component in a phase. Integration of Eq. (5.25) at constant T and P gives

$$G' = \sum \mu_i n_i$$

which, for the ice + water system, is written as

$$G' = \mu_{H_2O_{(s)}} n_{H_2O_{(s)}} + \mu_{H_2O_{(l)}} n_{H_2O_{(l)}} \tag{7.3}$$

Comparison of Eqs. (7.2) and (7.3) shows that $\mu_{H_2O} = G_{H_2O}$ or, in general, $\mu_i = G_i$; i.e., the chemical potential of a species in a particular state equals the molar Gibbs free energy of the species in the particular state.

This result could also have been obtained from a consideration of Eq. (5.16)

$$\left(\frac{\partial G'}{\partial n_i} \right)_{T,P} = \mu_i$$

In a one-component system, the chemical potential of the component equals the increase in the value of G' which occurs when 1 mole of the component is added to the system at constant T and P. That is, if the component is the species i,

$$\Delta G' = \mu_i$$

and as the increase in the value of G' for the one-component system is simply the molar Gibbs free energy of i, then

$$G_i = \mu_i$$

If the ice + water system is at 1 atm pressure and some temperature greater than 0°C, then the system is not stable and the ice spontaneously melts. This process decreases the Gibbs free energy of the system, and equilibrium is attained when all of the ice has melted. That is, for the change of state $H_2O_{(s)} \rightarrow H_2O_{(l)}$ at $T > 273$ K, and $P = 1$ atm,

$$\Delta G = G_{H_2O_{(l)}} - G_{H_2O_{(s)}} < 0$$

i.e.

$$G_{H_2O_{(l)}} < G_{H_2O_{(s)}}$$

The escaping tendency of H_2O from the solid phase is greater than the escaping tendency of H_2O from the liquid phase. Conversely, if, at $P = 1$ atm, the temperature is less than 0°C, then

$$G_{H_2O_{(l)}} > G_{H_2O_{(s)}}$$

The variations of $G_{H_2O_{(l)}}$ and $G_{H_2O_{(s)}}$ with temperature at constant pressure are shown in Fig. 7.1 and the variation of $\Delta G_{s \rightarrow l}$ with temperature at constant pressure is shown in Fig. 7.2.

Figs. 7.1 and 7.2 show that, at 1 atm pressure and temperatures greater than 0°C, the minimum Gibbs free energy occurs when all of the H_2O is in the liquid phase,

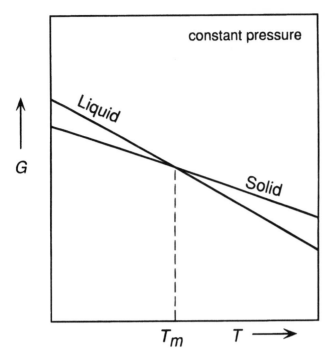

Figure 7.1 Schematic representation of the variations of the molar Gibbs free energies of solid and liquid water with temperature at constant pressure.

and at 1 atm pressure and temperatures lower than 0°C, the minimum Gibbs free energy occurs when all of the H_2O is in the solid phase. The slopes of the lines in Fig. 7.1 are obtained from Eq. (5.25) as

$$\left(\frac{\partial G}{\partial T}\right)_P = -S$$

and the curvatures of the lines are obtained from Eq. (6.12) as

$$\left(\frac{\partial^2 G}{\partial T^2}\right)_P = -\left(\frac{\partial S}{\partial T}\right)_P = -\frac{c_p}{T}$$

Similarly, the slope of the line in Fig. 7.2 is given as

$$\left(\frac{\partial \Delta G}{\partial T}\right)_P = -\Delta S$$

where ΔS is the change in the molar entropy which occurs as a result of the change of state. The slope of the line in Fig. 7.2 is negative, which shows that, at all temperatures,

$$S_{H_2O_{(l)}} > S_{H_2O_{(s)}}$$

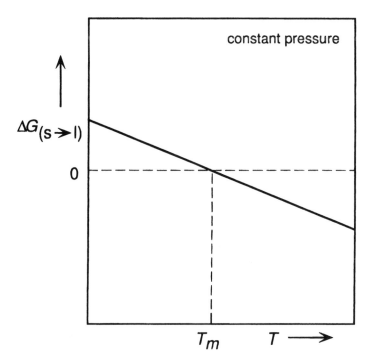

Figure 7.2 Schematic representation of the variation of the molar Gibbs free energy of melting of water with temperature at constant pressure.

as is to be expected in view of the fact that, at any temperature, the liquid phase is more disordered than is the solid phase.

The state in which the solid and liquid phases of a one-component system are in equilibrium with one another can be determined from consideration of the molar enthalpy H and the molar entropy S of the system. From Eq. (5.2),

$$G = H - TS$$

This can be written for both the solid and the liquid phases,

$$G_{(l)} = H_{(l)} - TS_{(l)}$$

and

$$G_{(s)} = H_{(s)} - TS_{(s)}$$

For the change of state solid → liquid, subtraction gives

$$\Delta G_{(s \to l)} = \Delta H_{(s \to l)} - T\Delta S_{(s \to l)}$$

where $\Delta H_{(s \to l)}$ and $\Delta S_{(s \to l)}$ are, respectively, the changes in the molar enthalpy and molar entropy which occur as a result of melting at the temperature T. From Eq. (7.1)

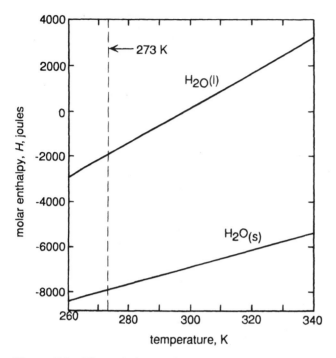

Figure 7.3 The variations, with temperature, of the molar enthalpies of solid and liquid water at 1 atm pressure. The molar enthalpy of liquid water at 298 K is arbitrarily assigned the value of zero.

equilibrium between the solid and the liquid phases occurs at that state at which $\Delta G_{(s \to l)} = 0$. This occurs at that temperature T_m at which

$$\Delta H_{(s \to l)} = T_m \Delta S_{(s \to l)} \tag{7.4}$$

For H_2O

$$\Delta H_m = \Delta H_{(s \to l)} = 6008 \text{ joules at 273 K}$$
$$S_{H_2O(l),\,298\text{ K}} = 70.08 \text{ J/K}$$
$$S_{H_2O(s),\,298\text{ K}} = 44.77 \text{ J/K}$$
$$c_{p,H_2O(l)} = 75.44 \text{ J/K}$$
$$c_{p,H_2O(s)} = 38 \text{ J/K}$$

Fig. 7.3 shows the variations of $H_{(s)}$ and $H_{(l)}$ at 1 atm pressure, in which, for convenience, $H_{(l),298\text{ K}}$ is arbitrarily assigned the value of zero, in which case

$$H_{(l),T} = \int_{298}^{T} c_{p,(l)} dT = 75.44(T - 298) \text{ J}$$

and

$$H_{(s),T} = \int_{298}^{273} c_{p,(l)}dT - \Delta H_m + \int_{273}^{T} c_{p,(s)}dT$$

$$= 75.44(273 - 298) - 6008 + 38(T - 273) \text{ J}$$

The molar enthalpy of melting at the temperature T, $\Delta H_{(s \to l),T}$ is the vertical separation between the two lines in Fig. 7.3.

Fig. 7.4 shows the variations of $S_{(s)}$ and $S_{(l)}$ with temperature at 1 atm pressure, where

$$S_{(l),T} = S_{(l),298} + \int_{298}^{T} c_{p,(l)}d \ln T$$

$$= 70.08 + 75.44 \ln \left(\frac{T}{298}\right) \text{ J/K}$$

and

$$S_{(s),T} = S_{(s),298} + \int_{298}^{T} c_{p,(s)}d \ln T = 44.77 + 38 \ln \left(\frac{T}{298}\right) \text{ J/K}$$

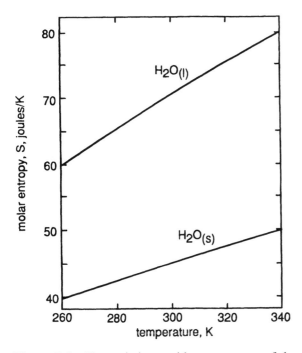

Figure 7.4 The variations, with temperature, of the molar entropies of solid and liquid water at 1 atm pressure.

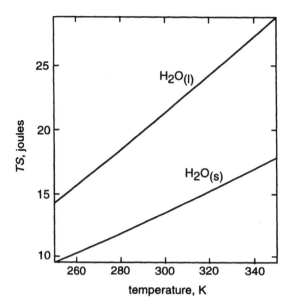

Figure 7.5 The variation, with temperature, of TS for solid and liquid water at 1 atm pressure.

The molar entropy of melting at the temperature T, $\Delta S_{(s \to l)}$ is the vertical separation between the two lines in Fig. 7.4. Fig. 7.5 shows the corresponding variations of $TS_{(s)}$ and $TS_{(l)}$ with temperature. Equilibrium between the solid and liquid phases occurs at that temperature at which the vertical separation between the two lines in Fig. 7.3 equals the vertical separation between the two lines in Fig. 7.5. This unique temperature is T_m, and, at this temperature,

$$\Delta H_{(s \to l)} = T_m \Delta S_{(s \to l)}$$

In Fig. 7.6, $\Delta H_{(s \to l)}$, $T \Delta S_{(s \to l)}$, and $\Delta G_{(s \to l)}$ are plotted as functions of temperature using the data in Figs. 7.3 and 7.5. This figure shows that $\Delta G_{(s \to l)} = 0$ at $T = T_m = 273$ K, which is thus the temperature at which solid and liquid water are in equilibrium with one another at 1 atm pressure.

Equilibrium between two phases thus occurs as the result of a compromise between enthalpy considerations and entropy considerations. Equilibrium requires that G' for the system have its minimum value at the fixed values of P and T, and Eq. (5.2) shows that minimization of G' requires that H be small and S be large. Fig. 7.3 shows that, at all temperatures, $H_{(l)} > H_{(s)}$, and thus, from consideration of the contribution of enthalpy to the Gibbs free energy, and in the absence of any other consideration, it would seem that the solid would always be stable with respect to the liquid. However Fig 7.4 shows that, at all temperatures, $S_{(l)} > S_{(s)}$. Thus from consideration of the contribution of entropy to the Gibbs free energy, in the absence of any other consideration, it would seem that the liquid phase is always stable with respect to the solid phase. However, as the contribution of the entropy, TS, to G is

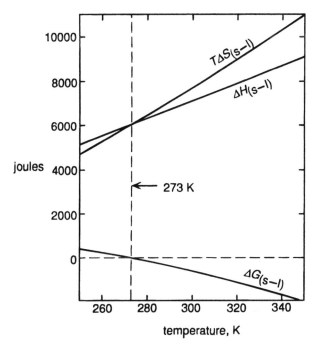

Figure 7.6 The variations, with temperature, of the molar Gibbs free energy of melting, the molar enthalpy of melting, and $T \times$ the molar entropy of melting of water at 1 atm pressure.

dependent on temperature, a unique temperature T_m occurs above which the contribution of the entropy outweighs the contribution of the enthalpy and below which the reverse is the case. The temperature T_m is that at which $H_{(l)} - T_m S_{(l)}$ equals $H_{(s)} - T_m S_{(s)}$ and hence is the temperature at which the molar Gibbs free energy of the solid has the same value as the molar Gibbs free energy of the liquid. This discussion is analogous to that presented in Sec. 5.3 where, at constant T and V, the equilibrium between a solid and its vapor was examined in terms of minimization of the Helmholtz free energy, A, of the system.

7.3 THE VARIATION OF GIBBS FREE ENERGY WITH PRESSURE AT CONSTANT TEMPERATURE

Consider the application of Le Chatelier's principle to ice and water, coexisting in equilibrium with one another at 0°C, when the pressure exerted on the system is increased to a value greater than 1 atm. Le Chatelier's principle states that, when subjected to an external influence, the state of a system at equilibrium shifts in that direction which tends to nullify the effect of the external influence. Thus when the pressure exerted on a system is increased, the state of the system shifts in the direction which causes a decrease in its volume. As ice at 0°C has a larger molar volume than has water at 0°C, the melting of ice is the change in state caused by an increase in pressure. The influence of an increase in pressure, at constant temperature, on the molar Gibbs free energies of the phases is given by Eq. (5.25) as

$$\left(\frac{\partial G_{(l)}}{\partial P}\right)_T = V_{(l)} \quad \text{and} \quad \left(\frac{\partial G_{(s)}}{\partial P}\right)_T = V_{(s)}$$

i.e., the rate of increase of G with increase in pressure at constant temperature equals the molar volume of the phase at the temperature T and the pressure, P. For the change of the state solid → liquid,

$$\left(\frac{\partial \Delta G_{(s \to l)}}{\partial P}\right)_T = \Delta V_{(s \to l)}$$

and as $\Delta V_{(s \to l)}$ for H_2O at 0°C is negative, the ice melts when the pressure is increased to a value greater than 1 atm. Thus, corresponding to Fig. 7.1, which showed the variation of $G_{(s)}$ and $G_{(l)}$ with T at constant P, Fig. 7.7 shows the variation of $G_{(s)}$ and $G_{(l)}$ with P at constant T. Water is anomalous in that, usually, melting causes an increase in volume.

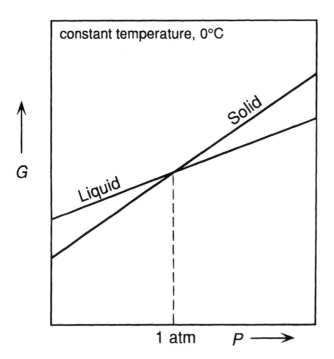

Figure 7.7 Schematic representation of the variations of the molar Gibbs free energies of solid and liquid water with pressure at constant temperature.

7.4 GIBBS FREE ENERGY AS A FUNCTION OF TEMPERATURE AND PRESSURE

Consideration of Figs. 7.1 and 7.7 shows that it is possible to maintain equilibrium between the solid and liquid phase by simultaneously varying the temperature and pressure in such a manner that $\Delta G_{(s\to l)}$ remains zero. For equilibrium to be maintained

$$G_{(l)} = G_{(s)}$$

or, for any infinitesimal change in T and P,

$$dG_{(l)} = dG_{(s)}$$

From Eq. (5.12)

$$dG_{(l)} = -S_{(l)}dT + V_{(l)}dP$$

and

$$dG_{(s)} = -S_{(s)}dT + V_{(s)}dP$$

Thus, for equilibrium to be maintained between the two phases,

$$-S_{(l)}dT + V_{(l)}dP = -S_{(s)}dT + V_{(s)}dP$$

or

$$\left(\frac{dP}{dT}\right)_{eq} = \frac{S_{(s)} - S_{(l)}}{V_{(s)} - V_{(l)}} = \frac{\Delta S_{(l\to s)}}{\Delta V_{(l\to s)}}$$

At equilibrium $\Delta G = 0$, and hence $\Delta H = T\Delta S$, substitution of which into the above equation gives

$$\left(\frac{dP}{dT}\right)_{eq} = \frac{\Delta H}{T\Delta V} \qquad (7.5)$$

Eq. (7.5), which is known as the Clapeyron equation, gives the required relationships between the variations of temperature and pressure which are required for the maintenance of equilibrium between the two phases.

The value of $\Delta V_{(s\to l)}$ for H_2O is negative and $\Delta H_{(s\to l)}$ for all materials is positive. Thus $(dP/dT)_{eq}$ for H_2O is negative, i.e., an increase in pressure decreases the equilibrium melting temperature, and it is for this reason that iceskating is possible. The pressure of the skate on the solid ice decreases its melting temperature, and, provided that the melting temperature is decreased to a value below the actual temperature of the ice, the ice melts to produce liquid water which acts as a lubricant for the skate on the ice. For most materials $\Delta V_{(s\to l)}$ is positive, which means that an increase in pressure increases the equilibrium melting temperature.

The thermodynamic states of the solid and liquid phase can be represented in a three-dimensional diagram with G, T, and P as coordinates; such a diagram, drawn schematically for H_2O, is shown in Fig. 7.8. In this figure each phase occurs on a

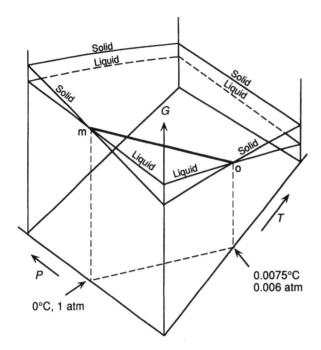

Figure 7.8 Schematic representation of the equilibrium surfaces of the solid and liquid phases of water in *G-T-P* space.

surface in *G-T-P* space, and the line along which the surfaces intersect represents the variation of *P* with *T* required for maintenance of the equilibrium between the two phases. At any state, which is determined by fixing the values of *T* and *P*, the equilibrium phase is that which has the lower value of *G*. Fig. 7.1, if drawn at $P = 0.006$ atm, corresponds to the right front face of Fig. 7.8, and Fig. 7.7, if drawn for $T = 0°C$, corresponds to the left front face of Fig. 7.8.

7.5 EQUILIBRIUM BETWEEN THE VAPOR PHASE AND A CONDENSED PHASE

If Eq. (7.5) is applied to an equilibrium between a vapor phase and a condensed phase then ΔV is the change in the molar volume accompanying evaporation or sublimation and ΔH is the corresponding change in the molar enthalpy, i.e., the molar latent heat of evaporation or sublimation. Thus

$$\Delta V = V_{vapor} - V_{condensed\ phase}$$

and as V_{vapor} is very much larger than $V_{condensed\ phase}$, then, with the introduction of an insignificant error,

$$\Delta V = V_{vapor}$$

Thus, for condensed phase-vapor equilibria, Eq. (7.5) can be written as

$$\left(\frac{dP}{dT}\right)_{eq} = \frac{\Delta H}{TV_{(v)}}$$

in which $V_{(v)}$ is the molar volume of the vapor. If it is further assumed that the vapor in equilibrium with the condensed phase behaves ideally, i.e., $PV = RT$, then

$$\left(\frac{dP}{dT}\right)_{eq} = \frac{P\Delta H}{RT^2}$$

rearrangement of which gives

$$\frac{dP}{P} = \frac{\Delta H}{RT^2} dT$$

or

$$d \ln P = \frac{\Delta H}{RT^2} dT \tag{7.6}$$

Eq. (7.6) is known as the Clausius-Clapeyron equation.

If ΔH is independent of temperature, i.e., if $c_p(\text{vapor}) = c_p(\text{condensed phase})$, integration of Eq. (7.6) gives

$$\ln P = -\frac{\Delta H}{RT} + \text{constant} \tag{7.7}$$

As equilibrium is maintained between the vapor phase and the condensed phase, the value of P at any T in Eq. (7.7) is the saturated vapor pressure exerted by the condensed phase at the temperature T. Eq. (7.7) thus shows that the saturated vapor pressure exerted by a condensed phase increases exponentially with increasing temperature, as was noted in Sec. 5.3. If Δc_p for the evaporation or sublimation is not zero, but is independent of temperature, then, from Eq. (6.9), ΔH_T in Eq. (7.6) is

$$\Delta H_T = \Delta H_{298} + \Delta c_p(T - 298)$$
$$= (\Delta H_{298} - 298\Delta c_p) + \Delta c_p T$$

in which case integration of Eq. (7.6) gives

$$\ln P = \left(\frac{298\Delta c_p - \Delta H_{298}}{R}\right)\frac{1}{T} + \frac{\Delta c_p}{R} \ln T + \text{constant}$$

which is normally expressed in the form

$$\ln P = -\frac{A}{T} + B \ln T + C \tag{7.8}$$

In Eq. (7.8),

$$\Delta H_T = AR + BRT$$

7.6 GRAPHICAL REPRESENTATION OF PHASE EQUILIBRIA IN A ONE-COMPONENT SYSTEM

In an equilibrium between a liquid and a vapor the normal boiling point of the liquid is defined as that temperature at which the saturated vapor pressure exerted by the liquid is 1 atm. Knowledge of the molar heat capacities of the liquid and vapor phases, the molar heat of evaporation at any one temperature, $\Delta H_{evap,T}$, and the normal boiling temperature allows the saturated vapor pressure-temperature to be determined for any material. For example, for H_2O

$$c_{p,H_2O(v)} = 30 + 10.7 \times 10^{-3}T + 0.33 \times 10^5 T^{-2} \text{ J/K}$$

in the range of temperature 298–2500 K and

$$c_{p,H_2O(l)} = 75.44 \text{ J/K}$$

in the range of temperature 273–373 K. Thus, for the change of state

$$H_2O_{(l)} \rightarrow H_2O_{(v)}$$

$$\Delta c_{p(l \rightarrow v)} = -45.44 + 10.7 \times 10^{-3}T + 0.33 \times 10^5 T^{-2} \text{ J/K}$$

At the normal boiling temperature of 373 K, $\Delta H_{evap} = 41,090$ J, and thus

$$\Delta H_{evapT} = \Delta H_{evap\,373} + \int_{373}^{T} \Delta c_{p(l \rightarrow v)} dT$$

$$= 41,090 - 45.44(T - 373) + 5.35 \times 10^{-3}(T^2 - 373^2)$$

$$- 0.33 \times 10^5 \left(\frac{1}{T} - \frac{1}{373} \right)$$

$$= 57,383 - 45.44T + 5.35 \times 10^{-3}T^2 - \frac{0.33 \times 10^5}{T} \text{ J}$$

Now

$$d \ln p = \frac{\Delta H_{evap}}{RT^2} dT$$

and so, with $R = 8.3144$ J/K·mole,

$$\ln p = -\frac{57,383}{RT} - \frac{45.44 \ln T}{R} + \frac{5.35 \times 10^{-3}T}{R} + \frac{0.33 \times 10^5}{2\,RT^2} + \text{constant}$$

At the boiling point of 373 K, $p = 1$ atm, and thus the integration constant is evaluated as 51.10. In terms of logarithms to base 10, this gives

$$\log p \text{ (atm)} = -\frac{57,383}{2.303RT} - \frac{45.44}{R} \log T + \frac{5.35 \times 10^{-3}T}{2.303 \times R}$$

$$+ \frac{0.33 \times 10^5}{2 \times 2.303RT^2} + \frac{51.10}{2.303} \tag{7.9}$$

$$= -\frac{2{,}997}{T} - 2.373 \log T + 0.279 \times 10^{-3}T + \frac{862}{T^2} + 21.75$$

which is thus the variation of the saturated vapor pressure of water with temperature in the range of temperature 273–373 K. Curve-fitting of experimentally measured vapor pressure of liquid water to an expression of the form

$$\log p \text{ (atm)} = \frac{A}{T} + B \log T + C$$

gives

$$\log p \text{ (atm)} = -\frac{2900}{T} - 4.65 \log T + 19.732 \qquad (7.10)$$

Eqs. (7.9) and (7.10) are shown in Fig. 7.9 as plots of $\log p$ (atm) vs. inverse temperature. The agreement between the two lines increases with increasing temperature. In Fig. 7.9 the slope of the line at any temperature equals $\Delta H_{\text{evap},T}/4.575$. The saturated vapor pressures of several of the more common elements are presented in Fig. 7.10, again as the variations of $\log p$ with inverse temperature.

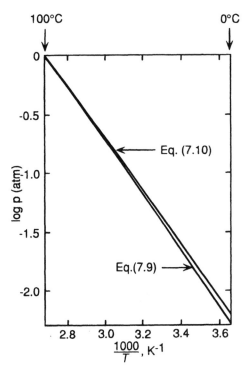

Figure 7.9   The saturated vapor pressure of water as a function of temperature.

Fig. 7.11 is a one-component phase diagram which uses T and P as coordinates. Line AOA' is a graphical representation of the integral of Eq. (7.5), which is the variation of pressure with temperature required for phase equilibrium between the solid and liquid phases. If ΔH_m is independent of temperature, integration of Eq. (7.5) gives an expression of the form

$$P = \frac{\Delta H}{\Delta V} \ln T + \text{constant} \tag{7.11}$$

By definition the normal melting temperature of the material is the melting temperature at a pressure of 1 atm, and in Fig. 7.11 the normal melting point is designated as the point m. The line BOB' is the line for equilibrium between the vapor and the liquid given by Eq. (7.7) or (7.9) in which ΔH_T is $\Delta H_{\text{evap},T}$. In the case of water the line BOB' represents the variation, with temperature, of the saturated vapor pressure of the liquid, or alternatively, the variation, with pressure, of the dew point of water vapor. The line BOB' passes through the normal boiling point (represented by the point b in the figure) and intersects the line AOA' at the *triple point, O*. The triple point is the state represented by the invariant values of P and T at which the solid, liquid, and vapor phases are in equilibrium with each other. Knowledge of the triple point, together with the value of $\Delta H_{\text{sublim},T}$, allows the variation of the saturated vapor pressure of the solid with temperature to be determined. This equilibrium line is drawn as COC' in Fig. 7.11.

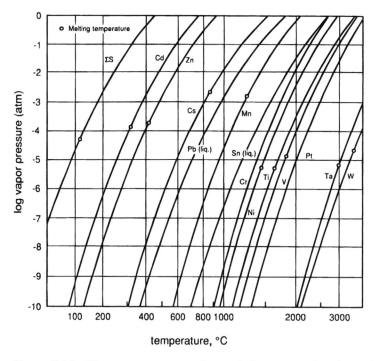

Figure 7.10   The vapor pressures of several elements as functions of temperature.

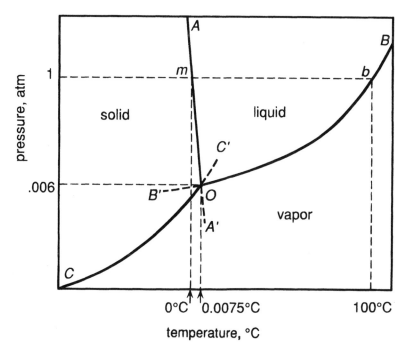

Figure 7.11 Schematic representation of part of the phase diagram for H_2O.

In the G-T-P surface for the states of existence of the vapor phase were included in Fig. 7.8, it would intersect with the solid-state surface along a line and would intersect with the liquid-state surface along a line. Projection of these lines, together with the line of intersection of the solid- and liquid-state surfaces, onto the two-dimensional P-T basal plane of Fig. 7.8 would produce Fig. 7.11. All three state surfaces in the re-drawn Fig. 7.8 would intersect at a point, projection of which onto the P-T basal plane gives the invariant point O. The dashed lines OA', OB', and OC' in Fig. 7.11 represent, respectively, metastable solid-liquid, metastable vapor-liquid, and metastable vapor-solid equilibria. The equilibria are metastable because, in the case of the line OB', the intersection of the liquid- and vapor-state surfaces in the redrawn Fig. 7.8 lies at higher values of G than does the solid-state surface for the same values of P and T. Similarly, the solid-liquid equilibrium OA' is metastable with respect to the vapor phase, and the solid-vapor equilibrium OC' is metastable with respect to the liquid phase.

Fig. 7.12*a* shows three isobaric sections of the redrawn Fig. 7.8 at $P_1 > P_{\text{triple point}}$, $P_2 = P_{\text{triple point}}$, and $P_3 < P_{\text{triple point}}$, and Fig. 7.12*b* shows three isothermal sections of the redrawn Fig. 7.8 at $T_1 < T_{\text{triple point}}$, $T_2 = T_{\text{triple point}}$, and $T_3 > T_{\text{triple point}}$. In Fig. 7.12*a,* the slopes of the lines in any isobaric section increase negatively in the order solid, liquid, vapor, in accordance with the fact that $S_{(s)} < S_{(l)} < S_{(v)}$. Similarly, in Fig. 7.12*b* the slopes of the lines in any isothermal section increase in the order liquid, solid, vapor in accordance with the fact that, for H_2O, $V_{(l)} < V_{(s)} < V_{(v)}$.

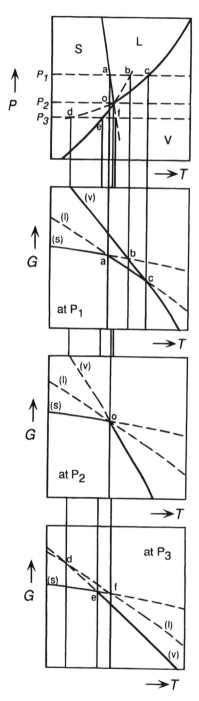

Figure 7.12 *(a)* schematic representation of the constant-pressure variations of the molar Gibbs free energies of solid, liquid, and vapor H_2O at pressures above, at, and below the triple-point pressure.

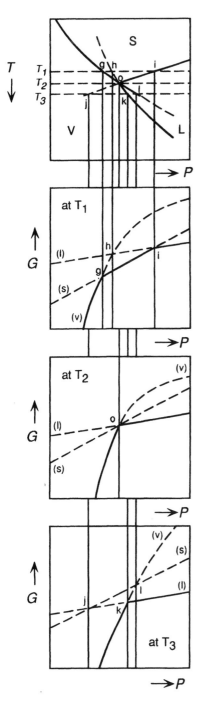

Figure 7.12 *(b)* Schematic representation of the constant-temperature variations of the molar Gibbs free energies of solid, liquid, and vapor H_2O at temperatures above, at, and below the triple-point temperature.

The lines *OA, OB,* and *OC* divide Fig. 7.11 into three areas within each of which only one phase is stable. Within these areas the pressure exerted on the phase and the temperature of the phase can be independently varied without upsetting the one-phase equilibrium. The equilibrium thus has two *degrees of freedom,* where the number of degrees of freedom that an equilibrium has is the maximum number of variables which may be independently varied without upsetting the equilibrium. The single-phase areas meet at the lines *OA, OB,* and *OC* along which two phases coexist in equilibrium, and for continued maintenance of any of these equilibria only one variable (either *P* or *T*) can be independently varied. Two-phase equilibria in a one-component system thus have only one degree of freedom. The three two-phase equilibrium lines meet at the triple point, which is the invariant state at which solid, liquid, and vapor coexist in equilibrium. The three-phase equilibrium in a one-component system thus has no degrees of freedom, and three is therefore the maximum number of phases which can coexist at equilibrium in a one-component system. The number of degrees of freedom, *F,* that a system containing *C* components can have when *P* phases are in equilibrium is given by

$$F = C - P + 2$$

This expression is called the Gibbs phase rule.

7.7 SOLID-SOLID EQUILIBRIA

Elements which can exist in more than one crystal form are said to exhibit allotropy, and chemical compounds which can exist in more than one solid form are said to exhibit polymorphism. The variation of pressure with temperature required to maintain equilibrium between two solids is given by Eq. (7.10) in which ΔH and ΔV are the changes in the molar enthalpy and the molar volume for the change of state solid I $\rightarrow$ solid II. The phase diagram for iron at relatively low pressures is shown in Fig. 7.13. Iron has body-centered crystal structures, the α and δ phases, at, respectively, low and high temperatures, and a face-centered crystal structure, the γ phase at intermediate temperatures; Fig. 7.13 shows three triple points involving two condensed phases and the vapor phase. As atoms in the face-centered crystal structure fill space more efficiently than do atoms in the body-centered structure, the molar volume of γ-Fe is less than those of α-Fe and δ-Fe, and consequently, the line for equilibrium between α and γ has a negative slope, and the line for equilibrium between γ and δ has a positive slope. With increasing pressure the slope of the γ-δ line becomes greater than that of the δ-liquid line, and the two lines meet at a triple point for the three-phase γ-δ-liquid equilibrium at $P = 14{,}420$ atm and $T = 1590°C$. The vapor pressure of liquid iron, which is given by

$$\log p \ (\text{atm}) = \frac{-19{,}710}{T} - 1.27 \log T + 10.39$$

reaches 1 atm at 3057°C, which is thus the normal boiling temperature of iron.

Fig. 7.14 shows a schematic representation of the variation, with temperature at constant pressure, of the molar Gibbs free energies of the bcc, fcc, liquid, and vapor

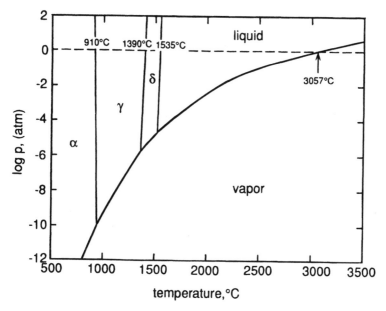

Figure 7.13 The phase diagram for iron.

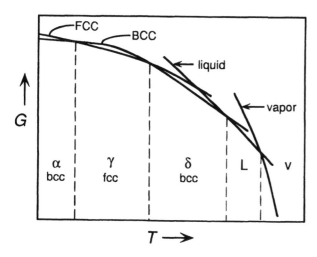

Figure 7.14 Schematic representation of the variation of the molar Gibbs free energies of the bcc, fcc, liquid, and vapor phases of iron with temperature at constant pressure.

phases of iron. The curvature of the bcc iron line is such that it intersects the fcc iron line twice, with the consequence that, at 1 atm pressure, bcc iron is stable relative to fcc iron at temperatures less than 910°C and at temperatures greater than 1390°C.

A schematic phase diagram for zirconia, ZrO_2, is shown in Fig. 7.15. Zirconia has monoclinic, tetragonal, and cubic polymorphs, and its existence in any of five phases (three polymorphs plus liquid and vapor) means that the phase diagram contains 5!/3! = 20 triple points, five of which are shown in Fig. 7.15. The states *a, b,* and *c* are stable triple points for, respectively, the three-phase equilibria monoclinic-tetragonal-vapor, tetragonal-cubic-vapor, and cubic-liquid-vapor, and the states *d* and *e* are metastable triple points. The state *d* is that at which the extrapolated vapor pressure lines of the monoclinic and the cubic lines meet in the phase field of stable tetragonal ZrO_2. The state *d* is thus the metastable triple point for the equilibrium between vapor, monoclinic, and cubic zirconia, which occurs at a higher value of molar Gibbs free energy than that of tetragonal zirconia at the same value of *P* and *T.* Similarly the state *e,* which is that at which the extrapolated vapor pressures of tetragonal and liquid zirconia intersect in the phase field of stable cubic zirconia, is the metastable triple point for equilibrium between liquid, vapor, and tetragonal zirconia.

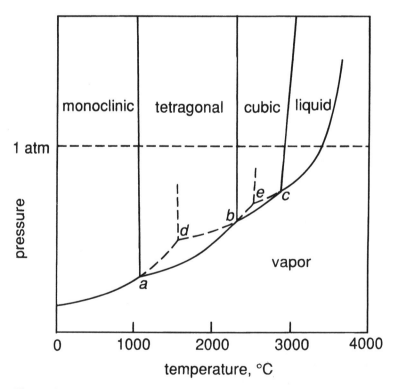

Figure 7.15 A schematic phase diagram for zirconia, ZrO_2.

7.8 SUMMARY

Knowledge of the dependencies, on temperature and pressure, of the changes in molar enthalpy and molar entropy caused by phase changes in a system allows determination of the corresponding change in the molar Gibbs free energy of the system. As a closed one-component system has only two independent variables, the dependence of G can be examined most simply by choosing T and P as the independent variables (these are the natural independent variables when G is the dependent variable). The phases in which the material can exist can thus be represented in a three-dimensional diagram using G, P, and T as coordinates, and in such a diagram the various states in which the material can exist occur as surfaces. In any state, which is determined by the values of P and T, the stable phase is that which has the lowest Gibbs free energy. The surfaces in the diagram intersect with one another along lines, and these lines represent the variations of P with T required for equilibrium between the two phases. The intersection of the surfaces for the solid and liquid phases gives the variation of the equilibrium melting temperature with pressure, and the intersection of the surfaces for the liquid and vapor phases gives the variation of the boiling temperature with pressure. Respectively, the normal melting and boiling points of the material occur on these intersections at $P = 1$ atm. Three surfaces intersect at a point in the diagram, and the values of P and T at which this intersection occurs are those of the invariant triple point at which an equilibrium occurs among three phases. In a one-component system no more than three phases can coexist in equilibrium with one another.

The three-dimensional G-T-P diagram illustrates the differences between stable, metastable, and unstable states and hence shows the difference between reversible and irreversible process paths. At any value of P and T the stable phase is that which has the lowest Gibbs free energy, and phases which have higher values of G at the same values of P and T are metastable with respect to the phase of lowest value of G. Phases with a value of G at any combination of P and T which do not lie on a surface in the diagram are unstable. A reversible process path involving a change in P and/or T lies on a phase surface, and the state of a phase is changed reversibly only when, during the change, the state of the system does not leave the surface of the phase. If the process path leaves the phase surface then the change of state, which necessarily passes through nonequilibrium states, is irreversible.

As the perspective representation, in two dimensions, of a three-dimensional diagram is difficult, it is normal practice to present the phase diagram for a one-component system as the basal plane of the G-T-P diagram, i.e., a P-T diagram, onto which are projected the lines along which two surfaces intersect (equilibrium between two phases) and the points at which three surfaces intersect (equilibrium among three phases). Such a diagram contains areas in which a single phase is stable, which are separated by lines along which two phases exist at equilibrium, and points at the intersection of three lines at which three phases coexist in equilibrium. The lines for equilibrium between a condensed phase and the vapor phase are called vapor pressure lines, and they are exponential in form. In view of the fact that saturated vapor pressures can vary over several orders of magnitude, phase diagrams can often be presented in more useful form as plots of $\log p$ vs. $1/T$ than as plots of P vs. T.

The development of phase diagrams for one-component systems demonstrate the use of the Gibbs free energy as a criterion for equilibrium when T and P are chosen as the independent variables.

7.9 NUMERICAL EXAMPLES

Example 1

The vapor pressure of solid NaF varies with temperature as

$$\ln p \text{ (atm)} = \frac{-34{,}450}{T} - 2.01 \ln T + 33.74$$

and the vapor pressure of liquid NaF varies with temperature as

$$\ln p \text{ (atm)} = \frac{-31{,}090}{T} - 2.52 \ln T + 34.66$$

Calculate:

a. The normal boiling temperature of NaF
b. The temperature and pressure at the triple point
c. The molar heat of evaporation of NaF at its normal boiling temperature
d. The molar heat of melting of NaF at the triple point
e. The difference between the constant-pressure molar heat capacities of liquid and solid NaF

The phase diagram is shown schematically in Fig. 7.16.

(a) The normal boiling temperature, T_b, is defined as that temperature at which the saturated vapor pressure of the liquid is 1 atm. Thus from the equation for the vapor pressure of the liquid, T_b, is

$$\ln(1) = 0 = -\frac{31{,}090}{T_b} - 2.52 \ln T_b + 34.66$$

which has the solution

$$T_b = 2006 \text{ K}$$

(b) The saturated vapor pressures for the solid and liquid phases intersect at the triple point. Thus at the temperature, T_{tp}, of the triple point

$$-\frac{34{,}450}{T_{tp}} - 2.01 \ln T_{tp} + 33.74 = -\frac{31{,}090}{T_{tp}} - 2.52 \ln T_{tp} + 34.66$$

which has the solution

$$T_{tp} = 1239 \text{ K}$$

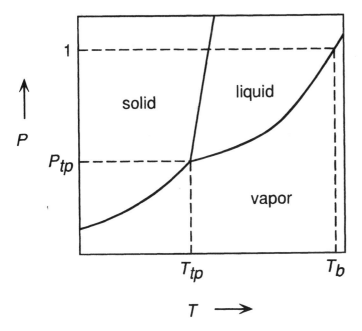

Figure 7.16 Schematic phase diagram for a one-component system.

The triple-point pressure is then calculated from the equation for the vapor pressure of the solid as

$$p = \exp\left(-\frac{34{,}450}{1239} - 2.01 \ln 1239 + 33.74\right) = 2.29 \times 10^{-4} \text{ atm}$$

or from the equation for the vapor pressure for the liquid as

$$p = \exp\left(-\frac{31{,}090}{1239} - 2.52 \ln 1239 + 34.66\right) = 2.29 \times 10^{-4} \text{ atm}$$

(c) For vapor in equilibrium with the liquid,

$$\ln p \text{ (atm)} = -\frac{31{,}090}{T} - 2.52 \ln T + 34.66$$

$$\frac{d \ln p}{dT} = \frac{\Delta H}{RT^2} = \frac{31{,}090}{T^2} - \frac{2.52}{T}$$

Thus

$$\Delta H_{(l \to v)} = (31{,}090 \times 8.3144) - (2.52 \times 8.3144)T = 258{,}500 - 20.95T$$

and, at the normal boiling temperature of 2006 K

$$\Delta H_{(l \to v)} = 258{,}500 - 20.95 \times 2006 = 216{,}500 \text{ J}$$

(d) For vapor in equilibrium with the solid

$$\ln p \text{ (atm)} = -\frac{34,450}{T} - 2.01 \ln T + 33.74$$

Thus

$$\Delta H_{(s \to v)} = (34,450 \times 8.3144) - (2.01 \times 8.3144)T$$
$$= 286,400 - 16.71T \text{ J}$$

At any temperature

$$\Delta H_{(s \to l)} + \Delta H_{(l \to v)} = \Delta H_{(s \to v)}$$

and thus

$$\Delta H_{(s \to l)} = 286,400 - 16.71T - 258,500 + 20.95T$$
$$= 27,900 + 4.24T$$

At the triple point

$$\Delta H_{(s \to l)} = 27,900 + (4.24 \times 1239) = 33,150 \text{ J}$$

(e) $\Delta H_{(s \to l)} = 27,900 + 4.24T$:

$$\frac{d\Delta H}{dT} = \Delta c_p = 4.24 \text{ J/K}$$
$$= c_{p,(l)} - c_{p,(s)}$$

Example 2

Carbon has three allotropes: graphite, diamond, and a metallic form called solid III. Graphite is the stable form of 298 K and 1 atm pressure, and increasing the pressure on graphite at temperatures less than 1440 K causes the transformation of graphite to diamond and then the transformation of diamond to solid III. Calculate the pressure which, when applied to graphite at 298 K, causes the transformation of graphite to diamond, given

$H_{298 \text{ K, (graphite)}} - H_{298 \text{ K, (diamond)}} = -1900 \text{ J}$
$S_{298 \text{ K, (graphite)}} = 5.74 \text{ J/K}$
$S_{298 \text{ K, (graphite)}} = 2.37 \text{ J/K}$
The density of graphite at 298 K is 2.22 g/cm^3
The density of diamond at 298 K is 3.515 g/cm^3

For the transformation graphite → diamond at 298 K,

$$\Delta G = \Delta H - T\Delta S$$
$$= 1900 - 298(2.37 - 5.74) = 2904 \text{ J}$$

For the transformation of graphite to diamond at any temperature T,

$$\left(\frac{\partial \Delta G_{\text{graphite}\rightarrow\text{diamond}}}{\partial P}\right)_T = \Delta V_{\text{graphite}\rightarrow\text{diamond}}$$

$$V_{\text{graphite}} = \frac{12}{2.22} = 5.405 \text{ cm}^3/\text{mole}$$

and

$$V_{\text{diamond}} = \frac{12}{3.515} = 3.415 \text{ cm}^3/\text{mole}$$

Thus

$$\Delta V = -1.99 \text{ cm}^3/\text{mole}$$

Equilibrium between graphite and diamond at 298 K requires that $\Delta G_{\text{graphite}\rightarrow\text{diamond}}$ be zero. As

$$\left(\frac{\partial \Delta G}{\partial P}\right)_T = \Delta V$$

then

$$\Delta G(P,T = 298) = \Delta G(P = 1, T = 298) + \int_1^P \Delta V dP$$

If the difference between the isothermal compressibilities of the two phases is negligibly small, i.e., if the influence of pressure on ΔV can be ignored, then, as

$$1 \text{ cm}^3 \cdot \text{atm} = \frac{8.3144}{82.057} = 0.1013 \text{ J}$$

$$\Delta G(P,T = 298) = 2904 + (-1.99 \times 0.1013)(P - 1)$$

and thus

$$(P - 1) = P = \frac{2904}{1.99 \times 0.1013} = 14{,}400 \text{ atm}$$

Transformation of graphite to diamond at 298 K requires the application of a pressure greater than 14,400 atm.

PROBLEMS

7.1 Using the vapor pressure-temperature relationships for $CaF_{2(\alpha)}$, $CaF_{2(\beta)}$, and liquid CaF_2, calculate:

a. The temperatures and pressures of the triple points for the equilibria $CaF_{2(\alpha)} - CaF_{2(\beta)} - CaF_{2(v)}$ and $CaF_{2(\beta)} - CaF_{2(l)} - CaF_{2(v)}$
b. The normal boiling temperature of CaF_2

    c. The molar latent heat of the transformation $CaF_{2(\alpha)} \rightarrow CaF_{2(\beta)}$

    d. The molar latent heat of melting of $CaF_{2(\beta)}$

7.2 Calculate the approximate pressure required to distill mercury at 100°C.

7.3 One mole of $SiCl_4$ vapor is contained at 1 atm pressure and 350 K in a rigid container of fixed volume. The temperature of the container and its contents is cooled to 280 K. At what temperature does condensation of the $SiCl_4$ vapor begin, and what fraction of the vapor has condensed when the temperature is 280 K?

7.4 The vapor pressures of zinc have been written as

$$\ln p \text{ (atm)} = -\frac{15,780}{T} - 0.755 \ln T + 19.25 \qquad (i)$$

and

$$\ln p \text{ (atm)} = -\frac{15,250}{T} - 1.255 \ln T + 21.79 \qquad (ii)$$

Which of the two equations is for solid zinc?

7.5 At the normal boiling temperature of iron, $T_b = 3330$ K, the rate of change of the vapor pressure of liquid iron with temperature is 3.72×10^{-3} atm/K. Calculate the molar latent heat of boiling of iron at 3330 K.

7.6 Below the triple point ($-56.2°C$) the vapor pressure of solid CO_2 is given as

$$\ln p \text{ (atm)} = -\frac{3116}{T} + 16.01$$

The molar latent heat of melting of CO_2 is 8330 joules. Calculate the vapor pressure exerted by liquid CO_2 at 25°C and explain why solid CO_2 is referred to as "dry ice."

7.7 The molar volumes of solid and liquid lead at the normal melting temperature of lead are, respectively, 18.92 and 19.47 cm^3. Calculate the pressure which must be applied to lead in order to increase its melting temperature by 20 centigrade degrees.

7.8 Nitrogen has a triple point at $P = 4650$ atm and $T = 44.5$ K, at which state the allotropes α, β, and γ coexist in equilibrium with one another. At the triple point $V_\beta - V_\alpha = 0.043$ cm^3/mole and $V_\alpha - V_\gamma = 0.165$ cm^3/mole. Also at the triple point $S_\beta - S_\alpha = 4.59$ J/K and $S_\alpha - S_\gamma = 1.25$ J/K. The state of $P = 1$ atm, $T = 36$ K lies on the boundary between the fields of stability of the α and β phases, and at this state, for the transformation of $\alpha \rightarrow \beta$, $\Delta S = 6.52$ J/K and $\Delta V = 0.22$ cm^3/mole. Sketch the phase diagram for nitrogen at low temperatures.

7.9 Measurements of the saturated vapor pressure of liquid $NdCl_5$ give 0.3045 atm at 478 K and 0.9310 atm at 520 K. Calculate the normal boiling temperature of $NdCl_5$.

Chapter 8

THE BEHAVIOR OF GASES

8.1 INTRODUCTION

Thus far frequent use has been made of the so-called ideal gas to illustrate the nature of changes in the thermodynamic state of a system. In this chapter the behavior of real gases is compared with ideal behavior, and the differences between the two are sought in the atomic or molecular properties of real gases. Although knowledge of the physical properties of a real gas is not required in a thermodynamic examination of the gas, an appreciation of the origin of physical properties provides a better understanding of the thermodynamic behavior.

8.2 THE *P-V-T* RELATIONSHIPS OF GASES

Experimental observation has shown that, for all real gases,

$$\frac{PV}{RT} \to 1_{\text{limit}P \to 0} \tag{8.1}$$

where

P is the pressure of the gas
V is the molar volume of the gas
R is the universal gas constant
T is the absolute temperature of the gas

Thus, as the pressure of the gas approaches zero, isotherms plotted on a P-V diagram approach the form of a rectangular hyperbola given by the equation

$$PV = RT \tag{8.2}$$

Eq. (8.2) is the equation of an ideal gas and is called the ideal gas law. A gas which obeys this law over a range of states is said to behave ideally in this range of states, and a gas which obeys this law in all states is called a perfect gas. The perfect gas is a convenient model with which the behavior of real gases can be compared.

The variation of V with P at several temperatures for a typical real gas is shown in Fig. 8.1. The figure shows that, as the temperature of the gas is decreased, the shape of the P-V isotherms changes, and, eventually, a value of $T = T_{\text{critical}}$ is reached at which, at some fixed pressure, P_{critical}, and fixed molar volume, V_{critical}, a horizontal inflexion occurs on the isotherm, i.e.,

$$\left(\frac{\partial P}{\partial V}\right)_{T_{cr}} = 0 \quad \text{and} \quad \left(\frac{\partial^2 P}{\partial V^2}\right)_{T_{cr}} = 0$$

At temperatures less than T_{cr} two phases can exist. For example, if 1 mole of vapor, initially in the state A (in Fig. 8.1), is isothermally compressed at T_8, the state of the vapor moves along the isotherm toward the state B. At B the pressure of the vapor is the saturated vapor pressure of the liquid at T_8, and further decrease in the volume of

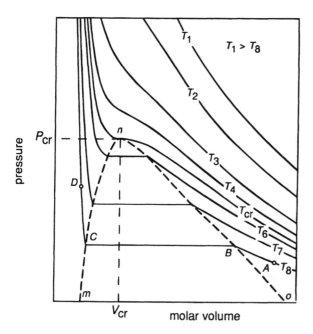

Figure 8.1 P-V isotherms for a typical real gas.

the system causes condensation of the vapor and consequent appearance of the liquid phase. The liquid phase, which is in equilibrium with the vapor, appears at the state C, and V_C is the molar volume of the liquid at P_C and T_8. Further decrease in the volume of the system causes further condensation, during which the states of the liquid and vapor phases remain fixed at C and B, respectively, and the total volume of the system, which is determined by the relative proportions of the liquid and vapor phases, moves along the horizontal line from B to C. Eventually condensation is complete, and the system exists as 100% liquid in the state C. Further increase in pressure moves the state of the system along the isotherm toward the state D. The large value of $-(\partial P/\partial V)_T$ in the range of liquid states and the small value of $-(\partial P/\partial V)_T$ in the range of vapor states indicate the low compressibility of the liquid phase and the high compressibility of the vapor phase.

Fig. 8.1 also shows that, as the temperature is increased up to T_{cr}, the molar volume of the liquid in equilibrium with the vapor (corresponding to the point C) progressively increases and the molar volume of the vapor in equilibrium with the liquid (corresponding to the point B) progressively decreases. Thus, as the temperature is increased toward T_{cr}, the vapor in equilibrium with liquid becomes more dense, and the liquid in equilibrium with the vapor becomes less dense. Eventually, when T_{cr} is reached, the molar volumes of the coexisting phases coincide at the state P_{cr}, T_{cr}. The critical point is thus the meeting point of the locus of the point C with temperature (the line mn) and the locus of the point B with temperature (the line on), and the complete locus line mno defines the field of vapor-liquid equilibrium.

At temperatures higher than T_{cr} distinct two-phase equilibrium (involving two phases separated by a boundary across which the properties of the system change abruptly) does not occur and thus the gaseous state cannot be liquified by isothermal compression at temperatures higher than T_{cr}. As the vapor can be condensed by isothermal compression at temperatures lower than T_{cr}, the critical isotherm provides a distinction between the gaseous and vapor states and defines the gaseous state phase field. The phase fields are shown in Fig. 8.2.

Liquefaction of a gas requires that the gas be cooled. Consider the process path $1 \rightarrow 2$ in Fig. 8.2. According to this path, which represents the cooling of the gas at constant pressure, the phase change gas $\rightarrow$ liquid occurs at the point a, at which the temperature falls below T_{cr}. In fact, at temperatures greater than T_{cr} the critical-temperature isotherm has no physical significance. In passing from the state 1 to the state 2 the molar volume of the system progressively decreases, and, hence, the density of the system progressively increases. No phase separation occurs between the states 1 and 2, and the system in the state 2 can equivalently be regarded as being a liquid of normal density or a gas of high density and, in state 1, can be regarded as being a gas of normal density or a liquid of low density. Physically, no distinction can be made between the liquid and gaseous states at pressures greater than P_{cr}, and consequently the system existing in these states is called a supercritical fluid. Thus, in the *P-T* phase diagram for the system (e.g., Fig. 7.10) the liquid-vapor equilibrium line (*OB* in Fig. 7.10) terminates at the critical point P_{cr}, T_{cr}.

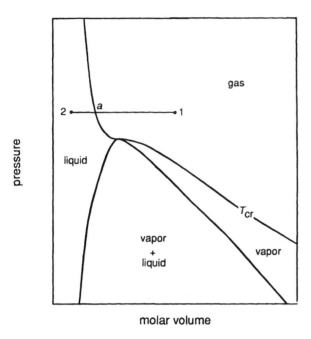

molar volume

Figure 8.2 The fields of phase stability of a typical real gas.

8.3 DEVIATION FROM IDEALITY AND EQUATIONS OF STATE FOR REAL GASES

The deviation of a real gas from ideal behavior can be measured as the deviation of the compressibility factor from unity. The compressibility factor, Z, is defined as

$$Z = \frac{PV}{RT} \tag{8.3}$$

which has the value of 1 for a perfect gas in all states of existence. Z itself is a function of the state of the system and, thus, is dependent on any two chosen dependent variables, e.g., $Z = Z(P,T)$. Fig. 8.3 shows the variation of Z with P at constant temperature for several gases. For all of the gases in Fig. 8.3 the Z is a linear function of P up to about 10 atm and, hence, can be expressed as

$$Z = mP + 1$$

or

$$\frac{PV}{RT} = mP + 1$$

which can be written as

$$P(V - mRT) = RT$$

or

$$P(V - b') = RT \tag{8.4}$$

where $b' = mRT$ and has the dimensions of volume.

Eq. (8.4) serves as the equation of state for the gases up to the pressures at which deviation from linear dependence of Z on P begins. Comparison with Eq. (8.4) shows that the deviations from ideal behavior, in the initial range of pressure, can be dealt with by making a correction to the volume term in the equation of state for an ideal gas. The need for such a correction is reasonable in view of the fact that an ideal gas is a system of noninteracting, volumeless particles, whereas the particles of real gases have small, but nevertheless finite, volumes. Thus, in a real gas, the volume available to the movement of Avogadro's number of particles is less than the molar volume of the gas by an amount equal to the volume excluded by the particles themselves, and the ideal gas equation should be corrected for this effect. At first sight it might appear that the constant b' in Eq. (8.4) is the volume excluded by the particles, but inspection of Fig. 8.3 shows that, with the exception of hydrogen, b' is a negative quantity. Thus the above interpretation of b' is incorrect, and Eq. (8.4) is a purely empirical equation which can be made to describe the behavior of real gases over a narrow range of low pressures in the vicinity of 0°C.

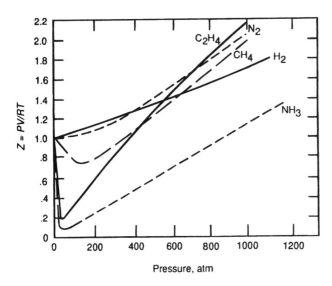

Figure 8.3 The variations, with pressure, of the compressibility factors of several gases at 0°C.

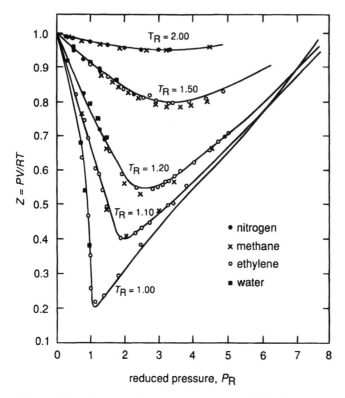

Figure 8.4 The variations of the compressibility factors of several gases with reduced pressure at several reduced temperatures.

If Fig. 8.3 is replotted as Z versus the reduced pressure, P_R (where $P_R = P/P_{cr}$) for fixed values of the reduced temperature, T_R ($= T/T_{cr}$), it is found that all gases lie on a single line. Fig. 8.4 shows a series of such plots. The behavior shown in Fig. 8.4 gives rise to the law of corresponding states, which states that all gases obey the same equation of state when expressed in terms of the reduced variables P_R, T_R, and V_R instead of P, T, and V. If the values of two reduced variables are identical for two gases then the gases have approximately equal values of the third reduced variable and are then said to be in corresponding states. Fig. 8.4 shows that the compressibility factor is the same function of the reduced variables for all gases (see Prob. 8.1).

8.4 THE VAN DER WAALS GAS

An ideal gas obeys the ideal gas law and has an internal energy, U, which is a function only of temperature. Consequently, an ideal gas is an assemblage of volumeless noninteracting particles, the energy of which is entirely the translational energy of motion of the constituent particles. Attempts to derive equations of state for real gases have attempted to modify the ideal gas equation by taking into consideration the facts that

1. The particles of a real gas occupy a finite volume and
2. The particles of a real gas are surrounded by force fields which cause them to interact with one another.

The magnitude of the importance of these two considerations depends on the state of the gas. For example, if the molar volume of the gas is large, then the volume fraction occupied by the particles themselves is small, and the magnitude of this effect on the behavior of the gas will be correspondingly small. Similarly, as the molar volume increases, the average distance between the particle increases, and thus the effect of interactions between particles on the behavior of the gas decreases. For a fixed number of moles of gas, an increase in the molar volume corresponds to a decrease in the density, n/V', and such states of existence occur at low pressure and high temperature, as can be seen from the ideal gas equation, i.e.,

$$\frac{n}{V'} = \frac{P}{RT}$$

Thus, approach toward ideal behavior is to be expected as the pressure is decreased and the temperature is increased.

The most celebrated equation of state for nonideal gases, which was derived from considerations 1 and 2 above, is the van der Waals equation, which, for 1 mole of gas, is written as

$$\left(P + \frac{a}{V^2}\right)(V - b) = RT$$

where P is the measured pressure of the gas, a/V^2 is a correction term for the interactions which occur among the particles of the gas, V is the measured volume of the gas, and b is a correction term for the finite volume of the particles.* The term b is determined by considering a collision between two spherical particles. Two particles, of radius r, collide when the distance between their centers decreases to a value less than $2r$, and, as is shown in Fig. 8.5a, at the point of collision the particles exclude a volume of

$$\frac{4}{3}\pi(2r)^3$$

to all other particles. The volume excluded per particles is thus

$$\frac{1}{2} \times \frac{4}{3}\pi(2r)^3 = 4 \times \frac{4}{3}\pi r^3$$

$$= 4 \times \text{the volume of one particle}$$

---

*For n moles of van der Waals gas, the equation of state is $\left(P + \dfrac{n^2 a}{V'^2}\right)(V' - nb) = nRT$ where $V' = nV$.

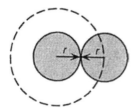

Figure 8.5 (a) Illustration of the volume excluded when two spherical atoms collide.

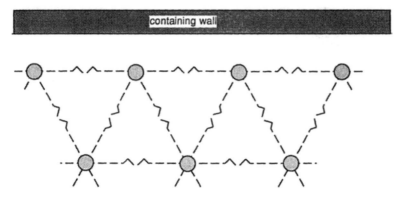

Figure 8.5 (b) The interactions among atoms in a gas phase.

The volume excluded is thus four times the volume of all of the particles present and has the value b. Thus in 1 mole of gas, the volume $(V - b)$ is that available for motion of the particles of the gas and is the molar volume which the gas would have were the gas ideal, i.e., if the particles were volumeless. The long-range attractive forces operating between the gas particles decrease the pressure exerted on the containing wall to a value less than that which would be exerted in the absence of the forces. van der Waals considered the following: The particles in the "layer" adjacent to the containing wall experience a net inward pull due to interaction with the particles in the next adjacent "layer." These attractive forces give rise to the phenomenon of "internal pressure," and the magnitude of the net inward pull (i.e., the decrease in the pressure exerted by the gas on the containing wall) is proportional to the number of particles in the "surface layer" and to the number of particles in the "next-to-the-surface layer." Both of these quantities are proportional to the density of the gas, n/V, and hence the net inward pull is proportional to the square of the density of the gas, or, for 1 mole of gas, equal to a/V^2, where a is a constant. Thus, if P is the measured pressure of the gas, $P + a/V^2$ is the pressure which the gas would exert on the containing wall if the gas were ideal, i.e., in the absence of interactions among the particles. The effect is illustrated in Fig. 8.5b.

   The van der Waals equation can be written as

$$PV^3 - (Pb + RT)V^2 + aV - ab = 0$$

which, being cubic in V, has three roots. Plotting V as a function of P for different values of T gives the family of isotherms shown in Fig. 8.6. As the temperature is increased the minimum and the maximum approach one another until, at T_{cr}, they coincide and produce a horizontal inflexion on the P-V curve. At this, the critical, point $T = T_{cr}$, $P = P_{cr}$, and $V = V_{cr}$, and the van der Waals equation gives

$$P_{cr} = \frac{RT_{cr}}{(V_{cr} - b)} - \frac{a}{V_{cr}^2}$$

$$\left(\frac{\partial P}{\partial V}\right)_{T_{cr}} = \frac{-RT_{cr}}{(V_{cr} - b)^2} + \frac{2a}{V_{cr}^3} = 0$$

$$\left(\frac{\partial^2 P}{\partial V^2}\right)_{T_{cr}} = \frac{2RT_{cr}}{(V_{cr} - b)^3} - \frac{6a}{V_{cr}^4} = 0$$

Solving these equations gives

$$T_{cr} = \frac{8a}{27bR}, \quad V_{cr} = 3b, \quad P_{cr} = \frac{a}{27b^2} \tag{8.5}$$

and hence the constants a and b for any gas can be evaluated from knowledge of the values of T_{cr} and P_{cr}. Alternatively, the values of a and b can be obtained by fitting the van der Waals equation to experimentally measured variations of V with T and P for real gases. The critical states, van der Waal constants, and values of Z at the critical point for several gases are listed in Table 8.1.

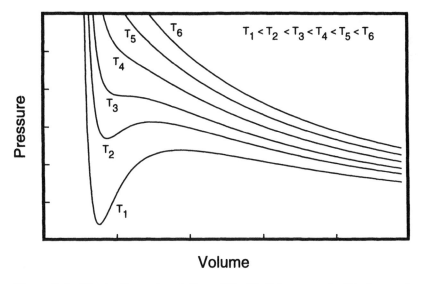

Figure 8.6 The isothermal variation of V with P for a van der Waals gas at several temperatures.

Table 8.1 The critical states, van der Waals constants, and values of Z at the critical points for several gases

Gas	T_{cr},K	P_{cr},atm	V_{cr},cm^3/mole	$a, \dfrac{l^2 \cdot atm}{mole^2}$	b,liters/mole	Z_{cr}
He	5.3	2.26	57.6	0.0341	0.0237	0.299
H$_2$	33.3	12.8	65.0	0.2461	0.0267	0.304
N$_2$	126.1	33.5	90.0	1.39	0.0391	0.292
CO	134.0	35.0	90.0	1.49	0.0399	0.295
O$_2$	153.4	49.7	74.4	1.36	0.0318	0.293
CO$_2$	304.2	73.0	95.7	3.59	0.0427	0.280
NH$_3$	405.6	111.5	72.4	4.17	0.0371	0.243
H$_2$O	647.2	217.7	45.0	5.46	0.0305	0.184

Consider the isothermal variation of V with P given by the van der Waals equation and shown in Fig. 8.7. Any increase in the pressure exerted on a system causes a decrease in the volume of the system, $(\partial P/\partial V)_T < 0$. This is a condition of intrinsic stability and, in Fig. 8.7, this condition is violated over the portion JHF, which means that this portion of the curve has no physical significance. The effect of pressure on the equilibrium state of the system can be obtained from a consideration of the variation of the Gibbs free energy with P along the isotherm. Eq. (5.12) gives the varia-

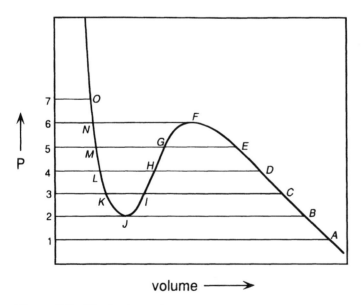

Figure 8.7 The isothermal variation, with pressure, of the volume of a van der Waals gas at a temperature below the critical temperature.

tion of G with P at constant T as $dG = VdP$, and integration of this equation between the state (P,T) and (P_A,T) gives

$$G(P,T) - G(P_A,T) = \int_{P_A}^{P} VdP$$

or

$$G = G_A + \int_{P_A}^{P} VdP \tag{8.5a}$$

If an arbitrary value is assigned to G_A, then graphical integration of the integral from Fig. 8.7 allows the variation of G with P, corresponding to the variation of V with P in Fig. 8.7, to be drawn. The values of the integrals are listed in Table 8.2, and the variation of G with P is shown in Fig. 8.8.

Fig. 8.8 shows that, as the pressure is increased from P_1, the value of G increases. At pressures greater than P_2 three states of existence become available to the system; for example, at P_3 the three states are given by the points I, K, and C. The stable, or equilibrium, state is that with the lowest Gibbs free energy, and hence over the range of pressure from P_2 to P_4 the stable states lie on the line BCD. As the pressure is increased above P_4 the state with the lowest Gibbs free energy no longer lies on the original line (the continuation of the line BCD) but lies on the line LMN. The change of stability at P_4 corresponds to a change of phase at this point, i.e., at pressures less than P_4 one phase is stable, and at pressures greater than P_4 another phase is stable. At low pressures ($P < P_4$), the system exists as a vapor, and at high pressures ($P > P_4$), it exists as a liquid. At P_4 G_D, which is the molar Gibbs free energy of the vapor phase, equals G_L, which is the molar Gibbs free energy of the liquid phase, and thus vapor and liquid coexist in equilibrium with one another at the state P_4,T. In Fig. 8.7 a tie-line connects the points D and L across a two-phase region. In

Table 8.2 Graphical integration of Fig. 8.7

$G_B = G_A + \int_{P_A}^{P_B} VdP$	$= G_A + $ area $1AB2$
G_C	$= G_A + $ area $1AC3$
G_D	$= G_A + $ area $1AD4$
G_E	$= G_A + $ area $1AE5$
G_F	$= G_A + $ area $1AF6$
G_G	$= G_A + $ area $1AE5 + $ area EFG
G_H	$= G_A + $ area $1AD4 + $ area DFH
G_I	$= G_A + $ area $1AC3 + $ area CFI
G_J	$= G_A + $ area $1AB2 + $ area BFJ
G_K	$= G_A + $ area $1AC3 + $ area $CFI - $ area IJK
G_L	$= G_A + $ area $1AD4 + $ area $DFH - $ area HJL
G_M	$= G_A + $ area $1AEF + $ area $EFG - $ area GJM
G_N	$= G_A + $ area $1AF6$ $\quad - $ area FJN
G_O	$= G_A + $ area $1AF6$ $\quad - $ area $FJN + $ area $6NO7$

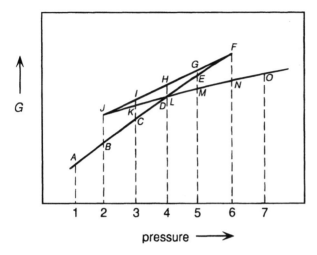

Figure 8.8 Schematic representation of the variation, with pressure, of the molar Gibbs free energy of a van der Waals gas at a constant temperature lower than the critical temperature.

Fig. 8.8 the lines *DF* and *LJ* represent, respectively, the metastable vapor and metastable liquid states. Thus, in the absence of nucleation of the liquid phase from the vapor phase at the state *D,* supersaturated vapor would exist along the line *DEF,* and, in the absence of nucleation of the vapor phase from the liquid phase at the state *L,* supersaturated liquid would exist along the line *LKJ.* In view of the violation of the criterion for intrinsic stability over the states path *JHF,* the states represented by this line in both Fig. 8.7 and Fig. 8.8 have no physical significance.

It is thus seen that the van der Waals equation predicts the phase change which occurs in the system at temperatures less than T_{cr}. At any temperature below T_{cr}, the value of *P* for equilibrium between the vapor and liquid phases, e.g., P_4 in Figs. 8.7 and 8.8, is that which the area *HFD* equals the area *LJH* in Fig. 8.7.

The measured values of T_{cr} and P_{cr} for carbon dioxide are, respectively, 31°C and 72.9 atm. Thus, from Eq. (8.5),

$$b = \frac{RT_{cr}}{8P_{cr}} = 0.0427 \text{ liters/mole}$$

and

$$a = 27b^2 P_{cr} = 3.59 \text{ liters}^2 \cdot \text{atm/mole}^2$$

in which case the van der Waals equation for CO_2 is given as

$$\left(P + \frac{3.59}{V^2}\right)(V - 0.0427) = RT$$

The variation of *P* with *V,* given by this equation, is shown at several temperatures in Fig. 8.9, in which it is seen that the 304 K isotherm exhibits a horizontal inflexion at

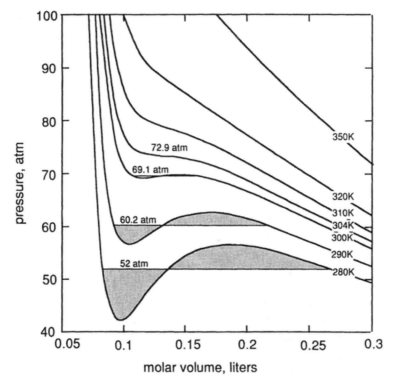

Figure 8.9 *P-V* isotherms for van der Waals carbon dioxide.

the critical point. At temperatures lower than 304 K the isotherms show the expected maxima and minima. The variation, with temperature, of the saturated vapor pressure of van der Waals liquid CO_2 can be determined by finding the tie-line on each isotherm which gives equal areas *DFH* and *LJH* as explained with reference to Fig. 8.7. Alternatively, the variation of the molar Gibbs free energy with pressure can be determined along each isotherm by graphical integration of the variation of *V* with *P*. These relationships are shown for several temperatures in Fig. 8.10, which shows the variation of the saturated vapor pressure of liquid CO_2 (the points *P*) with temperature. Fig. 8.10 also shows that, as the temperature increases toward the critical point, the range of nonphysical states (*J* to *F* in Fig. 8.8) diminishes and finally disappears at T_{cr}. At temperatures higher than T_{cr} the full-line indicates that one phase alone is stable over the entire range of pressure. As G_A in Eq. (8.5a) is a function of temperature, the positions of the isotherms in Fig. 8.10 with respect to one another are arbitrary; only the *P*-axis is quantitatively significant.

The variation of the saturated vapor pressure of liquid CO_2 with temperature, obtained from the van der Waals equation and plotted as the logarithm of *P* vs. the reciprocal of the absolute temperature, is shown in Fig. 8.11. Fig. 8.11 also shows the variation of the measured saturated vapor pressure with temperature. Comparison shows that the van der Waals equation predicts values of vapor pressure which

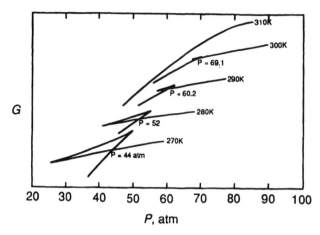

Figure 8.10   The variations of G with P for van der Waals carbon dioxide at several temperatures.

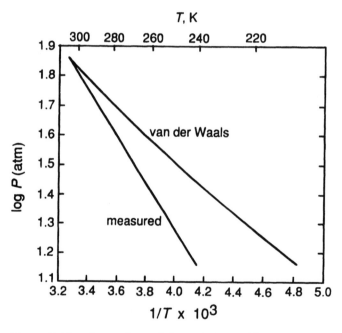

Figure 8.11   Comparison of the variation, with temperature, of the vapor pressure of van der Waals liquid carbon dioxide with the measured vapor pressures.

are higher than the measured values, although the difference between the two values decreases with increasing temperature. Consequently, the van der Waals equation predicts a value of the molar latent heat of evaporation of liquid CO_2 which is less than the measured value, with ΔH_{evap} being obtained as $-2.303\, R \times$ (the slope of the line) in Fig. 8.11. The molar latent heat of evaporation of a liquified van der Waals gas can be calculated as follows:

$$\Delta H_{evap} = H_v - H_l = U_v - U_l + P(V_v - V_l)$$

where V_v and V_l are, respectively, the molar volumes of the coexisting vapor and liquid phases, and P is the saturated vapor pressure, at the temperature T. From Eqs. (3.12) and (5.33),

$$\left(\frac{\partial U}{\partial V}\right)_T = T\left(\frac{\partial P}{\partial T}\right)_V - P$$

which, applied to the van der Waals gas, gives

$$\left(\frac{\partial U}{\partial V}\right)_T = T\left(\frac{R}{V - b}\right) - P = \frac{a}{V^2}$$

Integration gives

$$U = -\frac{a}{V} + \text{constant}$$

in which the integration constant is a function of temperature. Thus

$$\Delta H_{evap} = -\frac{a}{V_v} + \frac{a}{V_l} + P(V_v - V_l) \tag{8.6}$$

$$= -a\left(\frac{1}{V_v} - \frac{1}{V_l}\right) + P(V_v - V_l) \tag{8.7}$$

Eq. (8.7) thus correctly predicts that ΔH_{evap} for a van der Waals gas rapidly falls to zero as the temperature approaches T_{cr}, in which state $V_v = V_l$.

Although van der Waals developed his equation from a consideration of the physical factors causing nonideal behavior, the requirement that the pressure, volume, and temperature at the critical point be known for calculation of a and b means that the equation is empirical. This, however, does not detract from the usefulness of the equation in representing the behavior of a gas which exhibits a relatively small departure from ideality.

8.5 OTHER EQUATIONS OF STATE FOR NONIDEAL GASES

Other examples of derived equations of state for nonideal gases are the Dieterici equation

$$P(V - b')e^{a'/RTV} = RT$$

and the Berthelot equation

$$\left(P + \frac{A}{TV^2}\right)(V - B) = RT$$

Neither of these equations has a fundamental basis, and general empirical equations are normally used. Examples of such equations are the Beattie-Bridgeman equation, which contains five constants in addition to R and fits the P-V-T relationships over wide ranges of temperature and pressure, and the Kammerlingh-Onnes, or virial, equation of state. In the latter equation it is assumed that PV/RT is a power series of P or $1/V$, i.e.,

$$\frac{PV}{RT} = 1 + BP + CP^2 + \cdots$$

or

$$\frac{PV}{RT} = 1 + \frac{B'}{V} + \frac{C'}{V^2} + \cdots$$

The product PV is called the virial, B or B' is called the first virial coefficient, C or C' is called the second virial coefficient, etc., and the virial coefficients are functions of temperature. In both equations, as pressure approaches zero and volume approaches infinity, $PV/RT \rightarrow 1$. The virial equation converges rapidly in the gas phase, and thus the equation of state can be represented by the virial expansion over the entire range of densities and pressures. In practice, however, the virial equation is used only when the first few terms need to be retained. At low pressures or densities,

$$\frac{PV}{RT} = 1 + BP$$

or

$$\frac{PV}{RT} = 1 + \frac{B'}{V}$$

both of which are expressions of Eq. (8.4).

8.6 THE THERMODYNAMIC PROPERTIES OF IDEAL GASES AND MIXTURES OF IDEAL GASES

The variation of the molar Gibbs free energy of a closed system of fixed composition, with pressure at constant temperature, is given by the fundamental equation, Eq. (5.25) as

$$dG = VdP$$

For 1 mole of an ideal gas, this can be written as

$$dG = \frac{RT}{P} dP = RTd \ln P \tag{8.8}$$

and thus, for an isothermal change of pressure from P_1 to P_2 at T,

$$G(P_2,T) - G(P_1,T) = RT \ln \frac{P_2}{P_1} \tag{8.9}$$

As Gibbs free energies do not have absolute values (only changes in G can be measured), it is convenient to choose an arbitrary reference state from which the changes in Gibbs free energy can be measured. This reference state is called the *standard state* and is chosen as being the state of 1 mole of pure gas at 1 atm pressure and the temperature of interest. The Gibbs free energy of 1 mole of gas in the standard state $G(P = 1,T)$ is designated $G°(T)$, and thus, from Eq. (8.9), the Gibbs free energy of 1 mole of gas at any other pressure P is given as

$$G(P,T) = G°(T) + RT \ln P$$

or simply

$$G = G° + RT \ln P \tag{8.10}$$

In Eq. (8.10) notice that the logarithm of a dimensionless ratio, $P/1$, occurs in the right-hand term.

Mixtures of Ideal Gases

Before discussing the thermodynamic properties of mixtures of ideal gases, it is necessary to introduce the concepts of *mole fraction, partial pressure,* and *partial molar quantities.*

Mole fraction. When a system contains more than one component, i.e., when the composition of the system is variable, it is necessary to invent a means of expressing the composition. Several composition variables are in use, of which only one—the mole fraction—has any theoretical significance. The mole fraction, X_i, of the component i is defined as the ratio of the number of moles of i in the system to the total number of moles of all of the components in the system. For example, if the system contains n_A moles of A, n_B moles of B, and n_C moles of C, then

$$X_A = \frac{n_A}{n_A + n_B + n_C}$$

$$X_B = \frac{n_B}{n_A + n_B + n_C}$$

and

$$X_C = \frac{n_C}{n_A + n_B + n_C}$$

The convenience of the use of mole fraction as a composition variable lies in the fact that the sum of the mole fractions of all of the components in a system is unity. For example, in the above system $X_A + X_B + X_C = 1$.

Dalton's law of partial pressures. The pressure P exerted by a mixture of ideal gases is equal to the sum of the pressures exerted by each of the individual component gases, and the contribution made to the total pressure, P, by each individual gas is called the *partial pressure* of that gas. The partial pressure exerted by a component gas, p_i, is thus the pressure that it would exert if it alone were present. In a mixture of the ideal gases *A, B,* and *C,*

$$P = p_A + p_B + p_C$$

Consider a fixed volume V', at the temperature T, which contains n_A moles of an ideal gas A. The pressure exerted is thus

$$P = \frac{n_A RT}{V'} \tag{8.11}$$

If, to this constant volume containing n_A moles of gas A, n_B moles of ideal gas B are added, the pressure increases to

$$P = p_A + p_B = (n_A + n_B)\frac{RT}{V'} \tag{8.12}$$

Division of Eq. (8.11) by Eq. (8.12) gives

$$\frac{p_A}{p_A + p_B} = \frac{n_A}{n_A + n_B}$$

which, for the gas A in the mixture, can be written as

$$\frac{p_A}{P} = X_A$$

or

$$p_A = X_A P \tag{8.13}$$

Thus, in a mixture of ideal gases, the partial pressure of a component gas is the product of its mole fraction and the total pressure of the gas mixture. Eq. (8.13) is called Dalton's law of partial pressures.

Partial molar quantities. The molar value of any extensive state property of a component of a mixture is called the *partial molar value of the property.* This value is not necessarily equal to the value of the molar property of the pure component. The partial molar value of an extensive property Q of the component i in a mixture of components $i, j, k, \ldots$ is formally defined as

$$\bar{Q}_i = \left(\frac{\partial Q'}{\partial n_i}\right)_{T,P,n_j,n_k,\ldots} \tag{8.14}$$

where Q' is the value of the extensive property for an arbitrary quantity of the mixture.

$\bar{Q}_i$ is thus the rate of change of the value of Q' with n_i at constant temperature, pressure, and composition of the mixture, and, being a state property, is a function of temperature, pressure, and composition. The definition of $\bar{Q}_i$ can also be made as follows. If 1 mole of i is added, at constant temperature and pressure, to a quantity of mixture which is sufficiently large that the addition causes virtually no change in the composition of the mixture, the consequent increase in the value of Q' equals the value of $\overline{Q}_i$ in the mixture. In the case of the extensive property being the Gibbs free energy,

$$\overline{G}_i = \left(\frac{\partial G'}{\partial n_i}\right)_{T,P,n_j,n_k,\dots}$$

and, from Eq. (5.16), it is seen that

$$\overline{G}_i = \mu_i$$

i.e., the partial molar Gibbs free energy of a component in a mixture equals the chemical potential of the component in the mixture.

The relationships among the various state functions developed in the preceding chapters are applicable to the partial molar properties of the components of a system. For example, the fundamental equation, Eq. (5.25), at constant T and composition gives

$$\left(\frac{\partial G'}{\partial P}\right)_{T,\text{comp}} = V'$$

where G' is the Gibbs free energy of the system and V' is the volume of the system. For a variation in n_i, the number of moles of component i in the system, at constant T, P, and n_j,

$$\left[\frac{\partial}{\partial n_i}\left(\frac{\partial G'}{\partial P}\right)_{T,\text{comp}}\right]_{T,P,n_j} = \left(\frac{\partial V'}{\partial n_i}\right)_{T,P,n_j}$$

But, by definition

$$\left(\frac{\partial V'}{\partial n_i}\right)_{T,P,n_j} = \overline{V}_i$$

and, as G is a state function, in which case the order of partial differentiation has no influence on the result

$$\left[\frac{\partial}{\partial n_i}\left(\frac{\partial G'}{\partial P}\right)_{T,\text{comp}}\right]_{T,P,n_j} = \left[\frac{\partial}{\partial P}\left(\frac{\partial G'}{\partial n_i}\right)_{T,P,n_j}\right]_{T,\text{comp}}$$

$$= \left(\frac{\partial \overline{G}_i}{\partial P}\right)_{T,\text{comp}}$$

Hence

$$\left(\frac{\partial \overline{G}_i}{\partial P}\right)_{T,\text{comp}} = \overline{V}_i$$

which is simply the application of Eq. (5.25) to the component i in the system. Thus, for the ideal gas A in a mixture of ideal gases,

$$d\overline{G}_A = \overline{V}_A dP$$

The partial molar volume, $\overline{V}_A$, in a gas mixture is

$$\overline{V}_A = \frac{V'}{\sum n_i} = \frac{X_A RT}{p_A}$$

Differentiation of Eq. (8.13) at constant T and composition gives $dp_A = X_A dP$, and hence

$$d\overline{G}_A = \overline{V}_A dP = \frac{X_A RT}{p_A} \frac{dp_A}{X_A} = RTd(\ln p_A)$$

Integration between the limits $p_A = p_A$ and $p_A = 1$ gives

$$\overline{G}_A = G_A^\circ + RT \ln p_A \qquad (8.15)$$
$$= G_A^\circ + RT \ln X_A + RT \ln P$$

Eq. (8.15) could also have been obtained by integrating Eq. (8.8) from the standard state $p_A = P_A = 1$, $X_A = 1$, T to the state p_A, X_A, T.

The Heat of Mixing of Ideal Gases

For each component gas in a mixture of ideal gases

$$\overline{G}_i = G_i^\circ + RT \ln X_i + RT \ln P$$

where P is the total pressure of the gas mixture at the temperature T. Dividing by T and differentiating with respect to T at constant pressure and composition gives

$$\frac{\partial(\overline{G}_i/T)}{\partial T} = \frac{\partial(G_i^\circ/T)}{\partial T} \qquad (8.16)$$

But, from Eq. (5.37)

$$\left[\frac{\partial(G_i^\circ/T)}{\partial T}\right]_{P,comp} = -\frac{H_i^\circ}{T^2} \quad \text{and} \quad \left(\frac{\partial \overline{G}_i/T}{\partial T}\right)_{P,comp} = -\frac{\overline{H}_i}{T^2} \qquad (8.17)$$

and thus

$$\overline{H}_i = H_i^\circ \qquad (8.18)$$

i.e., the partial molar enthalpy of ideal gas i in a mixture of ideal gases equals the molar enthalpy of pure i, and thus the enthalpy of the gas mixture equals the sum of the enthalpies of the component gases before mixing, i.e.,

$$\Delta H'^{mix} = \sum_i n_i \overline{H}_i - \sum_i n_i H_i^\circ = 0 \qquad (8.19)$$

where $\Delta H'^{mix}$ is the change in the enthalpy caused by the process of mixing.

As G_i° is, by definition, a function only of temperature, then, from Eqs. (8.16) and (8.17) it is seen that $\overline{H}_i$ is a function only of temperature. Thus, in addition to being independent of composition, $\overline{H}_i$ is independent of pressure. The zero heat of mixing of ideal gases is a consequence of the fact that the particles of an ideal gas do not interact with one another.

The Gibbs Free Energy of Mixing of Ideal Gases

For each component gas i in a mixture of ideal gases,

$$\overline{G}_i = G_i^\circ + RT \ln p_i$$

and for each component gas before mixing

$$G_i = G_i^\circ + RT \ln P_i$$

where p_i is the partial pressure of i in the gas mixture and P_i is the pressure of the pure gas i before mixing.

The mixing process, being a change of state, can be written as

$$\text{unmixed components (state 1)} \rightarrow \text{mixed components (state 2)}$$

and

$$
\begin{aligned}
\Delta G(1 \rightarrow 2) &= G' \text{ (mixture)} - G' \text{ (unmixed components)} \\
&= \Delta G'^{\text{mix}} \\
&= \sum_i n_i \overline{G}_i - \sum_i n_i G_i \\
&= \sum_i n_i RT \ln \left(\frac{p_i}{P_i} \right)
\end{aligned}
\tag{8.20}
$$

The value of $\Delta G'^{\text{mix}}$ depends, thus, on the value of p_i and P_i for each gas. If, before mixing, the gases are all at the same pressure, i.e., if $P_i = P_j = P_k = \ldots$ and mixing is carried out at total constant volume such that the total pressure of the mixture, P_{mix}, equals the initial pressures of the gases before mixing, then, as $p_i/P_i = X_i$,

$$\Delta G'^{\text{mix}} = \sum_i n_i RT \ln X_i \tag{8.21}$$

As the values of X_i are less than unity, $\Delta G'^{\text{mix}}$ is a negative quantity, which corresponds with the fact that the mixing of gases is a spontaneous process.

The Entropy of Mixing of Ideal Gases

As $\Delta H'^{\text{mix}} = 0$ and

$$\Delta G'^{\text{mix}} = \Delta H'^{\text{mix}} - T\Delta S'^{\text{mix}}$$

then

$$\Delta S'^{\text{mix}} = -\sum_i n_i R \ln \left(\frac{p_i}{P_i} \right) \tag{8.22}$$

or, if $P_i = P_j = P_k = \ldots = P$ then

$$\Delta S'^{\text{mix}} = -\sum_i n_i R \ln X_i \tag{8.23}$$

which is seen to be positive, in accord with the fact that the mixing of gases is a spontaneous process.

8.7 THE THERMODYNAMIC TREATMENT OF NONIDEAL GASES

Eq. (8.10) showed that, at any temperature, the molar Gibbs free energy of an ideal gas is a linear function of the logarithm of the pressure of the gas. This property arises from the ideal gas law which was used in the derivation of Eq. (8.10), and thus, if the gas is not ideal, then the relationship between the logarithm of the pressure of the gas and its molar Gibbs free energy is not linear. However, in view of the simple form of Eq. (8.10), a function is invented which, when used in place of pressure in Eq. (8.10), gives a linear relationship between the molar Gibbs free energy of an nonideal gas and the logarithm of the function. This function is called the *fugacity, f,* and is partially defined by the equation

$$dG = RTd \ln f$$

The integration constant is chosen such that the fugacity approaches the pressure as the pressure approaches zero, i.e.,

$$\frac{f}{P} \rightarrow 1 \quad \text{as} \quad P \rightarrow 0$$

in which case

$$G = G^\circ + RT \ln f \tag{8.24}$$

where G° is the molar Gibbs free energy of the gas in its standard state, which is now defined as that state in which $f = 1$ at the temperature T. (The standard state for an ideal gas was defined as being $P = 1, T$.)

Consider a gas which obeys the equation of state

$$V = \frac{RT}{P} - \alpha$$

where α is a function only of temperature and is a measure of the deviation of the gas from ideality. Eq. (5.12) gives $dG = VdP$ at constant T, and Eq. (8.24) gives $dG = RTd \ln f$ at constant T. Thus, at constant T,

$$VdP = RTd \ln f$$

and hence

$$d \ln\left(\frac{f}{P}\right) = -\frac{\alpha}{RT}dP \tag{8.25}$$

Integration between the states $P = P$ and $P = 0$, at constant T gives

$$\ln\left(\frac{f}{P}\right)_{P=P} - \ln\left(\frac{f}{P}\right)_{P=0} = -\frac{\alpha P}{RT} \tag{8.26}$$

As $f/P = 1$ when $P = 0$, then $\ln (f/P) = 0$ when $P = 0$, and hence

$$\ln\left(\frac{f}{P}\right) = -\frac{\alpha P}{RT} \quad \text{or} \quad \frac{f}{P} = e^{-\alpha P/RT}$$

In order that α can be taken as being independent of pressure, the deviation of the gas from ideality must be small, in which case α is a small number. Thus,

$$e^{-\alpha P/RT} \approx 1 - \frac{\alpha P}{RT}$$

and hence

$$\frac{f}{P} = 1 - \frac{\alpha P}{RT} = 1 - \left(\frac{RT}{P} - V\right)\frac{P}{RT} = \frac{PV}{RT}$$

If the gas behaved ideally, then the ideal pressure, P_{id}, would be given as RT/V. Thus

$$\frac{f}{P} = \frac{P}{P_{id}} \qquad (8.27)$$

which shows that the actual pressure of the gas is the geometric mean of its fugacity and the pressure which it would exert if it behaved ideally. It is also seen that the percentage error involved in assuming that the fugacity is equal to the pressure is the same as the percentage departure from the ideal gas law.

Alternatively, the fugacity can be considered in terms of the compressibility factor Z. From Eq. (8.25)

$$d \ln\left(\frac{f}{P}\right) = -\frac{\alpha}{RT}dP = \left(\frac{V}{RT} - \frac{1}{P}\right)dP$$

But $Z = PV/RT$, and hence

$$d \ln\left(\frac{f}{P}\right) = \frac{Z-1}{P}dP$$

and

$$\ln\left(\frac{f}{P}\right)_{P=P} = \int_{P=0}^{P=P} \frac{Z-1}{P}dP \qquad (8.28)$$

This can be evaluated either by graphical integration of a plot of $(Z-1)P$ vs. P at constant T or by direct integration if Z is known as a function of P, i.e., if the virial equation of state of the gas is known.

For example, the variation of PV (cm^3·atm) with P in the range 0–200 atm for nitrogen gas at 0°C is represented by the equation

$$PV = 22{,}414.6 - 10.281P + 0.065189P^2 + 5.1955 \times 10^{-7}P^4$$
$$- 1.3156 \times 10^{-11}P^6 + 1.009 \times 10^{-16}P^8$$

Thus, dividing by $RT = 22{,}414.6$ at 0°C gives

$$\frac{PV}{RT} = Z = 1 - 4.5867 \times 10^{-4}P + 2.9083 \times 10^{-6}P^2 + 2.3179 \times 10^{-11}P^4$$
$$- 5.8694 \times 10^{-15}P^6 + 4.5015 \times 10^{-21}P^8$$

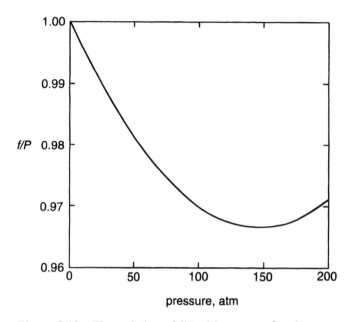

Figure 8.12   The variation of f/P with pressure for nitrogen gas at 0°C.

This variation of Z with P is shown graphically in Fig. 8.3. From integration of Eq. (8.28), $\ln\,(f/P)$ is obtained as

$$\ln\left(\frac{f}{P}\right) = -4.5867 \times 10^{-4}P + 1.4542 \times 10^{-6}P^2 + 5.7948 \times 10^{-12}P^4$$

$$- \, 0.9782 \times 10^{-16}P^6 + 5.627 \times 10^{-22}P^8$$

This variation of f/P with P is shown in Fig. 8.12.

The change in the molar Gibbs free energy of an nonideal gas caused by an isothermal change in pressure can be calculated from either

$$dG = VdP$$

or

$$dG = RTd \ln f$$

The correspondence between these two approaches is illustrated as follows. The virial equation of state of the gas is

$$\frac{PV}{RT} = Z = 1 + BP + CP^2 + DP^3 + \cdots$$

Then

$$V = RT\left(\frac{1}{P} + B + CP + DP^2 + \cdots\right)$$

and so, for the change of state of 1 mole of gas from (P_1, T) to (P_2, T),

$$\Delta G = \int_{P_1}^{P_2} V dP = RT \int_{P_1}^{P_2} \left(\frac{1}{P} + B + CP + DP^2 + \cdots \right) dP$$

$$= RT \left[\ln \frac{P_2}{P_1} + B(P_2 - P_1) + \frac{C}{2}(P_2^2 - P_1^2) + \frac{D}{3}(P_2^3 - P_1^3) + \cdots \right]$$

If the gas had been ideal, then

$$\Delta G = RT \ln \left(\frac{P_2}{P_1} \right)$$

and so the contribution to change in the molar Gibbs free energy arising from the nonideality of the gas is

$$RT \left[B(P_2 - P_1) + \frac{C}{2}(P_2^2 - P_1^2) + \frac{D}{3}(P_2^3 - P_1^3) + \cdots \right]$$

Alternatively, $dG = RT\, d \ln f$ where, from Eq. (8.29),

$$\ln \left(\frac{f}{P} \right) = \int_0^P \frac{Z - 1}{P} dP$$

$$= \int_0^P (B + CP + DP^2 + \cdots)dP$$

$$= BP + \frac{CP^2}{2} + \frac{DP^3}{3} + \cdots$$

Now

$$dG = RT\, d \ln f = RT\, d \ln \left(\frac{f}{P} \right) + RT\, d \ln P$$

and so

$$\Delta G = RT \left[B(P_2 - P_1) + \frac{C}{2}(P_2^2 - P_1^2) + \frac{D}{3}(P_2^3 - P_1^3) + \cdots \right] + RT \ln \left(\frac{P_2}{P_1} \right)$$

in agreement with the above.

Thus for 1 mole of nitrogen at 0°C, the difference between the Gibbs free energy at $P = 150$ atm and that at $P = 1$ atm is

$$\Delta G = RT \left[\ln \left(\frac{f}{P} \right)_{150} - \ln \left(\frac{f}{P} \right)_1 \right] + RT \ln 150$$

$$= 8.3144 \times 273(-0.034117 + 0.000457) + 8.3144 \times 273 \times 5.011 \text{ joules}$$

$$= -76 + 11{,}373$$

$$= 11{,}297 \text{ joules}$$

The contribution due to the nonideality of nitrogen is thus seen to be only 76 joules in almost 11,300 joules.

The number of terms which must be retained in the virial equation depends on the magnitude of the range of pressure over which it must be applied. For example, in the virial equation for nitrogen at 0°C only the first term is needed up to 6 atm and only the first two terms are needed up to 20 atm. When only the first term is needed the expression is

$$\frac{PV}{RT} = 1 + BP$$

or

$$V = \frac{RT}{P} + BRT$$

and hence $-BRT = \alpha$ in Eq. (8.25) and α is a function only of temperature.

Consider a nonideal gas which obeys the equation of state $PV = RT(1 + BP)$. The work done by this nonideal gas in a reversible, isothermal expansion from P_1 to P_2 is the same as that done when an ideal gas is reversibly and isothermally expanded from P_1 to P_2 at the same temperature. However, the work done by the nonideal gas in a reversible, isothermal expansion from V_1 to V_2 is greater than that done when an ideal gas is reversibly and isothermally expanded from V_1 to V_2 at the same temperature. Consider why this is so.

For the ideal gas $V = RT/P$, and for the nonideal gas $V = RT/P + BRT$. Thus, on a P-V diagram any isotherm for the nonideal gas is displaced from the isotherm for the ideal gas by the constant increment in volume BRT, as shown in Fig. 8.13. Because of the constant displacement the area under the isotherm for the ideal gas between P_1 and P_2 (the area *abcd*) is the same as the area under the isotherm for the nonideal gas between the same pressures (the area *efgh*). Thus the same amount of work is done by both gases in expanding isothermally from P_1 to P_2.

For the ideal gas,

$$w_{\text{ideal gas}} = \int_{V_1}^{V_2} P dV = RT \ln\left(\frac{V_2}{V_1}\right) = RT \ln\left(\frac{P_1}{P_2}\right)$$

and for the nonideal gas

$$w_{\text{nonideal gas}} = \int_{V_1}^{V_2} P dV$$

but, as $V = RT/P + BRT$ and hence, at constant T, $dV = -RT(dP/P^2)$, then

$$w_{\text{nonideal gas}} = -\int_{P_1}^{P_2} RT\left(\frac{dP}{P}\right) = RT \ln\left(\frac{P_1}{P_2}\right) = w_{\text{ideal gas}}$$

However, as any isotherm for the nonideal gas also lies above the isotherm for the ideal gas (for a positive value of B), the work done by the nonideal gas in expanding isothermally and reversibly from V_1 to V_2 (the area *aijd*) is greater than that done by

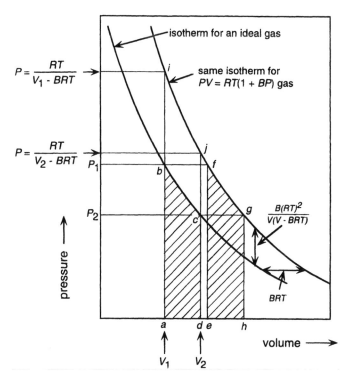

Figure 8.13   Isotherms for an ideal gas and a non-ideal gas.

the ideal gas in isothermally and reversibly expanding between V_1 and V_2 (the area *abcd*). The vertical separation between the two isotherms is

$$P_{\text{nonideal gas}} - P_{\text{ideal gas}} = \frac{RT}{V - BRT} - \frac{RT}{V} = \frac{B(RT)^2}{V(V - BRT)}$$

For the ideal gas, $w_{\text{ideal gas}} = RT \ln (V_2/V_1)$, and for the nonideal gas,

$$w = \int_{V_1}^{V_2} P dV$$

where

$$P = \frac{RT}{V - BRT}$$

such that

$$w_{\text{nonideal gas}} = RT \ln\left(\frac{V_2 - BRT}{V_1 - BRT}\right) > w_{\text{ideal gas}}$$

Consider the comparison of the behavior of hydrogen gas, for which $PV = RT(1 + 0.0064P)$, with that of an ideal gas in reversible isothermal expansions of 1 mole between $P_1 = 100$ atm and $P_2 = 50$ atm at 298 K:

$$V_{1(ideal, P_1 = 100 \text{ atm}, T = 298 \text{ K})} = \frac{RT}{P_1} = \frac{0.08206 \times 298}{100} = 0.2445 \text{ liters}$$

$$V_{1(H_2, P_1 = 100 \text{ atm}, T = 298 \text{ K})} = \frac{RT}{P_1} + RTB = 0.2445 + 0.08206 \times 298 \times 6.4 \times 10^{-4}$$

$$= 0.2445 + 0.0157 = 0.2602 \text{ liters}$$

$$V_{2(ideal, P_2 = 50 \text{ atm}, T = 298 \text{ K})} = \frac{0.08206 \times 298}{50} = 0.4890 \text{ liters}$$

$$V_{2(H_2, P_1 = 50 \text{ atm}, T = 298 \text{ K})} = 0.4890 + 0.0157 = 0.5047 \text{ liters}$$

Thus, for the change of state

$$(V_1 = 0.2445, T = 298 \text{ K}) \rightarrow (V_2 = 0.4890, T = 298 \text{ K})$$

$$w_{\text{ideal gas}} = RT \ln\left(\frac{V_2}{V_1}\right) = 8.3144 \times 298 \times \ln\left(\frac{0.4890}{0.2445}\right)$$

$$= 1717 \text{ joules}$$

and

$$w_{H_2} = RT \ln\left(\frac{V_2 - RTB}{V_1 - RTB}\right) = 8.3144 \times 298 \times \ln\left(\frac{0.4890 - 0.0157}{0.2445 - 0.0157}\right)$$

$$= 1801 \text{ joules}$$

At $V = 0.2445$ liters, $T = 298$ K, $P_{\text{ideal gas}} = 100$ atm, and

$$P_{H_2} = \frac{RT}{V - RTB} = \frac{0.08206 \times 298}{0.2445 - 0.0157}$$

$$= 106.4 \text{ atm}$$

and at $V = 0.489$ liters, $T = 298$ K, $P_{\text{ideal gas}} = 50$ atm, and

$$P_{H_2} = \frac{0.08206 \times 298}{0.4890 - 0.0157}$$

$$= 51.7 \text{ atm}$$

8.8 SUMMARY

An ideal gas is an assemblage of volumeless noninteracting particles which obeys the ideal gas law, $PV = RT$. The internal energy of an ideal gas arises solely from the translational motions of the gas particles and, hence, is a function only of temperature. The enthalpy of an ideal gas is also a function only of temperature. A consequence of the ideal gas law is that, at constant temperature, the Gibbs free energy of an ideal gas is a linear function of the logarithm of the pressure of the gas. As Gibbs free energies do not have absolute magnitudes (only differences can be measured) it is convenient to measure changes in Gibbs free energy from some arbitrary state. This state is chosen as $P = 1$ atm at the temperature of interest and is called the standard state. Thus, the difference between the molar Gibbs free energy in the state P, T and that in the standard state, $P = 1$ atm, T is $\Delta G = RT \ln P$.

The deviations of real gases from ideal behavior are caused by the atoms or molecules of real gases having finite volumes and by the interactions which occur among the atoms. Various attempts have been made to correct the ideal gas law for these effects, and the best-known derived equation is the van der Waals equation of state, which can be applied to gases which show small deviations from ideality. This equation predicts the condensation of vapor caused by compression at temperatures below the critical temperature, but does not give the correct dependence on temperature of the saturated vapor pressure of the liquid phase. Generally, measured variations of the molar volumes of gases with P and T are fitted to power series equations, in P or $1/V$, of the function PV. Such equations are called virial equations.

The compressibility factor, $Z = PV/RT$, of all real gases at constant reduced temperature, $T_R = T/T_{cr}$, is the same function of the reduced pressure, $P_R = P/P_{cr}$. This gives rise to the law of corresponding states, which states that when two gases have identical values of two reduced variables, they have almost identical values of the third reduced variable.

Consideration of the thermodynamic behavior of nonideal gases is facilitated by introduction of the fugacity, f, which is defined by the equation $dG = RT\, d \ln f$ and by the condition $f/P \to 1$ as $P \to 0$. Thus the standard state for a nonideal gas is that in which the fugacity is unity at the temperature of interest. For small deviations from ideality, the pressure of the gas is the geometric mean of its fugacity and P_{id}, the pressure which the gas would exert if it were ideal.

The composition of a mixture of gases is most conveniently expressed in terms of the mole fractions of its component gases, and if the mixture is ideal, the partial pressures exerted by the component gases are related to the total pressure P and the mole fraction X_i by $p_i = X_i P$. This equation is called Dalton's law of partial pressures. In a mixture of ideal gases the partial molar Gibbs free energy of a component gas is a linear function of the logarithm of its partial pressure, and in a mixture of nonideal gases is a linear function of the logarithm of its fugacity.

As the atoms in an ideal gas do not interact with one another no change in enthalpy occurs when different ideal gases are mixed, i.e., the enthalpy change of mixing of ideal gases, $\Delta H'^{\mathrm{mix}}$, is zero. The entropy change occurring when ideal gases are mixed arises solely from complete randomization of the different types of atoms in the available volume, and thus, as $\Delta H^{\mathrm{mix}} = 0$, $\Delta G^{\mathrm{mix}} = -T\Delta S^{\mathrm{mix}}$.

8.9 NUMERICAL EXAMPLES

Example 1

Assuming that nitrogen behaves as a van der Waals gas with $a = 1.39 \, l^2 \cdot atm/mole^2$ and $b = 39.1 \, cm^3/mole$, calculate the change in the Gibbs free energy and the change in entropy when the volume of 1 mole of nitrogen is increased from 1 to 2 liters at 400 K.

For a van der Waals gas

$$P = \frac{RT}{V - b} - \frac{a^2}{V^2}$$

and thus

$$
\begin{aligned}
dP &= -\frac{RTdV}{(V - b)^2} + \frac{RdT}{(V - b)} + \frac{2a^2 dV}{V^3} \\
&= \frac{RdT}{V - b} + \left[\frac{2a^2}{V^3} - \frac{RT}{(V - b)^2} \right] dV & \text{(i)} \\
&= \frac{RdT}{V - b} + \left[\frac{2a^2(V - b)^2 - RTV^3}{V^3(V - b)^2} \right] dV & \text{(ii)}
\end{aligned}
$$

At constant temperature

$$dG = VdP$$

which, from Eq. (i), gives

$$dG = \left[\frac{2a^2}{V^2} - \frac{RTV}{(V - b)^2} \right] dV$$

Integrating between V_2 and V_1 gives

$$
\begin{aligned}
\Delta G &= \int_{V_1}^{V_2} \left[\frac{2a^2}{V^2} - \frac{RTV}{(V - b)^2} \right] dV \\
&= \left[-\frac{2a^2}{V} + \frac{RTb}{V - b} - RT \ln(V - b) \right]_{V_1}^{V_2} \\
&= RTb \left(\frac{1}{V_2 - b} - \frac{1}{V_1 - b} \right) - RT \ln \frac{V_2 - b}{V_1 - b} - 2a^2 \left(\frac{1}{V_2} - \frac{1}{V_1} \right) \\
&= 0.08207 \times 400 \times 0.0391 \left(\frac{1}{2 - 0.0391} - \frac{1}{1 - 0.0391} \right) \\
&\quad - 0.082057 \times 400 \ln \frac{2 - 0.0391}{1 - 0.0394} - 2 \times 1.39^2 \left(\frac{1}{2} - \frac{1}{1} \right) \\
&= -0.68 - 23.41 - 1.93 \\
&= -26.02 \, l \cdot atm
\end{aligned}
$$

$$= -\frac{26.02 \times 8.3144}{0.082057} \text{ J}$$

$$= -2636 \text{ J}$$

From Eq. (6.17), at constant temperature,

$$dS = -\alpha V dP$$

where

$$\alpha = \frac{1}{V}\left(\frac{\partial V}{\partial T}\right)_P$$

From Eq. (ii), at constant pressure

$$\frac{dV}{dT} = \frac{R}{(V - b)} \frac{V^3(V - b)^2}{[2a^2(V - b)^2 - RTV^3]}$$

and thus, for a van der Waals gas

$$\alpha = \frac{RV^2(V - b)}{2a^2(V - b)^2 - RTV^3}$$

and

$$dS = -\left[\frac{RV^2(V - b)}{2a^2(V - b)^2 - RTV^3}\right]V\left[\frac{2a^2(V - b)^2 - RTV^3}{V^3(V - b)^2}\right]dV$$

$$= -\frac{RdV}{V - b}$$

Therefore,

$$\Delta S = R \ln\frac{V_2 - b}{V_1 - b}$$

$$= 8.3144 \ln\frac{2 - 0.0391}{1 - 0.0391}$$

$$= 5.93 \text{ J/K}$$

If the nitrogen had behaved as an ideal gas the changes in Gibbs free energy and entropy would have been

$$\Delta G = RT \ln\frac{V_1}{V_2}$$

$$= 8.3144 \times 400 \times \ln\frac{1}{2}$$

$$= -2305 \text{ J}$$

and

$$\Delta S = R \ln \frac{V_2}{V_1}$$
$$= 8.3144 \times \ln 2$$
$$= 5.76 \text{ J/K}$$

Example 2

The virial equation of state for n-butane at 460 K is $Z = 1 + A/V + B/V^2$ in which $A = -265 \text{ cm}^3/\text{g·mole}$ and $B = 30,250 \text{ cm}^6/\text{g·mole}^2$. Calculate the change in the Gibbs free energy when the volume of one mole of n-butane is decreased from 400 to 200 cm^3 at 460 K.

The equation of state is

$$PV = RT\left(1 - \frac{265}{V} + \frac{30,250}{V^2}\right)$$

$$\therefore P = RT\left(\frac{1}{V} - \frac{265}{V^2} + \frac{30,250}{V^3}\right)$$

and, at constant temperature

$$dP = RT\left(-\frac{1}{V^2} + \frac{2 \times 265}{V^3} - \frac{3 \times 30,250}{V^4}\right)dV$$

Thus,

$$dG = VdP = RT\left(-\frac{1}{V} + \frac{530}{V^2} - \frac{90,750}{V^3}\right)dV$$

and

$$\Delta G = RT\left[-\ln\frac{V_2}{V_1} - 530\left(\frac{1}{V_2} - \frac{1}{V_1}\right) + \frac{90,750}{2}\left(\frac{1}{V_2^2} - \frac{1}{V_1^2}\right)\right]$$

$$= 8.3144 \times 460\left[-\ln\frac{1}{2} - 530\left(\frac{1}{200} - \frac{1}{400}\right) + 45,375\left(\frac{1}{200^2} - \frac{1}{400^2}\right)\right]$$

$$= 8.3144 \times 460(0.693 - 1.325 + 0.851)$$

$$= 838 \text{ J}$$

PROBLEMS

8.1 Demonstrate the law of corresponding states by writing the van der Waals equation in terms of the reduced variables. Calculate the compressibility factor for a van der Waals gas at its critical point and compare the result with the values obtained for real gases at their critical points listed in Table 8.1 Calculate the value of $(\partial U/\partial V)_T$ for a van der Waals gas.

8.2 n moles of an ideal gas A and $(1 - n)$ moles of an ideal gas B, each at 1 atm pressure, are mixed at total constant pressure. What ratio of A to B in the mixture maximizes the decrease in the Gibbs free energy of the system? If the decrease in the Gibbs free energy is ΔG^M, to what value must the pressure be increased in order to increase the Gibbs free energy of the gas mixture by $\frac{1}{2}\Delta G^M$?

8.3 You are responsible for the purchase of oxygen gas which, before use, will be stored at a pressure of 200 atm at 300 K in a cylindrical vessel of diameter 0.2 meters and height 2 meters. Would you prefer that the gas behaved ideally or as a van der Waals gas? The van der Waals constants for oxygen are $a = 1.36$ liters^2·atm·mole^{-2} and $b = 0.0318$ liter/mole.

8.4 The virial equation of state for n-butane at 460 K is $Z = 1 + A/V + B/V^2$ in which $A = -265$ cm^3/g·mole and $B = 30{,}250$ cm^6/g·mole^2. Calculate the work required to reversibly compress 1 mole of n-butane from 50 to 100 atm at 460 K.

8.5 For sulfur dioxide, $T_{cr} = 430.7$ K and $P_{cr} = 77.8$ atm. Calculate

a   The critical van der Waals constants for the gas
b   The critical volume of van der Waals SO_2
c   The pressure exerted by 1 mole of SO_2 occupying a volume of 500 cm^3 at 500 K. Compare this with the pressure which would be exerted by an ideal gas occupying the same molar volume at the same temperature.

8.6 One hundred moles of hydrogen gas at 298 K are reversibly and isothermally compressed from 30 to 10 liters. The van der Waals constants for hydrogen are $a = 0.2461$ liters^2·atm mole^{-2} and $b = 0.02668$ 1/mole, and in the range of pressure 0–1500 atm, the virial equation for hydrogen is $PV = RT (1 + 6.4 \times 10^{-4} P)$. Calculate the work that must be done on the system to effect the required change in volume and compare this with the values that would be calculated assuming that (1) hydrogen behaves as a van der Waals gas and (2) hydrogen behaves as an ideal gas.

8.7 Using the virial equation of state for hydrogen at 298 K given in Prob. 8.6, calculate

a. The fugacity of hydrogen at 500 atm and 298 K
b. The pressure at which the fugacity is twice the pressure
c. The change in the Gibbs free energy caused by a compression of 1 mole of hydrogen at 298 K from 1 to 500 atm

What is the magnitude of the contribution to (c) caused by the nonideality of hydrogen?

Chapter 9

THE BEHAVIOR OF SOLUTIONS

9.1 INTRODUCTION

At high enough temperature and low enough pressure the interactions among the atoms or molecules in mixtures of gases are small enough that they can be neglected, in which case the mixture can be considered to be a mixture of ideal gases. Consequently, at high temperature and low pressure, all mixtures of gases have the same thermodynamic mixing properties. In contrast, strong interactions exist between atoms, molecules, or ions in condensed solutions, and the nature and magnitudes of these interactions have a significant influence on the thermodynamic behavior of the solution. The interactions, which are determined by such factors as atomic size, electronegativity, and electron-to-atom ratio, determine the extent to which a component is soluble in a solution and whether or not two or more components can react chemically to form a separate species. Solution thermodynamics is concerned with the vapor pressure-temperature-composition relationships of the components of a solution, and an examination of solution thermodynamics is made in the chapter.

9.2 RAOULT'S LAW AND HENRY'S LAW

If a quantity of pure liquid A is placed in a closed, initially evacuated vessel at the temperature T, it will spontaneously evaporate until the pressure in the vessel reaches the saturated vapor pressure of liquid A, p_A°, at the temperature T. In this state a dynamic equilibrium is established in which the rate of evaporation of liquid A equals the rate of condensation of A vapor. The rate of evaporation, $r_{e(A)}$, is determined by the magnitude of the energy of the bonds between the atoms of A at the surface of the liquid. The forces exerted between the atoms are such that each surface atom is located near the bottom of a potential energy well, and for an atom to leave the surface of the liquid and enter the vapor phase, it must acquire an activation energy, E^*. The intrinsic rate of evaporation, $r_{e(A)}$, is determined by the depth of the potential energy well, i.e., by the magnitude of E^*, and by the temperature T. On the other hand the rate of condensation, $r_{c(A)}$, is proportional to the number of A atoms in the vapor phase which strike the surface of the liquid in unit time. For a given temperature, this is proportional to the pressure of the vapor. Thus $r_{c(A)} = kp_A^\circ$, and at equilibrium

$$r_{e(A)} = kp_A^\circ \tag{9.1}$$

Eq. 9.1 illustrates why the saturated vapor pressures of liquids are exponential functions of temperature.* Similarly, when pure liquid B is placed in an initially evacuated vessel at the temperature T phase equilibrium between the liquid and its vapor occurs when

$$r_{e(B)} = k'p_B^\circ \tag{9.2}$$

Consider the effect of the addition of a small quantity of liquid B to liquid A. If the mole fraction of A in the solution is X_A and the atomic diameters of A and B are similar, then, assuming that the composition of the surface of the liquid is the same as that of the bulk liquid, the fraction of the surface sites occupied by A atoms is X_A. As A can only evaporate from surface sites occupied by A atoms, the rate of evaporation of A is decreased by the factor X_A, and as, at equilibrium, the rates of evaporation and condensation are equal to one another, the equilibrium vapor pressure of A exerted by the A–B solution is decreased from p_A° to p_A where

$$r_{e(A)}X_A = kp_A \tag{9.3}$$

Similarly, for liquid B containing a small amount of A,

$$r_{e(B)}X_B = k'p_B \tag{9.4}$$

Combination of Eqs. (9.1) and (9.3) gives

$$p_A = X_Ap_A^\circ \tag{9.5}$$

and combination of Eqs. (9.2) and (9.4) gives

$$p_B = X_Bp_B^\circ \tag{9.6}$$

---

*The energies of the atoms at the surface are quantized, and the distribution of the surface atoms among the available quantized energy levels is given by Eq. (4.13) as $n_i = n \exp(-E_i/kT)/P$, where n_i/n is the fraction of the atoms in the E_ith energy level, and P, the partition function, is given by $P = \sum_0^\infty \exp(-E_i/kT)$. If the quantized energy levels are spaced closely enough that the summation can be replaced by an integral, then $P = \int_0^\infty \exp(-E_i/kT)dE = kT$, which is thus the average energy per atom, and hence the fraction of surface atoms which have energies greater than the activation energy for evaporation, E^*, is $n_i^*/n = (1/kT) \int_{E^*}^\infty \exp(-E_i/kT)dE = \exp(-E^*/kT)$. From Eq. (9.1), as the evaporation rate, $r_{e(A)}$, is proportional to n_i^*/n, it is seen that the evaporation rate increases exponentially with increasing temperature and decreases exponentially with increasing value of E^*.

Eqs. (9.5) and (9.6) are expressions of *Raoult's law,* which states that the vapor pressure exerted by a component i in a solution is equal to the product of the mole fraction of i in the solution and the saturated vapor pressure of pure liquid i at the temperature of the solution. Raoult's law is shown in Fig. 9.1; the components of a solution which obeys Raoult's law are said to exhibit *Raoultian* behavior.

The derivations of Eqs. (9.3) and (9.4) required the assumption that the intrinsic rates of evaporation of A and B be independent of the composition of the solution. This requires that the magnitudes of the A–A, B–B, and A–B bond energies in the solution be identical, in which case the depth of the potential energy well of an atom at the surface is independent of the types of atoms which it has as nearest neighbors.

Consider the case in which the A–B bond energy is considerably more negative than the A–A and B–B bond energies, and consider a solution of A in B which is sufficiently dilute that every A atom on the surface of the liquid is surrounded only by B atoms. In this case the A atoms at the surface are each located in a deeper potential energy well than are the A atoms at the surface of pure A. Thus, in order to leave the surface and enter the vapor phase, the A atoms have to be lifted from deeper wells, and, consequently, the intrinsic rate of evaporation of A is decreased from $r_{e(A)}$ to $r'_{e(A)}$. Equilibrium between the condensed solution and the vapor phase occurs when

$$r'_{e(A)}X_A = kp_A \tag{9.7}$$

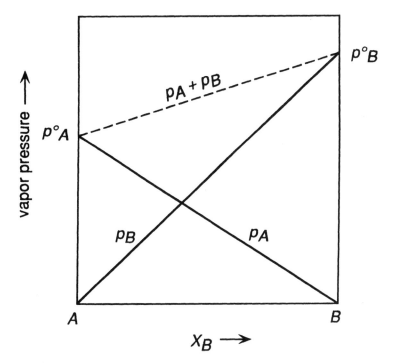

Figure 9.1 The vapor pressures exerted by the components of a binary Raoultian solution.

Combination of Eqs. (9.1) and (9.7) then gives

$$p_A = \frac{r'_{e(A)}}{r_{e(A)}} X_A p_A^\circ \qquad (9.8)$$

and as $r'_{e(A)} < r_{e(A)}$, p_A in Eq. (9.8) is a smaller quantity than p_A in Eq. (9.5). Eq. (9.8) can be written as

$$p_A = k_A X_A \qquad (9.9)$$

As the mole fraction of A in the A–B solution is increased, the probability that all of the A atoms on the surface of the liquid are surrounded only by B atoms decreases. The occurrence of a pair of neighboring A atoms on the surface decreases the depth of the potential wells in which they are located and hence increases the value of $r'_{e(A)}$. Beyond some critical mole fraction of A $r'_{e(A)}$ varies with composition, and, hence, Eq. (9.9) is no longer obeyed by A in solution. Consequently, Eq. (9.9) is obeyed only over an initial range of concentration of A in B, the extent of which is dependent on the temperature of the solution and on the relative magnitudes of the A–A, B–B, and A–B bond energies. A similar consideration of dilute solutions of B in A gives

$$p_B = k_B X_B \qquad (9.10)$$

which is obeyed over an initial range of concentration. Eqs. (9.9) and (9.10) are known as *Henry's law,* and, in the ranges of composition in which Henry's law is obeyed, the solutes are said to exhibit Henrian behavior. If the A–B bond energy is more negative than the A–A and B–B bond energies, then, as $r'_{e(A)} < r_{e(A)}$, the Henry's law line lies below the Raoult's law line. Conversely, if the A–B bond energy is less negative than the A–B and B–B bond energies, the solute atom, surrounded only by solvent atoms, is located in a shallower potential energy well than that which occurs in the pure solute. In this case $r'_{e(A)} > r_{e(A)}$, and hence, the Henry's law line lies above the Raoult's law line. Henrian behavior of the solute is illustrated in Figs. 9.2a and b.

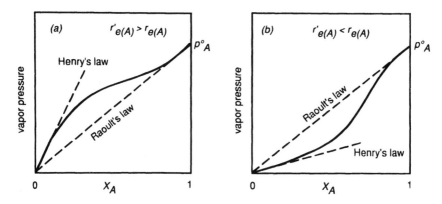

Figure 9.2 (*a*) The vapor pressure of a component of a binary solution which exhibits positive deviation from Raoultian behavior. (*b*) The vapor pressure of a component of a binary solution which exhibits negative deviation from Raoultian behavior.

9.3 THE THERMODYNAMIC ACTIVITY OF A COMPONENT IN SOLUTION

The thermodynamic activity of a component in any state at the temperature T is formally defined as being the ratio of the fugacity of the substance in that state to its fugacity in its standard state, i.e., for the species or substance i,

$$\text{activity of } i = a_i = \frac{f_i}{f^{\circ}_i} \tag{9.11}$$

In a condensed solution, f_i is the fugacity of the component i in the solution at the temperature T, and f°_i is the fugacity of pure i (the standard state) at the temperature T. If the vapor in equilibrium with the condensed solution is ideal, then $f_i = p_i$, in which case

$$a_i = \frac{p_i}{p^{\circ}_i} \tag{9.12}$$

i.e., the activity of i in a solution, with respect to pure i, is the ratio of the partial pressure of i exerted by the solution to the saturated vapor pressure of pure i at the same temperature. If the component i exhibits Raoultian behavior, then Eqs. (9.5) and (9.12) give

$$a_i = X_i \tag{9.13}$$

which is thus an alternative expression of Raoult's law. Fig. 9.3 shows Raoultian behavior in a binary solution in terms of the activities of the two components. The

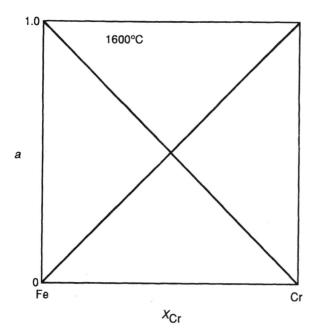

Figure 9.3   Activities in the binary system iron-chromium at 1600°C.

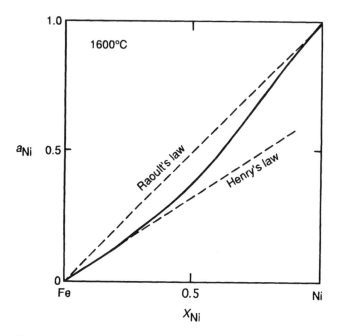

Figure 9.4   The activity of nickel in the system iron-nickel at 1600°C.

definition of activity thus normalizes the vapor pressure-composition relationship with respect to the saturated vapor pressure exerted in the standard state.

Similarly, over the composition range in which Henry's law is obeyed by the solute i, Eqs. (9.9) and (9.12) give

$$a_i = k_i X_i \tag{9.14}$$

which is an alternative expression of Henry's law. Henrian behavior, in terms of the activity of a component of a binary solution, is shown in Fig. 9.4.

9.4 THE GIBBS-DUHEM EQUATION

It is frequently found that the extensive thermodynamic properties of only one component of a binary (or multicomponent) solution are amenable to experimental measurement. In such cases the corresponding properties of the other component can be obtained from a general relationship between the values of the properties of both components. This relationship, which is known as the Gibbs-Duhem relationship, is introduced in this section, and some of its applications are discussed in Sec. 9.8.

The value of an extensive thermodynamic property of a solution is a function of the temperature, the pressure, and the numbers of moles of the components of the solution; i.e., if Q is an extensive molar property then

$$Q' = Q' (T, P, n_i, n_j, n_k, \dots)$$

Thus, at constant T and P the variation of Q with the composition of the solution is given as

$$dQ' = \left(\frac{\partial Q'}{\partial n_i}\right)_{T, P, n_j, n_k, \dots} dn_i + \left(\frac{\partial Q'}{\partial n_j}\right)_{T, P, n_i, n_k, \dots} dn_j + \left(\frac{\partial Q'}{\partial n_k}\right)_{T, P, n_i, n_j, \dots} dn_k + \cdots \quad (9.15)$$

In Chap. 8 the partial molar value of an extensive property of a component was defined as

$$\overline{Q}_i = \left(\frac{\partial Q'}{\partial n_i}\right)_{T, P, n_j, n_k, \dots}$$

in which case Eq. (9.15) can be written as

$$dQ' = \overline{Q}_i dn_i + \overline{Q}_j dn_j + \overline{Q}_k dn_k + \cdots \quad (9.16)$$

Also in Chap. 8 it was seen that $\overline{Q}_i$ is the increase in the value of Q' for the mixture or solution when 1 mole of i is added to a large quantity of the solution at constant T and P. (The stipulation that the quantity of solution be large is necessitated by the requirement that the addition of 1 mole of i to the solution should not cause a measurable change in its composition.) Thus, if $\overline{Q}_i$ is the value of Q per mole of i in the solution, then the value of Q' for the solution itself is

$$Q' = n_i \overline{Q}_i + n_j \overline{Q}_j + n_k \overline{Q}_k + \cdots \quad (9.17)$$

differentiation of which gives

$$dQ' = n_i d\overline{Q}_i + n_j d\overline{Q}_j + n_k d\overline{Q}_k + \cdots + \overline{Q}_i dn_i + \overline{Q}_j dn_j + \overline{Q}_k dn_k + \cdots \quad (9.18)$$

Comparison of Eqs. (9.16) and (9.18) shows that, at constant T and P,

$$n_i d\overline{Q}_i + n_j d\overline{Q}_j + n_k d\overline{Q}_k + \cdots = 0$$

or, generally,

$$\sum_i n_i d\overline{Q}_i = 0 \quad (9.19)$$

Division of Eq. (9.19) by n, the total number of moles of all the components of the solution, gives

$$\sum_i X_i d\overline{Q}_i = 0 \quad (9.20)$$

Eqs. (9.19) and (9.20) are equivalent expressions of the *Gibbs-Duhem* equation.

9.5 THE GIBBS FREE ENERGY OF FORMATION OF A SOLUTION

The Molar Gibbs Free Energy of a Solution and the Partial Molar Gibbs Free Energies of the Components of the Solution

In terms of the Gibbs free energy (as an extensive thermodynamic property), Eq. (9.17), for a binary A–B solution at fixed temperature and pressure, is

$$G' = n_A \overline{G}_A + n_B \overline{G}_B \tag{9.21}$$

where $\overline{G}_A$ and $\overline{G}_B$ are, respectively, the partial molar Gibbs free energies of A and B in the solution. Dividing both sides of Eq. (9.21) by $n_A + n_B$ gives the molar Gibbs free energy of the solution as

$$G = X_A \overline{G}_A + X_B \overline{G}_B \tag{9.22}$$

and differentiation gives

$$dG = X_A d\overline{G}_A + X_B d\overline{G}_B + \overline{G}_A dX_A + \overline{G}_B dX_B \tag{9.23}$$

However, as the Gibbs-Duhem equation gives $X_A d\overline{G}_A + X_B d\overline{G}_B = 0$, Eq. (9.23) becomes

$$dG = \overline{G}_A dX_A + \overline{G}_B dX_B \tag{9.24}$$

or

$$\frac{dG}{dX_A} = \overline{G}_A - \overline{G}_B \tag{9.25}$$

($X_A + X_B = 1$, and hence $dX_A = -dX_B$). Multiplication of Eq. (9.25) by X_B gives

$$X_B \frac{dG}{dX_A} = X_B \overline{G}_A - X_B \overline{G}_B \tag{9.26}$$

and addition of Eqs. (9.26) and (9.22) gives

$$G + X_B \frac{dG}{dX_A} = \overline{G}_A (X_A + X_B) \tag{9.27a}$$

or

$$\overline{G}_A = G + X_B \frac{dG}{dX_A}$$

Similarly

$$\overline{G}_B = G + X_A \frac{dG}{dX_B} \tag{9.27b}$$

These expressions relate the dependence on composition of the partial molar Gibbs free energies of the components of a binary solution and the molar Gibbs free energy of the solution.

The Change in Gibbs Free Energy Due to the Formation of a Solution

The pure component i, occurring in a condensed state at the temperature T, exerts an equilibrium vapor pressure p_i°, and when occurring in a condensed solution at the temperature T, it exerts a lower equilibrium pressure p_i. Consider the following isothermal three-step process:

a. The evaporation of 1 mole of pure condensed i to vapor i at the pressure p_i°
b. A decrease in the pressure of 1 mole of vapor i from p_i° to p_i
c. The condensation of 1 mole of vapor i from the pressure p_i to the condensed solution

The difference between the molar Gibbs free energy of i in the solution and the molar Gibbs free energy of pure i is given by the sum $\Delta G_{(a)} + \Delta G_{(b)} + \Delta G_{(c)}$. However, as steps (a) and (c) are processes conducted at equilibrium, $\Delta G_{(a)}$ and $\Delta G_{(c)}$ are both zero. The overall change in Gibbs free energy which accompanies the isothermal three-step process is thus $\Delta G_{(b)}$ which, from Eq. (8.9), is given as

$$\Delta G_{(b)} = RT \ln \left(\frac{p_i}{p_i^\circ} \right)$$

and, from Eq. (9.12), this can be written as

$$\Delta G_{(b)} = G_i(\text{in solution}) - G_i(\text{pure}) = RT \ln a_i$$

But G_i (in solution) is simply the partial molar Gibbs free energy of i in the solution, and G_i (pure) is the molar Gibbs free energy of pure i. The difference between the two is the change in the Gibbs free energy accompanying the solution of 1 mole of i in the solution. This quantity is designated as $\Delta \overline{G}_i^M$, the partial molar Gibbs free energy of the solution of i. Thus

$$\Delta \overline{G}_i^M = \overline{G}_i - G_i^\circ = RT \ln a_i \tag{9.28}$$

If n_A moles of A and n_B moles of B are mixed to form a solution at constant temperature and pressure,

$$\text{the Gibbs free energy before mixing} = n_A G_A^\circ + n_B G_B^\circ$$

and

$$\text{the Gibbs free energy after mixing} = n_A \overline{G}_A + n_B \overline{G}_B$$

The change in the Gibbs free energy caused by the mixing process, $\Delta G'^M$, referred to as the integral Gibbs free energy of mixing, is the difference between these quantities, i.e.,

$$\Delta G'^M = (n_A \overline{G}_A + n_B \overline{G}_B) - (n_A G_A^\circ + n_B G_B^\circ)$$
$$= n_A(\overline{G}_A - G_A^\circ) + n_B(\overline{G}_B - G_B^\circ)$$

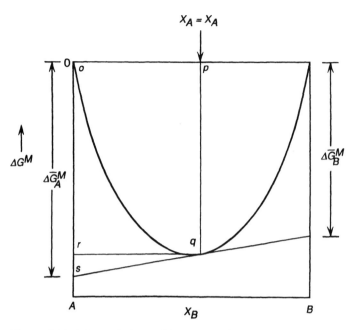

Figure 9.5   The variation, with composition, of the molar Gibbs free energy of formation of a binary solution.

Substitution from Eq. (9.28) gives

$$\Delta G'^M = n_A \Delta \overline{G}_A^M + n_B \Delta \overline{G}_B^M \tag{9.29}$$

or

$$\Delta G'^M = RT(n_A \ln a_A + n_B \ln a_B) \tag{9.30}$$

For 1 mole of solution, Eqs. (9.29) and (9.30), respectively, become

$$\Delta G^M = X_A \Delta \overline{G}_A^M + X_B \Delta \overline{G}_B^M \tag{9.31}$$

and

$$\Delta G^M = RT (X_A \ln a_A + X_B \ln a_B) \tag{9.32}$$

The variation of ΔG^M with composition, given by Eq. (9.32), is shown in Fig. 9.5.

The Method of Tangential Intercepts

In terms of the solution properties, Eqs. (9.27a) and (9.27b) can be written as

$$\Delta \overline{G}_A^M = \Delta G^M + X_B \frac{d\Delta G^M}{dX_A} \tag{9.33a}$$

and

$$\Delta \overline{G}_B^M = \Delta G^M + X_A \frac{d\Delta G^M}{dX_B} \tag{9.33b}$$

Consider the composition X_A (the point p in Fig. 9.5). At X_A

$$\Delta G^M = pq$$
$$X_B = rq$$

$$\frac{d\Delta G^M}{dX_A} = \text{the slope of the tangent to the } \Delta G^M \text{ curve at the composition } X_A$$

$$= \frac{rs}{rq}$$

From Eq. (9.33a),

$$\Delta \overline{G}_A^M = pq + rq\frac{rs}{rq} = pq + rs = or + rs = os$$
$$= \text{the tangential intercept at } X_A = 1$$

Similarly, at the composition X_A, $\Delta \overline{G}_B^M = $ the tangential intercept at $X_B = 1$.

The method of tangential intercepts can be used to obtain the partial molar values of any extensive property from the variation, with composition, of the integral value of the property.

9.6 THE PROPERTIES OF RAOULTIAN IDEAL SOLUTIONS

The components of a Raoultian ideal solution obey the relation

$$a_i = X_i$$

and hence, for an ideal binary A–B solution, Eq. (9.32) becomes

$$\Delta G^{M,\text{id}} = RT(X_A \ln X_A + X_B \ln X_B) \tag{9.34}$$

with

$$\Delta \overline{G}_A^{M,\text{id}} = RT \ln X_A \quad \text{and} \quad \Delta \overline{G}_B^{M,\text{id}} = RT \ln X_B$$

As discussed in Chap. 8, the general thermodynamic relationships between the state properties of a system are applicable to the partial molar properties of the components of a system. Thus, for the species i occurring in a solution,

$$\left(\frac{\partial \overline{G}_i}{\partial P}\right)_{T,\text{comp}} = \overline{V}_i \tag{9.35}$$

and, for pure i

$$\left(\frac{\partial G_i^\circ}{\partial P}\right)_{T,\text{comp}} = V_i^\circ \tag{9.36}$$

The Change in Volume Accompanying the Formation of an Ideal Solution

Subtraction of Eq. (9.36) from Eq. (9.35) gives

$$\left[\frac{\partial(\overline{G}_i - G_i^\circ)}{\partial P}\right]_{T,\text{comp}} = (\overline{V}_i - V_i^\circ)$$

or

$$\left(\frac{\partial \Delta \overline{G}_i^M}{\partial P}\right)_{T,\text{comp}} = \Delta \overline{V}_i^M \tag{9.37}$$

In an ideal solution, $\Delta \overline{G}_i^{M,\text{id}} = RT \ln X_i$, and, as X_i is not a function of pressure, then

$$\Delta \overline{V}_i^{M,\text{id}} = 0$$

The change in volume due to mixing, $\Delta V'^M$, is the difference between the volumes of the components in the solution and the volumes of the pure components, i.e., for a binary A–B solution containing n_A moles of A and n_B moles of B,

$$\begin{aligned}
\Delta V'^M &= (n_A \overline{V}_A + n_B \overline{V}_B) - (n_A V_A^\circ + n_B V_B^\circ) \\
&= n_A(\overline{V}_A - V_A^\circ) + n_B(\overline{V}_B - V_B^\circ) \\
&= n_A \Delta \overline{V}_A^M + n_B \Delta \overline{V}_B^M
\end{aligned}$$

For an ideal solution $\Delta \overline{V}_i^{M,\text{id}} = 0$, and thus it is seen that the change in volume accompanying an ideal solution is zero, i.e.,

$$\Delta V^{M,\text{id}} = 0 \tag{9.38}$$

The volume of an ideal solution is thus equal to the sum of the volumes of the pure components. The variation, with composition, of the molar volume of an ideal binary solution is shown in Fig. 9.6. At any composition the values of the partial molar volumes $\overline{V}_A^M$ and $\overline{V}_B^M$ are obtained as the intercepts of the tangents to the volume-

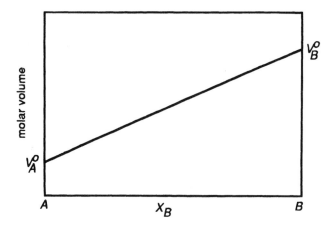

Figure 9.6 The variation, with composition, of the molar volume of a binary Raoultian solution.

composition line with the respective axes. As the molar volume of an ideal solution is a linear function of composition, then, trivially, the tangent at any point coincides with the straight line such that

$$\overline{V}_A = V_A^\circ \quad \text{and} \quad \overline{V}_B = V_B^\circ$$

The Heat of Formation of an Ideal Solution

For a component in a solution, the Gibbs-Helmholtz equation, Eq. (5.37), is

$$\left[\frac{\partial \overline{G}_i/T}{\partial T}\right]_{P,\text{comp}} = -\frac{\overline{H}_i}{T^2} \tag{9.39}$$

and, for the pure component,

$$\left[\frac{\partial(G_i^\circ/T)}{\partial T}\right]_{P,\text{comp}} = -\frac{H_i^\circ}{T^2} \tag{9.40}$$

where $\overline{H}_i$ and H_i° are, respectively, the partial molar enthalpy of i in the solution and the standard molar enthalpy of i. Subtraction of Eq. (9.40) from Eq. (9.39) gives

$$\left[\frac{\partial\left(\dfrac{\overline{G}_i - G_i^\circ}{T}\right)}{\partial T}\right]_{P,\text{comp}} = -\frac{(\overline{H}_i - H_i^\circ)}{T^2}$$

or

$$\left[\frac{\partial(\Delta \overline{G}_i^M/T)}{\partial T}\right]_{P,\text{comp}} = -\frac{\Delta \overline{H}_i^M}{T^2} \tag{9.41}$$

where $\Delta \overline{H}_i^M$ is the partial molar heat of solution of i.

In an ideal solution $\Delta \overline{G}_i^{M,\text{id}} = RT \ln X_i$, substituton of which into Eq. (9.41)

$$\frac{d(R \ln X_i)}{dT} = -\frac{\Delta \overline{H}_i^M}{T^2}$$

and, as X_i is not a function of temperature, it is seen that, for a component of an ideal solution,

$$\Delta \overline{H}_i^{M,\text{id}} = \overline{H}_i - H_i^\circ = 0$$

or

$$\overline{H}_i = H_i^\circ \tag{9.42}$$

The heat of formation of a solution (or the heat of mixing of the components) is the difference between the enthalpies of the components in solution and the enthalpies of the pure components before mixing. Thus for a mixture of n_A moles of A and n_B moles of B

$$\begin{aligned}
\Delta H'^M &= (n_A\overline{H}_A + n_B\overline{H}_B) - (n_A H_A^\circ + n_B H_B^\circ) \\
&= n_A(\overline{H}_A - H_A^\circ) + n_B(\overline{H}_B - H_B^\circ) \\
&= n_A\Delta \overline{H}_A^M + n_B\Delta \overline{H}_B^M
\end{aligned}$$

For an ideal solution $\Delta \bar{H}_i^M = 0$, and thus it is seen that the heat of formation of an ideal solution, $\Delta H^{M,\text{id}}$, is zero, i.e.,

$$\Delta H^{M,\text{id}} = 0 \tag{9.43}$$

The Entropy of Formation of an Ideal Solution

The fundamental equation, Eq. (5.25), gives

$$\left(\frac{\partial G}{\partial T}\right)_{P,\text{comp}} = -S$$

Thus, for the formation of a solution

$$\left(\frac{\partial \Delta G^M}{\partial T}\right)_{P,\text{comp}} = -\Delta S^M$$

For an ideal solution

$$\Delta G^{M,\text{id}} = RT(X_A \ln X_A + X_B \ln X_B)$$

and hence

$$\Delta S^{M,\text{id}} = -\left(\frac{\partial \Delta G^{M,\text{id}}}{\partial T}\right)_{P,\text{comp}} \tag{9.44}$$
$$= -R(X_A \ln X_A + X_B \ln X_B)$$

Eq. (9.44) shows that the entropy of formation of an ideal binary solution is independent of temperature.

Eq. (4.18) gave, for the mixing of N_A particles of A with N_B particles of B,

$$\Delta S'_{\text{conf}} = k \ln \frac{(N_A + N_B)!}{N_A! N_B!} \tag{4.17}$$
$$= k[\ln (N_A + N_B)! - \ln N_A! - \ln N_B!]$$

Application of Stirling's theorem* gives

$$\Delta S'_{\text{conf}} = k[(N_A + N_B) \ln (N_A + N_B)$$
$$- (N_A + N_B) - N_A \ln N_A + N_A - N_B \ln N_B + N_B]$$
$$= -k\left[N_A \ln \left(\frac{N_A}{N_A + N_B}\right) + N_B \left(\frac{N_B}{N_A + N_B}\right)\right]$$

---

*Stirling's theorem is $m! = \sqrt{2\pi m}\, m^m e^m$, and thus $\ln m! = \frac{1}{2} \ln (2\pi m) + m \ln m - m$, which, for large values of m, can be written as $\ln m! = m \ln m - m$.

Now,

$$\frac{N_A}{N_A + N_B} = \frac{n_A}{n_A + n_B} = X_A$$

and, similarly,

$$\frac{N_B}{N_A + N_B} = X_B$$

Also

$$N_A \text{ particles of } A = \frac{N_A}{N_O} \text{ moles of } A = n_A \text{ moles of } A$$

and

$$N_B \text{ particles of } B = \frac{N_B}{N_O} \text{ moles of } B = n_B \text{ moles of } B$$

where N_O is Avogadro's number. Thus

$$\Delta S'_{\text{conf}} = -kN_O(n_A \ln X_A + n_B \ln X_B)$$

But, as Boltzmann's constant (k) times Avogadro's number (N_O) equals the gas constant (R),

$$\Delta S'_{\text{conf}} = -R(n_A \ln X_A + n_B \ln X_B)$$

Division by $n_A + n_B$, the total number of moles, gives

$$\Delta S_{\text{conf}} = -R(X_A \ln X_A + X_B \ln X_B) \tag{9.45}$$

which is identical with Eq. (9.44). Thus the increase in entropy accompanying the formation of an ideal solution is a measure of the increase in the number of spatial configurations which become available to the system as a result of the mixing process. This is dependent only on the numbers of moles of the components in the solution and is independent of temperature. The variation of $\Delta S^{M,\text{id}}$ with composition in a binary A–B solution is shown in Fig. 9.7. As

$$\Delta S^M = X_A \Delta \bar{S}^M_A + X_B \Delta \bar{S}^M_B$$

it is seen that, in an ideal solution

$$\Delta \bar{S}^{M,\text{id}}_A = -R \ln X_A \quad \text{and} \quad \Delta \bar{S}^{M,\text{id}}_B = -R \ln X_B$$

For any solution

$$\Delta G^M = \Delta H^M - T\Delta S^M$$

and, for an ideal solution, as $\Delta H^{M,\text{id}} = 0$, then

$$\Delta G^{M,\text{id}} = -T\Delta S^{M,\text{id}}$$

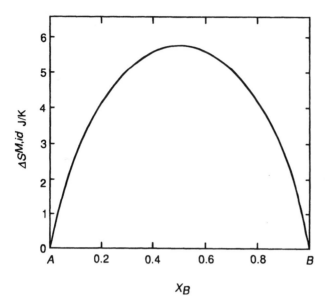

Figure 9.7 The variation, with composition, of the molar entropy of formation of a binary Raoultian solution.

9.7 NONIDEAL SOLUTIONS

A nonideal solution is one in which the activities of the components are not equal to their mole fractions. However, in view of the convenience of the concept of activity and the simplicity of Raoult's law, it is convenient to define an additional thermodynamic function called the *activity coefficient*, γ. The activity coefficient of a component of a solution is defined as the ratio of the activity of the component to its mole fraction, i.e., for the component i,

$$\gamma_i = \frac{a_i}{X_i} \tag{9.46}$$

The value of γ_i can be greater or less than unity ($\gamma_i = 1$ gives ideal Raoultian behavior). If $\gamma_i > 1$, then the component i is said to exhibit a positive deviation from Raoultian ideal behavior, and, if $\gamma_i < 1$, then the component i is said to exhibit a negative deviation from Raoult's law. Fig. 9.8 shows the variation of a_i with X_i for a component i which exhibits negative deviations, and Fig. 9.9 shows a system which exhibits positive deviations. The corresponding variations of γ_i are shown in Figs. 9.10 and 9.11.

 If γ_i varies with temperature, then $\Delta \overline{H}_i^M$ has a nonzero value, i.e., from Eq. (9.41)

$$\frac{\partial(\Delta \overline{G}_i^M / T)}{\partial T} = -\frac{\Delta \overline{H}_i^M}{T^2}$$

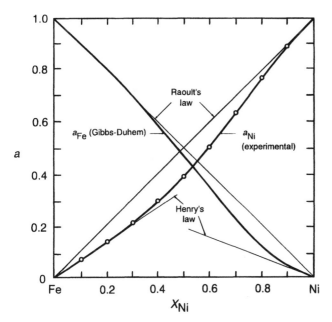

Figure 9.8 Activities in the system iron-nickel at 1600°C. (From G. R. Zellars, S. L. Payne, J. P. Morris, and R. L. Kipp, "The Activities of Iron and Nickel in Liquid Fe-Ni Alloys," *Trans. AIME* (1959), vol. 215, p. 181.)

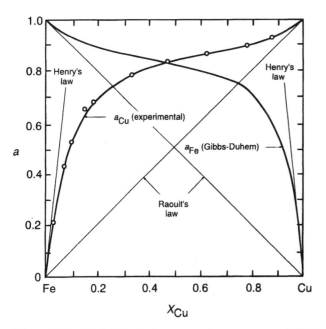

Figure 9.9 Activities in the system iron-copper at 1550°C. (From J. P. Morris and G. R. Zellars, "Vapor Pressure of Liquid Copper and Activities in Liquid Fe-Cu Alloys," *Trans. AIME* (1956), vol. 206, p. 1086.)

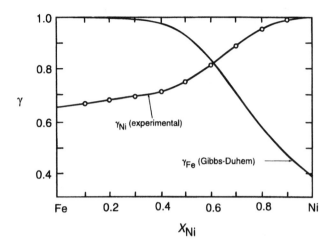

Figure 9.10 Activity coefficients in the system iron-nickel at 1600°C.

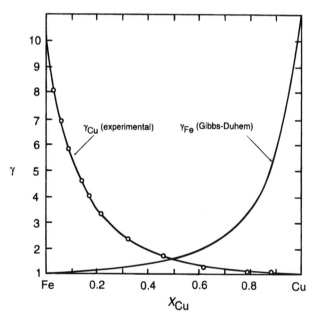

Figure 9.11 Activity coefficients in the system iron-copper at 1550°C.

and

$$\Delta \overline{G}_i^M = RT \ln a_i = RT \ln \gamma_i + RT \ln X_i$$

Thus

$$\frac{\partial (\Delta \overline{G}_i^M / T)}{\partial T} = \frac{\partial (R \ln \gamma_i)}{\partial T} = -\frac{\Delta \overline{H}_i^M}{T^2}$$

and as

$$d\left(\frac{1}{T}\right) = -\frac{dT}{T^2}$$

then

$$\frac{\partial(R \ln \gamma_i)}{\partial(1/T)} = \Delta \overline{H}_i^M \qquad (9.47)$$

In general, increasing the temperature of a nonideal solutions causes a decrease in the extent to which its components deviate from ideal behavior, i.e., if $\gamma_i > 1$ then an increase in temperature causes γ_i to decrease toward unity, and if $\gamma_i < 1$ an increase in temperature causes γ_i to increase toward unity. Thus in a solution, the components of which exhibit positive deviations from ideality, the values of the activity coefficients decrease with increasing temperature, and hence, from Eq. (9.47), the partial molar heats of solution of the components are positive quantities. Thus ΔH^M, the molar heat of formation of the solution is a positive quantity, which indicates that the mixing process is endothermic. ΔH^M is the quantity of heat absorbed from the thermostatting heat reservoir surrounding the solution per mole of solution formed at the temperature T. Conversely, in a solution, the components of which exhibit negative deviations from ideality, the activity coefficients increase with increasing temperature, and hence, the partial molar heats of mixing and the molar heat of mixing are negative. Such a solution forms exothermically, and ΔH^M is the heat absorbed by the thermostatting heat reservoir, per mole of solution formed, at the temperature T.

Exothermic mixing in an A–B binary system occurs when the A–B bond energy is more negative than both the A–A and B–B bond energies, and this causes a tendency toward "ordering" in the solution, in which the A atoms attempt to only have B atoms as nearest neighbors and vice versa. Exothermic mixing thus indicates a tendency toward the formation of a compound between the two components. Conversely, endothermic mixing occurs when the A–B bond energy is less negative than both the A–A and B–B bond energies, and this causes a tendency toward phase separation or "clustering" in the solution. The A atoms attempt to be coordinated only by A atoms, and the B atoms attempt to be coordinated only by B atoms. In both cases the equilibrium configuration of the solution is reached as a compromise between the enthalpy factors, which, being determined by the relative magnitudes of the bond energies, attempt to either completely order or completely unmix the solution, and the entropy factor which attempts to maximize the randomness of mixing of the atoms in the solution.

9.8 APPLICATION OF THE GIBBS-DUHEM RELATION TO THE DETERMINATION OF ACTIVITY

As was stated in Sec. 9.4, it is often found that the activity of only one component of a binary solution can be measured experimentally. In such cases the variation of the activity of the other component can be obtained from an application of the Gibbs-Duhem equation, Eq. (9.20),

$$\sum_i X_i d\overline{Q}_i$$

Applied to a binary A–B solution and using Gibbs free energy as the extensive property, Eq. (9.20) becomes

$$X_A d\Delta\overline{G}_A^M + X_B d\Delta\overline{G}_B^M = 0 \qquad (9.48)$$

and, as $\Delta\overline{G}_i^M = RT \ln a_i$, then

$$X_A d \ln a_A + X_B d \ln a_B = 0 \qquad (9.49)$$

or

$$d \log a_A = -\frac{X_B}{X_A} d \log a_B \qquad (9.50)$$

If the variation of a_B with composition is known, then integration of Eq. (9.50) from $X_A = 1$ to X_A gives the value of $\log a_A$ at X_A as

$$\log a_A \text{ at } X_A = -\int_{\log a_B \text{ at } X_A = 1}^{\log a_B \text{ at } X_A = X_A} (X_B/X_A) \, d \log a_B \qquad (9.51)$$

As an analytical expression for the variation of the activity of B is not usually computed, Eq. (9.51) is solved by graphical integration.

Fig. 9.12 shows a typical variation of $\log a_B$ with composition, and the value of $\log a_A$ at $X_A = X_A$ is equal to the shaded area under the curve. Two points are to be noticed in Fig. 9.12:

1. As $X_B \to 1$, $a_B \to 1$, $\log a_B \to 0$, and $X_B/X_A \to \infty$. Thus the curve exhibits a tail to infinity as $X_B \to 1$.
2. As $X_B \to 0$, $a_B \to 0$, and $\log a_B \to -\infty$. Thus the curve exhibits a tail to minus infinity as $X_B \to 0$.

Of these two points, point 2 is the more serious, as the calculation of $\log a_A$ at any composition involves the evaluation of the area under a curve which tails to minus infinity. This introduces an uncertainty into the calculation.

The tail to minus infinity can be eliminated by considering activity coefficients instead of activities in the Gibbs-Duhem equation. In the binary A–B solution

$$X_A + X_B = 1$$

Thus

$$dX_A + dX_B = 0 \qquad (9.52)$$

Therefore

$$X_A \frac{dX_A}{X_A} + X_B \frac{dX_B}{X_B} = 0$$

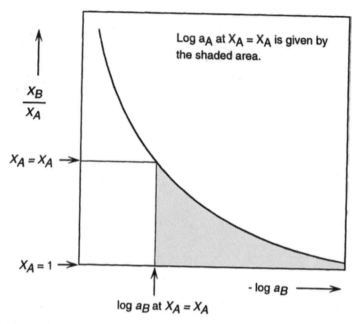

Figure 9.12 A schematic representation of the variation of log a_B with X_B/X_A in a binary solution, and illustration of the application of the Gibbs-Duhem equation to calculation of the activity of component A.

or

$$X_A \, d \log X_A + X_B \, d \log X_B = 0 \qquad (9.53)$$

Subtraction of Eq. (9.53) from Eq. (9.49) gives

$$X_A \, d \log \gamma_A + X_B \, d \log \gamma_B = 0$$

or

$$d \log \gamma_A = -\frac{X_B}{X_A} d \log \gamma_B \qquad (9.54)$$

Thus, if the variation of γ_B with composition is known, then integration of Eq. (9.54) gives the value of log γ_A at the composition $X_A = X_A$ as

$$\log \gamma_A \text{ at } X_A = -\int_{\log \gamma_B \text{ at } X_A = 1}^{\log \gamma_B \text{ at } X_A = X_A} (X_B/X_A) \, d \log \gamma_B \qquad (9.55)$$

A typical variation of log γ_B with composition in a binary A–B solution is shown in Fig. 9.13. The value of log γ_A at $X_A = X_A$ is given as the shaded area under the curve between the limits log γ_B at $X_A = X_A$ and log γ_B at $X_A = 1$.

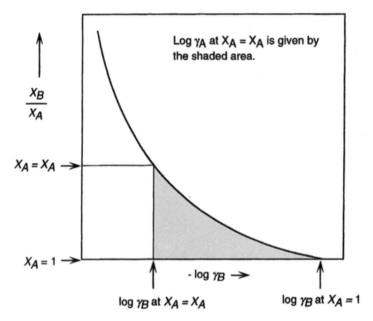

Figure 9.13 A schematic representation of the variation of log γ_B with X_B/X_A in a binary solution, and illustration of the application of the Gibbs-Duhem equation to calculation of the activity coefficient of component A.

The α-Function

The second tail to infinity is eliminated by the introduction of the α-function, which for the component i is defined as

$$\alpha_i = \frac{\ln \gamma_i}{(1 - X_i)^2} \tag{9.56}$$

The α-function is always finite by virtue of the fact that $\gamma_i \to 1$ as $X_i \to 1$. For the components of a binary A–B solution

$$\alpha_A = \frac{\ln \gamma_A}{X_B^2} \quad \text{and} \quad \alpha_B = \frac{\ln \gamma_B}{X_A^2}$$

or

$$\ln \gamma_A = \alpha_A X_B^2 \quad \text{and} \quad \ln \gamma_B = \alpha_B X_A^2 \tag{9.57}$$

Differentiation of $\ln \gamma_B = \alpha_B X_A^2$ gives

$$d \ln \gamma_B = 2\alpha_B X_A dX_A + X_A^2 d\alpha_B \tag{9.58}$$

and substitution of Eq. (9.58) into Eq. (9.54) gives

$$d \ln \gamma_A = -\frac{X_B}{X_A} 2\alpha_B X_A dX_A - \frac{X_B}{X_A} X_A^2 d\alpha_B \qquad (9.59)$$

$$= -2X_B \alpha_B dX_A - X_B X_A d\alpha_B$$

Integration of Eq. (9.59) gives

$$\ln \gamma_A = -\int_{X_A=1}^{X_A=X_A} 2X_B \alpha_B dX_A - \int_{\alpha_B \text{ at } X_A=1}^{\alpha_B \text{ at } X_A=X_A} X_B X_A d\alpha_B \qquad (9.60)$$

By virtue of the identity

$$\int d(x,y) = \int y dx + \int x dy$$

the second integral on the right-hand side of Eq. (9.60) can be written as

$$\int X_B X_A d\alpha_B = \int d(X_B X_A \alpha_B) - \int \alpha_B d(X_B X_A)$$

substitution of which into Eq. (9.60) gives

$$\ln \gamma_A = -\int 2X_B \alpha_B dX_A - \int d(X_B X_A \alpha_B) + \int \alpha_B d(X_B X_A)$$

$$= -\int 2X_B \alpha_B dX_A - X_B X_A \alpha_B + \int \alpha_B X_B dX_A + \int \alpha_B X_A dX_B$$

$$= -\int 2X_B \alpha_B dX_A - X_B X_A \alpha_B + \int \alpha_B X_B dX_A - \int \alpha_B X_A dX_A$$

$$= -X_B X_A \alpha_B - \int (2X_B - X_B + X_A)\alpha_B dX_A$$

$$= -X_B X_A \alpha_B - \int_{X_A=1}^{X_A=X_A} \alpha_B dX_A \qquad (9.61)$$

Thus $\ln \gamma_A$ at $X_A = X_A$ is obtained as $-X_B X_A \alpha_B$ minus the area under the plot of α_B vs. X_A from $X_A = X_A$ to $X_A = 1$, and, as α_B is everywhere finite, the integration does not involve a tail to infinity.

Fig. 9.8 shows the variation of a_{Ni} with composition in the system Fe-Ni at 1600°C as measured by Zellars et al.,[*] and Fig. 9.10 shows the corresponding variation of γ_{Ni} with composition. Extrapolation of γ_{Ni} to $X_{Ni} = 0$ in Fig. 9.10 gives the value of the Henry's law constant (k in Eq. (9.14)) as 0.66 for Ni in Fe at 1600°C. This, then, is the slope of the Henry's law line for Ni in Fe drawn in Fig. 9.8. The variation of γ_{Fe} with composition, shown in Fig. 9.10, is determined from consideration of either Fig. 9.14 or Fig 9.16. Fig. 9.14 shows the variation of log γ_{Ni} with X_{Ni}/X_{Fe}, graphical integration of which, according to Eq. (9.55), gives the variation of log γ_{Fe}

---

[*]G. R. Zellars, S. L. Payne, J. P. Morris, and R. L. Kip, "The Activities of Iron and Nickel in Liquid Fe-Ni Alloys," *Trans. AIME* (1959), vol. 215, p. 181.

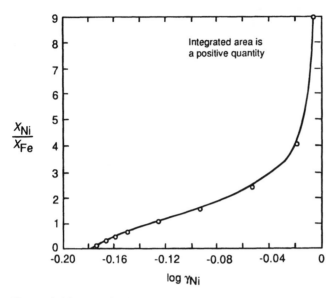

Figure 9.14 Application of the Gibbs-Duhem equation to determination of the activity of iron in the system iron-nickel.

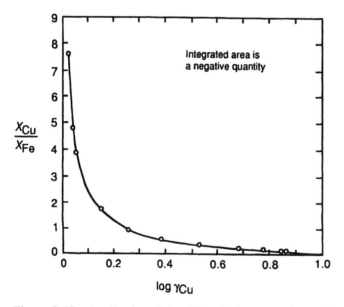

Figure 9.15 Application of the Gibbs-Duhem equation to determination of the activity of iron in the system iron-copper.

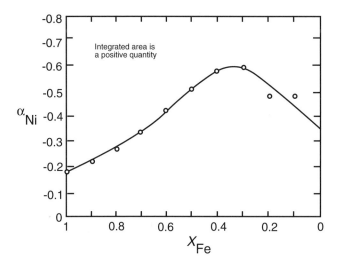

Figure 9.16 The variation of α_{Ni} with composition in the system iron-nickel.

with composition. In the graphical integration, as log γ_{Ni} increases with increasing X_{Ni}/X_{Fe}, the integrated area under the curve between $X_{Ni} = X_{Ni}$ and $X_{Ni} = 0$ is a positive quantity. Thus log γ_{Fe} is everywhere a negative quantity, and so Fe, like Ni, exhibits negative deviations from Raoult's law. The variation of α_{Ni} with composition is shown in Fig. 9.16. As α_{Ni} is everywhere negative, the integrated area from $X_{Fe} = X_{Fe}$ to $X_{Fe} = 1$ is a positive quantity. Fig. 9.9 shows the variation of a_{Cu} with composition in the Fe–Cu system at 1550°C measured by Morris and Zellars.* Fig. 9.11 shows the corresponding variation of γ_{Cu} with composition. Extrapolation of γ_{Cu} to $X_{Cu} = 0$ gives $k_{Cu} = 10.1$. Figures 9.15 and 9.17, respectively, show plots of log γ_{Cu} vs. X_{Cu}/X_{Fe} and α_{Cu} vs. X_{Fe}. As log γ_{Cu} decreases with increasing X_{Cu}/X_{Fe}, and α_{Cu} is everywhere positive, the integrated areas in both figures are negative quantities.

The Relationship between Henry's and Raoult's Laws

Henry's law for the solute B in a binary A–B solution is

$$a_B = k_B X_B$$

or, in terms of logarithms,

$$\ln a_B = \ln k_B + \ln X_B$$

differentiation of which gives

$$d \ln a_B = d \ln X_B$$

*J. P. Morris and G. R. Zellars, "Vapor Pressure of Liquid Copper and Activities in Liquid Fe-Cu Alloys," *Trans. AIME* (1956), vol. 206, p. 1086.

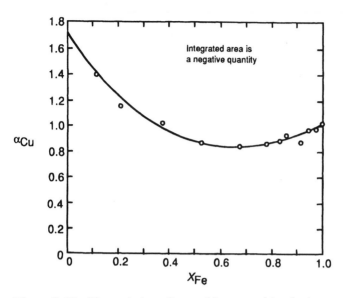

Figure 9.17 The variation of α_{Cu} with composition in the system iron-copper.

Inserting this into the Gibbs-Duhem equation, Eq. (9.49), gives

$$d \ln a_A = -\frac{X_B}{X_A} d \ln X_B = -\frac{X_B}{X_A} \frac{dX_B}{X_B} = -\frac{dX_B}{X_A}$$

$$= \frac{dX_A}{X_A} = d \ln X_A$$

Integration gives

$$\ln a_A = \ln X_A + \ln \text{ constant}$$

or

$$a_A = \text{constant} \times X_A$$

But, by definition, $a_i = 1$ when $X_i = 1$, and thus the integration constant equals unity. Consequently, in the range of composition over which the solute B obeys Henry's law, the solvent A obeys Raoult's law.

Direct Calculation of the Integral Molar Gibbs Free Energy of Mixing

Eq. (11.33b) gave

$$\Delta \bar{G}_A^M = \Delta G^M + X_B \frac{d\Delta G^M}{dX_A}$$

Rearranging and dividing by X_B^2 gives

$$\frac{\Delta \overline{G}_A^M dX_A}{X_B^2} = \frac{X_B d\Delta G^M - \Delta G^M dX_B}{X_B^2} = d\left(\frac{\Delta G^M}{X_B}\right)$$

or

$$d\left(\frac{\Delta G^M}{X_B}\right) = \frac{\Delta \overline{G}_A^M}{X_B^2} dX_A$$

Integrating between $X_A = X_A$ and $X_A = 0$ gives

$$\Delta G^M = X_B \int_0^{X_A} \frac{\Delta \overline{G}_A^M}{X_B^2} dX_A \qquad (9.62)$$

and, as $\Delta \overline{G}_A^M = RT \ln a_A$, the integral molar Gibbs free energy of mixing of A and B can be obtained directly from the variation of a_A with composition as

$$\Delta G^M = RTX_B \int_0^{X_A} \frac{\ln a_A}{X_B^2} dX_A \qquad (9.63)$$

The measured activities of Ni in Fe and Cu in Fe shown in Figs. 9.8 and 9.9 can be used to obtain

$$\Delta G^M \text{ (in the system Fe–Ni)} = RTX_{Fe} \int_0^{X_{Ni}} \frac{\ln a_{Ni}}{X_{Fe}^2} dX_{Ni}$$

and

$$\Delta G^M \text{ (in the system Cu–Fe)} = RTX_{Fe} \int_0^{X_{Cu}} \frac{\ln a_{Cu}}{X_{Fe}^2} dX_{Cu}$$

The graphical integrations of these equations are shown in Fig. 9.18, in which line (a) is $(\ln a_{Cu})/X_{Fe}^2$ vs. X_{Cu} and line (c) is $(\ln a_{Ni})/X_{Fe}^2$ vs. X_{Ni}. Line (b) shows the variation of $(\ln X_i)/(1 - X_i)^2$ with X_i, which is the variation of the function for a component i which exhibits Raoultian behavior. As is seen, some uncertainty is introduced into the integration by virtue of the fact that the function $(\ln a_i)/(1 - X_i)^2 \rightarrow -\infty$ as $X_i \rightarrow 0$. In Fig. 9.18 the shaded area (which is the value of the integral between $X_{Cu} = 0.5$ and $X_{Cu} = 0$), multiplied by the factor $2.303 \times 8.3144 \times 1823 \times 0.5$, gives the value of ΔG^M at $X_{Fe} = 0.5$.

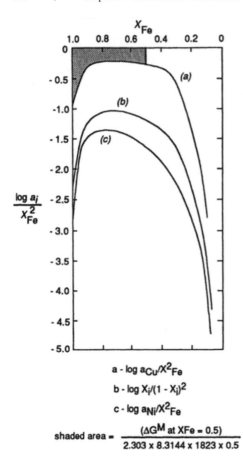

a - $\log a_{Cu}/X^2_{Fe}$

b - $\log X_i/(1 - X_i)^2$

c - $\log a_{Ni}/X^2_{Fe}$

shaded area = $\dfrac{(\Delta G^M \text{ at } X_{Fe} = 0.5)}{2.303 \times 8.3144 \times 1823 \times 0.5}$

Figure 9.18 Illustration of the direct calculation of the integral molar Gibbs free energies of mixing in the systems iron-copper at 1550°C and iron-nickel at 1600°C.

The variations of ΔG^M obtained from the graphical integrations are shown in Fig. 9.19. Applied to a solution which exhibits Raoultian behavior (line b), the integration gives

$$\Delta G^M = RT(1 - X_i) \int_0^{X_i} \frac{\ln X_i}{(1 - X_i)^2} dX_i$$

$$= RT(1 - X_i) \left[\frac{X_i \ln X_i}{1 - X_i} + \ln(1 - X_i) \right]$$

$$= RT \left[X_i \ln X_i + (1 - X_i) \ln(1 - X_i) \right]$$

in agreement with Eq. (9.34).

The uncertainty caused by the infinite tail as $X \to 0$ is eliminated if the equation is used the calculate the excess Gibbs free energy (see Sec. 9.9). Eq. (9.62) is a

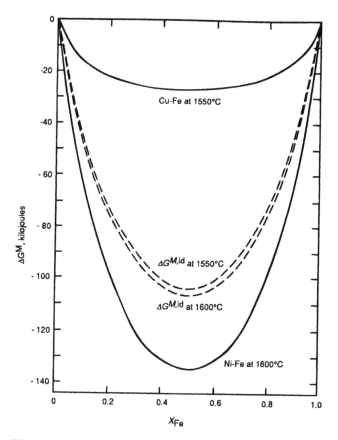

Figure 9.19 The integral molar Gibbs free energies of mixing in the systems iron-copper at 1550°C and iron-nickel at 1600°C.

general equation which relates the integral and partial molar values of any extensive thermodynamic function, e.g.,

$$\Delta H^M = X_B \int_0^{X_A} \frac{\Delta \overline{H}_A^M}{X_B^2} \, dX_A \qquad (9.64)$$

and

$$\Delta S^M = X_B \int_0^{X_A} \frac{\Delta \overline{S}_A^M}{X_B^2} \, dX_A \qquad (9.65)$$

9.9 REGULAR SOLUTIONS

Thus far two classes of solution have been identified:

1. Ideal or Raoultian solutions in which $a_i = X_i$, $\Delta \overline{H}_i^M = 0$, $\Delta \overline{V}_i^M = 0$, and $\Delta \overline{S}_i^M = -R \ln X_i$
2. Nonideal solutions in which $a_i \neq X_i$ and $\Delta \overline{H}_i^M \neq 0$

Attempts to classify nonideal solutions have involved the development of equations that describe the behavior of hypothetical solutions, and the simplest of these mathematical formalisms is that which generates what is known as "regular solution behavior."

In 1895 Margules* suggested that, at constant temperature, the activity coefficients, γ_A and γ_B, of the components of a binary solution could be represented by power series of the form

$$\ln \gamma_A = \alpha_1 X_B + \frac{1}{2} \alpha_2 X_B^2 + \frac{1}{3} \alpha_3 X_B^3 + \cdots \tag{9.66}$$

$$\ln \gamma_B = \beta_1 X_A + \frac{1}{2} \beta_2 X_A^2 + \frac{1}{3} \beta_3 X_A^3 + \cdots$$

and, by application of the Gibbs-Duhem equation, namely,

$$X_A \, d \ln \gamma_A = -X_B \, d \ln \gamma_B \tag{9.54}$$

he showed that if these equations are to hold over the entire range of composition of the solution, then $\alpha_1 = \beta_1 = 0$. This is proved by obtaining both sides of Eq. (9.54) as power series of X_A and X_B and equating the coefficients. By such comparison of the coefficients of the power series, Margules further demonstrated that if the variations of the activity coefficients can be represented by the quadratic terms alone, then

$$\alpha_2 = \beta_2 \tag{9.67}$$

In 1929 Hildebrand,[†] using an equation of van Laar[‡] which is based upon the van der Waals equation of state for mixtures, showed that if the value of the van der Waals "b" is the same for both components, then, in the binary A–B solution,

$$RT \ln \gamma_B = \alpha' X_A^2$$

and

$$RT \ln \gamma_A = \alpha' X_B^2 \tag{9.68}$$

---

*M. Margules "Über die Zusammensetzung der gesättigten Dampfe von Mischungen," *Stizungs-berichte. Akad. Wiss. Vienna* (1895), vol. 104, p. 1243.

[†]J. H. Hildebrand, "Solubility XII, Regular Solutions," *J. Am. Chem. Soc.* (1929), vol. 51, p. 66.

[‡]J. J. van Laar, "The Vapor Pressure of Binary Mixtures," *Z. Physik. Chem.* (1910), vol. 72, p. 723.

Hildebrand assigned the term "regular solution" to one obeying Eq. (9.68). Consideration of Eq. (9.61) shows that if the value of α for one component, say, component B, is independent of composition, then

$$\begin{aligned}
\ln \gamma_A &= -X_A X_B \alpha_B - \alpha_B(X_A - 1) \\
&= -X_A X_B \alpha_B + \alpha_B X_B \\
&= \alpha_B X_B(1 - X_A) \\
&= \alpha_B X_B^2
\end{aligned}$$

But, as Eq. (9.57) gave $\ln \gamma_A = \alpha_A X_B^2$, it is seen that

$$\alpha_A = \alpha_B = \alpha$$

From Eq. (9.68), α for a regular solution is an inverse function of temperature, i.e.,

$$\alpha = \frac{\alpha'}{RT} \tag{9.69}$$

Hildebrand defined a regular solution as one in which

$$\Delta \overline{H}_i^M \neq 0 \quad \text{and} \quad \Delta \overline{S}_i^M = \Delta \overline{S}_i^{M,\text{id}} = -R \ln X_i$$

The properties of a regular solution are best examined by means of the concept of excess functions. The excess value of an extensive thermodynamic solution property is simply the difference between its actual value and the value that it would have if the solution were ideal, e.g., as applied to the Gibbs free energy of the solution.

$$G = G^{\text{id}} + G^{\text{XS}} \tag{9.70}$$

where

G = the molar Gibbs free energy of the solution
G^{id} = the molar Gibbs free energy which the solution would have if it were ideal
G^{XS} = the excess molar Gibbs free energy of the solution

Subtraction of the Gibbs free energy of the unmixed components ($X_A G_A^{\circ} + X_B G_B^{\circ}$) from both sides of Eq. (9.70) gives

$$\Delta G^M = \Delta G^{M,\text{id}} + G^{\text{XS}} \tag{9.71}$$

As for any solution,

$$\Delta G^M = \Delta H^M - T \Delta S^M$$

and, for an ideal solution,

$$\Delta G^{M,\text{id}} = -T \Delta S^{M,\text{id}}$$

then

$$G^{\text{XS}} = \Delta G^M - \Delta G^{M,\text{id}} = \Delta H^M - T(\Delta S^M - \Delta S^{M,\text{id}}) \tag{9.72}$$

For a regular solution, $\Delta S^M = \Delta S^{M,\text{id}}$, and thus

$$G^{XS} = \Delta H^M \tag{9.73}$$

Now

$$\Delta G^M = RT(X_A \ln a_A + X_B \ln a_B)$$
$$= RT(X_A \ln X_A + X_B \ln X_B) + RT(X_A \ln \gamma_A + X_B \ln \gamma_B)$$

and as

$$\Delta G^{M,\text{id}} = RT(X_A \ln X_A + X_B \ln X_B)$$

then

$$G^{XS} = RT(X_A \ln \gamma_A + X_B \ln \gamma_B) \tag{9.74a}$$
$$= X_A \overline{G}_A^{XS} + X_B \overline{G}_B^{XS} \tag{9.74b}$$

For a regular solution, $\ln \gamma_A = \alpha X_B^2$, and $\ln \gamma_B = \alpha X_A^2$, substitution of which into Eq. (9.74a) gives

$$G^{XS} = RT\alpha X_A X_B \tag{9.75}$$

or, from Eq (9.69), gives

$$G^{XS} = \alpha' X_A X_B \tag{9.76}$$

It is thus seen that G^{XS} for a regular solution is independent of temperature. This can also been shown as follows:

$$\left(\frac{\partial G^{XS}}{\partial T} \right)_{P,\text{comp}} = -S^{XS}$$

and, as S^{XS} for a regular solution is zero, then G^{XS}, and hence ΔH^M, are independent of temperature. From Eqs. (9.74a), (9.74b), and (9.75), at any given composition,

$$\overline{G}_A^{XS} = RT_1 \ln \gamma_{A(T_1)} = RT_2 \ln \gamma_{A(T_2)} = \alpha' X_B^2$$

and hence, for a regular solution,

$$\frac{\ln \gamma_A \text{ at the temperature } T_2}{\ln \gamma_A \text{ at the temperature } T_1} = \frac{T_1}{T_2} \tag{9.77}$$

Eq. (9.77) is of considerable practical use in converting activity data for a regular solution at one temperature to activity data at another temperature.

Figs. 9.20 and 9.21, respectively, show the symmetrical variation, with composition, of the activities and activity coefficients in the system tin-thallium measured by Hildebrand and Sharma* at three temperatures, and Fig. 9.22 shows the linear

---

*J. H. Hildebrand and J. N. Sharma, "The Activities of Molten Alloys of Thallium with Tin and Lead," *J. Am. Chem. Soc.* (1929), vol. 51, p. 462.

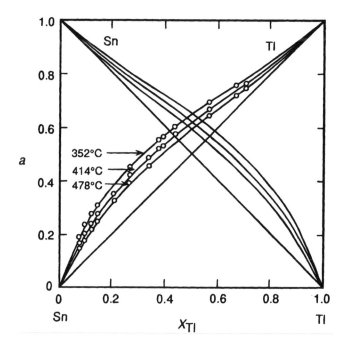

Figure 9.20 Activities in the system thallium-tin. (From J. H. Hildebrand and J. N. Sharma, "The Activities of Molten Alloys of Thallium with Tin and Lead," *J. Am. Chem. Soc.* (1929), vol. 51, p. 462.)

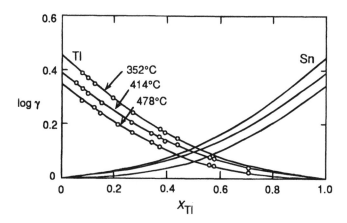

Figure 9.21 Activities coefficients in the system thallium-tin. (From J. H. Hildebrand and J. N. Sharma, "The Activities of Molten Alloys of Thallium with Tin and Lead," *J. Am. Chem Soc.* (1929), vol. 51, p. 462.)

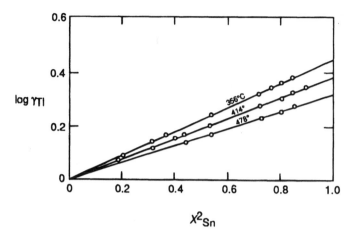

Figure 9.22   Log γ_{Tl} vs. X_{Sn}^2 in the system thallium-tin. (From J. H. Hildebrand and J. N. Sharma, "The Activities of Molten Alloys of Thallium with Tin and Lead," *J. Am. Chem. Soc* (1929), vol. 51, p. 462.)

variations of log γ_{Tl} with X_{Sn}^2, the slopes of which equal α at the given temperatures. The variation of γ_i with X_i is that of a regular solution, but Fig. 9.23 shows that αT, which for strict adherence to regular behavior should be independent of T, decreases slowly with increasing temperature. Fig. 9.24 shows the variations, with composition, of ΔG^M, ΔH^M, and $-T\Delta S^M$ for the system Tl–Sn 414°C. It is to be noted that a parabolic form for ΔH^M, or G^{XS} should not be taken as being a demonstration that the solution is regular, as it is frequently found that G^{XS} or ΔH^M can be adequately expressed by means of the relations

$$\Delta H^M = b X_A X_B \quad \text{or} \quad G^{XS} = b' X_A X_B$$

where b and b' are unequal, in which case from Eq. (9.72)

$$\Delta S^M \neq \Delta S^{M,\text{id}}$$

This type of behavior is found in melts in the system Au–Cu and Au–Ag. In the system Au–Cu at 1550 K, $G^{XS} = -24,060\ X_{Cu}X_{iAu}$ joules is parabolic, ΔH^M is asymmetric, and $S^{XS} \neq 0$. Conversely, in the system Au–Ag at 1350 K, $\Delta H^M = -20,590\ X_{Ag}X_{Au}$ joules is parabolic, G^{XS} is asymmetric, and $S^{XS} \neq 0$.

The molar excess Gibbs free energy can be obtained from a knowledge of the dependence, on composition, of the activity coefficient of one component by means of Eq. (9.63), written as

$$G^{XS} = RTX_B \int_0^{X_A} \frac{\ln \gamma_A}{X_B^2} dX_A$$

Thus, for a Raoultian solutions, as $\gamma_A = 1$, $G^{XS} = 0$, and for a regular solution, as $\ln \gamma_A/X_B^2 = \alpha$, $G^{XS} = RT\alpha X_A X_B$.

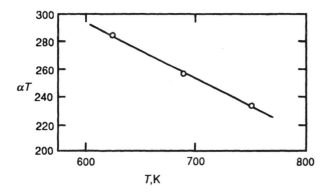

Figure 9.23 The variation of the product αT with T in the system Tl–Sn.

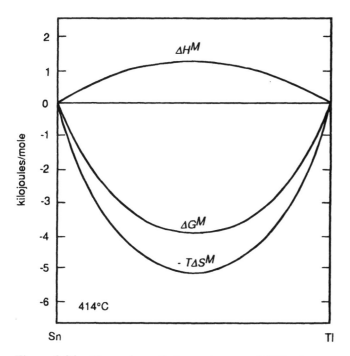

Figure 9.24 The molar enthalpy, entropy, and Gibbs free energy of mixing of thallium and tin at 414°C.

9.10 A STATISTICAL MODEL OF SOLUTIONS

Regular solution behavior can be understood by application of the statistical mixing model, introduced in Chap. 4, to two components which have equal molar volumes and which do not exhibit a change in molar volume when mixed. In both the pure state and in solution the interatomic forces exist only between neighboring atoms, in which case the energy of the solution is the sum of the interatomic bond energies.

Consider 1 mole of a mixed crystal containing N_A atoms of A and N_B atoms of B such that

$$X_A = \frac{N_A}{N_A + N_B} = \frac{N_A}{N_O} \quad \text{and} \quad X_B = \frac{N_B}{N_O}$$

where N_O is Avogadro's number. The mixed crystal, or solid solution, contains three types of atomic bond:

1. A–A bonds the energy of each of which is E_{AA}
2. B–B bonds the energy of each of which is E_{BB}
3. A–B bonds the energy of each of which is E_{AB}

By considering the relative zero of energy to be that when the atoms are infinitely far apart, the bond energies E_{AA}, E_{BB}, and E_{AB} are negative quantities. Let the coordination number of an atom in the crystal be z, i.e., each atom has z nearest neighbors. If, in the solution, there are P_{AA} A–A bonds, P_{B-B} B–B bonds, and P_{AB} A–B bonds, the energy of the solution, E, is obtained as the linear combination

$$E = P_{AA}E_{AA} + P_{BB}E_{BB} + P_{AB}E_{AB} \tag{9.78}$$

and the problem of calculating E becomes one of calculating the values of P_{AA}, P_{BB}, and P_{AB}.

The number of A atoms × the number of bonds per atom
= the number of A–B bonds + 2 × the number of A–A bonds.

(The factor 2 arises because each A–A bond involves two A atoms.) Thus

$$N_A z = P_{AB} + 2 P_{AA}$$

or

$$P_{AA} = \frac{N_A z}{2} - \frac{P_{AB}}{2} \tag{9.79}$$

Similarly, for B, $N_B z = P_{AB} + 2P_{BB}$ or

$$P_{BB} = \frac{N_B z}{2} - \frac{P_{AB}}{2} \tag{9.80}$$

Substitution of Eqs. (9.79) and (9.80) into Eq. (9.78) gives

$$E = \left(\frac{N_A z}{2} - \frac{P_{AB}}{2} \right) E_{AA} + \left(\frac{N_B z}{2} - \frac{P_{AB}}{2} \right) E_{BB} + P_{AB}E_{AB}$$

$$= \frac{1}{2} z N_A E_{AA} + \frac{1}{2} z N_B E_{BB} + P_{AB}[E_{AB} - \frac{1}{2}(E_{AA} + E_{BB})] \tag{9.81}$$

Consider now the energies of the pure components before mixing. With N_A atoms in pure A

2 × the number of A–A bonds
= the number of atoms × the number of bonds per atom

i.e.,

$$P_{AA} = \frac{1}{2} N_A z$$

and similarly, for N_B atoms in pure B

$$P_{BB} = \frac{1}{2} N_B z$$

Thus

$$\Delta E^M = \text{(the energy of the solution)} - \text{(the energy of the unmixed components)}$$

$$= P_{AB}[E_{AB} - \frac{1}{2}(E_{AA} + E_{BB})]$$

For the mixing process, from Eq. (5.10),

$$\Delta H^M = \Delta E^M - P\Delta V^M$$

and, as it has been stipulated that $\Delta V^M = 0$, then

$$\Delta H^M = \Delta E^M = P_{AB}[E_{AB} - \frac{1}{2}(E_{AA} + E_{BB})] \tag{9.82}$$

Eq. (9.82) shows that, for given values of E_{AA}, E_{BB}, and E_{AB}, ΔH^M depends on P_{AB} and further that, for the solution to be ideal, i.e., for $\Delta H^M = 0$,

$$E_{AB} = \frac{(E_{AA} + E_{BB})}{2} \tag{9.83}$$

Thus, contrary to the preliminary discussion in Sec. 9.2 which suggested that ideal mixing required the condition $E_{AB} = E_{AA} = E_{BB}$, it is seen that a sufficient condition is that E_{AB} be the average of E_{AA} and E_{BB}. If $|E_{AB}| > |\frac{1}{2}(E_{AA} + E_{BB})|$, then, from Eq. (9.82), ΔH^M is a negative quantity, corresponding to negative deviations from Raoultian ideal behavior, and, if $|E_{AB}| < |\frac{1}{2}(E_{AA} + E_{BB})|$, then ΔH^M is a positive quantity, corresponding to positive deviations from Raoultian ideality.

If $\Delta H^M = 0$, then the mixing of the N_A atoms with the N_B atoms of B is random, in which case Eq. (9.45) gives

$$\Delta S^M = \Delta S^{M,\text{id}} = -R(X_A \ln X_A + X_B \ln X_B)$$

In solutions which exhibit relatively small deviations from ideal behavior, that is $|\Delta H^M| \leq RT$, it can be assumed that the mixing of the atoms is also approximately random, in which case P_{AB} can be calculated as follows. Consider two neighboring lattice sites in the crystal labeled 1 and 2. The probability that site 1 is occupied by an A atom is

$$\frac{\text{the number of } A \text{ atoms in the crystal}}{\text{the number of lattice sites in the crystal}} = \frac{N_A}{N_O} = X_A$$

and similarly, the probability that site 2 is occupied by a B atom is X_B. The probability that site 1 is occupied by an A atom and site 2 is simultaneously occupied by a B atom is thus $X_A X_B$. But the probability that site 1 is occupied by a B atom and site 2 is simultaneously occupied by an A atom is also $X_A X_B$. Thus the probability that a neighboring pair of sites contains an A–B pair is $2X_A X_B$. By a similar argument, the probability that the neighboring sites contain an A–A pair is X_A^2 and that the neighboring sites contain a B–B pair is X_B^2. The probability that the neighboring sites contain an A–B pair or an A–A pair or a B–B pair is

$$X_A^2 + 2X_A X_B + X_B^2$$
$$= (X_A + X_B)^2$$
$$= 1$$

As the mole of crystal contains $\frac{1}{2} zN_O$ pairs of lattice sites, then

$$\text{the number of } A\text{–}B \text{ pairs}$$
$$= \text{the number of pairs of sites} \times \text{the probability of an } A\text{–}B \text{ pair}$$

i.e.,

$$P_{AB} = \frac{1}{2}zN_O \times 2X_A X_B = zN_O X_A X_B \tag{9.84}$$

Similarly

$$P_{AA} = \frac{1}{2}zN_O \times X_A^2 = \frac{1}{2}zN_O X_A^2$$

and

$$P_{BB} = \frac{1}{2}zN_O X_B^2$$

Substituting Eq. (9.84) into Eq. (9.82) gives

$$\Delta H^M = zN_O X_A X_B [E_{AB} - \frac{1}{2}(E_{AA} + E_{BB})]$$

and if

$$\Omega = zN_O [E_{AB} - \frac{1}{2}(E_{AA} + E_{BB})]$$

then

$$\Delta H^M = \Omega X_A X_B \tag{9.85}$$

which shows that ΔH^M is a parabolic function of composition. As random mixing is assumed, the statistical model corresponds to the regular solution model, i.e.,

$$\Delta H^M = G^{XS} = \Omega X_A X_B = RT\alpha X_A X_B \tag{9.86}$$

and thus

$$\alpha = \frac{\Omega}{RT} \tag{9.87}$$

Application of Eq. (9.33a) to the heat of mixing gives

$$\Delta \overline{H}_A^M = \Delta H^M + X_B \frac{\partial \Delta H^M}{\partial X_A}$$

and from Eq. (9.86)

$$\frac{\partial \Delta H^M}{\partial X_A} = \Omega(X_B - X_A)$$

Thus

$$\Delta \overline{H}_A^M = \Omega X_A X_B + X_B \Omega(X_B - X_A) = \Omega X_B^2 \tag{9.88a}$$

and

$$\Delta \overline{H}_B^M = \Omega X_A^2 \tag{9.88b}$$

As the mixing is random, then

$$\Delta \overline{S}_A^M = -R \ln X_A \quad \text{and} \quad \Delta \overline{S}_B^M = -R \ln X_B$$

and hence

$$\begin{aligned} \Delta \overline{G}_A^M &= \Delta \overline{H}_A^M - T\Delta \overline{S}_A^M \\ &= \Omega X_B^2 + RT \ln X_A \end{aligned} \tag{9.89}$$

But

$$\begin{aligned} \Delta \overline{G}_A^M &= RT \ln a_A \\ &= RT \ln \gamma_A + RT \ln X_A \end{aligned} \tag{9.90}$$

comparison of which with Eq. (9.89) indicates that

$$\ln \gamma_A = \frac{\Omega}{RT} X_B^2 = \alpha X_B^2 \tag{9.91}$$

The value of γ thus depends on the value of Ω, which, in turn, is determined by the values of the bond energies E_{AA}, E_{BB}, and E_{AB}. If Ω is negative, then $\gamma_A < 1$, and if Ω is positive, then $\gamma_A > 1$.

Henry's law requires that γ_A, and hence $\ln \gamma_A$, approach a constant value as X_B approaches unity. Thus, as $X_B \to 1$, $\ln \gamma_A \to \ln\gamma_A^\circ = \Omega/RT$, with this limiting value being approached asymptotically. Similarly, in view of the relationship between

Henry's and Raoult's laws, Raoult's law is approached asymptotically by the component i as $X_i \rightarrow 1$.

The applicability of the statistical model to real solutions decreases as the magnitude of Ω increases, i.e., if the magnitude of E_{AB} is significantly greater or less than the average of E_{AA} and E_{BB} then random mixing of the A and B atoms cannot be assumed. The equilibrium configuration of a solution at constant T and P is that which minimizes the Gibbs free energy G, where $G = H - TS$ is measured relative to the unmixed components. As has been seen, minimization of G occurs as a compromise between minimization of H and maximization of S. If $|E_{AB}| > |\frac{1}{2}(E_{AA} + E_{BB})|$ then minimization of H corresponds to maximization of the number of A–B pairs (complete ordering of the solution). On the other hand, maximization of S corresponds to completely random mixing. Minimization of G thus occurs as a compromise between maximization of P_{AB} (the tendency toward which increases with increasingly negative values of Ω) and random mixing (the tendency toward which increases with increasing temperature). The critical parameters are thus Ω and T, and, if Ω is appreciably negative and the temperature is not too high, then the value of P_{AB} will be greater than that for random mixing, in which case the assumption of random mixing is not valid.

Similarly, if $|E_{AB}| < |\frac{1}{2}(E_{AA} + E_{BB})|$, then minimization of H corresponds to minimization of the number of A–B pairs (complete clustering in the solution), and minimization of G occurs as a compromise between minimization of P_{AB} (the tendency toward which increases with increasingly positive values of Ω) and random mixing. Thus if Ω is appreciably positive and the temperature is not too high, then the value of P_{AB} will be less than that for random mixing, in which case the assumption of random mixing is again invalid.

In order for the statistical model, and hence the regular solution model, to be applicable, it is necessary that the above-mentioned compromise be such that the equilibrium solution configuration be not too distant from random mixing. As the entropy contribution to the Gibbs free energy is dependent on temperature, then

1. For any value of Ω, more nearly random mixing occurs as the temperature is increased, and
2. For any given temperature, more nearly random mixing occurs with smaller values of Ω.

The preceding discussion can be illustrated qualitatively by Fig. 9.25a and Fig. 9.25b. In these figures the x-axis represents the range of spatial configurations available to the atoms of a 50 mole percent $A - 50$ mole percent B solution, quantified as the probability of the occurrence of an A–B pair. The extreme configurations are complete clustering (A and B immiscible) in which the probability of the occurrence of an A–B pair is zero, and the completely ordered structure in which the probability of the occurrence of an A–B pair is unity. Random mixing occurs when the probability of the occurrence of an A–B pair is 0.5. Movement along the x-axis from 0 to 1 occurs as follows. A and B atoms are exchanged to form dilute solutions of A in B and B in A and this process is continued until a single homogeneous solution is formed in

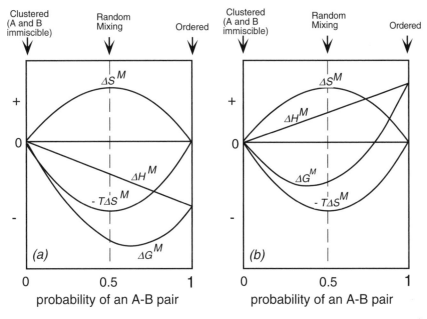

Figure 9.25 Illustration of the origins of deviation from regular solution behavior.

which the probability of the occurrence of an $A–B$ pair is less than 0.5. Thereafter the atoms are rearranged in such a manner as to continuously increase the probability of the occurrence of an $A–B$ pair to its limiting value of 1 in the completely ordered state. The change in the entropy of the system is given by the curve ΔS^M in Fig. 9.25a and Fig. 9.25b; it increases from zero in the immiscible configuration, passes through a maximum at the random mixing configuration and decreases to zero in the ordered configuration. The corresponding variations of $-T\Delta S^M$ are shown in both figures. Fig. 9.25a is drawn for an exothermic solution and Fig. 9.25b is drawn for an endothermic solution. Thus, in Fig. 25a, the heat of mixing line, identified as ΔH^M, begins at an arbitrary value of zero on the y-axis and decreases linearly with increasing probability of an $A–B$ pair. In Fig. 25b the ΔH^M, begins at zero and increases linearly with increasing probability of an $A–B$ pair. The variations of the Gibbs free energy with configuration, given as the sum of ΔH^M and $-T\Delta S^M$, are shown as the lines ΔG^M. The equilibrium configuration occurs at the position of the minimum on the ΔG^M curve, which is seen to be at a value of the probability of the occurrence of an $A–B$ pair of greater than 0.5 in the exothermically forming solution and at a value of less than 0.5 in the endothermically forming solution. It is seen that the random configuration is the equilibrium configuration only when $\Delta H^M = 0$ and that, as the magnitude of $|\Omega|$ for the system $A–B$ increases, then, at constant temperature, the position of the minimum in the ΔG^M curve moves further away from the random configuration. Similarly, for any given system (of fixed Ω), as T, and hence $|T\Delta S^M|$, increases, the position of the minimum in the ΔG^M curve moves toward the random configuration. Figs. 9.25a and b also illustrate that both extreme configurations are physically unrealizable as, in order to have the minimum in the ΔG^M curve coincide with either extreme, infinite values of ΔH^M would be required (negative for complete ordering and positive for complete clustering). Similarly, with a non-zero ΔH^M the random configuration becomes the equilibrium configuration only at infinite temperature.

9.11 SUBREGULAR SOLUTIONS

In the regular solution model, the constant value of Ω, which, via Eq. (9.85), gives a parabolic variation of ΔH^M, and the ideal entropy of mixing lead to variations of G^{XS} and ΔG^M which are symmetrical about the composition $X_A = 0.5$. The model can be made more flexible by arbitrarily allowing Ω to vary with composition such as

$$\Omega = a + bX_B + cX_B^2 + dX_B^3 + \cdots \tag{9.92}$$

and the so-called subregular solution model is that in which the values of all of the constants in Eq. (9.92), other than a and b, are zero. Thus, the subregular solution model gives the molar excess Gibbs free energy of formation of a binary A–B solution as

$$G^{XS} = (a + bX_B)X_AX_B \tag{9.93}$$

Eq. (9.93) is an empirical equation, i.e., the constants a and b have no physical significance and are simply parameters the values of which can be adjusted in an attempt to fit the equation to experimentally measured data. The application of Eqs. (9.27a) and (9.27b) to Eq. (9.93) gives the partial molar excess Gibbs free energies of the components A and B as

$$\overline{G}_A^{XS} = aX_B^2 + bX_B^2(X_B - X_A) \tag{9.94a}$$

and

$$\overline{G}_B^{XS} = aX_A^2 + 2bX_A^2X_B \tag{9.94b}$$

The variations of G^{XS} with X_B, with composition, for several combinations of a and b are shown in Fig. 9.26. The maxima and/or minima in the curves occur at

$$\frac{dG^{XS}}{dX_B} = 0$$

which, from Eq. (9.93), written as

$$G^{XS} = aX_B + 2(b - a)X_B^2 - bX_B^3$$

gives

$$\frac{dG^{XS}}{dX_B} = a + 2(b - a)X_B - 3bX_B^2 = 0$$

or

$$X_B = \frac{2(b - a) \pm 2\sqrt{b^2 + ab + a^2}}{6b}$$

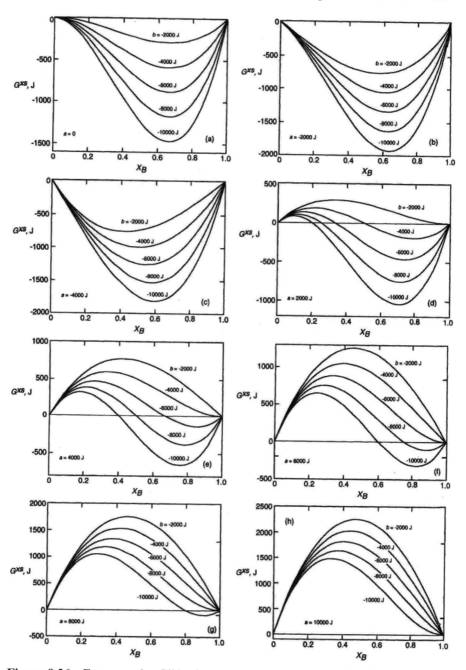

Figure 9.26 Excess molar Gibbs free energy curves generated by the subregular solution model.

Thus, as shown in Fig. 9.26a, with $a = 0$, the minimum in the curve occurs at $X_B = \frac{2}{3}$, and with $b = 0$ the solution behavior is regular. In Fig. 9.26e, with $a = 4000$ J and $b = -10,000$ J, a maximum occurs in the curve at $X_B = 0.17$ and a minimum occurs at $X_B = 0.76$.

The variation of Ω with composition in the system Ag–Au, obtained from the experimental measurements of Oriani at 1344 K, is shown in in Fig. 9.27.* Fitting these data to the subregular solution model with $a = -13,465$ J and $b = 5412.8$ gives the variation of ΔG^M with composition shown by the line in Fig. 9.28. The open circles are experimentally measured values of ΔG^M.

The influence of temperature on the behavior of subregular solutions is accommodated by introducing a third constant, τ, to give the molar excess Gibbs free energy of mixing as

$$G^{XS} = (a_0 + b_0 X_B) X_A X_B \left(1 - \frac{T}{\tau} \right) \tag{9.95}$$

The molar excess entropy of mixing is thus

$$S^{XS} = -\frac{\partial G^{XS}}{\partial T} \tag{9.96}$$

$$= \frac{(a_0 + b_0) X_A X_B}{\tau}$$

and the molar heat of mixing (which is also the molar excess heat of mixing) is given by

$$\Delta H^M = G^{XS} + T S^{XS} \tag{9.97}$$

$$= (a_0 + b_0 X_B) X_A X_B \left(2 - \frac{T}{\tau} \right)$$

9.12 SUMMARY

1. Raoult's law is $p_i = X_i p_i^\circ$, and a component of a solution that conforms with this law is said to exhibit Raoultian behavior. In all solutions, the behavior of the component i approaches Raoult's law as $X_i \to 1$.
2. Henry's law is $p_i = k' X_i$, and a component of a solution which conforms with this equation is said to exhibit Henrian behavior. In all solutions, the behavior of the component i approaches Henry's law as $X_i \to 0$. In a binary solution Henry's law is obeyed by the solute in that composition range over which Raoult's law is obeyed by the solvent.
3. The activity of the component i in a solution, with respect to a given standard state, is the ratio of the vapor pressure of i (strictly, the fugacity of i) exerted by the solution to the vapor pressure (the fugacity) of i in the given standard state. If the standard state is chosen as being pure i, then $a_i = p_i/p_i^\circ$. An activity is thus

---

*R. A. Oriani, "Thermodynamics of Liquid Ag-Au and Au-Cu Alloys and the Question of Strain Energy in Solid Solutions," *Acta Met.* (1956), vol. 4, p. 15.

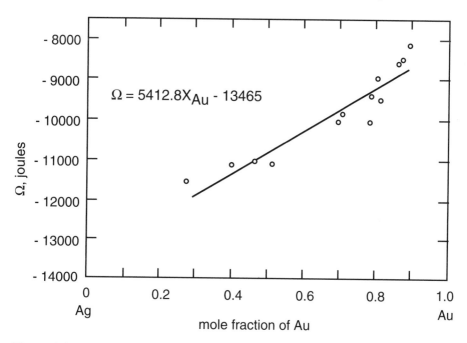

Figure 9.27 The variation, with composition, of Ω, calculated from experimental measurements of ΔG^M in the system Ag–Au at 1350 K.

a ratio, and its introduction effects a normalization of the vapor pressure exerted by the component i in the solution. In terms of activity, Raoult's law is $a_i = X_i$, and Henry's law is $a_i = kX_i$.

4. The difference between the value of an extensive thermodynamic property per mole of i in a solution, and the value of the property per mole of i in its standard state is called the partial molar property change of i for the solution process, i.e., if Q is any extensive thermodynamic property, the change in the property due to solution of 1 mole of i is $\Delta \overline{Q}_i^M = \overline{Q}_i - Q_i^\circ$. In the case of Gibbs free energy $\Delta \overline{G}_i^M = \overline{G}_i - G_i^\circ$. This difference in the molar Gibbs free energy is related to the activity of i in solution, with respect to the standard state, as $\Delta \overline{G}_i^M = RT \ln a_i$, and $\Delta \overline{G}_i^M$ is called the partial molar Gibbs free energy of solution of i.

   The change in the Gibbs free energy accompanying the formation of 1 mole of solution from the pure components i (called the integral Gibbs free energy change) is $\Delta G^M = \Sigma_i X_i \Delta \overline{G}_i^M$, so that, for the binary A–B, $\Delta G^M = X_A \Delta \overline{G}_A^M + X_B \Delta \overline{G}_B^M$. As $\Delta \overline{G}_A^M = RT \ln a_A$, then $\Delta G^M = RT(X_A \ln a_A + X_B \ln a_B)$. In a Raoultian solution, as $a_i = X_i$, then $\Delta G^M = RT(X_A \ln X_A + X_B \ln X_B)$. For any general extensive thermodynamic property $Q, \Delta Q^M = \Sigma_i X_i \Delta \overline{Q}_i^M$.

5. A Raoultian solution has the properties $a_i = X_i$, $\overline{V}_i = V_i^\circ$ (i.e., there is no change in volume when the components are mixed), $\overline{H}_i^M = H_i^\circ$ (i.e., there is zero heat of mixing), and $\Delta G^{M,\text{id}} = RT(X_A + X_B \ln X_B)$. As $\Delta S^{M,\text{id}} = -(\partial \Delta G^{M,\text{Bid}}/\partial T)$, $\Delta S^{M,\text{id}} = -R \Sigma_i X_i \ln X_i$, so that, in a Raoultian solution, $\Delta \overline{S}_i^M = -R \ln X_i$.

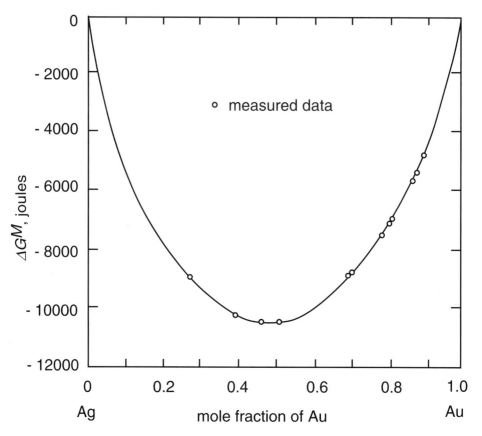

Figure 9.28 The subregular solution model fitted to experimental measurements of ΔG^M in the system Ag–Au at 1350 K as $\Delta G^M = RT (X_{Ag} \ln X_{Ag} + X_{Au} \ln X_{Au}) + (5{,}412.8 X_{Au} - 13{,}465)X_{Au}X_{Ag}$.

$\Delta S^{M,id}$ is thus independent of temperature and is simply an expression for the maximum number of spatial configurations available to the system.

6. The thermodynamic behavior of non-Raoultian solutions is dealt with by introducing the activity coefficient, γ, which for the component i is defined as $\gamma_i = a_i/X_i$. The coefficient γ_i, which, can have values of greater or less than unity, thus quantifies the deviation of i from Raoultian behavior. As $\ln a_i = \ln X_i + \ln \gamma_i$, $d \ln a_i/d(1/T) = \Delta \bar{H}_i^M/R = d \ln \gamma_i/d(1/T)$. Thus if $d\gamma_i/dT$ is positive, $\Delta \bar{H}_i^M$ is negative and vice versa. The magnitude of the heat of formation of a nonideal solution is determined by the magnitudes of the deviations of the components of the solution from Raoultian behavior. Nonideal components approach Raoultian behavior with increasing temperature. Thus if $\gamma_i < 1$, then $d\gamma_i/dT$ is positive and if $\gamma_i > 1$, $d\gamma_i/dT$ is negative. Solutions, the components of which exhibit negative deviations from Raoult's law, form exothermically, i.e., $\Delta H^M < 0$, and vice versa.

7. The Gibbs-Duhem relationship is $\Sigma_i X_i d\overline{Q}_i = 0$ at constant temperature and pressure, where $\overline{Q}_i$ is the partial molar value of the extensive thermodynamic function Q of the solution component i. The excess value of an extensive thermodynamic property of a solution is the difference between the actual value and the value that the property would have if the components obeyed Raoult's law. Thus, for the general function Q, $Q^{XS} = Q - Q^{id}$, or for the Gibbs free energy, $G^{XS} = G - G^{id}$, or $G^{XS} = \Delta G^M - \Delta G^{M,id}$. As $\gamma_i = a_i/X_i$, then $G^{XS} = RT \Sigma X_i \ln \gamma_i$.

8. A regular solution is one which has an ideal entropy of formation and a non-zero heat of formation from its pure components. The activity coefficients of the components of a regular solution are given by the expression $RT \ln \gamma_i = \alpha'(1 - X_i)^2$, where α' is a temperature-independent constant, the value of which is characteristic of the particular solution. Thus $\ln \gamma_i$ varies inversely with temperature, and, as $\overline{G}_i^{XS} = RT \ln \gamma_i$, then $\overline{G}^{XS} = \Delta \overline{H}_i^M$ is independent of temperature. Furthermore, the heat of formation of a regular solution, being equal to G^{XS}, is a parabolic function of composition, given by $\Delta H^M = G^{XS} = RT\alpha X_A X_B = \alpha' X_A X_B$.

9. Regular solution behavior is predicted by a statistical solution model in which it is assumed that the atoms mix randomly and that the energy of the solution is the sum of the individual interatomic bond energies in the solution. Random mixing can be assumed only if, in the system $A–B$, the $A–B$ bond energy is not significantly different from the average of the $A–A$ and $B–B$ bond energies in the pure components. For any such deviation the validity of the assumption of random mixing increases with increasing temperature. The statistical model predicts tendency toward Raoultian behavior and Henrian behavior as, respectively, $X_i \to 1$ and as $X_i \to 0$.

10. The subregular solution model is one in which the value of Ω is assumed to be a linear function of composition, being given by $\Omega = a + bX_B$. The variation of the molar excess Gibbs free energy of mixing is thus given by $G^{XS} = (a + bX_B)X_A X_B$. The constants a and b are curve-fitting parameters and have no physical significance.

9.13 NUMERICAL EXAMPLES

Example 1

Copper and gold form complete ranges of solid solution at temperatures between 410°C and 889°C, and, at 600°C, the excess molar Gibbs free energy of formation of the solid solutions is given by

$$G^{XS} = -28,280 X_{Au} X_{Cu} \text{ J}$$

Calculate the partial pressures of Au and Cu exerted by the solid solution of $X_{Cu} = 0.6$ at 600°C.

From Eq. (9.87), the solid solutions are regular with $\Omega = -28,280$ J. Therefore, from Eq. (9.91),

$$\ln \gamma_{Cu} = \frac{\Omega}{RT} X_{Au}^2 = -\frac{28,280 \times 0.4^2}{8.3144 \times 873} = -0.624$$

Thus,

$$\gamma_{Cu} = 0.536 \quad \text{and} \quad a_{Cu} = \gamma_{Cu}X_{Cu} = 0.536 \times 0.6 = 0.322$$

Similarly,

$$\ln \gamma_{Au} = -\frac{28,280 \times 0.6^2}{8.3144 \times 873} = -1.403$$

Thus

$$\gamma_{Au} = 0.246 \quad \text{and} \quad a_{Au} = 0.246 \times 0.4 = 0.098$$

The saturated vapor pressure of solid copper is given by

$$\ln p^{\circ}_{Cu} \text{ (atm)} = -\frac{40,920}{T} - 0.86 \ln T + 21.67$$

and the saturated vapor pressure of solid gold is given by

$$\ln p^{\circ}_{Au} \text{ (atm)} = -\frac{45,650}{T} - 0.306 \ln T + 10.81$$

Therefore, at 873 K,

$$p^{\circ}_{Cu} = 3.35 \times 10^{-14} \text{ atm}$$

and

$$p^{\circ}_{Au} = 1.52 \times 10^{-16} \text{ atm}$$

From Eq. (9.12), $a_i = p_i/p^{\circ}_i$, and thus the partial pressures exerted by the alloy are

$$p_{Cu} = 0.322 \times 3.35 \times 10^{-14} = 1.08 \times 10^{-14} \text{ atm}$$

and

$$p_{Au} = 0.098 \times 1.52 \times 10^{-16} = 1.50 \times 10^{-16} \text{ atm}$$

Example 2

At 700 K, the activity of Ga in a liquid Ga–Cd solution of composition $X_{Ga} = 0.5$ has the value 0.79. On the assumption that liquid solutions of Ga and Cd exhibit regular solution behavior, estimate the energy of the Ga–Cd bond in the solution. The molar enthalpies of evaporation of liquid Ga and liquid Cd at their melting temperatures are, respectively, 270,000 and 100,000 J.

With $a_{Ga} = 0.79$ at $X_{Ga} = 0.5$,

$$\gamma_{Ga} = \frac{a_{Ga}}{X_{Ga}} = \frac{0.79}{0.5} = 1.59$$

Therefore, from Eq. (9.91),

$$\ln 1.59 = \frac{\Omega \times 0.5^2}{8.3144 \times 700}$$

which gives

$$\Omega = \frac{0.464 \times 8.3144 \times 700}{0.5^2} = 10,795 \text{ J}$$

At their melting temperatures, the coordination numbers of liquid Cd and liquid Ga are, respectively, 8 and 11. It will thus be assumed that the coordination number in the 50–50 solution is the average of 8 and 11, namely, 9.5. The bond energy, E_{Ga-Ga}, is obtained from the molar enthalpy of evaporation, ΔH_{evap}, according to

$$\Delta H_{evap,Ga-Ga} = -\frac{1}{2}zN_0E_{Ga-Ga}$$

The negative sign is required to conform with the convention that bond energies are negative quantities. Thus

$$E_{Ga-Ga} = -\frac{270,000 \times 2}{11 \times 6.023 \times 10^{23}} = -8.15 \times 10^{-20} \text{ J}$$

and similarly,

$$E_{Cd-Cd} = -\frac{100,000 \times 2}{8 \times 6.023 \times 10^{23}} = -4.15 \times 10^{-20} \text{ J}$$

The bond energy, E_{Cd-Ga}, is obtained from

$$\Omega = zN_0\left[E_{Cd-Ga} - \frac{1}{2}(E_{Cd-Cd} + E_{Ga-Ga})\right]$$

i.e.,

$$10,795 = 9.5 \times 6.023 \times 10^{23}\left[E_{Cd-Ga} - \frac{1}{2}(-4.15 \times 10^{-20} - 8.15 \times 10^{-20})\right]$$

as

$$E_{Cd-Ga} = -5.96 \times 10^{-20} \text{ J}$$

PROBLEMS

9.1 One mole of solid Cr_2O_3 at 2500 K is dissolved in a large volume of a liquid Raoultian solution of Al_2O_3 and Cr_2O_3 in which $X_{Cr_2O_3} = 0.2$ and which is also at 2500 K. Calculate the changes in enthalpy and entropy caused by the addition. The normal melting temperature of Cr_2O_3 is 2538 K, and it can be assumed that the $\Delta S_{m,Al_2O_3} = \Delta S_{m,Cr_2O_3}$.

9.2 When 1 mole of argon gas is bubbled through a large volume of an Fe–Mn melt of $X_{Mn} = 0.5$ at 1863 K evaporation of Mn into the Ar causes the mass of the melt to decrease by 1.50 g. The gas leaves the melt at a pressure of 1 atm. Calculate the activity coefficient of Mn in the liquid alloy.

9.3 The variation, with composition, of G^{XS} for liquid Fe–Mn alloys at 1863 K is listed below.

    a. Does the system exhibit regular solution behavior?
    b. Calculate $\bar{G}^{XS}_{Fe}$ and $\bar{G}^{XS}_{Mn}$ at $X_{Mn} = 0.6$.
    c. Calculate ΔG^M at $X_{Mn} = 0.4$.
    d. Calculate the partial pressures of Mn and Fe exerted by the alloy of $X_{Mn} = 0.2$.

X_{Mn}	0.1	0.2	0.3	0.4	0.5	0.6	0.7	0.8	0.9
G^{XS} joules	395	703	925	1054	1100	1054	925	703	395

9.4 Calculate the heat required to form a liquid solution at 1356 K starting with 1 mole of Cu and 1 mole of Ag at 298 K. At 1356 K the molar heat of mixing of liquid Cu and liquid Ag is given by $\Delta H^M = -20{,}590 X_{Cu} X_{Ag}$.

9.5 Melts in the system Pb–Sn exhibit regular solution behavior. At 473°C $a_{Pb} = 0.055$ in a liquid solution of $X_{Pb} = 0.1$. Calculate the value of Ω for the system and calculate the activity of Sn in the liquid solution of $X_{Sn} = 0.5$ at 500°C.

9.6 The activities of Cu in liquid Fe–Cu alloys at 1550°C have been determined as

X_{Cu}	1	0.9	0.8	0.7	0.6	0.5	0.4	0.3	0.2	0.1	0.05
a_{Cu}	1.0	0.935	0.895	0.865	0.850	0.830	0.810	0.780	0.720	0.575	0.40

Using, separately, Eqs. (9.55) and (9.61), calculate the variation of a_{Fe} with composition in the system at 1550°C.

9.7 The activities of Ni in liquid Fe–Ni alloys at 1600°C have been determined as

X_{Ni}	1	0.9	0.8	0.7	0.6	0.5	0.4	0.3	0.2	0.1
a_{Ni}	1	0.89	0.766	0.62	0.485	0.374	0.283	0.207	0.136	0.067

Using, separately, Eqs. (9.55) and (9.61), calculate the variation of a_{Fe} with composition in the system at 1600°C.

9.8 Tin obeys Henry's law in dilute liquid solutions of Sn and Cd and the Henrian activity coefficient of Sn, γ°_{Sn}, varies with temperature as

$$\ln \gamma^{\circ}_{Sn} = -\frac{840}{T} + 1.58$$

Calculate the change in temperature when 1 mole of liquid Sn and 99 moles of liquid Cd are mixed in an adiabatic enclosure. The molar constant pressure heat capacity of the alloy formed is 29.5 J/K.

9.9 Use the Gibbs-Duhem equation to show that, if the activity coefficients of the components of a binary solution can be expressed as

$$\ln \gamma_A = \alpha_1 X_B + \frac{1}{2}\alpha_2 X_B^2 + \frac{1}{3}\alpha_3 X_B^3 + \cdots$$

and

$$\ln \gamma_B = \beta_1 X_A + \frac{1}{2}\beta_2 X_A^2 + \frac{1}{3}\beta_3 X_A^3 + \cdots$$

over the entire range of composition, then $\alpha_1 = \beta_1 = 0$, and that, if the variation can be represented by the quadratic terms alone, then $\alpha_2 = \beta_2$.

9.10 The activity coefficient of Zn in liquid Zn–Cd alloys at 435°C can be represented as

$$\ln \gamma_{Zn} = 0.875 \, X_{Cd}^2 - 0.30 \, X_{Cd}^3$$

Derive the corresponding expression for the dependence of $\ln \gamma_{Cd}$ on composition and calculate the activity of cadmium in the alloy of $X_{Cd} = 0.5$ at 435°C.

9.11 The molar excess Gibbs free energy of formation of solid solutions in the system Au–Ni can be represented by

$$G^{XS} = X_{Ni}X_{Au}(24,140 \, X_{Au} + 38,280 \, X_{Ni} - 14,230 \, X_{Au}X_{Ni})\left(1 - \frac{T}{2660}\right) J$$

Calculate the activities of Au and Ni in the alloy of $X_{Au} = 0.5$ at 1100 K.

Chapter 10

GIBBS FREE ENERGY COMPOSITION AND PHASE DIAGRAMS OF BINARY SYSTEMS

10.1 INTRODUCTION

It has been seen that, at constant temperature and pressure, the stable state of existence of a system is that which has the minimum possible value of Gibbs free energy. Thus, phase stability in a system, as normally presented on an isobaric phase diagram, can be determined from knowledge of the variations of the Gibbs free energies of the various possible phases with composition and temperature. When a liquid solution is cooled, a liquidus temperature is eventually reached, at which point a solid phase begins to separate from the liquid solution. This solid phase could be a virtually pure component, a solid solution of the same or different composition from the liquid, or a chemical compound formed by reaction between two or more of the components. In all possible cases the composition of the solid phase which is in equilibrium with the liquid solution is that which minimizes the Gibbs free energy. If liquid solutions are stable over the entire range of composition, then the Gibbs free energies of the liquid states are lower'than those of any possible solid state, and conversely, if the temperature of the system is lower than the lowest solidus temperature, then the Gibbs free energies of the solid states are everywhere lower than those of liquid states. At intermediate temperatures, the variation of Gibbs free energy with composition will identify ranges of composition over which liquid states are stable, ranges over which solid states are stable, and intermediate ranges in which solid and liquid phases coexist in equilibrium with one another. Thus, by virtue of the facts that (1) the state of lowest Gibbs free energy is the stable state and (2) when phases coexist in equilibrium $\bar{G}_i$ has the same value in all of the coexisting phases, there must exist a quantitative correspondence between Gibbs free energy-composition diagrams and "phase diagrams." This correspondence is examined in this chapter, in which it will be seen that "normal" phase diagrams are generated by, and are simply representations of, Gibbs free energy–composition diagrams.

10.2 GIBBS FREE ENERGY AND THERMODYNAMIC ACTIVITY

The Gibbs free energy of mixing of the components A and B to form a mole of solution is given by

$$\Delta G^M = RT(X_A \ln a_A + X_B \ln a_B)$$

and ΔG^M is the difference between the Gibbs free energy of a mole of the homogeneous solution and the Gibbs free energy of the corresponding numbers of moles of the unmixed components. As only changes in Gibbs free energy can be measured, the Gibbs free energies of the pure unmixed components are assigned the value of zero. If the solution is ideal, i.e., if $a_i = X_i$, then the molar Gibbs free energy of mixing, given by

$$\Delta G^{M,\text{id}} = RT(X_A \ln X_A + X_B \ln X_B)$$

has the characteristic shape shown, at the temperature T, as curve I in Fig. 10.1. As $\Delta H^{M,\text{id}} = 0$, then $\Delta G^{M,\text{id}} = -T \Delta S^{M,\text{id}}$, and hence curve I in Fig. 10.1 is obtained as $-T \times$ (the curve drawn in Fig. 9.7). It is thus seen that the shape of the variation of $\Delta G^{M,\text{id}}$ with composition depends only on temperature.

If the solution exhibits a slight positive deviation from ideal mixing, i.e., if $\gamma_i > 1$ and $a_i > X_i$, then, at the temperature T, the Gibbs free energy of mixing curve is typically as shown by curve II in Fig. 10.1; and if the solution shows a slight negative deviation from ideal mixing, i.e., if $\gamma_i < 1$ and $a_i < X_i$, the Gibbs free energy of mixing curve is typically as shown by curve III in Fig. 10.1. From Eqs. (9.33a and b) the tangent drawn to the ΔG^M curve at any composition intersects the $X_A = 1$ and $X_B = 1$ axes at $\Delta \overline{G}_A^M$ and $\Delta \overline{G}_B^M$, respectively, and, as $\Delta \overline{G}_i^M = RT \ln a_i$, a correspondence is provided between the ΔG^M-composition and activity-composition curves. In Fig. 10.1, at the composition Y, tangents drawn to curves I, II, and III intersect the $X_B = 1$ axis at a, b, and c, respectively. Thus

$$|Bb = \Delta \overline{G}_B^M = RT \ln a_B \text{ (in system II)}| < |Ba = \Delta \overline{G}_B^M = RT \ln X_B|$$
$$< |Bc = \Delta \overline{G}_B^M = RT \ln a_B \text{ (in system III)}|$$

from which it is seen that

$$\gamma_B \text{ in system II} > 1 > \gamma_B \text{ in system III}$$

The variation, with composition, of the tangential intercepts generates the variations of activity with composition shown in Fig. 10.2.

As $X_i \to 0$, $a_i \to 0$, and hence the tangential intercept $\Delta \overline{G}_i^M = RT \ln a_i \to -\infty$, which indicates that all Gibbs free energy of mixing curves have vertical tangents at their extremities. Similarly, by virtue of being logarithmic, the entropy of the mixing curve shown in Fig. 9.7 has vertical tangents at its extremities.

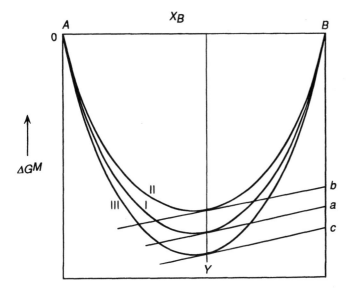

Figure 10.1 The molar Gibbs free energies of mixing in binary systems exhibiting ideal behavior (I), positive deviation from ideal behavior (II), and negative deviation from ideal behavior (III).

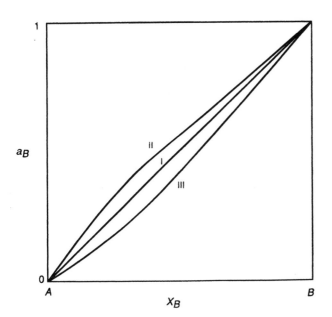

Figure 10.2 The activities of component B obtained from lines I, II, and III in Fig. 10.1.

10.3 THE GIBBS FREE ENERGY OF FORMATION OF REGULAR SOLUTIONS

If curves II and III in Fig. 10.1 are drawn for regular solutions, then deviation of ΔG^M from $\Delta G^{M,\text{id}}$ is due only to the nonzero heat of mixing and the difference between the two curves, $\Delta G^M - \Delta G^{M,\text{id}}$

$$= G^{\text{XS}} = RT\alpha X_A X_B = \Omega X_A X_B = \Delta H^M$$

For curve II, $|\Delta G^M| < |\Delta G^{M,\text{id}}|$, and thus ΔH^M is a positive quantity (α and Ω are positive quantities). It is of interest to consider the effect of increasingly positive values of α on the shape of the Gibbs free energy of mixing curve for a regular solution. In Fig. 10.3, curve I is drawn as $- \Delta S^{M,\text{id}}/R = X_A \ln X_A + X_B \ln X_B$. This curve represents $\Delta G^{M,\text{id}}/RT$. Curves for $\Delta H^M/RT = \alpha X_A X_B$ are drawn for $\alpha = 0, +0.5, +1.0, +1.5, +2.0, +2.5,$ and $+3.0$, and the corresponding $\Delta G^M/RT$ curves are drawn as the sum of the particular $\Delta H^M/RT$ and $-\Delta S^{M,\text{id}}/R$ curves. As the magnitude of α is increased it is seen that the shape of the $\Delta G^M/RT$ curve continuously changes from a shape typi-

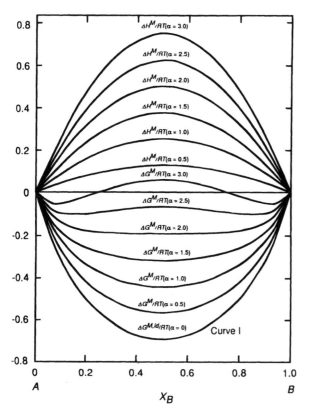

Figure 10.3 The effect of the magnitude of α on the integral molar heats and integral molar Gibbs free energies of formation of a binary regular solution.

fied by $\alpha = 0$ to a form typified by $\alpha = 3$. Before discussing the consequences of this change of shape on the behavior of the solutions, it is pertinent to examine the significance of the shape of the curve. Curve I from Fig. 10.1 is reproduced in Fig. 10.4a. This curve is "convex downwards" at all compositions. Thus the homogeneous solution formed from any mixture of A and B is the stable state, as this state has the lowest possible Gibbs free energy. Consider, further, two separate solutions, say, a and b in Fig. 10.4a. Before mixing of these two solutions, the Gibbs free energy of the two-solution system, with respect to pure A and pure B, lies on the straight line joining a and b, with the exact position being determined, via the lever rule, by the relative proportions of the separate solutions. If the solutions a and b are present in equal amounts then the Gibbs free energy of the system is given by the point c. When mixed, the two solutions form a new homogeneous solution, as thereby the Gibbs free energy of the system is decreased from c and d, the minimum Gibbs free energy which it can have. Consider now Fig. 10.4b in which the $\Delta G^M/RT$ curve for $\alpha = +3.0$ is reproduced from Fig. 10.3. This curve is "convex downwards" only between a and n and between p and B and is "convex upwards" between n and p. The Gibbs free energy of a system of composition between m and q is minimized when the system occurs as two solu-

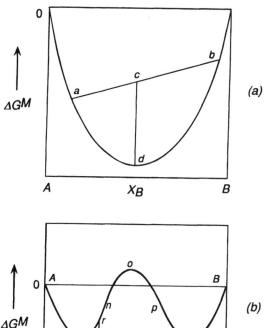

Figure 10.4 (*a*) The molar Gibbs free energies of mixing of binary components which form a complete range of solutions. (*b*) The molar Gibbs free energies of mixing of binary components in a system which exhibits a miscibility gap.

tions, one of composition m and the other of composition q; e.g., if the homogeneous solution of composition r separates into the two coexisting solutions m and q, the Gibbs free energy of the system is decreased from r to s. The equilibrium coexistence of two separate solutions at the temperature T and pressure P requires that

$$\overline{G}_A \text{ (in solution } m) = \overline{G}_A \text{ (in solution } q) \tag{i}$$

and

$$\overline{G}_B \text{ (in solution } m) = \overline{G}_B \text{ (in solution } q) \tag{ii}$$

Subtracting G_A° from both sides of Eq. (i) gives

$$RT \ln a_{A(\text{in solution } m)} = RT \ln a_{A(\text{in solution } q)}$$

or

$$a_{A(\text{in solution } m)} = a_{A(\text{in solution } q)} \tag{iii}$$

Similarly

$$a_{B(\text{in solution } m)} = a_{B(\text{in solution } q)} \tag{iv}$$

Equations (iii) and (iv) are the criteria for equilibrium coexistence of two solutions (or phases) at constant T and P. As $\Delta \overline{G}_A^M$ (in m) $= \Delta \overline{G}_A^M$ (in q), and $\Delta \overline{G}_B^M$ (in m) $= \Delta \overline{G}_B^M$ (in q), then it is seen that the tangent to the curve at the point m is also the tangent to the curve at the point q. The positioning of this double tangent defines the positions of the points m and q on the Gibbs free energy of mixing curve.

The A–B system, as represented in Fig. 10.4b, is one in which, at the temperature T, the value of α is sufficiently positive that the consequent tendency toward clustering of like atoms is great enough to cause phase separation. A homogeneous solution (phase I) is formed when B is initially added to A and saturation of phase I with B occurs at the composition m. Further addition of B causes the appearance of a second solution (phase II) of composition q (which is phase II saturated with A), and continued addition of B causes an increase in the ratio of phase II to phase I occurring, until the overall composition of the two-phase system reaches q, at which point phase I disappears. A homogeneous solution (phase II) occurs between the compositions q and B. The curve mn represents the Gibbs free energy of phase I supersaturated with B, and the curve qp represents the Gibbs free energy of phase II supersaturated with B. As the line $AmqB$ represents the equilibrium states of the system, then this line alone has physical significance, and the line is the isobaric, isothermal section of the system as it occurs in G-T-P-composition space.

10.4 CRITERIA FOR PHASE STABILITY IN REGULAR SOLUTIONS

For a given temperature it is obvious that a critical value of α occurs below which a homogeneous solution is stable over the entire range of composition and above which phase separation occurs. The criteria used to determine this critical value are illustrated in Fig. 10.5. Fig. 10.5a, b, and c show the variations of ΔG^M, $\partial \Delta G^M / \partial X_B$, $\partial^2 \Delta G^M / \partial X_B^2$, and $\partial^3 \Delta G^M / \partial X_B^3$ with composition for $\alpha < \alpha_{\text{critical}}$, $\alpha = \alpha_{\text{critical}}$, and $\alpha > \alpha_{\text{critical}}$ respectively. The critical value of α is seen to be that which makes $\partial^2 \Delta G^M / \partial X_B^2$ and $\partial^3 \Delta G^M / \partial X_B^3$ simultaneously equal to zero at that composition at which immiscibility becomes imminent. For a regular solution,

$$\Delta G^M = RT(X_A \ln X_A + X_B \ln X_B) + RT\alpha X_A X_B$$

$$\frac{\partial \Delta G^M}{\partial X_B} = RT \left[\ln \frac{X_B}{X_A} + \alpha(X_A - X_B) \right]$$

$$\frac{\partial^2 \Delta G^M}{\partial X_B^2} = RT \left(\frac{1}{X_A} + \frac{1}{X_B} - 2\alpha \right)$$

and

$$\frac{\partial^3 \Delta G^M}{\partial X_B^3} = RT \left(\frac{1}{X_A^2} - \frac{1}{X_B^2} \right)$$

The third derivative, $\partial^3 \Delta G^M / \partial X_B^3 = 0$ at $X_A = X_B = 0.5$, and thus the second derivative, $\partial^2 \Delta G^M / X_B^2 = 0$ at $X_A = 0.5$, when $\alpha = 2$, which is thus the critical value of α above which phase separation occurs. As α is an inverse function of temperature, given by

$$\Omega = RT\alpha$$

a critical temperature occurs in any regular system with a positive value of Ω, above which $\alpha < 2$ and below which $\alpha > 2$. The critical temperature, T_{cr}, is

$$T_{\text{cr}} = \frac{\Omega}{2R} \tag{10.1}$$

Fig. 10.6a shows the variation, with temperature, of the Gibbs free energy of mixing curve for a regular solution which has a positive molar heat of mixing ($\Omega = 16,630$ joules) and a critical temperature of $T_{\text{cr}} = 16,630/2R = 1000$ K. The Gibbs free energy expression contains a negative logarithmic term, the magnitude of which is proportional to temperature, and a positive parabolic term which is independent of temperature. At high enough temperature, the logarithmic contribution predominates and the Gibbs free energy of mixing is convex downwards at all compositions. However, with decreasing temperature, the contribution of the logarithmic term decreases, and eventually the positive parabolic term predominates and produces a range of composition centered on $X_B = 0.5$ over which the Gibbs free energy curve is convex upwards. The logarithmic term still requires that the tangents to the curve at $X_A = 1$ and $X_B = 1$ be vertical. Fig. 10.6b shows the phase diagram for the system, in which the miscibility curve bounding the two-phase region is simply the locus of the double tangent compositions in Fig. 10.6a. The influence of temperature on the variations of the activity of component B with composition is shown in Fig. 10.6c. The activities are obtained from the intercepts, with the $X_B = 1$ axis, of tangents

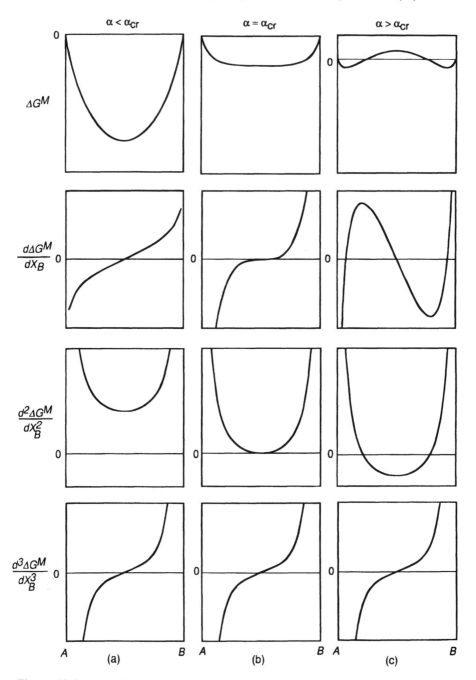

Figure 10.5 The effect of the magnitude of α on the first, second, and third derivatives of the integral Gibbs free energy of mixing with respect to composition.

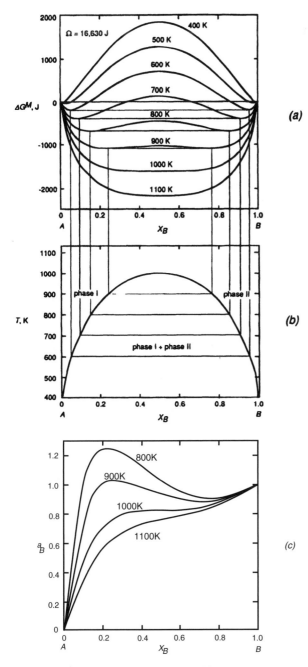

Figure 10.6 (*a*) The effect of temperature on the molar Gibbs free energy of mixing a binary regular solution for which $\Omega = 16,630$ joules. (*b*) The loci of the double tangent points in (*a*), which generate the phase diagram for the system. (*c*) The activities of component B derived from (*a*).

drawn to the free energy curves as $\Delta \overline{G}_B^M = RT \ln a_B$. At T_{cr} the activity exhibits a horizontal inflexion at $X_B = 0.5$, as is seen from the following. From Eq. (9.33b),

$$\Delta \overline{G}_B^M = \Delta G^M + X_A \left(\frac{\partial \Delta G^M}{\partial X_B} \right) = RT \ln a_B$$

Thus

$$\frac{\partial \Delta \overline{G}_B^M}{\partial X_B} = X_A \frac{\partial^2 \Delta G^M}{\partial X_B^2} = \frac{RT}{a_B} \frac{\partial a_B}{\partial X_B} \tag{10.2}$$

and

$$\frac{\partial^2 \Delta \overline{G}_B^M}{\partial X_B^2} = X_A \left(\frac{\partial^3 \Delta G^M}{\partial X_B^3} \right) - \left(\frac{\partial^2 \Delta G^M}{\partial X_B^2} \right) = \frac{RT \partial^2 a_B}{a_B \partial X_B^2} - \frac{RT}{a_B^2} \left(\frac{\partial a_B}{\partial X_B} \right)^2 \tag{10.3}$$

At T_{cr} and $X_B = 0.5$ both the second and third derivatives of ΔG^M with respect to X_B are zero, and thus, from Eqs. (10.2) and (10.3), the first and second derivatives of a_B with respect to X_B are zero, which produces a horizontal inflexion point on the activity curve at $X_B = 0.5$ and T_{cr}. At $T < T_{cr}$ the activity curve has a maximum and a minimum, which occur at the spinodal compositions (where $\partial^2 \Delta G^M / \partial X_B^2$, and hence $\partial a_B / \partial X_B$, are zero), e.g., the points n and p in Fig. 10.4b and the points b and c on the activity curve at 800 K shown in Fig. 10.7. The portion of the curve given by ab in Fig. 10.7 represents the activity of B in phase I which is supersaturated with B and the portion of the activity curve given by cd represents the activity of B in phase II

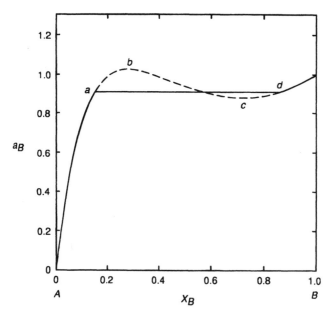

Figure 10.7 The activity of B at 800 K derived from Fig. 10.6a.

which is supersaturated with A. The value of $\partial a_B/\partial X_B$ is negative between b and c, and this violates an intrinsic criterion for stability which requires that $\partial a_i/\partial X_i$ always be positive [cf. $(\partial P/\partial V)_T > 0$ over the portion JHF in Fig. 8.7]. Thus the derived activity curve between b and c, and, consequently, the Gibbs free energy of mixing curve between the spinodal compositions, have no physical significance. The horizontal line drawn between a and d in Fig. 10.7 represents the actual constant activity of B in the two-phase region, and the compositions a and d are those of the double tangents to the Gibbs free energy of mixing curve.

10.5 LIQUID AND SOLID STANDARD STATES

Thus far the standard state of a component of a condensed system has been chosen as being the pure component in its stable state at the particular temperature and pressure of interest. At 1 atm pressure (the pressure normally considered), the stable state is determined by whether or not the temperature of interest is above or below the normal melting temperature of the component. In the discussion of condensed binary solutions, it has been tacitly assumed that the temperature of interest is above or below the melting temperatures of both components, i.e., Fig. 10.7 could be drawn for liquid immiscibility, in which case the standard states are the two pure liquids, or it could be drawn for solid immiscibility, in which case the standard states are the two pure solids. As the standard state of a component is simply a reference state to which the component in any other state is compared, it follows that any state can be chosen as the standard state, and the choice is normally made purely on the basis of convenience.

Consider the binary system A–B at a temperature T which is below $T_{m(B)}$, the melting temperature of B, and above $T_{m(A)}$, the melting temperature of A. Consider, further, that this system forms Raoultian ideal liquid solutions and Raoultian ideal solid solutions. The phase diagram for the system and the temperature of interest, T, are shown in Fig. 10.8a. Fig. 10.8b shows the two Gibbs free energy of mixing curves of interest, curve I drawn for liquid solutions and curve II drawn for solid solutions. At the temperature T, the stable states of pure A and B are located at $\Delta G^M = 0$, with pure liquid A located at $X_A = 1$ (the point a) and pure solid B located at $X_B = 1$ (the point b). The point c represents the molar Gibbs free energy of solid A relative to that of liquid A at the temperature T, and $T > T_{m(A)}$, then $G^\circ_{A(s)} - G^\circ_{A(l)}$ is a positive quantity which is equal to the negative of the molar Gibbs free energy of melting of A at the temperature T. That is,

$$G^\circ_{A(s)} - G^\circ_{A(l)} = -\Delta G^\circ_{m(A)} = -(\Delta H^\circ_{m(A)} - T\Delta S^\circ_{m(A)})$$

and if $c_{p,A(s)} = c_{p,A(l)}$, that is, if $\Delta H^\circ_{m(A)}$ and $\Delta S^\circ_{m(A)}$ are independent of temperature, then

$$\Delta G^\circ_{m(A)} = \Delta H^\circ_{m(A)} \left(\frac{T_{m(A)} - T}{T_{m(A)}} \right) \tag{10.4}$$

Similarly, the point d represents the molar Gibbs free energy of liquid B relative to that of solid B at the temperature T, and, as $T < T_{m(B)}$, then $G^\circ_{B(l)} - G^\circ_{B(s)}$ is a positive

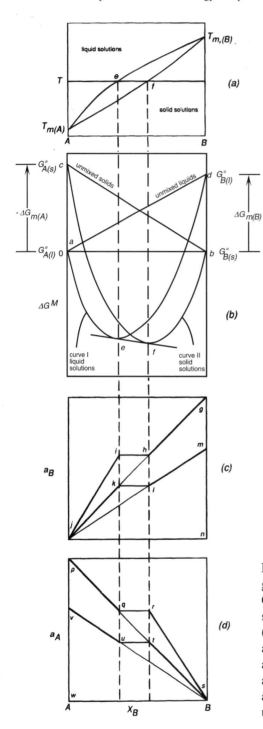

Figure 10.8 (*a*) The phase diagram for the system *A*–*B*. (*b*) The Gibbs free energies of mixing in the system *A*–*B* at the temperature *T*. (*c*) The activities of *B* at the temperature *T* and comparison of the solid and liquid standard states. (*d*) The activities of *A* at the temperature *T*, and comparison of the solid and liquid standard states.

quantity, equal to $\Delta G^\circ_{m(B)}$. The differences between $\Delta G^\circ_{B(l)}$ and $\Delta G^\circ_{B(s)}$, and $\Delta G^\circ_{A(s)}$ and $\Delta G^\circ_{A(l)}$, and their dependence on temperature are shown in Figs. 7.1 and 7.2. The straight line in Fig. 10.8a joining a and d represents the Gibbs free energy of unmixed liquid A and liquid B relative to that of the standard state of unmixed liquid A and solid B, and the straight line joining c and b represents the Gibbs free energy of unmixed solid A and solid B relative to that of the standard state. The straight line cb has the equation

$$\Delta G = -X_A \Delta G^\circ_{m(A)}$$

and the equation for the straight line ad is

$$\Delta G = X_B \Delta G^\circ_{m(B)}$$

At any composition the formation of a homogeneous liquid solution from pure liquid A and pure solid B can be considered as being a two-step process involving

1. The melting of X_B moles of B, which involves the change in Gibbs free energy $\Delta G = X_B \Delta G^\circ_{m(B)}$, and
2. The mixing of X_B moles of liquid B and X_A moles of liquid A to form an ideal liquid solution, which involves the change in Gibbs free energy,

$$\Delta G = \Delta G^{M,\text{id}} = RT(X_A \ln X_A + X_B \ln X_B)$$

Thus, the molar Gibbs free energy of formation of an ideal liquid solution, $\Delta G^M_{(l)}$, from liquid A and solid B is given by

$$\Delta G^M_{(l)} = RT(X_A \ln X_A + X_B \ln X_B) + X_B \Delta G^\circ_{m(B)} \tag{10.5}$$

which is the equation of curve I in Fig. 10.8b. Similarly, at any composition, the formation of an ideal solid solution from liquid A and solid B involves a change in Gibbs free energy of

$$\Delta G^M_{(s)} = RT(X_A \ln X_A + X_B \ln X_B) - X_A \Delta G^\circ_{m(A)} \tag{10.6}$$

which is the equation of curve II in Fig. 10.8b.

At the composition e, the tangent to the curve for the liquid solutions is also the tangent to the solid solution at the composition f. Thus, at the temperature T, liquid of composition e is in equilibrium with solid of composition f, i.e., e is the liquidus composition and f is the solidus composition, as seen in Fig. 10.8a. As the temperature is varied, say, lowered, consideration of Figs. 7.1 and 7.2 shows that the magnitude of ca decreases and the magnitude of db increases. The consequent movement of the positions of curves I and II relative to one another is such that the positions e and f of the double tangent to the curves shift to the left. Correspondingly, if the temperature is increased, the relative movement of the Gibbs free energy curves is such that e and f shift to the right. The loci of e and f with change in temperature trace out the liquidus and solidus lines, respectively.

For equilibrium between the solid and liquid phases,

$$\Delta \overline{G}^M_A \text{ (in the solid solution)} = \Delta \overline{G}^M_A \text{ (in the liquid solution)} \tag{10.7}$$

and

$$\Delta \overline{G}_B^M \text{ (in the liquid solution)} = \Delta \overline{G}_B^M \text{ (in the solid solution)} \qquad (10.8)$$

At any temperature T, these two conditions fix the solidus and liquidus compositions, i.e., the position of the points of double tangency. From Eq. (10.5)

$$\frac{\partial G_{(l)}^M}{\partial X_{A(l)}} = RT (\ln X_{A(l)} - \ln X_{B(l)}) - \Delta G_{m(B)}^\circ$$

Thus

$$X_{B(l)} \frac{\partial \Delta G_{(l)}^M}{\partial X_{A(l)}} = RT(X_{B(l)} \ln X_{A(l)} - X_{B(l)} \ln X_{B(l)}) - X_{B(l)} \Delta G_{m(B)}^\circ \qquad (10.9)$$

From Eq. (9.33a)

$$\Delta \overline{G}_A^M \text{ (in liquid solutions)} = \Delta G_{(l)}^M + X_{B(l)} \frac{\partial \Delta G_{(l)}^M}{\partial X_{A(l)}}$$

and thus, adding Eq. (10.5) and (10.9) gives

$$\Delta \overline{G}_A^M \text{ (in liquid solutions)} = RT \ln X_{A(l)} \qquad (10.10)$$

From Eq. (10.6)

$$\frac{\partial \Delta G_{(s)}^M}{\partial X_{A(s)}} = RT (\ln X_{A(s)} - \ln X_{B(s)}) - \Delta G_{m(A)}^\circ$$

and thus

$$X_{B(s)} \frac{\partial \Delta G_{(s)}^M}{\partial X_{A(s)}} = RT(X_{B(s)} \ln X_{A(s)} - X_{B(s)} \ln X_{B(s)}) - X_{B(s)} \Delta G_{m(A)}^\circ \qquad (10.11)$$

Adding Eqs. (10.6) and (10.11) gives

$$\Delta \overline{G}_A^M \text{ (in solid solutions)} = \Delta G_{(s)}^M + X_{B(s)} \frac{\partial \Delta G_{(s)}^M}{\partial X_{A(s)}} = RT \ln X_{A(s)} - \Delta G_{m(A)}^\circ \qquad (10.12)$$

Thus, from Eqs. (10.7), (10.10), and (10.12)

$$RT \ln X_{A(l)} = RT \ln X_{A(s)} - \Delta G_{m(A)}^\circ \qquad (10.13)$$

Similarly, from Eqs. (10.5) and (9.33b)

$$\Delta \overline{G}_B^M \text{ (in liquid solutions)} = \Delta G_{(l)}^M + X_{A(l)} \frac{\partial \Delta G_{(l)}^M}{\partial X_{B(l)}} = RT \ln X_{B(l)} + \Delta G_{m(B)}^\circ \qquad (10.14)$$

and from Eqs. (10.6) and (9.33b)

$$\Delta \overline{G}_B^M \text{ (in solid solutions)} = \Delta G_{(s)}^M + X_{A(s)} \frac{\partial \Delta G_{(s)}}{\partial X_{B(s)}} = RT \ln X_{B(s)} \qquad (10.15)$$

Thus, from Eqs. (10.8), (10.14), and (10.15),

$$RT \ln X_{B(l)} + \Delta G^\circ_{m(B)} = RT \ln X_{B(s)} \tag{10.16}$$

The solidus and liquidus compositions are thus determined by Eqs. (10.13) and (10.16) as follows. Eq. (10.13) can be written as

$$X_{A(l)} = X_{A(s)} \exp\left(\frac{-\Delta G^\circ_{m(A)}}{RT}\right) \tag{10.17}$$

and, noting that $X_B = 1 - X_A$, Eq. (10.16) can be written as

$$(1 - X_{A(l)}) = (1 - X_{A(s)}) \exp\left(\frac{-\Delta G^\circ_{m(B)}}{RT}\right) \tag{10.18}$$

Combination of Eqs (10.17) and (10.18) gives

$$X_{A(s)} = \frac{1 - \exp(-\Delta G^\circ_{m(B)}/RT)}{\exp(-\Delta G^\circ_{m(A)}/RT) - \exp(-\Delta G^\circ_{m(B)}/RT)} \tag{10.19}$$

and

$$X_{A(l)} = \frac{[1 - \exp(-\Delta G^\circ_{m(B)}/RT)] \exp(-\Delta G^\circ_{m(A)}/RT)}{\exp(-\Delta G^\circ_{m(A)}/RT) - \exp(-\Delta G^\circ_{m(B)}/RT)} \tag{10.20}$$

Thus, if $c_{p,i(s)} = c_{p,i(l)}$, in which case

$$\Delta G^\circ_{m(i)} = \Delta H^\circ_{m(i)} \left[\frac{T_{m(i)} - T}{T_{m(i)}}\right] \tag{10.4}$$

it is seen that the phase diagram for a system which forms ideal solid and liquid solutions is determined only by the melting temperatures and the molar heats of melting of the components.

Example

The system Ge–Si exhibits complete ranges of liquid and solid solutions. (1) Calculate the phase diagram for the system assuming that the solid and liquid solutions are Raoultian in their behavior and (2) calculate the temperature at which the liquidus (and hence solidus) composition exerts its maximum vapor pressure.

Silicon melts at 1685 K, and its standard Gibbs free energy change on melting is

$$\Delta G^\circ_{m,Si} = 50{,}200 - 29.8\, T \text{ J}$$

The saturated vapor pressure of solid Si is

$$\log p^\circ_{Si(s)} \text{ (atm)} = -\frac{23{,}550}{T} - 0.565 \log T + 9.47$$

Germanium melts at 1213 K and its standard Gibbs free energy change on melting is

$$\Delta G^\circ_{m,Ge} = 36{,}800 - 30.3\, T \text{ J}$$

The saturated vapor pressure of liquid Ge is

$$\log p^\circ_{Ge(l)} \text{ (atm)} = -\frac{18{,}700}{T} - 0.565 \log T + 9.99$$

The equation of the liquidus line is then obtained from Eq. (10.20) as

$$X_{Ge,(liquidus),T} = \frac{\exp\left(\dfrac{-\Delta G^\circ_{m,Ge}}{RT}\right)\left[1 - \exp\left(\dfrac{-\Delta G^\circ_{m,Si}}{RT}\right)\right]}{\exp\left(\dfrac{-\Delta G^\circ_{m,Ge}}{RT}\right) - \exp\left(\dfrac{-\Delta G^\circ_{m,Si}}{RT}\right)}$$

and the equation of the solidus line is obtained from Eq. (10.19) as

$$X_{Si,(solidus),T} = \frac{1 - \exp\left(\dfrac{-\Delta G^\circ_{m,Ge}}{RT}\right)}{\exp\left(\dfrac{-\Delta G^\circ_{m,Si}}{RT}\right) - \exp\left(\dfrac{-\Delta G^\circ_{m,Ge}}{RT}\right)}$$

The calculated liquidus and solidus lines are shown in comparison with the measured lines in Fig. 10.9a.

The partial pressure of Si exerted by the solidus composition (and hence by the corresponding liquidus melt) at the temperature T is

$$p_{Si,T} = X_{Si,(solidus),T} \times p^\circ_{Si,(s),T} \tag{i}$$

and the partial pressure of Ge exerted by the liquidus melt composition (and hence by the corresponding solidus) is

$$p_{Ge,T} = X_{Ge,(liquidus),T} \times p^\circ_{Ge,(l),T} \tag{ii}$$

Eqs. (i) and (ii), together with the sum of the partial pressures, are shown in Fig. 10.9b. In Eq. (i) the values of both $X_{Si,(solidus),T}$ and $p^\circ_{Si,(s),T}$ increase with increasing liquidus temperature, and thus the partial pressure of Si exerted by the liquidus composition increases from zero at 1213 K to the saturated vapor pressure of pure solid Si (log $p^\circ_{Si,(s),1683\ K} = -6.33$) at 1685 K. In contrast, in Eq. (ii), increasing the liquidus temperature causes an increase in $p^\circ_{Ge,(l),T}$ and a decrease in $X_{Ge,(liquidus),T}$, and Fig. 10.9b shows that, at lower liquidus temperatures, the influence of $p^\circ_{Ge,(l),T}$ on the partial pressure of Ge predominates and the partial pressure initially increases with increasing liquidus temperature. However, with continued increase in temperature along the liquidus line the relative influence of the dilution of Ge increases, and the partial pressure of Ge passes through a maximum at the liquidus state $X_{Ge} = 0.193$, $T = 1621$ K before decreasing rapidly to zero at 1685 K. The maximum in the partial pressure of Ge causes a maximum in the total vapor pressure to occur at the liquidus state $X_{Ge} = 0.165$, $T = 1630$ K.

Fig. 10.10 shows the Gibbs free energy of mixing curves for a binary system A–B which forms ideal solid solutions and ideal liquid solutions, drawn at a temperature of 500 K, which is lower than $T_{m,(B)}$ and higher than $T_{m,(A)}$. At 500 K, $\Delta G^\circ_{m,(A)} = -1500$ J and $\Delta G^\circ_{m,(B)} = 1000$ J. Fig. 10.10a shows the curves when liquid A and solid B are chosen as the standard states, located at $\Delta G^M = 0$, Fig. 10.10b shows the curves when

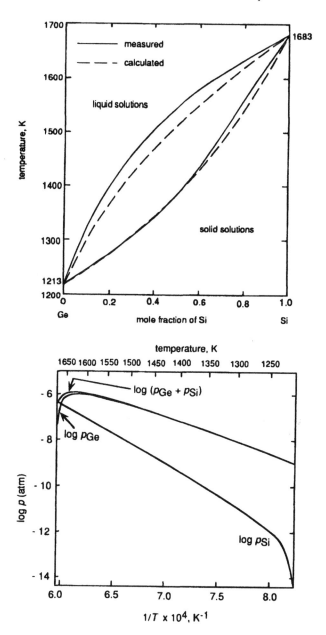

Figure 10.9 (*a*) The calculated phase diagram for the system Ge–Si assuming Raoultian behavior of the solid and liquid solutions. (*b*) The variations, with temperature, of the partial pressures of Ge and Si (and their sum) with composition along the liquidus line.

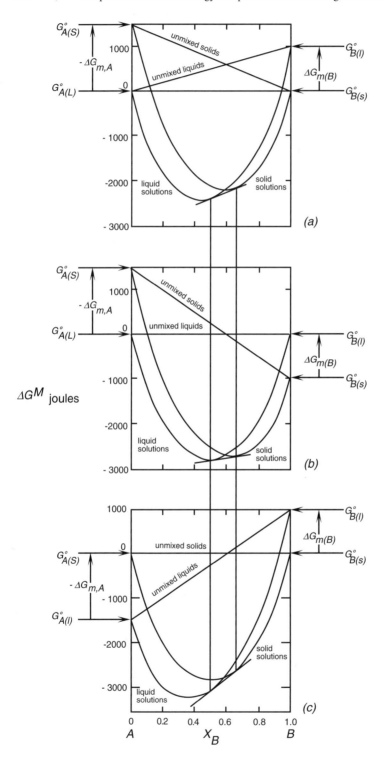

liquid A and liquid B are chosen as the standard states, and Fig. 10.10c shows the curves when solid A and solid B are chosen as the standard states. Comparison among the three shows that, because of the logarithmic nature of the Gibbs free energy curves, the positions of the points of double tangency are not influenced by the choice of standard state; they are determined only by the temperature T and by the magnitude of the difference between $G^{\circ}_{(l)}$ and $G^{\circ}_{(s)}$ for both components at the temperature T.

The activity-composition relationships for component B are shown in Fig. 10.8c. As two standard states are available, the point b for solid B and the point d for liquid B, the lengths of the tangential intercepts with the $X_B = 1$ axis can be measured from b, in which case the activities of B are obtained with respect to solid B as the standard state, or the lengths can be measured from d, which gives the activities with respect to liquid B as the standard state.

If pure solid B is chosen as the standard state and is located at the point g in Fig. 10.8c, then the length gn is, by definition, unity, and this defines the solid standard state activity scale. The line $ghij$ then represents a_B in the solutions with respect to solid B having unit activity at g. The line is obtained from the variation of the tangential intercepts from the curve $aefb$ to the $X_B = 1$ axis, measured from the point b. On this activity scale Raoult's law is given by jg, and the points i and h represent, respectively, the activity of B in the coexisting liquid solution e and solid solution f. The point m represents the activity of pure liquid B measured on the solid standard state activity scale of B. This activity is less than unity, being given by the ratio mn/gn. For B in any state along the $aefb$ Gibbs free energy curve, in which state the partial molar Gibbs free energy of B is $\overline{G}_B$, the following relations hold:

$$\overline{G}_B = G^{\circ}_{B(l)} + RT \ln (a_B \text{ with respect to liquid } B)$$

and

$$\overline{G}_B = G^{\circ}_{B(s)} + RT \ln (a_B \text{ with respect to solid } B)$$

Thus

$$G^{\circ}_{B(l)} - G^{\circ}_{B(s)} = \Delta G^{\circ}_{m(B)} = RT \ln \left(\frac{a_B \text{ with respect to solid } B}{a_B \text{ with respect to liquid } B} \right) \quad (10.21)$$

As $T < T_{m(B)}$, $\Delta G^{\circ}_{m(B)}$ is a positive quantity, and thus the activity of B in any solution with respect to solid B as the standard state is less than the activity of B with respect to liquid B as the standard state, where both activities are measured on the same

Figure 10.10 The Gibbs free energy of mixing curves for a binary system A–B which forms ideal solid solutions and ideal liquid solutions, at a temperature which is higher than $T_{m(A)}$ and lower than $T_{M(B)}$. (a) Liquid A and solid B chosen as standard states located at $\Delta G^M = 0$. (b) Liquid A and liquid B chosen as standard states located at $\Delta G^M = 0$. (c) Solid A and solid B chosen as standard states located at $\Delta G^M = 0$. The positions of the points of double-tangency are not influenced by the choice of standard state.

(solid or liquid) activity scale. For pure B $a_{B(s)} > a_{B(l)}$ i.e., $gn > mn$ in Fig. 10.8c, and, if $gn = 1$, then $mn = \exp(-\Delta G^{\circ}_{m(B)}/RT)$. Equation (10.21) simply states that the length of the tangential intercept from any point on the curve $aefb$, measured from b + the length bd = the length of the tangential intercept from the same point on the curve measured from d, which is a restatement of Eq. (10.21).

If pure liquid B is chosen as the standard state and is located at the point m, then the length mn is, by definition, unity, and this defines the liquid standard state activity scale. Raoult's law on this scale is given by the line jm, and the activities of B in solution, with respect to pure liquid B having unit activity, are represented by the line $mlkj$. The activity of solid B, located at g, is greater than unity on the liquid standard state activity scale, being equal to $\exp(\Delta G^{\circ}_{m(B)}/RT)$. When measured on one or the other of the two activity scales, the lines $jihg$ and $jklm$ vary in the constant ratio $\exp(\Delta G^{\circ}_{m(B)}/RT)$, but $jihg$ measured on the solid standard state activity scale is identical with $jklm$ measured on the liquid standard state activity scale.

The variation of a_A with composition is shown in Fig. 10.8d. In this case, as $T > T_{m(A)}$, $\Delta G^{\circ}_{m(A)}$ is a negative quantity, and hence, from Eq. (10.3) applied to component A,

$$a_A(\text{with respect to liquid } A) > a_A (\text{with respect to solid } A)$$

when measured on the same activity scale. If pure liquid A is chosen as the standard state and is located at the point p, then the length of pw is, by definition, unity, and the line $pqrs$ represents the activity of A in the solution with respect to the liquid standard state. On the liquid standard state activity scale, the activity of pure solid A, located at the point v, has the value $\exp(\Delta G^{\circ}_{m(A)}/RT)$. If, on the other hand, pure solid A is chosen as the standard state, then the length of vw is, by definition, unity, and Raoult's law is given by vs. The line $vuts$ represents the activities of A in the solutions with respect to pure solid A. On the solid standard state activity scale, liquid A, located at the point p, has the value $\exp(-\Delta G^{\circ}_{m(A)}/RT)$. Again, the two lines, measured on one or the other of the two activity scales, vary in the constant ratio $\exp(\Delta G^{\circ}_{m(A)}/RT)$, and when measured on their respective scales, are identical.

If the temperature of the system is decreased to a value less than T indicated in Fig. 10.8a, then the length of ac, being equal to $|\Delta G^{\circ}_{m(A)}|$ at the temperature of interest, decreases, and, correspondingly, the magnitude of $|\Delta G^{\circ}_{m(B)}|$, and hence the length of bd, increase. The consequent change in the positions of the Gibbs free energy of mixing curves I and II in Fig. 10.8b causes the double tangent points e and f to shift to the left toward A. The effect on the activities is as follows. In the case of both components,

$$\frac{a_i \text{ with respect to solid } i}{a_i \text{ with respect to liquid } i} = \exp\left(\frac{\Delta G^{\circ}_{m(i)}}{RT}\right)$$

which, from Eq. (10.4),

$$= \exp\left[\Delta H^{\circ}_{m(i)}\left(\frac{T_{m(i)} - T}{RTT_{m(i)}}\right)\right] \tag{10.22}$$

With respect to component B, if the temperature, which is less than $T_{m(B)}$, is decreased, the ratio $a_{B(solid)}/a_{B(liquid)}$, which is greater than unity, increases. Thus, in Fig. 10.8c, the ratio gn/mn increases. With respect to the component A, if the temperature, which is higher than $T_{m(A)}$, is decreased, then the ratio $a_{A(solid)}/a_{A(liquid)}$, which is less than unity, increases. Thus the ratio vw/pw in Fig. 10.8d increases. At the temperature $T_{m(B)}$, solid and liquid B coexist in equilibrium, $\Delta G^{\circ}_{m(A)} = 0$, and the points p and v coincide. Similarly, at the temperature $T^{\circ}_{m(B)}$ the points m and g coincide.

10.6 PHASE DIAGRAMS, GIBBS FREE ENERGY, AND THERMODYNAMIC ACTIVITY

Complete mutual solid solubility of the components A and B requires that A and B have the same crystal structures, be of comparable atomic size, and have similar electronegativities and valencies. If any one of these conditions is not met, then a miscibility gap will occur in the solid state. Consider the system A–B, the phase diagram of which is shown in Fig. 10.11a, in which A and B have differing crystal structures. Two terminal solid solutions, α and β, occur. The molar Gibbs free energy of mixing curves, at the temperature T_1, are shown in Fig. 10.11b. In this figure, a and c, located at $\Delta G^M = 0$, represent, respectively, the molar Gibbs free energies of pure solid A and pure liquid B, and b and d represent, respectively, the molar Gibbs free energies of pure liquid A and pure solid B. The curve aeg (curve I) is the Gibbs free energy of mixing of solid A and solid B to form homogeneous α solid solutions which have the same crystal structure as has A. This curve intersects the $X_B = 1$ axis at the molar Gibbs free energy which solid B would have if it had the same crystal structure as has A. Similarly, the curve dh (curve II) represents the Gibbs free energy of mixing of solid B and solid A to form homogeneous β solid solutions which have the same crystal structure as has B. This curve intersects the $X_A = 1$ axis at the molar Gibbs free energy which A would have if it had the same crystal structure as B. The curve bfc (curve III) represents the molar Gibbs free energy of mixing of liquid A and liquid B to form a homogeneous liquid solution. As curve II lies everywhere above curve III, solid β solutions are not stable at the temperature T_1. The double tangent to the curves I and III identifies the α solidus composition at the temperature T_1 as e and the liquidus composition as f. Fig. 10.11c shows the activity-composition relationships of the components at the temperature T_1, drawn with respect to solid as the standard state for A and liquid as the standard state for B. These relationships are drawn in accordance with the assumption that the liquid solutions exhibit Raoultian ideality and the solid solutions show positive deviations from Raoult's law.

As the temperature decreases below T_1 the length of ab increases and the length of cd decreases until, at $T = T_{m(B)}$, the points c and d coincide at $\Delta G^M = 0$. At $T_2 < T_{m(B)}$ the point c (liquid B) lies above d in Fig. 10.12b, and, as curve II lies partially below curve III, two double tangents can be drawn: one to the curves I and III, which defines the compositions of the solidus α and its conjugate liquidus, and one to the curves II and III, which defines the compositions of the solidus β and its conjugate liquidus. The activity-composition curves at T_2 are shown in Fig. 10.11c, in which the solid is the standard state for both components.

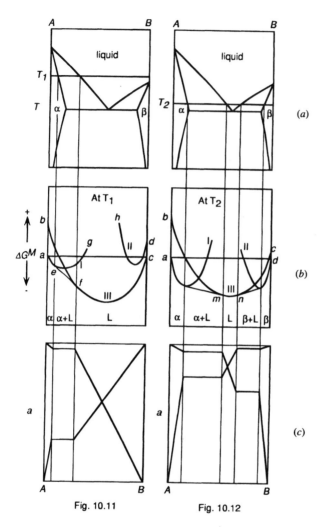

Figures 10.11–10.14   The effect of temperature on the molar Gibbs free energies of mixing and the activities of the components of the system A–B.

With further decrease in temperature the two liquidus compositions, m and n in Fig. 10.12b, approach one another and, at the unique temperature, T_E, the eutectic temperature, they coincide, which means that the two double tangents merge to form the triple tangent to the three curves shown in Fig. 10.13b. At compositions between o and p in Fig. 10.13b a doubly saturated eutectic liquid coexists in equilibrium with α and β solid solutions. From the Gibbs phase rule discussed in Sec. 7.6, this three-phase equilibrium has one degree of freedom, which is used to specify the pressure of the system. Thus, at the specified pressure, the three-phase equilibrium is invariant. Fig. 10.13c shows the activities of A and B at T_E. At $T_3 < T_E$ curve III lies above the double tangent to curves I and II, and thus the liquid phase is not stable. This behavior and the corresponding activity-composition relationships are shown, respectively, in Figs. 10.14b and c.

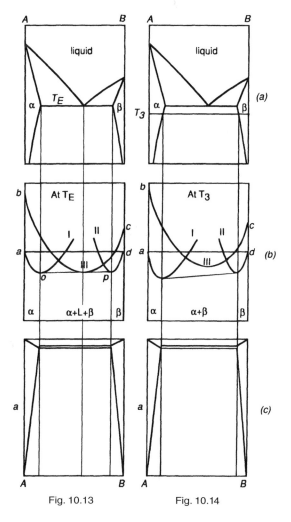

Fig. 10.13   Fig. 10.14

If the ranges of solid solubility in the α and β phases are immeasurably small, then, as a reasonable approximation, it can be said that A and B are insoluble in one another in the solid state. The phase diagram for such a system is shown in Fig. 10.15a. As all Gibbs free energy of mixing curves have vertical tangents at their extremities, any pure substance presents an infinite chemical sink to any other substance, or conversely, it is impossible to obtain an absolutely pure substance. As the range of solid solubility in Fig. 10.15a is so small that it may be neglected on the scale of Fig. 10.15a, then also the Gibbs free energy curves for formation of α and β (curves I and II in Figs. 10.11–10.14) are so compressed toward the $X_A = 1$ and $X_B = 1$ axes, respectively, that on the scale of Figs. 10.11–10.14, they coincide with the vertical axes. The sequence in Fig. 10.16 shows how, as the solubility of B in α decreases, the Gibbs free energy curve for α is compressed against the $X_A = 1$ axis. The Gibbs free energy of formation of the liquid solutions in the system A–B at the temperature T is shown in Fig. 10.15b. The "double tangent" to the α solid solution and

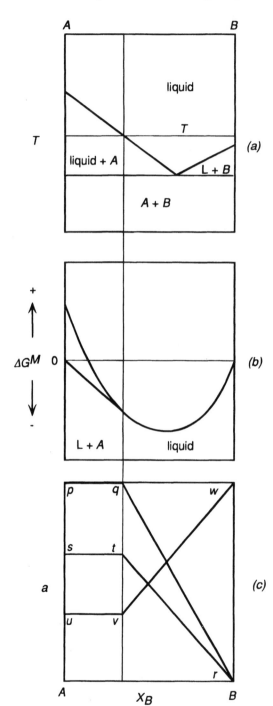

Figure 10.15 The molar Gibbs free energy of mixing and the activities in a binary eutectic system that exhibits complete liquid miscibility and vitually complete solid immiscibility.

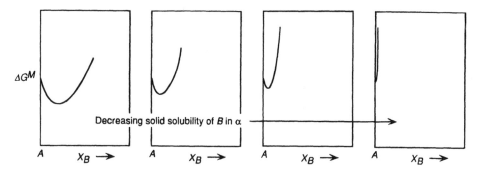

Figure 10.16 The effect of decreasing solid solubility on the molar Gibbs free energy of mixing curve.

liquid solution curves is reduced to a tangent drawn from the point on the $X_A = 1$ axis which represents pure solid A to the liquid solutions curve. The corresponding activity-composition relations are shown in Fig. 10.15c. Again these are drawn in accordance with the supposition the the liquid solutions are ideal. In Fig. 10.15c pqr is the activity of A with respect to pure solid A at p, s is the activity of pure liquid A with respect to solid A at p, str is the activity of A with respect to liquid A having unit activity at s, and $Auvw$ is the activity of B with respect to liquid B having unit activity at w.

In a binary system which exhibits complete miscibility in the liquid state and virtually complete immiscibility in the solid state, e.g., Fig. 10.15a, the variations of the activities of the components of the liquid solutions can be obtained from consideration of the liquidus curves. At any temperature T (Fig. 10.15a), the system with a composition between pure A and the liquidus composition exists as virtually pure solid A in equilibrium with a liquid solution of the liquidus composition. Thus, at T

$$G^{\circ}_{A(s)} = \overline{G}^A_{(l)}$$
$$= G^{\circ}_{A(l)} + RT \ln a_A$$

in which a_A is with respect to liquid A as the standard state. Thus

$$\Delta G^{\circ}_{m(A)} = -RT \ln a_A \tag{10.23}$$

or, if the liquid solutions are Raoultian,

$$\Delta G^{\circ}_{m(A)} = -RT \ln X_A \tag{10.24}$$

It can be noted that Eq. (10.24) is simply Eq. (10.17) with $X_{A(s)} = 1$.

Consider the application of Eq. (10.24) to calculation of the liquidus lines in a binary eutectic system. In the system Cd–Bi, the phase diagram for which is shown in Fig. 10.17, cadmium is virtually insoluble in solid bismuth, and the maximum solubility of bismuth in solid cadmium is 2.75 mole percent at the eutectic temperature of 419 K. If the liquidus solutions are ideal, the Bi liquidus is obtained from Eq. (10.24) as

$$\Delta G^{\circ}_{m(\text{Bi})} = -RT \ln X_{\text{Bi(liquidus)}}$$

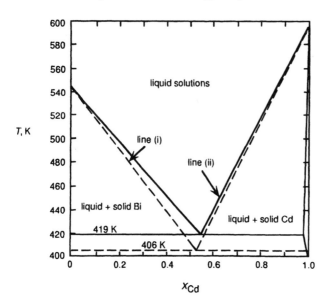

Figure 10.17 The phase diagram for the system Bi–Cd. The full lines are the measured liquidus lines, and the broken lines are calculated assuming no solid solution and ideal mixing in the liquid solutions.

$\Delta H^{\circ}_{m(\text{Bi})} = 10,900$ J at $T_{m(\text{Bi})} = 554$ K, and thus

$$\Delta S^{\circ}_{m(\text{Bi})} = \frac{10,900}{544} = 20.0 \text{ J/K at } 544 \text{ K}$$

The molar constant pressure heat capacities of solid and liquid bismuth vary with temperature as

$$c_{p,\text{Bi}(s)} = 18.8 + 22.6 \times 10^{-3} T \text{ J/K}$$
$$c_{p,\text{Bi}(l)} = 20 + 6.15 \times 10^{-3}T + 21.1 \times 10^{5}T^{-2} \text{ J/K}$$

Thus

$$c_{p,\text{Bi}(l)} - c_{p,\text{Bi}(s)} = \Delta c_{p,\text{Bi}} = 1.2 - 16.45 \times 10^{-3}T + 21.1 \times 10^{5}T^{-2} \text{ J/K}$$

and

$$\Delta G^{\circ}_{m(\text{Bi})} = \Delta H^{\circ}_{m(\text{Bi}),544} + \int_{544}^{T} \Delta c_{p,\text{Bi}} \, dT - T\left(\Delta S^{\circ}_{m(\text{Bi}),544} + \int_{544}^{T} \frac{\Delta c_{p,\text{Bi}}}{T} \, dT\right)$$
$$= 16,560 - 23.79 \, T - 1.2T \ln T + 8.225 \times 10^{-3}T^{2} - 10.55 \times 10^{5}T^{-1}$$
$$= -RT \ln X_{\text{Bi(liquidus)}} \tag{10.25}$$

or

$$\ln X_{\text{Bi(liquidus)}} = \frac{-1992}{T} + 2.861 + 0.144 \ln T - 9.892 \times 10^{-4}T + 1.269 \times \frac{10^5}{T^2}$$

This equation is drawn as the broken line (i) in Fig. 10.17.

Similarly, if the small solid solubility of Bi in Cd is ignored,

$$\Delta G^{\circ}_{m(\text{Cd})} = -RT \ln X_{\text{Cd(liquidus)}}$$

$\Delta H^{\circ}_{m(\text{Cd})} = 6400$ J at $T_{m,\text{Cd}} = 594$ K, and thus $\Delta S^{\circ}_{m(\text{Cd})} = 6400/594 = 10.77$ J/K at 594 K. The constant pressure molar heat capacities are

$$c_{p,\text{Cd}(s)} = 22.2 + 12.3 \times 10^{-3}T \text{ J/K}$$

and

$$c_{p,\text{Cd}(l)} = 29.7 \text{ J/K}$$

Thus

$$c_{p,\text{Cd}(l)} - c_{p,\text{Cd}(s)} = \Delta c_{p,\text{Cd}} = 7.5 - 12.3 \times 10^{-3}T \text{ J/K}$$

$$\Delta G^{\circ}_{m(\text{Cd})} = \Delta H^{\circ}_{m(\text{Cd}),594} + \int_{594}^{T} \Delta c_{p,\text{Cd}} \, dT - T\left(\Delta S^{\circ}_{m(\text{Cd}),594} + \int_{594}^{T} \frac{\Delta c_{p,\text{Cd}}}{T} \, dT\right)$$

$$= 4155 + 37.32T - 7.5T \ln T + 6.15 \times 10^{-3}T^2 \text{ J} \qquad (10.26)$$

$$= -RT \ln X_{\text{Cd(liquidus)}}$$

or

$$\ln X_{\text{Cd(liquidus)}} = \frac{-495}{T} - 4.489 + 0.90 \ln T - 7.397 \times 10^{-4}T$$

which is drawn as the broken line (ii) in Fig. 10.17. Lines (i) and (ii) intersect at the composition of the Raoultian liquid which is simultaneously saturated with Cd and Bi and at 406 K, which would be the eutectic temperature if the liquids were ideal. The actual liquidus lines lie above those calculated, and the actual eutectic temperature is 419 K. From Eq. (10.25), $\Delta G^{\circ}_{m(\text{Bi}),419 \text{ K}} = 2482$ J, and from Eq. (10.26), $\Delta G^{\circ}_{m(\text{Cd}),419 \text{ K}} = 1898$ J. Thus, from Eq. (10.23), in the actual eutectic melt,

$$a_{\text{Bi}} = \exp\left(\frac{-2482}{8.3144 \times 419}\right) = 0.49$$

and

$$a_{\text{Cd}} = \exp\left(\frac{-1898}{8.3144 \times 419}\right) = 0.58$$

The actual eutectic composition is $X_{\text{Cd}} = 0.55$, $X_{\text{Bi}} = 0.45$, and thus the activity coefficients are

$$\gamma_{\text{Bi}} = \frac{0.49}{0.45} = 1.09$$

and

$$\gamma_{Cd} = \frac{0.58}{0.55} = 1.05$$

Thus, positive deviations from Raoultian ideality cause an increase in the liquidus temperatures.

It is now of interest to examine what happens to the liquidus line as the magnitude of the positive deviation from Raoultian behavior in the liquids increases, i.e., as G^{XS} becomes increasingly positive. Assuming regular solution behavior, Eq. (10.23), written in the form

$$-\Delta G^\circ_{m(A)} = RT \ln X_A + RT \ln \gamma_A$$

becomes

$$-\Delta G^\circ_{m(A)} = RT \ln X_A + RT\alpha(1 - X_A)^2$$

or, from Eq. (9.90),

$$-\Delta G^\circ_{m(A)} = RT \ln X_A + \Omega(1 - X_A)^2 \qquad (10.27)$$

Consider a hypothetical system $A–B$ in which $\Delta H^\circ_{m(A)} = 10$ kJ at $T_{m,A} = 2000$ K. Thus, for this system

$$-10{,}000 + 5T = RT \ln X_A + \Omega(1 - X_A)^2$$

where X_A is the composition of the A liquidus at the temperature T. The A liquidus lines, drawn for $\Omega = 0$, 10, 20, 25.3, 30, 40, and 50 kJ, are shown in Fig. 10.18. As Ω exceeds some critical value (which is 25.3 kJ in this case), the form of the liquidus line changes from a monotonic decrease in liquidus temperature with decreasing X_A to a form which contains a maximum and a minimum. At the critical value of Ω the maximum and minimum coincide at $X_A = 0.5$ to produce a horizontal inflexion in the liquidus curve. It is apparent that, when Ω exceeds the critical value, isothermal tie-lines cannot be drawn between pure solid A and all points on the liquidus lines, which, necessarily, means that the calculated liquidus lines are impossible.

From Eq. (10.21)

$$\ln a_A = \frac{-\Delta H^\circ_{m(A)}}{RT} + \frac{\Delta H^\circ_{m(A)}}{RT_{m(A)}}$$

Thus

$$d \ln a_A = \frac{da_A}{a_A} = \frac{\Delta H^\circ_{m(A)}}{RT^2} dT$$

or

$$\frac{dT}{dX_A} = \frac{RT^2}{\Delta H^\circ_{m(A)} a_A} \frac{da_A}{dX_A} \qquad (10.28)$$

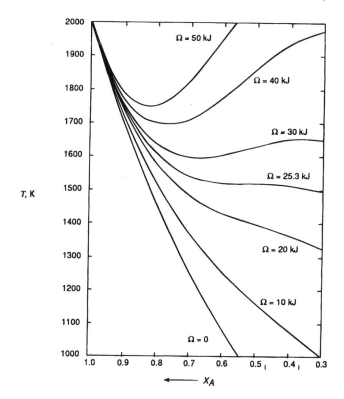

Figure 10.18 Calculated liquidus lines assuming regular solution behavior in the liquid solutions and no solid solubility.

and also

$$\frac{d^2T}{dX_A^2} = \frac{2\,RT}{\Delta H^\circ_{m(A)}a_A}\frac{da_A}{dX_A}\frac{dT}{dX_A} - \frac{RT^2}{\Delta H^\circ_{m(A)}a_A^2}\left(\frac{da_A}{dX_A}\right)^2 + \frac{RT^2}{\Delta H^\circ_{m(A)}a_A}\frac{d^2a_A}{dX_A^2} \quad (10.29)$$

In Eqs. (10.2) and (10.3) it was seen that $da_A/dX_A = d^2a_A/dX_A^2 = 0$ at the state of imminent immiscibility. Thus in Eqs. (10.28) and (10.29), $dT/dX_A = d^2T/dX_A^2 = 0$ at the state of imminent immiscibility. In Fig. 10.18, $\Omega_{cr} = 25.3$ kJ and the horizontal inflexion in the critical liquidus curve occurs at $X_A = 0.5$, $T = 1413$ K. Thus from Eq. (9.90)

$$\alpha_{cr} = \frac{\Omega_{cr}}{RT_{cr}} = \frac{25{,}390}{8.3144 \times 1413} = 2$$

which is in accord with Eq. (10.1). The phase equilibria generated when $\Omega > \Omega_{cr}$ are shown in Fig. 10.19 which shows the immiscibility in regular liquid solutions with $\Omega = 30{,}000$ J and the A-liquidus for $\Omega = 30{,}000$ J shown in Fig. 10.18. The liquid immiscibility curve and the A-liquidus curve intersect at 1620 K to produce a three-phase monotectic equilibrium between A and liquidus L_1 and L_2. The liquid immiscibility curve is metastable at temperatures less than 1620 K, and the calculated A-liquidus is physically impossible between the compositions of L_1 and L_2 at 1620 K.

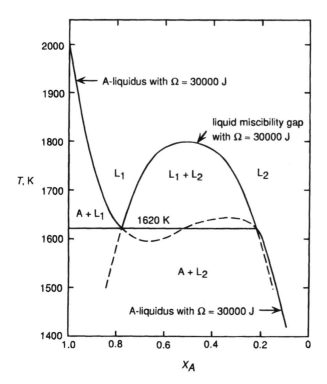

Figure 10.19 The monotectic equilibrium in a binary system in which the liquid solutions exhibit regular solution behavior with $\Omega = 30{,}000$ J.

10.7 THE PHASE DIAGRAMS OF BINARY SYSTEMS THAT EXHIBIT REGULAR SOLUTION BEHAVIOR IN THE LIQUID AND SOLID STATES

Consider the binary system A–B which forms regular liquid solutions and regular solid solutions. The melting temperatures of A and B are, respectively, 800 and 1200 K, and the molar Gibbs free energies of melting are

$$\Delta G_{m,(A)} = 8000 - 10T$$
$$\Delta G_{m,(B)} = 12{,}000 - 10T$$

Consider the system in which $\Omega_l = -20{,}000$ J in the liquid solutions and $\Omega_s = 0$ in the solid solutions. The Gibbs free energy of mixing curves at 1000 K are shown in Fig. 10.20a. As $T_{m,(A)} < 1000$ K $< T_{m,(B)}$, liquid is chosen as the standard state for A, and solid is chosen as the standard state for B. With reference to these standard states, the Gibbs free energies of mixing are

$$\Delta G_l^M = X_B \Delta G_{m,(B)}^\circ + RT(X_A \ln X_A + X_B \ln X_B) + \Omega_l X_A X_B$$

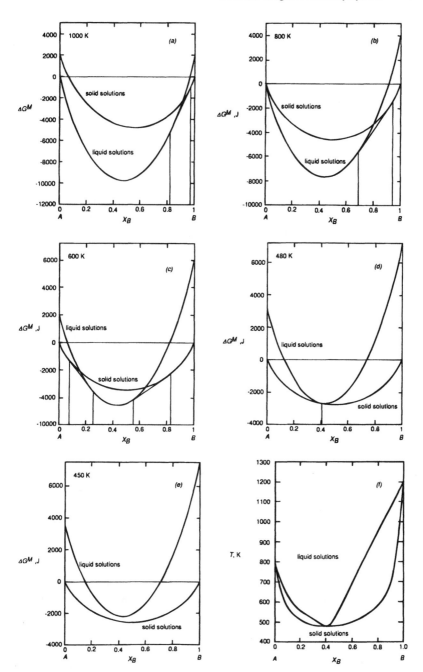

Figure 10.20 The Gibbs free energy of mixing curves at various temperatures, and the phase diagram for a binary system which forms regular solid solutions in which $\Omega_s = 0$ and regular liquid solutions in which $\Omega_l = -20{,}000$ J.

for the liquid solutions and

$$\Delta G_s^M = -X_A \Delta G_{m,(A)}^\circ + RT(X_A \ln X_A + X_B \ln X_B) + \Omega_s X_A X_B$$

for the solid solutions. Inserting the numerical data gives

$$\Delta G_l^M = (12{,}000 - 10T)X_A + 8.3144T(X_A \ln X_A + X_B \ln X_B) - 20{,}000X_A X_B$$

and

$$\Delta G_s^M = -(8000 - 10T)X_A + 8.3144T(X_A \ln X_A + X_B \ln X_B)$$

The double tangent to the curves in Fig. 10.20a gives the liquidus composition of $X_B = 0.82$ and the solidus composition of $X_B = 0.97$. Decreasing the temperature causes the Gibbs free energies of the liquids to increase relative to those of the solids. As shown in Fig. 10.20b, the Gibbs free energies of pure solid A and pure liquid B are equal at the melting temperature of A, and the double tangent gives liquidus and solidus compositions of $X_B = 0.69$ and $X_B = 0.94$, respectively. At temperatures lower than $T_{m,(A)}$ and $T_{m,(B)}$ solid is chosen as the standard state for both components and the Gibbs free energies of mixing are written as

$$\Delta G_l^M = (12{,}000 - 10T)X_B + (8000 - 10T)X_A$$
$$+ 8.3144T(X_A \ln X_A + X_B \ln X_B) - 20{,}000X_A X_B$$

and

$$\Delta G_s^M = RT(X_A \ln X_A + X_B \ln X_B)$$

The curves at 600 K, shown in Fig. 10.20c, contain two double tangents and, as the temperature is further decreased, the curve for the liquids is raised relative to the curve for the solids until, at 480 K, the two double tangents collapse to a single point of contact between the curves at $X_B = 0.41$. At temperatures less than 480 K the curve for the liquids lies above that for the solids, and thus solid solutions are stable over the entire range of composition. The variations, with temperature, of the compositions of the double tangents give the phase diagram shown in Fig. 10.20f.

The behavior of the system in which $\Omega_l = -2000$ J and $\Omega_s = 10{,}000$ J is shown in Fig. 10.21. As seen in Figs. 10.21a–c, the behavior is similar to that shown in Fig. 10.20. However, with a positive value of Ω_s, a critical temperature exists below which immiscibility occurs in the solid state. With $\Omega_s = 10{,}000$ J, the critical temperature is $10{,}000/8.3144R = 601$ K, and the mixing curves at 601 K are shown in Fig. 10.21d. The phase diagram is shown in Fig. 10.21f. With increasingly negative values of Ω_l and increasingly positive values of Ω_s the temperature of the point of contact of the liquidus curve with the solidus curve decreases and the critical temperature in the solid state increases, which eventually produces a eutectic system.

The behavior of a system in which $\Omega_l = 20{,}000$ J and $\Omega_s = 30{,}000$ J is shown in Fig. 10.22. The critical temperatures for the liquid and solid solutions are, respec-

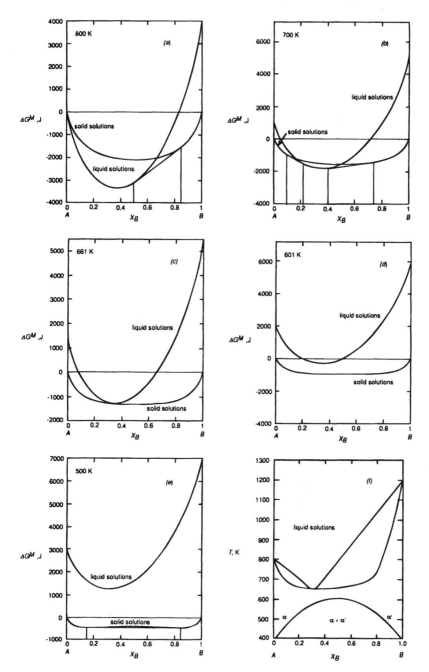

Figure 10.21 The Gibbs free energy of mixing curves at various temperatures, and the phase diagram for a binary system which forms regular solid solutions in which $\Omega_s = 10{,}000$ J and regular liquid solutions in which $\Omega_l = -2000$ J.

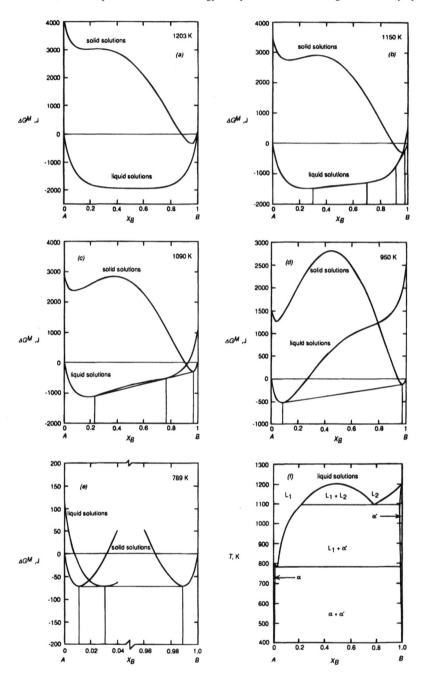

Figure 10.22 The Gibbs free energy of mixing curves at various temperatures, and the phase diagram for a binary system which forms regular solid solutions in which $\Omega_s = 30{,}000$ J and regular liquid solutions in which $\Omega_l = 20{,}000$ J.

tively, 1203 and 1804 K, and the Gibbs free energy of mixing curves at 1203 K are shown in Fig. 10.22a. At $T \geq 1203$ K homogeneous liquids are the stable state, and at temperatures lower than 1203 K immiscibility occurs in the liquid state. The curves shown in Fig. 10.22b at 1150 K contain two double tangents, one joining the conjugate liquid solutions L_1 and L_2 and one connecting the liquidus L_2 with the solidus α'. With decreasing temperature the compositions of the conjugate liquid L_2 and the liquidus L_2 approach one another until, at 1090 K, the two double tangents merge to form a triple tangent to liquid compositions at $X_B = 0.23$ and 0.77 and α' at $X_B = 0.98$. This is a monotectic equilibrium. Further cooling produces a single double tangent between L_1 and α', as shown in Fig. 10.22c, and at 789 K another triple tangent occurs between α at $X_B = 0.01$, L_1 at $X_B = 0.03$ and $\alpha' = 0.99$. At temperatures lower than the eutectic temperature of 789 K the liquid phase is not stable, and, depending on its composition, the system exists as α, $\alpha + \alpha'$, or α'. The monotectic and eutectic equilibria are shown in the phase diagram in Fig. 10.22f.

The influence of systematic changes in the values of Ω_l and Ω_s on the phase relationships which occur in the binary system A–B which forms regular solid and liquid solutions is shown in Fig. 10.23.* In moving from the bottom of any column to the top, the value of Ω_s becomes more positive at constant Ω_l, and in moving from left to right along any row the value of Ω_l becomes more positive at constant Ω_s. In the sequence Figs. 10.23a–e, the liquid solutions become increasing less stable relative to the solid phases with the consequence that the eutectic temperature is increased and, in the transition from Fig. 10.23d–e, the A-liquidus becomes unstable and a monotectic equilibrium occurs. Similarly in the sequence Fig. 10.23h–i the temperature at which the three-phase equilibrium occurs is increased from 633 to 799 K with the consequence that the eutectic equilibrium in Fig. 10.23h becomes a peritectic equilibrium in Fig. 10.23i, and in Fig. 10.23j the immiscibility in the solid state disappears at a temperature below that at which a peritectic equilibrium could occur. Also, with $\Omega_l = 20$ kJ in Fig. 10.23j, liquid immiscibility occurs at temperatures lower than $20,000/(2 \times 8.3144) = 1202$ K, and hence a monotectic equilibrium occurs at 1190 K. In Fig. 10.23k the three-phase L_1-L_2-α equilibrium occurs at 1360 K, which, being higher than $T_{m,(B)}$, produces a syntectic equilibrium in which the composition of the α phase lies between the compositions of the two liquids. In the sequence Figs. 10.23 ($p \rightarrow l \rightarrow f$) the solid phase becomes increasingly less stable than the liquid phase, which deepens the depression of the liquidus and solidus curves and eventually forms a eutectic. Figs. 10.20 and 10.22, respectively, show the Gibbs free energy relations in Figs. 10.23p and e, and Fig. 10.21 shows the phase equilibria occurring between those in Fig. 10.23l and those in Fig. 10.23f.

---

*A. D. Pelton and W. T. Thompson, *Prog. Solid State Chem.* (1975), vol. 10, part 3, p. 119.

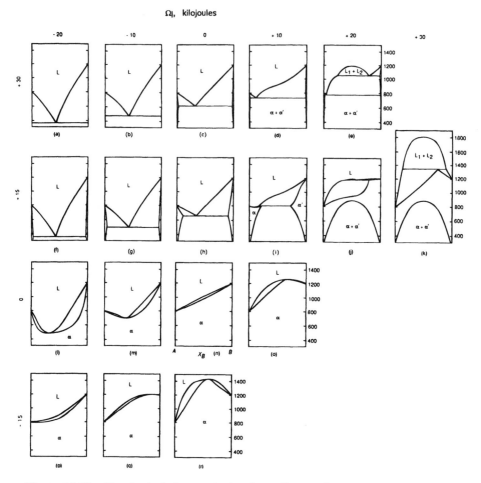

Figure 10.23 Topological changes in the phase diagram for a system A–B with regular solid and liquid solutions, brought about by systematic changes in the values of Ω_s and Ω_l. The melting temperatures of A and B are, respectively, 800 and 1200 K, and the molar entropies of melting of both components are 10 J/K. (From A. D. Pelton and W. T. Thompson, *Prog. Solid State Chem.* (1975), vol. 10, part 3, p. 119).

10.8 SUMMARY

1. The molar Gibbs free energy of formation of binary solution A–B is given by

$$\Delta G^M = RT(X_A \ln a_A + X_B \ln a_B)$$

2. For a regular solution,

$$\Delta G^M - \Delta G^{M,\text{id}} = G^{\text{xs}} = RT\alpha X_A X_B = \Omega X_A X_B = \Delta H^M$$

3. The criteria for equilibrium between the phases α and β in the binary system A–B are

$$a_A \text{ (in } \alpha) = a_A \text{ (in } \beta)$$

and

$$a_B \text{ (in } \alpha) = a_B \text{ (in } \beta)$$

4. Immiscibility becomes imminent in a regular solution at the critical value of $\alpha = 2$. The critical temperature, below which immiscibility occurs in a regular system, is given by $T_{cr} = \Omega/2R$.

5. In a binary system A–B which forms ideal liquid solutions and ideal solid solutions, the solidus is given by

$$X_{A(s)} = \frac{1 - \exp(-\Delta G^\circ_{m(B)}/RT)}{\exp(-\Delta G^\circ_{m(A)}/RT) - \exp(-\Delta G^\circ_{m(B)}/RT)}$$

and the liquidus line is given by

$$X_{A(l)} = \frac{[1 - \exp(-\Delta G^\circ_{m(B)}/RT]\exp(-\Delta G^\circ_{m(A)}/RT)}{\exp(-\Delta G^\circ_{m(A)}/RT) - \exp(-\Delta G^\circ_{m(B)}/RT)}$$

6. In a binary system A–B which contains a eutectic equilibrium and in which the extent of solid solution is negligibly small, the liquidus lines are determined by the conditions

$$\Delta G^\circ_{m(A)} = -RT \ln a_A \text{ (in the } A\text{–liquidus melt)}$$

and

$$\Delta G^\circ_{m(B)} = -RT \ln a_B \text{ (in the } B\text{–liquidus melt)}$$

7. Thus, if the liquid solutions are ideal, the A-liquidus compositions are given by

$$-\Delta G^\circ_{m(A)} = RT \ln X_A$$

and if the liquid solutions are regular, the A-liquidus compositions are given by

$$-\Delta G^\circ_{m(A)} = RT \ln X_A + \Omega(1 - X_A)^2$$

10.9 NUMERICAL EXAMPLE

The phase diagram for the system Cs-Rb is shown in Fig. 10.24. Examine the extent to which the phase diagram can be reproduced assuming that the liquid solutions are ideal and that the solid solutions are regular.

For Cs

$$\Delta G^\circ_{m,Cs} = 2100 - 6.95T \text{ J}$$

and for Rb

$$\Delta G^\circ_{m,Rb} = 2200 - 7.05T$$

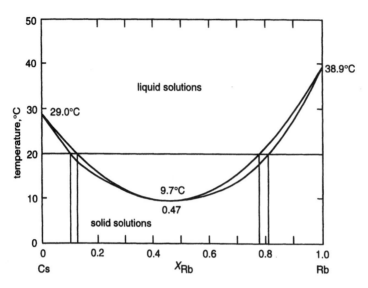

Figure 10.24 The phase diagram for the system Cs–Rb.

Initially, draw the Gibbs free energy of mixing curves for the solid and liquid solutions at 9.7°C (282.7 K) and determine if some value of Ω gives Gibbs free energy of mixing curves similar to those shown in Fig. 10.21c, i.e., the curve for the liquid solutions lies above the curve for the solid solutions except at a single composition at which the two curves touch one another.

Relative to the pure solids as standard states at the temperature T,

$$\Delta G^M \text{ (liquid solutions)} = X_{Rb}\Delta G^{\circ}_{m,Rb} + X_{Cs}\Delta G^{\circ}_{m,Cs} \qquad (i)$$
$$+ RT(X_{Rb} \ln X_{Rb} + X_{Cs} \ln X_{Cs})$$

which, at 282.7 K, becomes

$$\Delta G^M \text{(liquid solutions)} = 205X_{Rb} + 133(1 - X_{Rb}) + 8.3144 \times 282.7$$
$$\times [X_{Rb} \ln X_{Rb} + (1 - X_{Rb}) \ln (1 - X_{Rb})]$$

Also,

$$\Delta G^M \text{(solid solutions)} = RT(X_{Rb} \ln X_{Rb} + X_{Cs} \ln X_{Cs}) + \Omega X_{Rb}X_{Cs} \qquad (ii)$$

which at 282.7 K becomes

$$\Delta G^M \text{(solid solutions)} = 8.3144 \times 282.7 \times (X_{Rb} \ln X_{Rb} + X_{Cs} \ln X_{Cs}) + \Omega X_{Rb}X_{Cs}$$

The Gibbs free energy of mixing curves, drawn with $\Omega = 668$ J, are shown in Fig. 10.25, which shows that the curves touch one another at $X_{Rb} = 0.47$, in exact agreement with Fig. 10.24.

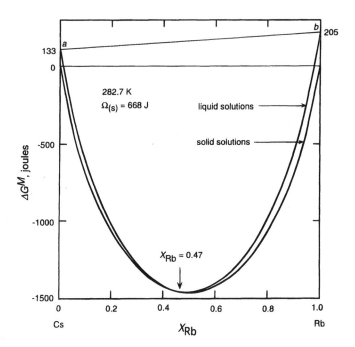

Figure 10.25 The molar Gibbs free energy of mixing curves for ideal liquid solutions and regular solid solutions (with $\Omega = 668$ J) at 282.7 K, drawn using the Gibbs free energies of melting of Cs and Rb.

The Gibbs free energies of mixing, given by Eqs. (i) and (ii), with $\Omega = 668$ J and $T = 20°C$, 292 K, are shown in Fig. 10.26. The double tangent at the Cs-rich side of the diagram gives the solidus composition as $X_{Rb} = 0.10$ and the liquidus composition as $X_{Rb} = 0.13$, which is in excellent agreement with the phase diagram. At the Rb-rich side of the diagram the double tangent gives the solidus and liquidus compositions as, respectively, $X_{Rb} = 0.81$, and $X_{Rb} = 0.75$, which are in good agreement with the phase diagram values of 0.80 and 0.77.

It is thus seen that the phase diagram is reproduced by assuming that the liquid solutions are ideal and the solid solutions are regular with $\Omega = 668$ J.

PROBLEMS

10.1 CaF_2 and MgF_2 are mutually insoluble in the solid state and form a simple binary eutectic system. Calculate the composition and temperature of the eutectic melt assuming that the liquid solutions are Raoultian. The actual eutectic occurs at $X_{CaF_2} = 0.45$ and $T = 1243$ K.

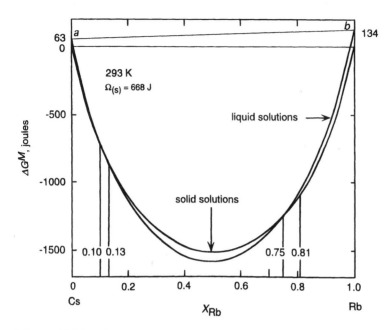

Figure 10.26 The molar Gibbs free energy of mixing curves for ideal liquid solutions and regular solid solutions (with $\Omega = 668$ J) at 293 K, drawn using the Gibbs free energies of melting of Cs and Rb.

10.2 Gold and silicon are mutually insoluble in the solid state and form a eutectic system with a eutectic temperature of 636 K and a eutectic composition of $X_{Si} = 0.186$. Calculate the Gibbs free energy of the eutectic melt relative to (a) unmixed liquid Au and liquid Si, and (b) unmixed solid Au and solid Si.

At 636 K $\Delta G^\circ_{m,Au} = 12{,}600 \left[\dfrac{1338 - 636}{1338} \right] = 6{,}611$ J

and $\Delta G^\circ_{m,Si} = 50{,}200 \left[\dfrac{1658 - 636}{1658} \right] = 30{,}943$ J

a. $\Delta G^M = ab = -(0.186 \times 30{,}943 + 0.814 \times 6{,}611)$ J $= -11{,}140$ J
b. $\Delta G^M = 0$

10.3 Al_2O_3, which melts at 2324 K, and Cr_2O_3, which melts at 2538 K form complete ranges of solid and liquid solutions. Assuming that $\Delta S^\circ_{m,Cr_2O_3} = \Delta S^\circ_{m,Al_2O_3}$, and that the solid and liquid solutions in the system Al_2O_3–Cr_2O_3 behave ideally, calculate

a. The temperature at which equilibrium melting begins when an alloy of $X_{Al_2O_3} = 0.5$ is heated
b. The composition of the melt which first forms

c. The temperature at which equilibrium melting is complete

d. The composition of last-formed solid

10.4 $Na_2O \cdot B_2O_3$ and $K_2O \cdot B_2O_3$ form complete ranges of solid and liquid solutions and the solidus and liquidus show a common minimum at the equimolar composition and $T = 1123$ K. Calculate the molar Gibbs free energy of formation of the equimolar solid solution from solid $Na_2O \cdot B_2O_3$ at 1123 K, assuming that the liquid solutions are ideal.

10.5 SiO_2, which melts at 1723°C, and TiO_2, which melts at 1842°C, are immiscible in the solid state, and the SiO_2–TiO_2 binary system contains a monotectic equilibrium at 1794°C, at which temperature virtually pure TiO_2 is in equilibrium with two liquids containing mole fractions of SiO_2 of 0.04 and 0.76. If, for the purpose of simple calculation, it is assumed that the compositions of the two liquids are $X_{SiO_2} = 0.24$ and $X_{SiO_2} = 0.76$ and that the liquid solutions are regular in their behavior, what is the value of Ω_l and at what temperature does the liquid immiscibility gap disappear?

10.6 The binary system Ge–Si contains complete solid and liquid solutions. The melting temperatures are $T_{m,Si} = 1685$ K and $T_{m,Ge} = 1210$ K, and $\Delta H^\circ_{m,Si} = 50{,}200$ J. At 1200°C the liquidus and solidus compositions are, respectively, $X_{Si} = 0.32$ and $X_{Si} = 0.665$. Calculate the value of $\Delta H^\circ_{m,Ge}$ assuming

a. That the liquid solutions are ideal.

b. That the solid solutions are ideal.

Which assumption gives the better estimate? The actual value of $\Delta H^\circ_{m,Ge}$ at $T_{m,Ge}$ is 36,900 J.

10.7 CaO and MgO form a simple eutectic system with limited ranges of solid solubility. The eutectic temperature is 2370°C. Assuming that the solutes in the two solid solutions obey Henry's law with γ°_{CaO} in MgO = 12.88 and γ°_{MgO} in CaO = 6.23 at 2300°C, calculate the solutibility of CaO in MgO and the solubility of MgO in CaO at 2300°C.

Chapter 11

REACTIONS INVOLVING GASES

11.1 INTRODUCTION

In Chap. 8 it was seen that the absence of interatomic forces between the atoms of ideal gases causes the heat of mixing of ideal gases to be zero. This situation represents one extreme of a range of possible situations. Toward the other extreme of this range is the situation in which gases which exhibit marked chemical affinity for one another are mixed. For example, considerable heat is released when gaseous hydrogen and oxygen are mixed in the presence of a catalyst. The thermodynamics of such a system can be treated in either of two ways. The mixture can be considered to be a highly nonideal mixture of H_2 and O_2, the thermodynamic equilibrium state of which, at given temperature and pressure, can be defined in terms of the fugacities of the components H_2 and O_2; or it can be considered that the H_2 and O_2 have reacted with one another to some extent to produce the product species H_2O. In the latter case, if the pressure of the system is low enough, the equilibrium state at the given temperature can be defined in terms of the partial pressures exerted by the three species, H_2, O_2, and H_2O, occurring in the system. Although both treatments are thermodynamically equivalent, the latter, by virtue of its correspondence with physical reality, is by far the more convenient and practical.

As with any constant-pressure, constant-temperature system, the equilibrium state is that in which the Gibbs free energy of the system has its minimum possible value. If the gases, present initially in the system, react to form distinct product species, the total change in the Gibbs free energy of the system comprises a contribution arising from the change in Gibbs free energy due to the chemical reaction and a contribution arising from the mixing of the appearing product gases with the remaining reactant gases. Knowledge of the variation of this change in Gibbs free energy with composition (which ranges from the pure unmixed reactant gases to the pure unmixed product gases) allows determination of the equilibrium state in any system of reactive gases. This determination is facilitated by the introduction of the equilibrium constant for the reaction, and it will be seen that the relation between this constant and the standard Gibbs free energy change for the reaction is one of the more important relationships in reaction equilibrium thermodynamics.

11.2 REACTION EQUILIBRIUM IN A GAS MIXTURE AND THE EQUILIBRIUM CONSTANT

Consider the reaction

$$A_{(g)} + B_{(g)} = 2C_{(g)}$$

occurring at constant temperature and constant pressure P. At any instant during the reaction, the Gibbs free energy of the system is

$$G' = n_A \overline{G}_A + n_B \overline{G}_B + n_C \overline{G}_C \tag{11.1}$$

where n_A, n_B, and n_C are, respectively, the numbers of moles of A, B, and C present in the reaction system at that instant, and $\overline{G}_A$, $\overline{G}_B$, and $\overline{G}_C$ are, respectively, the partial molar Gibbs free energies of A, B, and C in the gas mixture which occurs at that instant. The problem is to determine the values of n_A, n_B, and n_C which minimize the value of G' in Eq. (11.1), as this state of minimum Gibbs free energy is the equilibrium state of the system at the given temperature and pressure. That is, once chemical reaction between A and B has proceeded to the extent that the Gibbs free energy of the system has been minimized, the reaction stops. On the microscopic level reaction equilibrium occurs when the rate of the chemical reaction proceeding from right to left equals that of the reaction proceeding from left to right.

The stoichiometry of the reaction allows the numbers of moles of all the species present at any instant to be expressed in terms of the number of moles of any one of the species. Starting with 1 mole of A and 1 mole of B (i.e., 2 moles of gas), as 1 atom of A reacts with 1 atom of B to produce 2 molecules of C, then, at any time during the reaction

$$n_A = n_B$$

and

$$n_C = 2 - n_A - n_B = 2(1 - n_A)$$

Eq. (11.1) can thus be written as

$$G' = n_A \overline{G}_A + n_A \overline{G}_B + 2(1 - n_A)\overline{G}_C$$

From Eq. (8.15),

$$\overline{G}_i = G_i^\circ + RT \ln P + RT \ln X_i$$

and

$$X_A = \frac{n_A}{2}, \quad X_B = \frac{n_A}{2}, \quad X_C = \frac{2(1 - n_A)}{2} = (1 - n_A)$$

substitution of which gives

$$G' = n_A(G_A^\circ + G_B^\circ - 2G_C^\circ) + 2G_C^\circ + 2RT \ln P$$

$$+ 2 RT \left[n_A \ln \left(\frac{n_A}{2} \right) + (1 - n_A) \ln (1 - n_A) \right]$$

or

$$G' - 2G_C^\circ = n_A(-\Delta G^\circ) + 2 RT \ln P$$

$$+ 2RT \left[n_A \ln \left(\frac{n_A}{2} \right) + (1 - n_A) \ln (1 - n_A) \right] \quad (11.2)$$

where $\Delta G^\circ = 2G_C^\circ - G_A^\circ - G_B^\circ$ is the *standard Gibbs free energy change* for the chemical reaction at the temperature T. The standard Gibbs free energy change for any reaction is the difference between the sum of the Gibbs free energies of the reaction products in their standard states and the sum of the Gibbs free energies of the reactants in their standard states. In the present case ΔG° is the difference between the Gibbs free energy of 2 moles of C at 1 atm pressure and the temperature T, and 1 mole of A and 1 mole of B, each at 1 atm pressure and the temperature T. If the total pressure of the system is 1 atm, then Eq. (11.2) simplifies to

$$G' - 2G_C^\circ = n_A(-\Delta G^\circ) + 2RT \left[n_A \ln \left(\frac{n_A}{2} \right) + (1 - n_A) \ln (1 - n_A) \right] \quad (11.3)$$

The left-hand side of Eq. (11.3), which is the difference between the Gibbs free energy of the 2 mole system when $n_A = n_A$ and the Gibbs free energy of the system when it consists of 2 moles of C, is determined by two factors:

1. The change in the Gibbs free energy due to the chemical reaction, i.e., due to the disappearance of the reactants and the appearance of the products, given by the first term on the right-hand side of Eq. (11.3).
2. The decrease in the Gibbs free energy caused by mixing of the gases, given by the second term on the right-hand side of Eq. (11.3).

Fig. 11.1 is drawn for the reaction

$$A_{(g)} + B_{(g)} = 2C_{(g)}$$

at 500 K and 1 atm pressure. ΔG_{500}° for the reaction is taken as being -5000 joules. If the reference for the Gibbs free energy is arbitrarily chosen as being $(G_A^\circ + G_B^\circ) = 0$, then $2G_C^\circ = -5000$ joules. In Fig. 11.1 the ordinate $\Delta G'$ is plotted as the difference between the Gibbs free energy of the system containing n_A moles of A and the

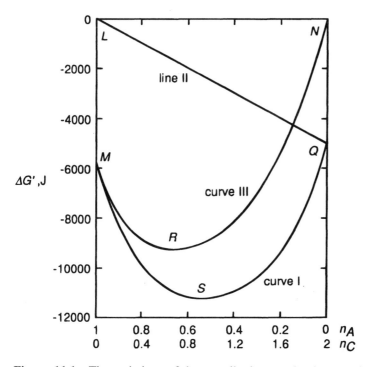

Figure 11.1 The variations of the contributions to the decrease in the Gibbs free energy due to chemical reaction (line II), the contribution to the decrease in the Gibbs free energy due to gas mixing (curve III) with the extent of the reaction $A_{(g)} + B_{(g)} = 2C_{(g)}$ for which $\Delta G^\circ = -5000$ joules at 500 K.

Gibbs free energy of the system comprising 1 mole of A and 1 mole of B before mixing of A and B occurs. Thus the point $L(n_A = 1, n_B = 1$ before mixing) is located at $\Delta G' = 0$, and the point Q ($n_C = 2$) is located at $\Delta G' = -5000$ joules. The point M represents the decrease in the Gibbs free energy due to mixing of 1 mole of A and 1 mole of B before any chemical reaction between the two occurs, i.e., from Eq. (8.20),

$$\Delta G(L \rightarrow M) = \sum_i n_i RT \ln \left(\frac{p_i}{P_i} \right)$$

$$= RT \left[n_A \ln \left(\frac{p_A}{P_A} \right) + n_B \ln \left(\frac{p_B}{P_B} \right) \right]$$

But $n_A = n_B = 1$, and

$$P_A = P_B = P_{\text{mixture}} = 1$$

in which case

$$p_A = p_B = \frac{1}{2}$$

Thus

$$\Delta G(L \rightarrow M) = 8.3144 \times 500 \times 2 \times \ln 0.5$$
$$= -5763 \text{ J}$$

or

$$G' + 5000 = 1 \times 5000 - 5763 \text{ J}$$

and hence

$$G' = -5763 \text{ J}$$

and so

$$\Delta G' = G' - (G_A^\circ + G_B^\circ) = -5763 \text{ J}$$

Curve I, which represents the variation of G' with n_A, is obtained as the sum of line II (given by the first term on the right-hand side of Eq. (11.3)—the decrease in Gibbs free energy due to chemical reaction) and curve III (given by the second term on the right-hand side of Eq. (11.3)—the decrease in Gibbs free energy due to gas mixing). As can be seen, the magnitude of the chemical reaction contribution to the decrease in the Gibbs free energy of the system increases linearly with increasing n_C, but the magnitude of the contribution to the change in the total Gibbs free energy due to gas mixing passes through a maximum at the composition R. R occurs at that composition of the gas mixture which permits maximum randomization of the system. Further chemical reaction, which takes the composition of the system beyond R, decreases the magnitude of the gas-mixing contribution as further increase in n_C, at the expense of n_A and n_B, decreases the randomness of the system. Eventually the composition S is reached in which state the sum of the two contributions to the decrease in the Gibbs free energy is a maximum. If chemical reaction continued beyond S, then, as the decrease in line II is smaller than the increase in curve III, the total Gibbs free energy of the system would increase. The composition S is thus that at which the Gibbs free energy of the system has its minimum value and hence is the equilibrium state.

The position of the minimum in curve I is fixed by the criterion that, at the minimum

$$\left(\frac{\partial G'}{\partial n_A} \right)_{T,P} = 0$$

and, as

$$G' = n_A \overline{G}_A + n_A \overline{G}_B + 2(1 - n_A)\overline{G}_C$$

then*

$$\left(\frac{\partial G'}{\partial n_A} \right) = \overline{G}_A + \overline{G}_B - 2\overline{G}_C = 0$$

---

*From the Gibbs-Duhem equation, Eq. (9.19), $n_A d\overline{G}_A + n_B d\overline{G}_B + n_C d\overline{G}_C = 0$.

i.e., the criterion for reaction equilibrium is

$$\overline{G}_A + \overline{G}_B = 2\overline{G}_C \tag{11.4}$$

Eq. (11.4) can be written as

$$G_A^\circ + RT \ln p_A + G_B^\circ + RT \ln p_B = 2G_C^\circ + 2RT \ln p_C \tag{11.5}$$

where p_A, p_B, and p_C are, respectively, the partial pressures of A, B, and C which occur at reaction equilibrium. Rearrangement of Eq. (11.5) gives

$$2G_C^\circ - G_A^\circ - G_B^\circ = -RT \ln \frac{p_C^2}{p_A p_B}$$

or

$$\Delta G^\circ = -RT \ln \left(\frac{p_C^2}{p_A p_B} \right) \tag{11.6}$$

The quotient of the equilibrium partial pressure of the reactants and products occurring as the logarithmic term in Eq. (11.6) is termed the *equilibrium constant for the reaction*, K_p, i.e.,

$$\left(\frac{p_C^2}{p_A p_B} \right)_{eq} = K_p \tag{11.7}$$

and hence

$$\Delta G^\circ = -RT \ln K_p \tag{11.8}$$

As ΔG° is a function only of temperature, it follows from Eq. (11.8) that K_p is a function only of temperature. For the example used in Fig. 11.1,

$$\ln K_p = -\frac{\Delta G^\circ}{RT} = \frac{5000}{8.3144 \times 500} = 1.203$$

Therefore

$$K_p = 3.329$$

Now

$$K_p = \frac{p_C^2}{p_A p_B} = \frac{X_C^2 P^2}{X_A P X_B P} = \frac{X_C^2}{X_A X_B}$$

$$= \frac{(1 - n_A)^2}{n_A^2/4} = 3.329$$

and hence

$$n_A = 0.523 \quad \text{(the other solution, } n_A = 11.4 \text{, is nonphysical)}$$

i.e., reaction equilibrium occurs with $n_A = n_B = 0.523$ and $n_C = 0.954$, or when the chemical reaction $A + B = 2C$ has proceeded to 47.7% completion. The minimum in curve I in Fig. 11.1 occurs at $n_A = 0.523$.

If the temperature T was such that $\Delta G°$ for the reaction was zero, there would be no chemical reaction contribution to $\Delta G'$, and the variation of $\Delta G'$ with n_A would be given by curve III in Fig. 11.1, i.e., the criterion for reaction equilibrium would be maximization of the randomness of the system, which occurs at the composition R. From Eq. (11.8), if $\Delta G° = 0$, then $K_p = 1$, and thus

$$1 = \frac{4(1 - n_A)^2}{n_A^2} \quad \text{and} \quad n_A = \frac{2}{3}$$

The minimum in curve III occurs at $n_A = n_B = n_C = \tfrac{2}{3}$, and thus maximum randomness in the system occurs when all three species are present in equal amount.

It is important to note that the minimum in curve I in Fig. 11.1, in representing the equilibrium state of the system at $P = 1$ atm and 500 K, is the only point on curve I which has any significance within the scope of classical thermodynamics. The point S is the only state on the curve I which lies on the equilibrium surface in P-T-composition space for the fixed values of $P = 1$ atm and $T = 500$ K. Any change in P and/or T moves the equilibrium composition over the equilibrium surface in P-T-composition space.

11.3 THE EFFECT OF TEMPERATURE ON THE EQUILIBRIUM CONSTANT

The position of the minimum in curve I in Fig. 11.1 is determined by the difference between the lengths of LM and NQ. The length LM is the decrease in the Gibbs free energy caused by mixing of the gases before the reaction begins, and the length NQ is the standard Gibbs free energy change, $\Delta G°$, for the reaction. The lengths of both of these lines are dependent on temperature, i.e., $LM = 2RT \ln 0.5$ and $NQ = \Delta G°$. The effect of temperature on the composition at which the minimum occurs in curve I (and hence on the value of K_p) thus depends on the relative effects of temperature on the lengths of LM and NQ. For given reactants, the length of LM increases linearly with temperature, and the variation of the length NQ with temperature is determined by the sign and magnitude of the standard entropy change for the reaction according to

$$\left(\frac{\partial \Delta G°}{\partial T} \right)_P = -\Delta S°$$

If $\Delta S°$ is negative, an increase in temperature increases the length LM and decreases the length NQ. Thus the position of the minimum in the curve shifts to the left, which indicates that K_p decreases with increasing temperature.

The exact variation of K_p with temperature is obtained from consideration of the Gibbs-Helmholtz equation, Eq. (5.37a),

$$\left[\frac{\partial (\Delta G°/T)}{\partial T} \right]_P = -\frac{\Delta H°}{T^2}$$

As $\Delta G^\circ = -RT \ln K_p$, then

$$\frac{\partial \ln K_p}{\partial T} = \frac{\Delta H^\circ}{RT^2} \tag{11.9}$$

or

$$\frac{\partial \ln K_p}{\partial \left(\dfrac{1}{T}\right)} = -\frac{\Delta H^\circ}{R} \tag{11.10}$$

Eq. (11.10), which is known as the *van't Hoff equation*, shows that the effect of temperature on K_p is determined by the sign and magnitude of ΔH° for the reaction. If ΔH° is positive, i.e., if the reaction is endothermic, then K_p increases with increasing temperature, and, conversely, if ΔH° is negative, i.e., if the reaction is exothermic, then K_p decreases with increasing temperature. Integration of Eq. (11.10) requires knowledge of the temperature dependence of ΔH°, which, as was seen in Chap. 6, depends on the value of Δc_p for the reaction.

The direction of the variation of K_p with temperature can be obtained from application of Le Chatelier's principle. If heat is added to a system at reaction equilibrium then the equilibrium is displaced in that direction which involves the absorption of heat. Consider the simple gaseous reaction,

$$Cl_2 = 2Cl$$

This reaction is endothermic and hence has a positive value of ΔH°. Thus the equilibrium constant, $K_p = p_{Cl}^2/p_{Cl_2}$, increases with increasing temperature, i.e., the equilibrium shifts in that direction which involves the absorption of heat. Conversely, if the reaction was written as

$$2Cl = Cl_2$$

then, as the reaction has a negative value of ΔH°, $K_p = p_{Cl_2}/p_{Cl}^2$ decreases with increasing temperature, i.e., the equilibrium shifts in that direction which involves the absorption of heat. In both cases an increase in temperature increases p_{Cl} and decreases p_{Cl_2}.

Eq. (11.10) shows that, if ΔH° is independent of temperature, then $\ln K_p$ varies linearly with $1/T$.

11.4 THE EFFECT OF PRESSURE ON THE EQUILIBRIUM CONSTANT

The equilibrium constant, K_p, as defined by Eq. (11.7) is independent of pressure. This is a consequence of the fact that ΔG°, being the difference between the Gibbs free energies of the pure products, each at unit pressure, and the Gibbs free energies of the pure reactants, each at unit pressure, is, by definition, independent of pressure.

However, reaction equilibrium expressed in terms of the numbers of moles of species present rather than in terms of the partial pressures of the species present is dependent on the total pressure if the chemical reaction involves a change in the total number of moles present.

Consider, again, the reaction

$$Cl_2 = 2Cl$$

Completion of this reaction causes a doubling of the number of moles present, and the effect of a change in pressure can again be obtained by application of Le Chatelier's principle. If the pressure exerted on a system at reaction equilibrium is increased, then the equilibrium shifts in that direction which tends to decrease the pressure exerted by the system, i.e., shifts in that direction which decreases the number of moles present. Thus if the pressure exerted on the Cl–Cl_2 system is increased, the equilibrium will shift toward the Cl_2 side, as, thereby, the total number of moles present will be decreased to accommodate the increased pressure. Specifically, the effect of pressure on the reaction equilibrium expressed in terms of numbers of moles present (or in terms of mole fractions) can be seen as follows:

$$K_p = \frac{p_{Cl}^2}{p_{Cl_2}} \quad \text{which is independent of pressure}$$

$$= \frac{X_{Cl}^2 P^2}{X_{Cl_2} P} = \frac{X_{Cl}^2 P}{X_{Cl_2}} = K_x P$$

where K_x is the equilibrium constant expressed in terms of the mole fractions. Thus if the pressure is increased, then K_x decreases in order to maintain K_p constant, and the decrease in K_x is achieved by the reaction equilibrium shifting toward the Cl_2 side such that X_{Cl} decreases and X_{Cl_2} increases.

In the case of the reaction $A + B = 2C$, the system at all times contains 2 moles of gas, and thus the reaction equilibrium, expressed in terms of the mole fractions, is independent of pressure, i.e.,

$$K_p = \frac{p_C^2}{p_A p_B} = \frac{X_C^2 P^2}{X_A P X_B P} = \frac{X_C^2}{X_A X_B} = K_x$$

This can also be seen from Eq. (11.2), as, if $P \neq 1$, then the effect of the nonzero term $2RT \ln P$ is the raising or lowering of curve I in Fig. 11.1 without affecting the position of the minimum with respect to the composition axis.

The magnitude of the effect of a change in pressure on the value of K_x depends on the magnitude of the change in the number of moles present in the system, occurring as a result of the chemical reaction. For the general reaction

$$aA + bB = cC + dD$$

$$K_p = \frac{p_C^c p_D^d}{p_A^a p_B^b} = \frac{X_C^c X_D^d}{X_A^a X_B^b} \frac{P^c P^d}{P^a P^b} = K_x P^{(c+d-a-b)}$$

which shows that K_x is independent of the total pressure only if $c + d - a - b = 0$.

11.5 REACTION EQUILIBRIUM AS A COMPROMISE BETWEEN ENTHALPY AND ENTROPY

The Gibbs free energy of a system is defined as

$$G = H - TS$$

and hence low values of G are obtained with low values of H and large values of S. It was seen, in the discussion of one-component systems, that equilibrium occurs as the result of a compromise between enthalpy and entropy considerations. Similar discussions can be made concerning chemical reaction equilibria.

Consider again the reaction $Cl_2 = 2Cl$. This reaction has a positive value of $\Delta H°$ ($\Delta H°$ is the thermal energy required to break Avogadro's number of Cl–Cl bonds) and has a positive value of $\Delta S°$ (2 moles of chlorine atoms are produced from 1 mole of chlorine molecules). Thus the system occurring as Cl atoms has a high value of H and a high value of S, and the system occurring as Cl_2 molecules has a low value of H and a low value of S. The minimum value of G thus occurs somewhere between the two extreme states. This compromise between the enthalpy and the entropy is analogous to the compromise between the contributions of the chemical reaction and the mixing of gases to the decrease in Gibbs free energy shown in Fig. 11.1.

For the reaction $A + B = 2C$, Eq. (11.3) can be written as

$$G' - 2G_C^\circ = n_A(-\Delta H°) + n_A(T\Delta S°) + 2RT\left[n_A \ln\left(\frac{n_A}{2}\right) + (1 - n_A)\ln(1 - n_A)\right]$$

$$= [n_A(-\Delta H°)] + T\left\{n_A\Delta S° + 2R\left[n_A \ln\left(\frac{n_A}{2}\right) + (1 - n_A)\ln(1 - n_A)\right]\right\}$$

$$(11.11)$$

The term in the first brackets is the enthalpy contribution to the change in the Gibbs free energy, and the second term on the right-hand side is the entropy contribution, with $n_A\Delta S°$ being the entropy change caused by the chemical reaction and $2R[n_A \ln(n_A/2) + (1 - n_A)\ln(1 - n_A)]$ being the entropy change caused by gas mixing. In the previous example $\Delta G°$ equaled -5000 joules at 500 K. Let it be that $\Delta H° = -2500$ joules and $\Delta S° = 5$ J/K, in which case Fig. 11.2 can be drawn from Fig. 11.1. In Fig. 11.2 the $\Delta H'$ line is the first term on the right-hand side of Eq. (11.11) and the $T\Delta S'$ line is the second term. The sum of these two gives $G' - 2G_C^\circ$, the scale of which is given on the left-hand edge of Fig. 11.2. The scale on the right-hand edge of the figure is $\Delta G'$ where, as before, the reference zero of Gibbs free energy is chosen $G_A^\circ + G_B^\circ = 0$ such that $\Delta G' = G'$ (i.e., the scale is displaced by $2G_C^\circ = -5000$ joules). On this scale the $\Delta G'$ curve in Fig. 11.2 is identical with curve I in Fig. 11.1. As can be seen, the minimum in the $\Delta G'$ curve is determined as the compromise between the minimum value of H' at $n_A = 0$ and the maximum value of $T\Delta S'$ at $n_A = 0.597$ (the point M in Fig. 11.2). If the temperature is increased then the $T\Delta S'$ term becomes relatively more important, and, hence, the

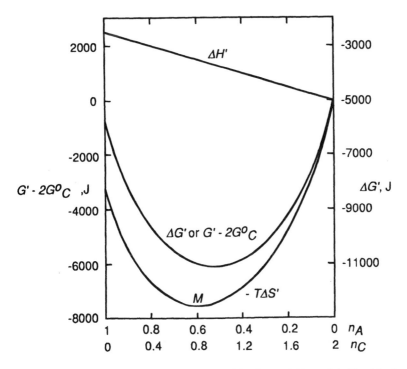

Figure 11.2 The variations at 500 K of $\Delta H'$, $-T\Delta S'$, and $\Delta G'$ with the extent of reaction $A(g) + B(g) = 2C(g)$ for which $\Delta G° = -2500 - 5T$.

equilibrium value of n_A increases (the minimum in the $\Delta G'$ curve shifts to the left). Thus, K_p decreases with increasing temperature in accordance with Eq. (11.9) ($\Delta H°$ is negative).

The effect of temperature on the $\Delta G'$ curve is illustrated in Fig. 11.3, in which curves are drawn for the reaction $A + B = 2C$ at 500, 1000, and 1500 K. It is assumed that Δc_p for the reaction is zero, in which case $\Delta H°$ and $\Delta S°$ are independent of temperature. As $\Delta G° = \Delta H° - T\Delta S°$, then

$$\Delta G°_{500} = -2500 - (500 \times 5) = -5000 \text{ J} \qquad K_{p,500} = 3.329$$

$$\Delta G°_{1000} = -2500 - (1000 \times 5) = -7500 \text{ J} \qquad K_{p,1000} = 2.465$$

$$\Delta G°_{1500} = -2500 - (1500 \times 5) = -10{,}000 \text{ J} \qquad K_{p,1500} = 2.229$$

The equilibrium values of n_A at 500, 1000, and 1500 K are thus, respectively, 0.523, 0.560, and 0.572.

Although K_p is constant at constant temperature, note that an infinite set of partial pressures of reactants and products correspond to the fixed value of K_p. If the reaction involves three species, then arbitrary choice of the partial pressures of two of the species uniquely fixes the equilibrium partial pressure of the third.

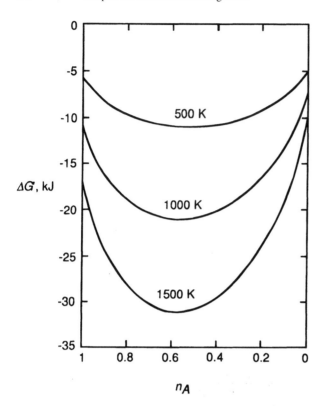

Figure 11.3 The effect of temperature on the equilibrium state of the reaction $A_{(g)} + B_{(g)} = 2C_{(g)}$ for which $\Delta G° = -2500 - 5T$ joules.

11.6 REACTION EQUILIBRIUM IN THE SYSTEM $SO_{2(g)}$–$SO_{3(g)}$–$O_{2(g)}$

Consider the equilibrium

$$SO_{2(g)} + \frac{1}{2}O_{2(g)} = SO_{3(g)} \qquad (11.12)$$

The standard Gibbs free energy change for this reaction is

$$\Delta G° = -94,600 + 89.37T \text{ J}$$

Thus, at 1000 K,

$$\Delta G°_{1000} = -5230 \text{ J}$$

$$\ln K_p = \frac{5230}{8.3144 \times 1000} = 0.629$$

and

$$K_p = 1.876 = \frac{p_{SO_3}}{p_{SO_2} p_{O_2}^{1/2}}$$

Consider the reaction between 1 mole of SO_2 gas at 1 atm pressure and ½ mole of O_2 gas at 1 atm pressure to form an equilibrium mixture of SO_2, SO_3, and O_2 at 1 atm pressure and 1000 K. From the stoichiometry of the chemical reaction given by Eq. (11.12), x moles of SO_3 are formed from reaction of x moles of SO_2 and ½x moles of O_2, and any reacting mixture contains x moles of SO_3, $(1 - x)$ moles of SO_2, and ½$(1 - x)$ moles of O_2, i.e.,

SO_2	$+$	$\dfrac{1}{2} O_2$	$\rightarrow$ SO_3
initially 1		$\dfrac{1}{2}$	0
upon reaction $1 - x$		$\dfrac{1}{2} - \dfrac{1}{2}x$	x

The total number of moles in the system, n_T, is

$$n_T = 1 - x + \frac{1}{2} - \frac{1}{2}x + x = \frac{1}{2}(3 - x)$$

and, as

$$p_i = \frac{n_i}{n_T} P$$

then

$$p_{SO_2} = \frac{2(1 - x)P}{(3 - x)}, \; p_{O_2} = \frac{(1 - x)P}{(3 - x)}, \; p_{SO_3} = \frac{2xP}{(3 - x)}$$

Thus

$$K_p^2 = \frac{p_{SO_3}^2}{p_{SO_2}^2 p_{O_2}} = \frac{(3 - x)x^2}{(1 - x)^3 P}$$

or

$$(1 - PK_p^2)x^3 + (3PK_p^2 - 3)x^2 - 3PK_p^2 x + PK_p^2 = 0 \qquad (11.13)$$

which, with $P = 1$ and $K_p = 1.876$, gives $x = 0.463$. Thus, at equilibrium, there are 0.537 moles of SO_2, 0.269 moles of O_2, and 0.463 moles of SO_3, such that

$$p_{SO_2} = \frac{2(1 - 0.463)}{(3 - 0.463)} = 0.423 \text{ atm}$$

$$p_{O_2} = \frac{(1 - 0.463)}{(3 - 0.463)} = 0.212 \text{ atm}$$

and

$$p_{SO_3} = \frac{2 \times 0.463}{(3 - 0.463)} = 0.365 \text{ atm}$$

As a check,

$$K_p = \frac{0.365}{0.423 \times 0.212^{1/2}} = 1.874$$

The composition of the equilibrium gas is thus 42.3% SO_2, 21.2% O_2, and 36.5% SO_3.

The Effect of Temperature

As $\Delta H°$ for the reaction given by Eq. (11.12) is negative ($-94,600$ J), and Le Chatelier's principle states that a decrease in temperature at constant pressure shifts the equilibrium in that direction which involves an evolution of heat, decreasing the temperature at constant pressure causes the equilibrium to shift toward the SO_3 side. At 900 K

$$\Delta G°_{900} = -14,167 \text{ J}$$

$$\ln K_p = \frac{14,167}{8.3144 \times 900} = 1.893$$

and thus

$$K_p = 6.64$$

Substituting $K_p = 6.64$ and $P = 1$ atm into Eq. (11.13) gives $x = 0.704$, and thus

$$p_{SO_2} = 0.258 \text{ atm}, \quad p_{O_2} = 0.129 \text{ atm}, \quad p_{SO_3} = 0.613 \text{ atm}$$

As a check,

$$K_p = \frac{0.613}{0.258 \times 0.129^{1/2}} = 6.64$$

Thus, of the total number of moles present,

25.8% are SO_2, which is a decrease from 42.3% at 1000 K
12.9% are O_2, which is a decrease from 21.2% at 1000 K
61.3% are SO_3, which is an increase from 36.5% at 1000 K

Thus it is seen that a decrease in temperature has shifted the equilibrium toward the SO_3 side.

The Effect of Pressure

Although K_p is independent of pressure, Le Chatelier's principle states that an increase in total pressure at constant temperature shifts the equilibrium in that direction which involves a decrease in the number of moles in the system, i.e., toward the SO_3 side. Consider the equilibrium mixture at $P = 10$ atm and $T = 1000$ K. Substituting $P = 10$ and $K_p = 1.876$ into Eq. (11.13) gives $x = 0.686$, in which case $n_{SO_2} = 0.314$, $n_{O_2} = 0.157$, and $n_{SO_3} = 0.686$. Thus

$$p_{SO_2} = \frac{2(1 - 0.686) \times 10}{(3 - 0.686)} = 2.714 \text{ atm}$$

$$p_{O_2} = \frac{(1 - 0.686) \times 10}{(3 - 0.686)} = 1.357 \text{ atm}$$

and

$$p_{SO_3} = \frac{2 \times 0.686 \times 10}{(3 - 0.686)} = 5.929 \text{ atm}$$

As a check,

$$K_p = \frac{5.929}{2.714 \times 1.357^{1/2}} = 1.875$$

Of the total number of moles present

27.1% are SO_2, which is a decrease from 42.3% at $P = 1$ atm
13.75% are O_2, which is a decrease from 21.2% at $P = 1$ atm
59.29% are SO_3, which is an increase from 36.5% at $P = 1$ atm

Thus it is seen that an increase in pressure has shifted the equilibrium toward the SO_3 side.

The Effect of Changes in Temperature and Pressure

In order to simultaneously vary the temperature and the pressure in such a manner that the numbers of moles of the three gaseous species present remain constant, Eq. (11.13) indicates that the variation must be such that the term PK_p^2 remains constant, where

$$K_p = e^{\frac{94,600}{8.3144T}} \times e^{\frac{-89.37}{8.3144}}$$

It is apparent that, by mixing of SO_2 gas and SO_3 gas, an equilibrium mixture with a known partial pressure of O_2 can be produced. For example, consider that it was required to have an SO_3–SO_2–O_2 mixture at 1 atm total pressure in which $p_{O_2} = 0.1$ atm. To obtain this gas mixture, SO_3 and SO_2, both at 1 atm pressure, would be mixed in the molar ratio $SO_2/SO_3 = a$ and allowed to equilibrate. If a moles of SO_2 and 1 mole of SO_3 are mixed, then from the stoichiometry of Eq. (11.12), x moles of SO_3 would decompose to form x moles of SO_2 and $\frac{1}{2}x$ moles of O_2 such that, at equilibrium, the numbers of moles present would be

$$SO_2 \qquad + \qquad \frac{1}{2}O_2 \qquad \rightarrow \qquad SO_3$$

initially $\qquad\qquad a \qquad\qquad\qquad 0 \qquad\qquad\qquad 1$

upon reaction $\qquad a + x \qquad\qquad \frac{1}{2}x \qquad\qquad 1 - x$

with

$$n_T = a + x + \frac{1}{2}x + 1 - x = \frac{1}{2}(2a + 2 + x)$$

In this gas mixture

$$p_{O_2} = \frac{n_{O_2}}{n_T}P = \frac{x}{2a + 2 + x}P$$

which, for $P = 1$ and $p_{O_2} = 0.1$, gives $a = 4.5x - 1$. Also

$$K_p^2 = \frac{p_{SO_3}^2}{p_{SO_2}^2 p_{O_2}} = \frac{(1 - x)^2(2a + 2 + x)}{(a + x)^2 xP}$$

At 1000 K, $K_p = 1.876$. Thus, substituting for a in terms of x and taking $K_p^2 = 3.519$ gives

$$96.45x^3 - 18.709x^2 - 6.481x = 0$$

or, since $x \neq 0$,

$$96.45x^2 - 18.709x - 6.481 = 0$$

which has the solution $x = 0.374$. Thus

$$a = (4.5 \times 0.374) - 1 = 0.683$$

$$p_{O_2} = \frac{xP}{2a + 2 + x} = \frac{0.374}{(2 \times 0.683) + 2 + 0.374} = 0.1 \text{ atm}$$

$$p_{SO_3} = \frac{2(1 - x)}{2a + 2 + x} = \frac{2(1 - 0.374)}{(2 \times 0.683) + 2 + 0.374} = 0.335 \text{ atm}$$

and

$$p_{SO_2} = \frac{2(a + x)}{2a + 2 + x} = \frac{2(0.683 + 0.374)}{(2 \times 0.683) + 2 + 0.374} = 0.565 \text{ atm}$$

In the equilibrium mixture, $p_{SO_2}/p_{SO_3} = 1.7$, compared with $p_{SO_2}/p_{SO_3} = a = 0.683$ in the initial mixture. As a check,

$$K_p = \frac{0.355}{0.565 \times 0.1^{1/2}} = 1.875$$

If it had been required to have $p_{O_2} = 0.212$ atm (the equilibrium value of p_{O_2} in an equilibrated gas at 1 atm pressure of initial composition 1 mole of SO_2 + ½ mole of O_2), then solution of the stoichiometric condition would have given $a = 0$, which indicates that pure SO_3 at 1000 K and 1 atm pressure decomposes to the same equilibrium gas mixture as that formed from an initial mixture of 1 mole of SO_2 + ½ mole of O_2 at 1 atm. Thus, 0.212 atm is the maximum value of p_{O_2} which can be produced in an SO_2–SO_3 mixture at 1000 K and 1 atm total pressure.

11.7 EQUILIBRIUM IN H_2O–H_2 AND CO_2–CO MIXTURES

H_2O–H_2 and CO_2–CO gas mixtures are used when it is required that the partial pressure of oxygen in a gas phase be fixed at a very low value. For example, if it were required to have a gaseous atmosphere containing a partial pressure of oxygen of 10^{-10} atm, then such an oxygen potential can be obtained with relative ease by establishing the equilibrium

$$H_2 + \frac{1}{2}O_2 = H_2O$$

for which

$$\Delta G° = -247{,}500 + 55.85T \text{ J}$$

From Eq. (11.8),

$$\ln K_p = \frac{247{,}500}{8.3144T} - \frac{55.85}{8.3144}$$

If it is required to have an atmosphere containing $p_{O_2} = 10^{-10}$ atm at $T = 2000$ K, then, at this temperature,

$$\ln K_p = \frac{247{,}500}{8.4144 \times 2000} - \frac{55.85}{8.3144}$$

$$= 8.167$$

Therefore

$$K_p = 3.521 \times 10^3 = \frac{p_{H_2O}}{p_{H_2}p_{O_2}^{1/2}}$$

and with $p_{O_2} = 10^{-10}$ atm,

$$\frac{p_{H_2O}}{p_{H_2}} = 3.521 \times 10^3 \times 10^{-5} = 3.521 \times 10^{-2}$$

Thus if, in the H_2–H_2O gas mixture, $p_{H_2} = 1$ atm, then P_{H_2O} must be 0.0352 atm. From Eq. (7.10) 0.0352 atm is the saturated vapor pressure of liquid water at 27.0°C, and thus the required gas mixture can be produced by bubbling hydrogen gas at 1 atm pressure through pure liquid water at 27.0°C to saturate it with water vapor. The establishment of reaction equilibrium at 2000 K gives $p_{O_2} = 10^{-10}$ atm in the gas.

Similarly the partial pressure of oxygen in a gaseous atmosphere can be determined by establishing the reaction

$$CO + \frac{1}{2}O_2 = CO_2$$

The standard Gibbs free energy change for the reaction $CO_{(g)} = \frac{1}{2}O_{2(g)} + C_{(s)}$ is

$$\Delta G° = 111,700 + 87.65T \text{ J}$$

and that for the reaction $C_{(s)} + O_{2(g)} = CO_{2(g)}$ is

$$\Delta G° = -394,100 - 0.84T \text{ J}$$

Summation of the standard Gibbs free energy changes gives

$$\Delta G° = -282,400 + 86.81T \text{ J}$$

for the reaction $CO_{(g)} + \frac{1}{2}O_{2(g)} = CO_{2(g)}$. Thus

$$\ln K_p = \frac{282,400}{8.3144T} - \frac{86.81}{8.3144}$$

If it were required to have $p_{O_2} = 10^{-20}$ atm at 1000 K, then

$$\ln K_p = \frac{282,400}{8.3144 \times 1000} - \frac{86.81}{8.3144} = 23.52$$

Therefore

$$K_p = 1.646 \times 10^{10} = \frac{p_{CO_2}}{p_{CO}p_{O_2}^{1/2}} = \frac{p_{CO_2}}{p_{CO} \times 10^{-10}}$$

and so

$$\frac{p_{CO_2}}{p_{CO}} = 1.646$$

If the total pressure $P = 1$ atm, then, as $p_{CO_2} + p_{CO} = 1$, $p_{CO_2} = 1.646(1 - p_{CO_2}) = 0.622$ atm and $p_{CO} = 0.378$. The required mixture is produced by mixing CO_2 and CO in the volume ratio 1.561/1, i.e., 62.2 volume percent CO_2 and 37.8 volume percent CO.

In both of the above equilibria the oxygen pressure in the equilibrated gas is so small that the p_{H_2}/p_{H_2O} and p_{CO_2}/p_{CO} ratios in the equilibrated gases are negligibly different from the corresponding ratios in the initial mixtures. The equality of the ratio of the volume percentages with the ratio of the partial pressures in a gas mixture can be demonstrated as follows. Consider a cm^3 of gas A at 1 atm pressure and b cm^3 of gas B at 1 atm pressure being mixed at constant pressure (and hence at constant total volume, $a + b$). The number of moles of $A = n_A = (1 \times a)/RT = a/RT$, and the number of moles of $B = n_B = (1 \times b)/RT = b/RT$. Therefore, in the mixture,

$$p_A = \frac{n_A RT}{V} = \frac{n_A RT}{(a + b)}$$

and

$$p_B = \frac{n_B RT}{(a + b)}$$

Thus

$$\frac{p_A}{p_B} = \frac{n_A}{n_B} = \frac{a}{b} = \frac{\text{volume percentage of } A}{\text{volume percentage of } B}$$

11.8 SUMMARY

1. The equilibrium state of the reaction $aA + bB = cC + dD$ is that in which $a\overline{G}_A + b\overline{G}_B = c\overline{G}_C + d\overline{G}_D$. The equilibrium state of the reaction is thus determined by the value of $\Delta G°$ for the reaction and is qualified by the equilibrium constant K_p, where $K_p = (p_C^c p_D^d / p_A^a p_B^b)_{eq}$ and the standard state for each reactant and product gas is the pure gas at 1 atm pressure. $\Delta G°$ and K_p are related as $\Delta G° = -RT \ln K_p$. This relation is one of the more powerful equations in chemical thermodynamics and will be used extensively in subsequent chapters. For increasingly negative values of $\Delta G°$, K_p becomes increasingly greater than unity, and, conversely, for increasingly positive values of $\Delta G°$, K_p becomes increasingly less than unity.

2. As $\Delta G°$ is a function only of temperature, then K_p is a function only of temperature, and the dependence of K_p on temperature is determined by the value of $\Delta H°$ for the reaction, i.e.,

$$\Delta G° = \Delta H° - T\Delta S° = -RT \ln K_p$$

and hence

$$\ln K_p = -\frac{\Delta H°}{RT} + \frac{\Delta S°}{R}$$

or

$$\frac{\partial \ln K_p}{\partial T} = \frac{\Delta H°}{RT^2}$$

or

$$\frac{\partial \ln K_p}{\partial (1/T)} = -\frac{\Delta H°}{R}$$

Thus, for an exothermic reaction, K_p decreases with increasing temperature, and for an endothermic reaction, K_p increases with increasing temperature.

3. As $p_i = X_i P$ in an ideal gas mixture, the equilibrium constant can be written in terms of the mole fractions, i.e.,

$$K_p = \frac{p_C^c p_D^d}{p_A^a p_B^b} = \frac{X_C^c X_D^d}{X_A^a X_B^b} P^{(c+d-a-b)} = K_x P^{(c+d-a-b)}$$

Although K_p is, by definition, independent of pressure, K_x is independent of pressure only if $(c + d - a - b) = 0$, that is, if the gas mixture contains a constant number of moles in all states along the reaction coordinate. If forward progression of the reaction decreases the number of moles present, i.e., if $(c + d - a - b) < 0$, then an increase in pressure increases the value of K_x, and, conversely, if $(c + d - a - b) > 0$, an increase in pressure decreases the value of K_x. The dependence of K_x on pressure and the dependence of K_p on temperature are examples of Le Chatelier's principle.

11.9 NUMERICAL EXAMPLES

Example 1

Consider the partial decomposition of gaseous P_4 according to

$$P_{4(g)} = 2P_{2(g)}$$

and calculate

1. The temperature at which $X_{P_4} = X_{P_2} = 0.5$ at total pressure of 1 atm
2. The total pressure at which $X_{P_4} = X_{P_2} = 0.5$ at 2000 K

1. For the reaction $P_{4(g)} = 2P_{2(g)}$,

$$\Delta G° = 225,400 + 7.90T \ln T - 209.4T \text{ J}$$

and hence

$$\ln K_p = \frac{-27,109}{T} - 0.95 \ln T + 25.18 \tag{i}$$

Partial decomposition of 1 mole of P_4 produces $(1 - x)$ moles of P_4 and $2x$ moles of P_2. Thus, for $X_{P_4} = (1 - x)/(1 + x) = 0.5$, $x = 1/3$ and hence

$$p_{P_4} = \left(\frac{1 - \dfrac{1}{3}}{1 + \dfrac{1}{3}} \right) P = 0.5P$$

and

$$p_{P_2} = \left(\frac{2 \times \dfrac{1}{3}}{1 + \dfrac{1}{3}} \right) P = 0.5P$$

Thus with $X_{P_4} = X_{P_2} = 0.5$ and the pressure P

$$K_p = \frac{p_{P_2}^2}{p_{P_4}} = 0.5P \tag{ii}$$

With $P = 1$ atm, Eq. (i) gives

$$\ln K_p = \ln (0.5) = \frac{-27,109}{T} - 0.95 \ln T + 25.18$$

which has the solution $T = 1429$ K. Thus at $P = 1$ atm and $T = 1429$ K, $X_{P_2} = X_{P_4} = 0.5$ in a P_4-P_2 mixture.

    2. From Eqs. (i) and (ii)

$$K_{p,2000\,K} = 81.83 = 0.5P$$

which gives $P = 163.6$ atm. Thus, at $P = 163.6$ atm and $T = 2000$ K, $X_{P_4} = X_{P_2} = 0.5$ in the mixture. The variations of X_{P_4} and X_{P_2} with temperature in an equilibrated P_4-P_2 mixture at $P = 1$ atm and at $P = 163.3$ atm are shown in Fig. 11.4. As the dissociation of P_4 is endothermic, increasing the temperature at constant pressure shifts the equilibrium toward P_2, and as the dissociation increases the number of moles present, increasing the pressure at constant temperature shifts the equilibrium toward P_4.

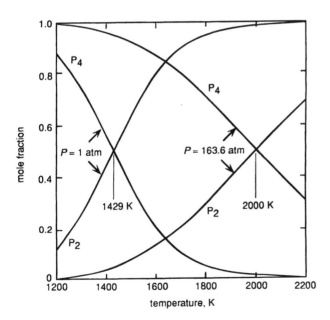

Figure 11.4 The variations of the mole fractions of P_2 and P_4 with temperature in an equilibrated P_2-P_4 mixture at $P = 1$ atm and at $P = 163.6$ atm.

Example 2

Consider the cracking of gaseous ammonia according to the reaction

$$2NH_{3(g)} = N_{2(g)} + 3H_{2(g)} \tag{i}$$

under conditions of

1. Constant total pressure and
2. Constant volume at 400°C

1. The standard Gibbs free energy change for the reaction given by Eq. (i) is

$$\Delta G° = 87,030 - 25.8T \ln T - 31.7T \text{ J}$$

Therefore,

$$\Delta G°_{673\,K} = -47,370 \text{ J} \quad \text{and} \quad K_{p,673\,K} = 4748$$

From the stoichiometry of the reaction, partial decomposition of 1 mole of NH_3 produces $3x$ moles of H_2, x moles of N_2, and $(1 - 2x)$ moles of NH_3. Thus for

	$2NH_3$	$\rightarrow$	N_2	$+$	$3H_2$
initially	1		0		0
upon decomposition	$1 - 2x$		x		$3x$

and $n_T = 1 - 2x + x + 3x = 1 + 2x$ moles. Thus,

$$p_{H_2} = \frac{3x}{1 + 2x}P, \; p_{N_2} = \frac{x}{1 + 2x}P, \quad \text{and} \quad p_{NH_3} = \frac{1 - 2x}{1 + 2x}P$$

such that

$$K_{p,673\,K} = \frac{p_{H_2}^3 p_{N_2}}{p_{NH_3}^2} = \frac{27x^4 P^2}{(1 + 2x)^2(1 - 2x)^2} \tag{ii}$$

Using the identity $(1 - y)(1 + y) = 1 - y^2$, Eq. (ii) can be written as

$$K_p = \frac{27x^4 P^2}{[1 - (2x)^2]^2}$$

or

$$K_p^{1/2} = \frac{5.196x^2 P}{(1 - 4x^2)}$$

Thus, for a constant total pressure of 1 atm,

$$(4748)^{1/2}(1 - 4x^2) = 5.196x^2$$

which has the solution $x = 0.4954$. Thus, at equilibrium

$$p_{H_2} = \frac{3x}{1 + 2x} = 0.7465 \text{ atm}$$

$$p_{N_2} = \frac{x}{1 + 2x} = 0.2488 \text{ atm}$$

and

$$p_{NH_3} = \frac{1 - 2x}{1 + 2x} = 0.0047 \text{ atm}$$

in which state 99.08% of the NH_3 has decomposed.

An alternative approach to the problem is as follows. From the stoichiometry of the reaction, at all times

$$p_{H_2} = 3p_{N_2} \tag{iii}$$

and

$$P = p_{NH_3} + p_{N_2} + p_{H_2} \tag{iv}$$

Eliminating p_{NH_3} and p_{H_2} from Eqs. (iii) and (iv) and substituting into Eq. (ii) gives

$$K_p = \frac{27p_{N_2}^4}{(P - 4p_{N_2})^2}$$

or

$$K_p^{1/2} = \frac{(27)^{1/2}p_{N_2}^2}{(P - 4p_{N_2})}$$

with $P = 1$ atm,

$$(4784)^{1/2}(1 - 4p_{N_2}) = (27)^{1/2}p_{N_2}^2$$

which gives

$$p_{N_2} = 0.2488 \text{ atm}$$

$$p_{H_2} = 3p_{N_2} = 0.7464 \text{ atm}$$

and

$$p_{NH_3} = 1 - p_{H_2} - p_{N_2} = 0.0048 \text{ atm}$$

2. Now consider that the decomposition occurs at constant volume. As the decomposition reaction increases the number of moles of gas from 1 to $(1 + 2x)$, the reaction at constant P increases the volume of the gas by the factor $(1 + 2x)$. From Le Chatelier's principle, an increase in pressure shifts the equilibrium in that direction which decreases the number of moles of gas present, i.e., in the direction $3H_2 + N_2 \rightarrow 2NH_3$. Thus the extent of equilibrium decomposition of ammonia at constant volume will be less than that occurring at constant pressure and will cause an increase in pressure.

As before,

$$p_{H_2} = \frac{3x}{1 + 2x}P'$$

$$p_{N_2} = \frac{x}{1 + 2x}P'$$

and

$$p_{NH_3} = \frac{1 - 2x}{1 + 2x}P' \qquad (v)$$

where P' is the pressure of the reacting mixture. Before decomposition begins, the 1 mole of NH_3 obeys the relation $PV = RT$. The decomposition reaction, at constant V and T, increases the number of moles of gas to $(1 + 2x)$ and hence increases the total pressure of the gas to P', where

$$P'V = (1 + 2x)RT$$

Therefore,

$$V = \text{constant} = \frac{RT}{P} = \frac{(1 + 2x)RT}{P'}$$

such that, in Eq. (v), the original pressure of NH_3 at volume V before decomposition started is equal to $P'/(1 + 2x)$. Thus, for an original pressure of $P = 1$ atm, at equilibrium

$$p_{H_2} = 3x$$
$$p_{N_2} = x$$

and

$$p_{NH_3} = 1 - 2x$$

Hence,

$$K_p = \frac{27x^4}{(1 - 2x)^2}$$

or

$$K_p^{1/2} = (4784)^{1/2} = \frac{(27)^{1/2}x^2}{(1 - 2x)}$$

which has the solution $x = 0.4909$. Thus

$$p_{H_2} = 3x = 1.4727 \text{ atm}$$
$$p_{N_2} = x = 0.4909 \text{ atm}$$
$$p_{NH_3} = 1 - 2x = 0.0182 \text{ atm}$$

$$P = \sum_i p_i = 1.9819 = 1 + 2x$$

in which state 98.18% of the NH_3 has decomposed.

It can be shown that the yield of NH_3 formed by reaction between H_2 and N_2 is maximized when the reactants H_2 and N_2 are mixed in the molar ratio 3:1. In the mixture, $P = p_{H_2} + p_{N_2} + p_{NH_3}$, and, at reaction equilibrium, let it be that

$$p_{H_2} = ap_{N_2}$$

Thus

$$p_{NH_3} = P - (a + 1)p_{N_2}$$

or

$$p_{N_2} = \frac{P - p_{NH_3}}{a + 1} \quad \text{and} \quad p_{H_2} = \frac{a(P - p_{NH_3})}{(a + 1)}$$

and hence

$$K_p = \frac{p_{NH_3}^2}{p_{H_2}^3 p_{N_2}} = \frac{p_{NH_3}^2}{[a(P - p_{NH_3})/(a + 1)]^3[(P - p_{NH_3})/(a + 1)]} \quad \text{(vi)}$$
$$= \frac{p_{NH_3}^2(a + 1)^4}{a^3(P - p_{NH_3})^4}$$

It now must be shown that p_{NH_3} has its maximum value when $a = 3$, i.e., that

$$\frac{dp_{NH_3}}{da} = 0 \quad \text{when} \quad a = 3$$

The derivative is most easily obtained by taking the logarithms of Eq. (vi),

$$\ln K_p + 3 \ln a + 4 \ln (P - p_{NH_3}) = 2 \ln p_{NH_3} + 4 \ln (a + 1)$$

and differentiating to obtain

$$\left[\frac{3}{a} - \frac{4}{(1 + a)}\right] da = \left[\frac{2}{p_{NH_3}} + \frac{4}{(P - p_{NH_3})}\right] dp_{NH_3}$$

Thus $dp_{NH_3}/da = 0$ requires that $3/a = 4/(1 + a)$, or $a = 3$.

The stoichiometry of the reaction shows that, for the ratio p_{H_2}/p_{N_2} to have the value 3 in the equilibrium mixture, the reactants H_2 and N_2 must be mixed in the ratio 3:1.

Example 3

Determine the state of equilibrium when CH_4 and H_2O are mixed in the molar ratio 1:2 and allowed to react at 1000 K and 1 atm to form H_2 and CO according to

$$CH_4 + H_2O = 3H_2 + CO$$

For $CH_{4(g)} = C_{(s)} + 2H_{2(g)}$

$$\Delta G° = 69,120 - 22.25T \ln T + 65.35T \text{ J}$$

For $H_2O_{(g)} = H_{2(g)} + \frac{1}{2} O_{2(g)}$

$$\Delta G° = 247{,}500 - 55.85T \text{ J}$$

and for $C_{(s)} + \frac{1}{2} O_{2(g)} = CO_{(g)}$

$$\Delta G° = -111{,}700 - 87.65T \text{ J}$$

Summation of the three standard Gibbs free energy changes gives

$$\Delta G° = 204{,}920 - 22.25T \ln T - 78.15T \text{ J}$$

For the reaction

$$CH_{4(g)} + H_2O_{(g)} = 3H_{2(g)} + CO_{(g)} \tag{i}$$

Thus $\Delta G°_{1000\ K} = -26{,}927$ J and $K_{P,(i)} = 25.5$. For

	CH_4	$+$	H_2O	$=$	$3H_2$	$+$	CO
Initially the numbers of moles are	1		2		0		0
and, on reaction, are	$1 - x$		$2 - x$		$3x$		x

which gives $n_T = 3 + 2x$ moles. Thus

$$p_{CH_4} = \frac{(1 - x)P}{3 + 2x}$$

$$p_{H_2O} = \frac{(2 - x)P}{3 + 2x}$$

$$p_{H_2} = \frac{3xP}{3 + 2x}$$

and

$$p_{CO} = \frac{xP}{3 + 2x}$$

and thus, at equilibrium

$$K_{p,(i)} = 25.5 = \frac{p_{H_2}^3 p_{CO}}{p_{CH_4} p_{H_2O}} = \frac{27x^4 P^2}{(3 + 2x)^2 (1 - x)(2 - x)} \tag{ii}$$

which, with $P = 1$ atm, gives $x = 0.9637$. Thus the equilibrium state is

$$p_{CH_4} = \frac{1 - 0.9637}{3 + (2 \times 0.9637)} = 0.00737 \text{ atm}$$

$$p_{H_2O} = \frac{2 - 0.9637}{3 + (2 \times 0.9637)} = 0.2103 \text{ atm}$$

$$p_{CO} = \frac{0.9637}{3 + (2 \times 0.9637)} = 0.1956 \text{ atm}$$

$$p_{H_2} = 3p_{CO} = 0.5867 \text{ atm}$$

As a check

$$K_{p,(i)} = \frac{0.5867^3 \times 0.1956}{0.00737 \times 0.2103} = 25.5$$

Consider, now, that the product CO reacts with the reactant H_2O to establish the separate equilibrium

$$CO + H_2O = H_2 + CO_2$$

For $H_2O_{(g)} = H_{2(g)} + \frac{1}{2}O_{2(g)}$

$$\Delta G^\circ = 247,500 - 55.85T \text{ (J)}$$

for $CO_{(g)} = C_{(s)} + \frac{1}{2}O_{2(g)}$

$$\Delta G^\circ = 111,700 + 87.65T \text{ J}$$

and for $C_{(s)} + O_{2(g)} = CO_{2(g)}$

$$\Delta G^\circ = -394,100 - 0.84T \text{ J}$$

Summation of these three standard Gibbs free energy changes gives

$$\Delta G^\circ = -34,900 + 30.96T \text{ J}$$

for the reaction

$$CO_{(g)} + H_2O_{(g)} = H_{2(g)} + CO_{2(g)} \tag{iii}$$

Thus

$$\Delta G^\circ_{1000\,K} = -3,940 \text{ J}$$

and

$$K_{p,(iii)} = 1.606 = \frac{p_{CO_2} p_{H_2}}{p_{H_2O} p_{CO}} \tag{iv}$$

Thus complete reaction equilibrium within the system at 1000 K and $P = 1$ atm requires that the partial pressure of the species CH_4, CO_2, CO, H_2, and H_2O simultaneously satisfy Eqs. (ii) and (iv).

Following the original approach, equilibrium in reaction (i) is established when the system contains $(1-x)$ moles of CH_4, $(2-x)$ moles of H_2O, $3x$ moles of H_2, and x moles of CO, i.e.

CH_4	$+$	H_2O	$=$	H_2	$+$	CO
$1-x$		$2-x$		$3x$		x

Then, with respect to reaction (iii)

	CO	$+$	H_2O	$=$	H_2	$+$	CO_2
initially	x		$2-x$		$3x$		0
and, on reaction	$x-y$		$2-x-y$		$3x+y$		y

Thus, at complete equilibrium,

$$n_{CH_4} = 1 - x, \; n_{H_2O} = 2 - x - y, \; n_{CO} = x - y, \; n_{H_2} = 3x + y, \; n_{CO_2} = y$$

and $n_T = 3 + 2x$.

Thus, the partial pressures are

$$p_{CH_4} = \frac{1 - x}{3 + 2x} P$$

$$p_{CO_2} = \frac{y}{3 + 2x} P$$

$$p_{H_2} = \frac{3x + y}{3 + 2x} P$$

$$p_{CO} = \frac{x - y}{3 + 2x} P$$

$$p_{H_2O} = \frac{2 - x - y}{3 + 2x} P$$

Thus, in Eq. (i)

$$K_{p,(i)} = 25.5 = \frac{(3x + y)^3 (x - y) P^2}{(1 - x)(2 - x - y)(3 + 2x)^2} \tag{v}$$

and, in Eq. (iv)

$$K_{p,(iii)} = 1.61 = \frac{y(3x + y)}{(x - y)(2 - x - y)} \tag{vi}$$

solution of which, with $P = 1$ atm, gives $x = 0.9555$ and $y = 0.2714$. Thus

$$p_{CH_4} = 0.0091 \text{ atm}$$
$$p_{CO_2} = 0.0553 \text{ atm}$$
$$p_{H_2} = 0.6390 \text{ atm}$$
$$p_{CO} = 0.1393 \text{ atm}$$
$$p_{H_2O} = 0.1574 \text{ atm}$$
$$\text{and } P_{total} = 1 \text{ atm}$$

When substituted into Eqs. (ii) and (iv), these partial pressure give $K_{p,(i)} = 25.5$ and $K_{p,(iii)} = 1.33$.

An alternative approach to the solution of the problem can be made by consideration of the mole balance in the system. The system contains the species CH_4, CO_2, CO, H_2O, and H_2 and thus

n_C, the number of moles of $C =$

$$n_{CH_4} + n_{CO_2} + n_{CO} \tag{vii}$$

n_O, the number of moles of oxygen, $=$

$$2n_{CO_2} + n_{CO} + n_{H_2O} \qquad (viii)$$

and n_H, the number of moles of H, $=$

$$2n_{H_2} + 2n_{H_2O} + 4n_{CH_4} \qquad (ix)$$

Initially, 1 mole of CH_4 was mixed with 2 moles of H_2O, and thus, in the closed system,

$$\frac{n_C}{n_O} = 0.5 \qquad (x)$$

and

$$\frac{n_H}{n_O} = 4 \qquad (xi)$$

From Eqs. (x), (vii), and (viii)

$$2n_{CH_4} + 2n_{CO} + 2n_{CO_2} = n_{CO} + 2n_{CO_2} + n_{H_2O}$$

or, as at constant total pressure, p_I is proportional to n_I

$$2p_{CH_4} = p_{H_2O} - p_{CO} \qquad (xii)$$

From Eqs. (xi), (ix), and (viii)

$$2n_{H_2} + 2n_{H_2O} + 4n_{CH_4} = 4n_{CO} + 8n_{CO_2}$$

or

$$2p_{CH_4} = 4p_{CO_2} + 2p_{CO} + p_{H_2O} - p_{H_2} \qquad (xiii)$$

Also

$$P = p_{CH_4} + p_{CO} + p_{CO_2} + p_{H_2O} + p_{H_2} \qquad (xiv)$$

Eliminating p_{CH_4} from Eqs. (xii) and (xiii) gives

$$p_{H_2} = 3p_{CO} + 4p_{CO_2} \qquad (xv)$$

and eliminating P_{CH_4} from Eqs. (xii) and (xiv) gives

$$2p_{H_2} = 2P - 3p_{H_2O} - p_{CO} - 2p_{CO_2} \qquad (xvi)$$

Eliminating p_{H_2} from Eqs. (xv) and (xvi) gives

$$p_{H_2O} = \frac{2}{3}P - \frac{7}{3}p_{CO} - \frac{10}{3}p_{CO_2} \qquad (xvii)$$

Combination of Eqs. (xii) and (xvii) gives

$$p_{CH_4} = \frac{1}{3}P - \frac{5}{3}p_{CO} - \frac{5}{3}p_{CO_2} \qquad (xviii)$$

and thus all five partial pressures are expressed in terms of the total pressure, P, and the partial pressures of CO and CO_2, namely

$$p_{H_2} = 3p_{CO} + 4p_{CO_2}$$

$$p_{H_2O} = \frac{2}{3}P - \frac{7}{3}p_{CO} - \frac{10}{3}p_{CO_2}$$

$$p_{CH_4} = \frac{1}{3}P - \frac{5}{3}p_{CO} - \frac{5}{3}p_{CO_2}$$

Thus Eqs. (ii) and (iv) can be written as

$$K_{p,(i)} = \frac{p_{H_2}^3 p_{CO}}{p_{CH_4} p_{H_2O}} = 25.5 = \frac{(3p_{CO} + 4p_{CO_2})p_{CO}}{\left(\dfrac{P}{5} - \dfrac{5}{3}p_{CO} - \dfrac{5}{3}p_{CO_2}\right)\left(\dfrac{2}{3}P - \dfrac{7}{3}p_{CO} - \dfrac{10}{3}p_{CO_2}\right)}$$

(xix)

and

$$K_{p,(iii)} = \frac{p_{CO_2} p_{H_2}}{p_{H_2O} p_{CO}} = 1.61 = \frac{p_{CO_2}(3p_{CO} + 4p_{CO_2})}{\left(\dfrac{2}{3}P - \dfrac{7}{3}p_{CO} - \dfrac{10}{3}p_{CO_2}\right)p_{CO}}$$

(xx)

Solution of Eqs. (xix) and (xx) with $P = 1$ atm gives $p_{CO} = 0.139$ atm and $p_{CO_2} = 0.055$ atm. The equilibrium state is thus

$$p_{H_2} = 0.637 \text{ atm}$$
$$p_{CO} = 0.139 \text{ atm}$$
$$p_{CO_2} = 0.055 \text{ atm}$$
$$p_{H_2O} = 0.159 \text{ atm}$$
$$p_{CH_4} = 0.01 \text{ atm}$$
$$P_{total} = 1 \text{ atm}$$

These partial pressures return $K_{p,(i)} = 22.6$ and $K_{p,(iii)} = 1.59$.

The partial pressure of oxygen in the gas can be calculated from consideration of either the CO_2/CO ratio or the H_2/H_2O ratio in the equilibrium state. For $CO + \frac{1}{2} O_2 = CO_2$,

$$\Delta G° = -282,400 + 86.81T \text{ J}$$

Thus $\Delta G°_{1000 K} = -195,550$ J and

$$K_{p,1000 K} = 1.64 \times 10^{10} = \frac{p_{CO_2}}{p_{CO} p_{O_2}^{1/2}} = \frac{0.0618}{0.1236 p_{O_2}^{1/2}}$$

which gives $p_{O_2} = 5.8 \times 10^{-22}$ atm. Alternatively, for $H_2 + \frac{1}{2} O_2 = H_2O$

$$\Delta G° = -247,500 + 55.85T \text{ J}$$

Thus $\Delta G^{\circ}_{1000\ K} = -191{,}700$ J and

$$K_{p,1000\ K} = 1.03 \times 10^{10} = \frac{p_{H_2O}}{p_{H_2}p_{O_2}^{1/2}} = \frac{0.1511}{0.6452 \times p_{O_2}^{1/2}}$$

which gives $p_{O_2} = 5.3 \times 10^{-22}$ atm.

The occurrence of the reaction (iii) is indicated if an attempt is made to calculate the value of p_{O_2} in the originally considered CH_4–CO–CO_2–H_2 mixture, i.e., the existence of the equilibrium

$$CO + \frac{1}{2}O_2 = CO_2$$

requires the production of H_2O to establish the equilibrium

$$H_2 + \frac{1}{2}O_2 = H_2O$$

PROBLEMS

11.1 A gas mixture of 50% CO, 25% CO_2, and 25% H_2 (by volume) is fed to a furnace at 900°C. Determine the composition of the equilibrium CO–CO_2–H_2–H_2O gas if the total pressure of the gas in the furnace is 1 atm.

11.2 How much heat is evolved when 1 mole of SO_2 and ½ mole of O_2, each at 1 atm pressure, react to form the equilibrium SO_3–SO_2–O_2 mixture at 1000 K and 1 atm pressure?

11.3 A CO_2–CO–H_2O–H_2 gas mixture at a total pressure of 1 atm exerts a partial pressure of oxygen of 10^{-7} atm at 1600°C. In what ratio were the CO_2 and H_2 mixed to produce the gas with this oxygen pressure?

11.4 Lithium bromide vapor dissociates according to $LiBr_{(g)} \rightarrow Li_{(g)} + \frac{1}{2}Br_{2,(g)}$. At what temperature does the partial pressure of Li reach the value of 10^{-5} atm when the gas is heated at a constant total pressure of 1 atm?

11.5 When SO_3 is decomposed at the constant pressure P and $T = 1000$ K, the partial pressure of O_2 in the equilibrium gas is 0.05 atm. What is the pressure P? If the pressure of this equilibrated gas is increased to 1 atm, to what value must the temperature be decreased to produce a gas mixture in which $p_{O_2} = 0.05$ atm?

11.6 For the dissociation of nitrogen according to

$$N_2 = 2N$$
$$\Delta G^{\circ} = 945{,}000 - 114.9T \text{ J}$$

Calculate

a. The equilibrium partial pressure of N in nitrogen gas at 3000 K and a total pressure of 1 atm

b. The total pressure of the gas, at 3000 K, at which the partial pressure of N_2 is 90% of the total pressure

11.7 Ammonia gas is heated to 300°C. At what total pressure is the mole fraction of N_2 in the equilibrium gas mixture equal to 0.2. Calculate the standard enthalpy change and the standard entropy change for the reaction

$$\frac{3}{2}H_{2(g)} + \frac{1}{2}N_{2(g)} = NH_{3(g)}$$

at 300°C.

11.8 By establishing the equilibrium

$$PCl_{5(g)} = PCl_{3(g)} + Cl_{2(g)}$$

at 500 K in a mixture of PCl_5 and PCl_3 a gas is obtained at 1 atm total pressure in which the partial pressure of Cl_2 is 0.1 atm. In what ratio were PCl_5 and PCl_3 mixed to obtain this equilibrium gas?

11.9 Air and hydrogen are mixed in the ratio 1:4 and are heated to 1200 K. Calculate the partial pressures of H_2 and O_2 in the equilibrium gas at 1 atm pressure and at 10 atm pressure. Air contains 21 volume percent O_2.

11.10 One mole of each of hydrogen, iodine vapor, and HI gas are allowed to react at 1500 K and $P = 1$ atm. Calculate the mole fractions of H_2, I_2, and HI in the equilibrium mixture. The temperature is then changed to that value at which p_{HI} in the equilibrated gas is five times p_{H_2}. What is this temperature?

Chapter 12

REACTIONS INVOLVING PURE CONDENSED PHASES AND A GASEOUS PHASE

12.1 INTRODUCTION

The criterion for equilibrium in a gaseous reaction system was discussed in Chapter 11. The question now is: How is the situation changed if one or more of the reactants or products of the reaction occurs as a condensed phase? As an introduction it will first be considered that the condensed phases are those of pure species, i.e., are of fixed composition. Many practical systems occur in this category, e.g., the reaction of pure metals with gaseous elements to form pure metal oxides, sulfides, halides, etc. Questions of interest include what is the maximum oxygen pressure which can be tolerated in a gaseous atmosphere without oxidation of a given metal occurring at a given temperature, or, to what temperature must a given carbonate be heated in a gaseous atmosphere of given partial pressure of carbon dioxide to cause decomposition of the carbonate? The first question is of interest in the bright annealing of copper, and the second is of interest in the production of lime from limestone.

In such systems, complete equilibrium entails the establishment of (1) phase equilibrium between the individual condensed phases and the gas phase, and (2) reaction equilibrium among the various species present in the gas phase. As phase equilibrium is established when the pure condensed phases exert their saturated vapor pressures, which are uniquely fixed when the temperature of the system is fixed, then the only pressures which can be varied at constant temperature are those of the species which exist only in the gas phase. The unique variations, with temperature, of the saturated vapor pressures of pure condensed phase species and the relative insensitivity of the Gibbs free energies of condensed phases to changes in pressure considerably simplify the thermodynamic treatment of reaction equilibria in systems containing both gaseous and pure condensed phases.

12.2 REACTION EQUILIBRIUM IN A SYSTEM CONTAINING PURE CONDENSED PHASES AND A GAS PHASE

Consider the reaction equilibrium between a pure solid metal M, its pure oxide MO, and oxygen gas at the temperature T and the pressure P:

$$M_{(s)} + \frac{1}{2}O_{2(g)} = MO_{(s)}$$

It is considered that oxygen is insoluble in the solid metal and that MO is stoichiometric. Both the metal M and the oxide MO exist as vapor species in the gas phase, as is required by the criterion for phase equilibrium, i.e.,

$$\overline{G}_M \text{ (in the gas phase)} = G_M \text{ (in the solid metal phase)}$$

and

$$\overline{G}_{MO} \text{ (in the gas phase)} = G_{MO} \text{ (in the solid oxide phase)}$$

and thus the reaction equilibrium

$$M_{(g)} + \frac{1}{2}O_{2(g)} = MO_{(g)}$$

is established in the gas phase. From Eq. (11.6) the criterion for this reaction equilibrium to occur at the temperature T is

$$G^{\circ}_{MO(g)} - \frac{1}{2}G^{\circ}_{O_2(g)} - G^{\circ}_{M(g)} = -RT\ln\frac{p_{MO}}{p_M p_{O_2}^{1/2}} \qquad (12.1)$$

or

$$\Delta G^{\circ} = -RT\ln\frac{p_{MO}}{p_M p_{O_2}^{1/2}}$$

where ΔG° is the difference between the Gibbs free energy of 1 mole of *gaseous* MO *at 1 atm pressure* and the sum of the Gibbs free energies of $\frac{1}{2}$ mole of oxygen gas at 1 atm pressure and 1 mole of *gaseous* M *at 1 atm pressure* all at the temperature T. As M and MO are present in the system as pure solids, phase equilibrium requires that p_{MO} in Eq. (12.1) be the saturated vapor pressure of solid MO at the temperature T, and that p_M be the saturated vapor pressure of solid M at the temperature T. Thus the values of p_{MO} and p_M in the gas phase are uniquely fixed by the temperature T, and so the value of p_{O_2} in Eq. (12.1), at which reaction equilibrium is established, is fixed at the temperature T. As has been stated, phase equilibrium in the system requires that

$$\overline{G}_M \text{ (in the gas)} = G_M \text{ (in the solid metal phase)} \tag{12.2}$$

and

$$\overline{G}_{MO} \text{ (in the gas phase)} = G_{MO} \text{ (in the solid oxide phase)} \tag{12.3}$$

Eq. (12.2) can be written as

$$G^\circ_{M(g)} + RT \ln p_{M(g)} = G^\circ_{M(s)} + \int_{P=1}^{P=p_{M(g)}} V_{M(s)} \, dP \tag{12.4}$$

and Eq. (12.3) can be written as

$$G^\circ_{MO(g)} + RT \ln p_{MO(g)} = G^\circ_{MO(s)} + \int_{P=1}^{P=p_{MO(g)}} V_{MO} \, dP \tag{12.5}$$

Consider the implications of Eq. (12.4). $G^\circ_{M(s)}$ is the molar Gibbs free energy of solid M under a pressure of 1 atm at the temperature T. The integral

$$\int_1^P V_{M(s)} \, dP$$

(where $V_{M(s)}$ is the molar volume of the solid metal at the pressure P and temperature T) is the effect, on the value of the molar Gibbs free energy of solid M at the temperature T, of a change in pressure from $P = 1$ atm to $P = P$. Consider iron as a typical metal at a temperature of 1000°C. The saturated vapor pressure of solid iron at 1000°C is 6×10^{-10} atm, and thus the term $RT \ln p_{M(g)}$ has the value $8.3144 \times 1273 \times \ln (6 \times 10^{-10}) = -224{,}750$ joules. The molar volume of solid iron at 1000°C is 7.34 cm^3, which in the range 0–1 atm, is independent of pressure. The value of the integral for $P = 6 \times 10^{-10}$ atm is -7.34×1 cm^3·atm $= -0.74$ joules. It is thus seen that $G^\circ_{Fe(g)}$ at 1000°C is much larger in value than $G^\circ_{Fe(s)}$ at 1000°C, which is to be expected in view of the large metastability, with respect to the solid, of iron vapor at 1 atm pressure and a temperature of 1000°C. Secondly, it is to be noted that the value of the integral, -0.74 joules, is small enough to be considered negligible, in which case Eq. (12.4) can be written as

$$G^\circ_{M(g)} + RT \ln p_{M(g)} = G^\circ_{M(s)}$$

As a consequence of the negligible effect of pressure on the Gibbs free energy of a condensed phase (when the pressure is in the range 0–1 atm), the standard state of a species occurring as a condensed phase can be defined as being the *pure species at the temperature T*, i.e., the specification that the pressure be 1 atm is no longer required, and $G^\circ_{M(s)}$ is now simply the molar Gibbs free energy of pure solid M at the temperature T. Similarly, Eq. (12.5) can be written as

$$G^\circ_{MO(g)} + RT \ln p_{MO(g)} = G^\circ_{MO(s)}$$

and thus Eq. (12.1) can be written as

$$G^\circ_{MO(s)} - \frac{1}{2}G^\circ_{O_2(g)} - G^\circ_{M(s)} = -RT \ln\left(\frac{1}{p_{O_2}^{1/2}}\right)$$

or

$$\Delta G^\circ = -RT \ln K \tag{12.6}$$

where $K = 1/p_{O_2}^{1/2}$, and ΔG° is the *standard Gibbs free energy change for the reaction*

$$M_{(s)} + \frac{1}{2}O_{2(g)} = MO_{(s)}$$

Thus in the case of a reaction equilibrium involving only pure condensed phases and a gas phase, the equilibrium constant K can be written solely in terms of those species which occur only in the gas phase. Again, as ΔG° is a function only of temperature, then K is a function only of temperature, and thus, at any fixed temperature, the establishment of reaction equilibrium occurs at a unique value of $p_{O_2} = p_{O_2(eq.T)}$. The equilibrium thus has one degree of freedom, as can be seen from application of the phase rule. $P = 3$ (two pure solids and a gas phase), $C = 2$ (metal M + oxygen), and thus $F = C + 2 - P = 2 + 2 - 3 = 1$.

If, at any temperature T, the actual partial pressure of oxygen in a closed metal–metal oxide–oxygen system is greater than $p_{O_2(eq.T)}$, spontaneous oxidation of the metal will occur, thus consuming oxygen and decreasing the oxygen pressure in the gas phase. When the actual oxygen pressure has thus been decreased to $p_{O_2(eq.T)}$, then, provided that both solid phases are still present, the oxidation ceases and reaction equilibrium is reestablished. Similarly, if the oxygen pressure in the closed vessel was originally less than $p_{O_2(eq.T)}$, spontaneous reduction of the oxide would occur until $p_{O_2(eq.T)}$ was reached.

Extraction metallurgical processes involving the reduction of oxide ores depend on the achievement and maintenance of an oxygen pressure less than $p_{O_2(eq.T)}$ in the reaction vessel. For example, the standard Gibbs free energy change for the reaction

$$4Cu_{(s)} + O_{2(g)} = 2Cu_2O_{(s)}$$

is

$$\Delta G^\circ = -324,400 + 138.5T \text{ J}$$

in the temperature range 298–1200 K. Thus,

$$-\ln K = \ln p_{O_2(eq.T)} = \frac{\Delta G^\circ}{RT}$$

or

$$\log p_{O_2(eq.T)} = -\frac{324,400}{2.303 \times 8.3144T} + \frac{138.5}{2.303 \times 8.3144}$$

$$= -\frac{16,940}{T} + 7.23$$

This variation of log $p_{O_2(eq.T)}$ with $1/T$ is drawn as the line ab in Fig. 12.1a, and all points on the line represent the unique oxygen pressure, $p_{O_2(eq.T)}$ required for equilibrium between solid Cu, solid Cu_2O, and oxygen gas at the particular temperature T. Thus ab divides the diagram into two regions. Above ab (where $p_{O_2} > p_{O_2(eq.T)}$) the metal phase is not stable, and thus the system exists as $Cu_2O_{(s)} + O_{2(g)}$, and below ab (where $p_{O_2} < p_{O_2(eq.T)}$) the oxide is not stable, and hence the system exists as $Cu_{(s)} + O_{2(g)}$.

Other equilibria among two condensed pure phases and a gas phase include the formation of carbonates and hydroxides. For example, at the temperature T the equilibrium

$$MO_{(s)} + H_2O_{(g)} = M(OH)_{2(s)}$$

occurs when

$$G^\circ_{MO(s)} + G^\circ_{H_2O(g)} + RT \ln p_{H_2O} = G^\circ_{M(OH)_{2(s)}}$$

i.e., when

$$\Delta G^\circ = -RT \ln K = RT \ln p_{H_2O(eq.T)}$$

and similarly, the equilibrium

$$MO_{(s)} + CO_{2(g)} = MCO_{3(s)}$$

occurs when

$$G^\circ_{MO(s)} + G^\circ_{CO_2(g)} + RT \ln p_{CO_2(eq.T)} = G^\circ_{MCO_3(s)}$$

i.e., when

$$\Delta G^\circ = -RT \ln K = RT \ln p_{CO_2(eq.T)}$$

For the reaction

$$MgO_{(s)} + CO_{2(g)} = MgCO_{3(s)}$$

$$\Delta G^\circ = -117,600 + 170T \text{ J}$$

in the temperature range 298 to 1000 K, and thus

$$\log p_{CO_2(eq.T)} = -\frac{117,600}{2.303 \times 8.3144T} + \frac{170}{2.303 \times 8.3133}$$

$$= -\frac{6141}{T} + 8.88$$

This variation is shown in Fig. 12.1b as the line cd, which again divides the diagram into two regions: one in which $MgO_{(s)} + CO_{2(g)}$ are stable, and one in which $MgCO_{3(s)} + CO_{2(g)}$ are stable.

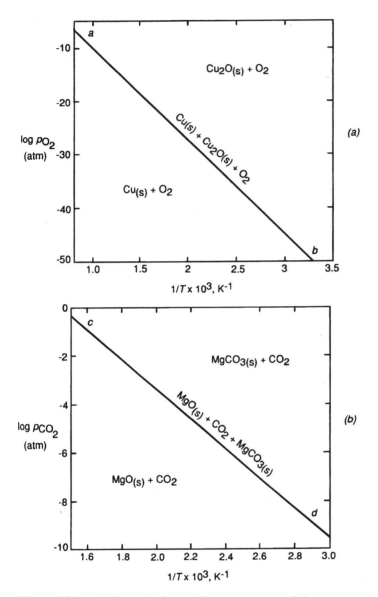

Figure 12.1 (*a*) the variation, with temperature, of the oxygen pressure required for maintenance of the equilibrium $4Cu_{(s)} + O_2 = 2Cu_2O$. (*b*) The variation, with temperature, of the carbon dioxide pressure required for maintenance of the equilibrium $MgO_{(s)} + CO_2 = MgCO_{3(s)}$.

12.3 THE VARIATION OF THE STANDARD GIBBS FREE ENERGY CHANGE WITH TEMPERATURE

For any chemical reaction, combination of Eqs. (6.11) and (6.12) gives $\Delta G°$ for the reaction as a function of temperature:

$$\Delta G_T^° = \Delta H_T^° - T\Delta S_T^° \tag{12.7}$$

$$= \Delta H_{298}^° + \int_{298}^{T} \Delta c_p dT - T\Delta S_{298}^° - T \int_{298}^{T} \frac{\Delta c_p}{T} dT$$

from which it is seen that the deviation from linearity between $\Delta G°$ and T depends on the sign and the magnitude of Δc_p for the reaction. Generally, however, the variation of $\Delta G°$ with T is considered as follows. For each of the individual reactants and products of the reaction, the molar heat capacity, c_p, is expressed, over a stated range of temperature, in the form

$$c_p = a + bT + cT^{-2}$$

Hence, for the reaction, again within the stated range of temperature,

$$\Delta c_p = \Delta a + \Delta bT + \Delta cT^{-2}$$

Kirchhoff's equation is

$$\left(\frac{\partial \Delta H°}{\partial T}\right)_P = \Delta c_p = \Delta a + \Delta bT + \Delta cT^{-2}$$

where $\Delta H°$ is the standard enthalpy change for the reaction. Integration gives

$$\Delta H_T^° = \Delta H_0 + \Delta aT + \frac{\Delta bT^2}{2} - \frac{\Delta c}{T} \tag{12.8}$$

where ΔH_0 is an integration constant which would be equal to the standard enthalpy of the reaction at 0 K only if the analytical expression for Δc_p as a function of T was valid down to 0 K. ΔH_0 is normally evaluated by substituting a known value of $\Delta H_T^°$ into Eq. (12.8).

For the reaction, the Gibbs-Helmholtz equation is

$$\frac{\partial \left(\frac{\Delta G°}{T}\right)}{\partial T} = -\frac{\Delta H°}{T^2} = -\frac{\Delta H_0}{T^2} - \frac{\Delta a}{T} - \frac{\Delta b}{2} + \frac{\Delta c}{T^3}$$

integration of which gives

$$\frac{\Delta G°}{T} = I + \frac{\Delta H_0}{T} - \Delta a \ln T - \frac{\Delta bT}{2} - \frac{\Delta c}{2T^2}$$

or

$$\Delta G° = IT + \Delta H° - \Delta aT \ln T - \frac{\Delta bT^2}{2} - \frac{\Delta c}{2T} \tag{12.9}$$

where I is an integration constant.

As $\Delta G° = -RT \ln K$, then Eq. (12.9) gives

$$\ln K = -\frac{\Delta H_0}{RT} - \frac{I}{R} + \frac{\Delta a \ln T}{R} + \frac{\Delta bT}{2R} + \frac{\Delta c}{2RT^2} \qquad (12.10)$$

The value of I can be determined if K is known at any temperature T. For the reaction

$$4Cu_{(s)} + O_{2(g)} + 2Cu_2O_{(s)}$$

$$\Delta H_{298}° = -324,400 \text{ J}$$

$$\Delta S_{298}° = -138.5 \text{ J/K}$$

and thus

$$\Delta G_{298}° = -324,400 + (298 \times 138.5) = -283,100 \text{ J}$$

In the range 298 to 1356 K,

$$c_{p,Cu(s)} = 22.6 + 6.3 \times 10^{-3}T \text{ J/K}$$

In the range 298 to 1200 K

$$c_{p,Cu_2O(s)} = 62.34 + 24 \times 10^{-3}T \text{ J/K}$$

and in the range 298 to 3000 K

$$c_{p,O_2(g)} = 30 + 4.2 \times 10^{-3}T - 1.7 \times 10^5T^{-2} \text{ J/K}$$

such that in the temperature range 298 to 1200 K

$$\Delta c_p = 2c_{p,Cu_2O(s)} - 4c_{p,Cu(s)} - c_{p,O_2(g)}$$
$$= 4.28 + 18.6 \times 10^{-3}T + 1.7 \times 10^5T^{-2} \text{ J/K}$$

Thus

$$\Delta H_T° = \Delta H_0 + 4.28T + 9.3 \times 10^{-3}T^2 - 1.7 \times 10^5T^{-1} \text{ J}$$

Substitution of $\Delta H_{298}° = -324,400 \text{ J}$ gives $\Delta H_0 = -325,900 \text{ J}$. Dividing by $-T^2$, integrating with respect to T, and multiplying through by T gives

$$\Delta G_T° = -325,900 - 4.28T \ln T - 9.3 \times 10^{-3}T^2 - 0.85 \times 10^5T^{-1} + IT$$

Substitution of $\Delta G_{298}° = -283,100 \text{ J}$ gives $I = 171.7$, and hence

$$\Delta G_T° = -325,900 - 4.28T \ln T - 9.3 \times 10^{-3}T^2 - 0.85 \times 10^5T^{-1} + 171.7 \text{ J} \quad (i)$$

and

$$-\ln K = \frac{\Delta G_T°}{RT} = \ln p_{O_2(eq.T)}$$

$$= -\frac{39,200}{T} - 0.515 \ln T - 1.1 \times 10^{-3}T - \frac{1.0 \times 10^4}{T^3} + 20.65$$

The variation of ΔG_T°, as calculated from the experimentally measured variation of $p_{O_2(eq.T)}$ with temperature, can be fitted to an equation of the form

$$\Delta G^\circ = A + BT \ln T + CT$$

For the oxidation of $4Cu_{(s)}$ to $2Cu_2O_{(s)}$, this gives

$$\Delta G^\circ = -338,900 - 14.2T \ln T + 247T \text{ J} \tag{ii}$$

which can be approximated in linear form by

$$\Delta G^\circ = -333,000 + 141.3T \text{ J} \tag{iii}$$

in the temperature range 298 to 1200 K. At 300 K Eqs. (i), (ii), and (iii) give, respectively, -282.8, -289.1, and -290.6 kJ, and, at 1200 K, give -169.7, -163.3, and -163.4 kJ.

It can be noticed that Eq. (12.10) is similar to the vapor pressure equation, Eq. (7.8). The relationship between the two can be seen as follows. Consider the evaporation of A,

$$A_{(l)} = A_{(v)}$$

Equilibrium occurs at the temperature T when

$$G_{A(l)}^\circ = G_{A(v)}^\circ + RT \ln p_A$$

i.e., when

$$\Delta G^\circ = -RT \ln p_A = -RT \ln K$$

If the liquid and the vapor have the same molar heat capacity, then

$$\ln p_A = -\frac{\Delta G^\circ}{RT} = -\frac{\Delta H^\circ}{RT} + \frac{\Delta S^\circ}{R}$$

This is to be compared with Eq. (7.7), which gave

$$\ln p_A = -\frac{\Delta H_{evap}}{RT} + \text{constant}$$

If the vapor behaves ideally, then, at constant temperature, $H_{(v)}$ is independent of pressure, and thus

$$\Delta H^\circ = H_{(v)}^\circ - H_{(l)}^\circ = H_{(v)} - H_{(l)} = \Delta H_{evap}$$

However, from Eq. (6.15), for the vapor

$$S(T,p_A) = S_T^\circ - R \ln p_A$$

and thus

$$\begin{aligned} \Delta S_{(l \to v)}^\circ &= S_{(v)}^\circ - S_{(l)}^\circ \\ &= S_{(v)} - S_{(l)}^\circ + R \ln p_A \\ &= \Delta S_{evap} + R \ln p_A \end{aligned}$$

The constant in Eq. (7.7) thus has the value $\Delta S_{evap}/R + \ln p_A$, and Eq. (7.7) becomes

$$\ln p_A = -\frac{\Delta H_{evap}}{RT} + \frac{\Delta S_{evap}}{R} + \ln p_A$$

indicating, thus, that when the vapor is in equilibrium with the liquid at the temperature T,

$$\Delta S_{evap} = \frac{\Delta H_{evap}}{T}$$

or, as is required,

$$\Delta G_{evap} = 0$$

12.4 ELLINGHAM DIAGRAMS

Ellingham* plotted the experimentally determined variations of $\Delta G°$ with T for the oxidation and sulfidation of a series of metals and found that, in spite of the terms involving $\ln T$, T^2, and T^{-1} in Eq. (12.9), the relationship approximated to straight lines over ranges of temperature in which no change of state occurred. The relations could thus be expressed by means of the simple equation

$$\Delta G° = A + BT \tag{12.11}$$

in which the constant A is identified with the temperature-independent standard enthalpy change for the reaction, $\Delta H°$, and the constant B is identified with the negative of the temperature-independent standard entropy change for the reaction, $-\Delta S°$.

The variation of $\Delta G°$ with T for the oxidation reaction

$$4Ag_{(s)} + O_{2(g)} = 2Ag_2O_{(s)}$$

is shown in Fig. 12.2, and Fig. 12.2 is known as an Ellingham diagram. From Eq. (12.11), the length of the intercept of the line with the $T = 0$ K axis gives $\Delta H°$, and $\Delta S°$ is given by the negative of the slope of the line. As $\Delta S°$ is a negative quantity (the reaction involves the disappearance of a mole of gas), the line has a positive slope. $\Delta G° = 0$ at $T = 462$ K, and thus, at this temperature pure solid silver and oxygen gas at 1 atm pressure are in equilibrium with pure solid silver oxide. From Eq. (12.6), $\Delta G° = -RT \ln K = RT \ln p_{O_2(eq.T)} = 0$ at 462 K, and, therefore $p_{O_2(eq.462\,K)} = 1$. If the temperature of the system (pure Ag_2O, pure $Ag_{(s)}$ and oxygen gas at 1 atm pressure) is decreased to T_1, then, as $\Delta G°$ for the oxidation reaction becomes negative, the metal phase becomes unstable relative to silver oxide and oxygen gas at 1 atm pressure and, hence, is spontaneously oxidized. The value of $p_{O_2(eq.T_1)}$ is calculated from $\Delta G°_{T_1} = RT_1 \ln p_{O_2(eq.T_1)}$, and, as $\Delta G°_{T_1}$ is a negative quantity, $p_{O_2(eq.T_1)} < 1$ atm.

---

*H. J. T. Ellingham, "Reducibility of Oxides and Sulfides in Metallurgical Processes," *J. Soc. Chem. Ind.* (1944), vol. 63, p. 125.

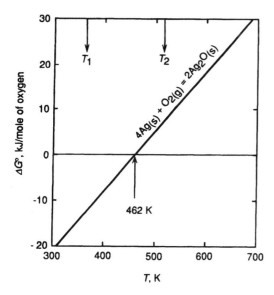

Figure 12.2 The Ellingham line for the oxidation of silver.

Similarly, if the temperature of the system is increased from 462 K to T_2, then, as $\Delta G°$ for the oxidation becomes positive, the oxide phase becomes unstable relative to silver metal and oxygen gas at 1 atm and, hence, spontaneously dissociates. As ΔG_{T_2} is a positive quantity, $p_{O_2(eq.T_2)}$ is greater than 1 atm. The value of $\Delta G°$ for an oxidation is thus a measure of the chemical affinity of the metal for oxygen, and the more negative the value of $\Delta G°$ at any temperature, the more stable the oxide.

For the oxidation reaction $A_{(s)} + O_{2(g)} = AO_{2(s)}$,

$$\Delta S° = S°_{AO_2(s)} - S°_{O_2(g)} - S°_{A(s)}$$

and, as generally, in the temperature range in which A and AO_2 are solid, $S°_{O_2}$ is considerably greater than both $S°_A$ and $S°_{AO_2}$ (see, e.g., Fig. 6.15), then,

$$\Delta S° \sim S°_{O_2}$$

Thus the standard entropy changes for oxidation reactions involving solid phases have almost the same value, which corresponds with the disappearance of 1 mole of oxygen gas initially at 1 atm pressure. As the slopes of the lines in an Ellingham diagram are equal to $-\Delta S°$, then the lines are more or less parallel to one another, as will be seen in Fig. 12.13.

$\Delta G°$ at any temperature is the sum of the enthalpy contribution $\Delta H°$ (which is independent of T if $\Delta c_p = 0$) and the entropy contribution $-T\Delta S$ (which, if $\Delta c_p = 0$, is a linear function of temperature). The two contributions are illustrated in Fig. 12.3 for the oxidation reactions,

$$2Co_{(s)} + O_{2(g)} = 2CoO_{(s)}$$

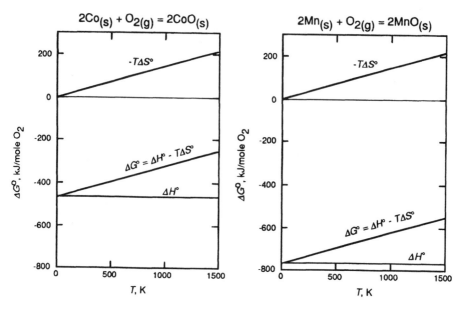

Figure 12.3 Illustration of the effect of the magnitude of $\Delta H°$ on the $\Delta G°$- T relationships for reactions of the type $2M_{(s)} + O_{2(g)} = 2MO_{(s)}$.

for which

$$\Delta G° = -467{,}800 + 143.7T \text{ J}$$

in the temperature range 298 to 1763 K, and

$$2Mn_{(s)} + O_{2(g)} = 2MnO_{(s)}$$

for which

$$\Delta G° = -769{,}400 + 145.6T \text{ J}$$

in the temperature range 298 to 1500 K.

As the values of $\Delta S°$ for these two reactions are virtually equal to one another, Fig. 12.3 shows that the relative stabilities of the oxides CoO and MnO are determined by their values of $\Delta H°$, in that, the more negative the value of $\Delta H°$, the more negative the value of $\Delta G°$ and, hence, the more stable the oxide. As

$$\ln K = -\frac{\Delta H°}{RT} + \frac{\Delta S°}{R} = \ln\left(\frac{1}{p_{O_2(eq.T)}}\right)$$

then

$$p_{O_2(eq.T)} = \exp\frac{\Delta H°}{RT}\exp\frac{-\Delta S°}{R} = \text{constant} \times \exp\frac{\Delta H°}{RT}$$

i.e., as $\Delta H°$ is a negative quantity, $p_{O_2(eq.T)}$ increases exponentially with increasing temperature and, at any temperature, decreases as $\Delta H°$ becomes more negative.

Consider two oxidation reactions, the Ellingham lines of which intersect one another, e.g.,

$$2A + O_2 = 2AO \qquad (i)$$

and

$$B + O_2 = BO_2 \qquad (ii)$$

which are shown in Fig. 12.4. From Fig. 12.4 it is seen that $\Delta H°_{(ii)}$ is more negative than $\Delta H°_{(i)}$ and that $\Delta S°_{(ii)}$ is more negative than $\Delta S°_{(i)}$. Subtraction of reaction (i) from reaction (ii) gives

$$B + 2AO = 2A + BO_2 \qquad (iii)$$

for which the variation of $\Delta G°$ with T is as shown in Fig. 12.5. At temperatures less than T_E, A and BO_2 are stable with respect to B and AO, and at temperatures higher than T_E the reverse is the case. At T_E, A, B, AO, and BO_2, occurring in their standard states, are in equilibrium with one another. The equilibrium at T_E (as with any equilibrium) occurs as the result of a compromise between enthalpy and entropy considerations. As $\Delta H°_{(iii)}$ is negative (being equal to $\Delta H°_{(ii)} - \Delta H°_{(i)}$) and $\Delta S°_{(iii)}$ is negative (being given by $\Delta S°_{(ii)} - \Delta S°_{(i)}$), the system A + B + O_2 has a minimum enthalpy when it occurs as A + BO_2 and has a maximum entropy when it occurs as B + AO. At T_E, $\Delta H°_{(iii)}$ equals $-T_E \Delta S°_{(iii)}$, and thus $\Delta G°_{(iii)} = 0$. At temperatures less than T_E the enthalpy contribution to $\Delta G°_{(iii)}$ outweighs the entropy contribution, and thus, as $\Delta G°_{(iii)}$ is negative, A + BO_2 is the stable state. At temperatures higher than T_E, the reverse is the case, $\Delta G°_{(iii)}$ is negative and B + AO is the stable state. Fig. 12.5 thus shows that if pure A were to be used as a reducing agent to reduce pure BO_2, to form pure B and pure AO, then the reduction would have to be conducted at temperatures

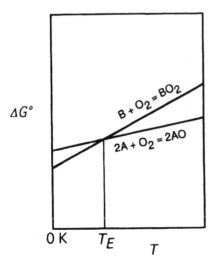

Figure 12.4 Intersecting Ellingham lines for two hypothetical oxidation reactions.

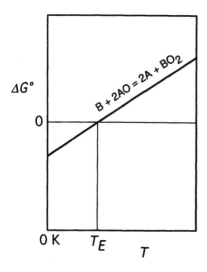

Figure 12.5 The variation of $\Delta G°$ with T for $B + 2AO = 2A + BO_2$ from Fig. 12.4.

higher than T_E. The foregoing discussion also illustrates that, in order to compare the stabilities of different oxides, the Ellingham diagrams must be drawn for oxidation reactions involving the consumption of the same number of moles of oxygen. The units of $\Delta G°$ for the oxidation reaction thus must be energy per mole of O_2, e.g., joules per mole of oxygen.

In order to avoid having to calculate the value of $p_{O_2(eq.T)}$ for any oxidation reaction Richardson* added a nomographic scale to the Ellingham diagram. This scale is constructed as follows. At any temperature T, the standard Gibbs free energy change for an oxidation reaction, $\Delta G_T°$, is given by Eq. (12.6) as $RT \ln p_{O_2(eq.T)}$. However, from Eq. (8.10), $G = G° + RT \ln P$, $\Delta G_T°$ is seen to be numerically equal to the decrease in the Gibbs free energy of 1 mole of oxygen gas when its pressure is decreased from 1 atm to $p_{O_2(eq.T)}$ atm at the temperature T. Consider the variation of ΔG with T in Eq. (8.10). For a decrease in the pressure of 1 mole of ideal gas from 1 atm to P atm, ΔG vs. T is a straight line with a slope of $R \ln P$, and, as $P < 1$, the line has a negative slope. Similarly for an increase in the pressure of 1 mole of ideal gas from 1 to P atm, the variation of ΔG with T is linear with a positive slope of $R \ln P$. Thus a series of lines can be drawn for given pressure changes (from 1 to P atm) as a function of temperature. These lines radiate from the point $\Delta G = 0$, $T = 0$ as shown in Fig. 12.6. Superimposition of Fig. 12.6 with a typical Ellingham diagram is shown in Fig. 12.7. In Fig. 12.7,

At T_1, $\Delta G_{T_1}° = ab$ = the decrease in Gibbs free energy when p_{O_2} is decreased from 1 atm to 10^{-20} atm at T_1.

At T_2, $\Delta G_{T_2}° = cd$ = the decrease in Gibbs free energy when p_{O_2} is decreased from 1 atm to 10^{-8} atm at T_2.

*F. D. Richardson and J. H. E. Jeffes, "The Thermodynamics of Substances of Interest in Iron and Steel Making from 0°C to 2400°C: I—Oxides," *J. Iron and Steel Inst.* (1948), vol. 160, p. 261.

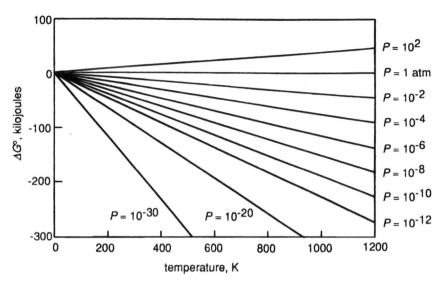

Figure 12.6 The variation, with temperature, of the difference between the Gibbs free energy of 1 mole of ideal gas in the state ($P = P$ atm, T) and the Gibbs free energy of 1 mole of ideal gas in the state ($P = 1$ atm, T).

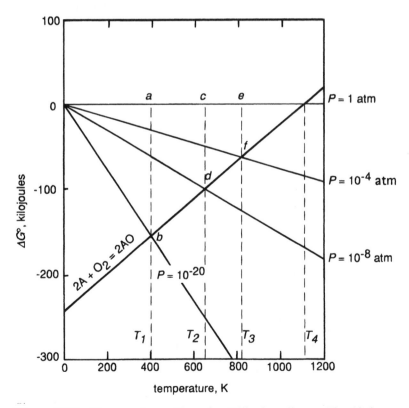

Figure 12.7 The superimposition of an Ellingham line on Fig. 12.6.

At T_3, $\Delta G^\circ_{T_3} = ef$ = the decrease in Gibbs free energy when p_{O_2} is decreased from 1 atm to 10^{-4} atm at T_3.

At T_4, $\Delta G = 0$, which corresponds to no change of p_{O_2} from 1 atm. Thus $p_{O_2(eq.T_4)} = 1$ atm.

The $p_{O_2(eq.T)}$ nomographic scale is thus added to the Ellingham diagram along the right-hand edge and along the bottom edge. The value of $p_{O_2(eq.T)}$ for any metal–metal oxide equilibrium is read off the graph as that value on the scale which is collinear with the points $\Delta G^\circ = 0$, $T = 0$, and ΔG°_T, $T = T$.

The reactions (i) and (ii) shown in Fig. 12.4 can be reexamined using the $p_{O_2(eq.T)}$ nomographic scale. Fig. 12.4 is reproduced with a nomographic oxygen pressure scale in Fig. 12.8. At any temperature lower than T°_E (say, T_1), it is seen that

$$p_{O_2} \text{ [eq. for reaction (ii) at } T_1] < p_{O_2} \text{ [eq. for reaction (i) at } T_1]$$

Thus, if metal A and metal B are placed in a closed system in an atmosphere of oxygen at $P = 1$ atm, both metals spontaneously oxidize, and, as a consequence of the consumption of oxygen to form the oxides, the pressure of oxygen decreases. Oxidation of both metals continues until the oxygen pressure is decreased to the value p_{O_2} [eq. for reaction (i) at T_1] at which point the oxidation of A ceases. However, as B + O_2 at p_{O_2} [eq. for reaction (i) at T_1] is still unstable with respect to BO_2, the oxidation

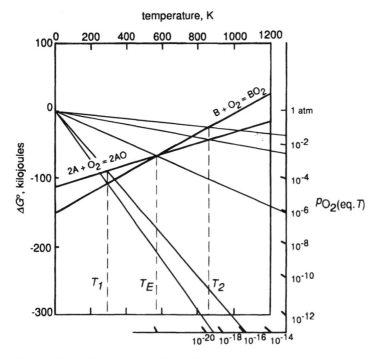

Figure 12.8 Illustration of the addition of the Richardson oxygen pressure nomographic scale to an Ellingham diagram.

of B continues until the oxygen pressure has been decreased to p_{O_2} [eq. for reaction (ii) at T_1]. As the oxygen pressure is decreased below p_{O_2} [eq. from reaction (i) at T_1] then AO becomes unstable with respect to A and O_2 at the prevailing pressure, and, hence, AO decomposes. When complete equilibrium is attained, the state of the system is A + BO$_2$ + O$_2$ at p_{O_2} [eq. for reaction (ii) at T_1]. At any temperature higher than T_E, (say, T_2),

$$p_{O_2} \text{ [eq. for reaction (i) at } T_2] < p_{O_2} \text{ [eq. for reaction (ii) at } T_2]$$

and an argument similar to that above shows that the equilibrium state of the closed system containing, initially A + B + O$_2$ at 1 atm pressure, is B + AO + O$_2$ at p_{O_2} (eq. for reaction (i) at T_2). It is thus obvious that A, B, AO, BO$_2$, and an oxygen atmosphere are in equilibrium only at that temperature T at which

$$p_{O_2(eq.(i)T)} = p_{O_2(eq.(ii)T)}$$

Fig. 12.8 shows that this unique temperature is T_E, the temperature at which the Ellingham lines intersect one another.

12.5 THE EFFECT OF PHASE TRANSFORMATIONS

In the previous section it was stated that the variation of $\Delta G°$ with temperature can be approximated by a straight line only over ranges of temperature in which no change of phase of a reactant or product occurs. However, as the enthalpy of a high-temperature phase (e.g., the liquid phase) exceeds that of a low-temperature phase (e.g., the solid phase) by the latent heat of the phase change, and similarly, the entropy of the higher-temperature phase exceeds that of the lower-temperature phase, then an "elbow" occurs on an Ellingham line at the temperature of a phase change in a reactant or product of the reaction.

Consider the reaction

$$A_{(s)} + O_{2(g)} = AO_{2(s)}$$

for which the $\Delta H°$ is the standard enthalpy change and $\Delta S°$ is the standard entropy change. At $T_{m,A}$, the melting temperature of A, the change of phase

$$A_{(s)} \rightarrow A_{(l)}$$

occurs, for which the standard enthalpy change (the latent heat of melting) is $\Delta H°_{m,A}$ and the corresponding change in entropy is $\Delta S°_{m,A}$. Thus for the reaction

$$A_{(l)} + O_{2(g)} = AO_{2(s)}$$

the standard enthalpy change is $\Delta H° - \Delta H°_{m,A}$ and the standard entropy change is $\Delta S° - \Delta S°_{m,A}$. As $\Delta H°_{m,A}$ and $\Delta S°_{m,A}$ are positive quantities (melting is an endothermic process), then $\Delta H° - \Delta H°_{m,A}$ is a larger negative quantity than is $\Delta H°$ and $\Delta S° - \Delta S°_{m,A}$ is a larger negative quantity than is $\Delta S°$. Consequently, the Ellingham line for the oxidation of liquid A to form solid AO$_2$ has a greater slope than the corresponding line for the oxidation of solid A, and the line contains an "elbow upwards" at

$T_{m,A}$. This is shown in Fig. 12.9a. The line does not contain a discontinuity as, at $T_{m,A}$, $G^\circ_{A(s)} = G^\circ_{A(l)}$. If the melting temperature of the oxide, T_{m,AO_2}, is lower than the melting temperature of the metal, then, at T_{m,AO_2}, the change of phase

$$AO_{2(s)} \rightarrow AO_{2(l)}$$

occurs, for which the standard enthalpy and entropy changes are, respectively, $\Delta H^\circ_{m,AO_2}$ and $\Delta S^\circ_{m,AO_2}$. Thus, for the reaction

$$A_{(s)} + O_{2(g)} = AO_{2(l)}$$

the standard enthalpy change is $\Delta H^\circ + \Delta H^\circ_{m,AO_2}$ and the standard entropy change is $\Delta S^\circ + \Delta S^\circ_{m,AO_2}$, both of which are less than the corresponding quantities ΔH° and ΔS°. In this case the Ellingham line for oxidation of the solid metal to produce the liquid oxide has a lower slope than the line for oxidation of the solid metal to the solid oxide, and thus, as shown in Fig. 12.9b, the Ellingham line has an elbow downwards at T_{m,AO_2}. If $T_{m,A} < T_{m,AO_2}$, the Ellingham line is as shown in Fig. 12.10a, and, if $T_{m,A} > T_{m,AO_2}$, the line is as shown in Fig. 12.10b.

Copper is a metal which melts at a lower temperature than its lowest oxide Cu_2O. The standard Gibbs free energy change for the oxidation of solid copper to form solid cuprous oxide in the range of temperature 298 K to $T_{m,Cu}$ is

$$\Delta G^\circ = -338,900 - 14.2T \ln T + 247T \text{ J} \tag{i}$$

and, for the oxidation of liquid copper to form solid cuprous oxide in the range of temperature $T_{m,Cu}$ to 1503 K is

$$\Delta G^\circ = -390,800 - 14.2T \ln T + 285.3T \text{ J} \tag{ii}$$

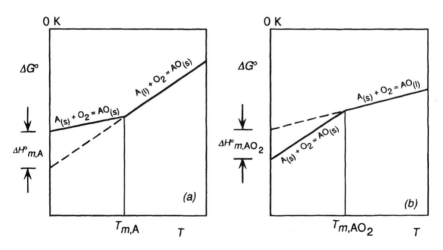

Figure 12.9 (a) The effect of melting of the metal on the Ellingham line for oxidation of the metal. (b) The effect of melting of the metal oxide on the Ellingham line for the oxidation of the metal.

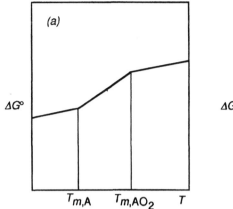
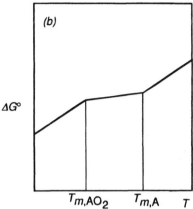

Figure 12.10 Illustration of the effects of phase changes of the reactants and products of a reaction on the Ellingham line for the reaction.

These two lines, which are drawn in Fig. 12.11, intersect at 1356 K, which is thus the melting temperature of copper. $\Delta G^\circ_{(i)} - \Delta G^\circ_{(ii)}$ gives

$$\Delta G = 51{,}900 - 38.3T \text{ J}$$

for the phase change

$$4Cu_{(s)} = 4Cu_{(l)}$$

or, for the melting of 1 mole of Cu,

$$\Delta G^\circ_{m,Cu} = 12{,}970 - 9.58T \text{ J}$$

from which

$$\Delta H^\circ_{m,Cu} = 12{,}970 \text{ J}$$

and

$$\Delta S^\circ_{m,Cu} = 9.58 \text{ J/K}$$

Thus, at $T_{m,Cu}$, the slope of the Ellingham line for the oxidation of Cu increases by 9.58 J/K.

As $FeCl_2$ boils at a lower temperature than the melting temperature of Fe, the Ellingham diagram for the chlorination of Fe shows "elbows downwards" at the melting temperature of $FeCl_2$ and at the boiling temperature of $FeCl_2$. For

$$Fe_{(s)} + Cl_{2(g)} = FeCl_{2(s)}$$

$$\Delta G^\circ = -346{,}300 - 12.68T \ln T + 212.9T \text{ J} \qquad \text{(iii)}$$

in the range 298 K to $T_{m,FeCl_2}$. For

$$Fe_{(s)} + Cl_{2(g)} = FeCl_{2(l)}$$

$$\Delta G^\circ = -286{,}400 + 63.68T \text{ J} \qquad \text{(iv)}$$

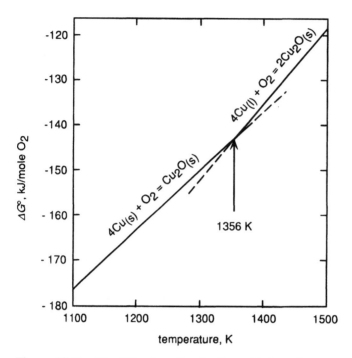

Figure 12.11 The Ellingham line for the oxidation of copper.

in the range T_{m,FeCl_2} to T_{b,FeCl_2} and for

$$\text{Fe}_{(s)} + \text{Cl}_{2(g)} = \text{FeCl}_{2(g)}$$

$$\Delta G^\circ = -105,600 + 41.8T \ln T - 375.1T \text{ J} \qquad (\text{v})$$

in the range T_{b,Cl_2} to $T_{m,\text{Fe}}$. Lines (iii), (iv), and (v) are shown in Fig. 12.12, which shows that

$$T_{m,\text{FeCl}_2} = 969 \text{ K} \quad \text{and} \quad T_{b,\text{FeCl}_2} = 1298 \text{ K}$$

for $\text{FeCl}_{2(s)} \rightarrow \text{FeCl}_{2(l)}$, $\Delta G^\circ_{(\text{iv})} - \Delta G^\circ_{(\text{iii})}$ gives

$$\Delta G^\circ_{m,\text{FeCl}_2} = 59,900 + 12.68T \ln T - 149.0T \text{ J}$$

Thus

$$\Delta H^\circ_{m,\text{FeCl}_2} = -T^2 \left[\frac{\partial(\Delta G^\circ_{m,\text{FeCl}_2}/T)}{\partial T} \right] = 59,900 - 12.68T \text{ J}$$

which, at 969 K gives $\Delta H^\circ_{m,\text{FeCl}_2} = 47,610$ J.

$$\Delta S^\circ_{m,\text{FeCl}_2} = -\frac{\partial \Delta G^\circ_{m,\text{FeCl}_2}}{\partial T}$$

$$= -12.68 \ln T - 12.68 + 149.0 \text{ J/K}$$

$$= 49.13 \text{ J/K at 969 K}$$

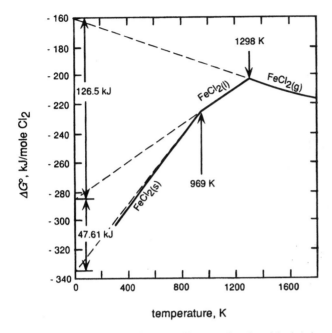

Figure 12.12 The Ellingham diagram for the chlorination of iron.

or, alternatively,

$$\Delta S^\circ_{m,\text{FeCl}_2} = \frac{\Delta H^\circ_{m,\text{FeCl}_2}}{T_{m,\text{FeCl}_2}} = \frac{47{,}610}{969} = 49.13 \text{ J/K}$$

Thus the difference in slope between lines (iii) and (iv) at 969 K is 49.13 J/K, and the difference between the tangential intercepts of the slopes of the two lines at 969 K, with the $T = 0$ axis, is 47,610 J. Similarly, $\Delta G^\circ_{(v)} - \Delta G^\circ_{(iv)}$ gives

$$\Delta G^\circ_{b,\text{FeCl}_2} = 180{,}800 + 41.8T \ln T - 438.8T \text{ J}$$

Thus

$$\Delta H^\circ_{b,\text{FeCl}_2} = -T^2 \left[\frac{\partial(\Delta G^\circ_{b,\text{FeCl}_2}/T)}{\partial T} \right]$$

$$= 180{,}800 - 41.8T \text{ J}$$

$$= 126{,}500 \text{ J at 1298 K}$$

and

$$\Delta S^\circ_{b,\text{FeCl}_2} = \frac{126{,}500}{1298} = 97.46 \text{ J/K at 1298 K}$$

Thus the change in slope between lines (iv) and (v) at 1298 K is 97.46 J/K, and the difference between the tangential intercepts is 126,500 J.

12.6 THE OXIDES OF CARBON

Carbon forms two gaseous oxides, CO and CO_2, according to

$$C_{(gr)} + O_{2(g)} = CO_{2(g)} \tag{i}$$

for which $\Delta G_{(i)}^{\circ} = -394,100 - 0.84T$ J and

$$2C_{(gr)} + O_{2(g)} = 2CO_{(g)} \tag{ii}$$

for which $\Delta G_{(ii)}^{\circ} = -223,400 - 175.3T$ J. Combination of reactions (i) and (ii) gives

$$2CO_{(g)} + O_{2(g)} = 2CO_{2(g)} \tag{iii}$$

for which $\Delta G_{(iii)}^{\circ} = 2\Delta G_{(i)}^{\circ} - \Delta G_{(ii)}^{\circ} = -564,800 + 173.62T$ J. The Ellingham lines for reactions (i), (ii), and (iii) are included in Fig. 12.13, in which it is seen that:

The line for reaction (iii) has a positive slope (2 moles of gas produced from 3 moles of gas, $\Delta S_{(iii)}^{\circ} = -173.62$ J/K).

The line for reaction (i) has virtually no slope (1 mole of gas produced from 1 mole of gas, $\Delta S_{(i)}^{\circ} = 0.84$ J/K),

The line for reaction (ii) has a negative slope (2 moles of gas produced from 1 mole of gas, $\Delta S_{(ii)}^{\circ} = 175.3$ J/K).

Consider the equilibrium

$$C_{(gr)} + CO_{2(g)} = 2CO_{(g)} \tag{iv}$$

for which $\Delta G_{(iv)}^{\circ} = \Delta G_{(ii)}^{\circ} - \Delta G_{(i)}^{\circ} = 170,700 - 174.5T$. $\Delta G_{(iv)}^{\circ} = 0$ at $T = 978$ K (705°C), the temperature at which the Ellingham lines for reactions (i) and (ii) intersect one another. At this temperature, CO and CO_2, in their standard states, i.e., both at 1 atm pressure, are in equilibrium with solid graphite, and the total pressure of the system is 2 atm. As reaction equilibria are normally considered for systems under a total pressure of 1 atm, it is instructive to calculate the temperature at which CO and CO_2, each at a pressure of 0.5 atm, are in equilibrium with solid C. Consideration of Le Chatelier's principle indicates whether this temperature is higher or lower than 978 K. For reaction (iv), $\Delta G_{978\,K}^{\circ} = 0 = -RT \ln K_p = -RT \ln (p_{CO}^2/p_{CO_2})$, i.e., at 978 K, $K_p = 1$, and hence $p_{CO_2} = p_{CO} = 1$ and $P_{total} = 2$ atm. If the pressure of the system is decreased to 1 atm, then, as K_p is independent of pressure, and thus remains equal to unity, p_{CO} becomes greater than p_{CO_2}, i.e., the equilibrium shifts toward the CO side, as is predicted by Le Chatelier's principle. As $\Delta H_{(iv)}^{\circ} = +170,700$ J, reaction (iv) is endothermic, and thus, from Le Chatelier's principle, as a decrease in temperature shifts the equilibrium in that direction which involves an evolution of heat, a decrease in temperature shifts the equilibrium towards the C + CO_2 side. Thus, if it is required to decrease the pressure of the system from 2 to 1 atm, and, at the same time maintain $p_{CO} = p_{CO_2}$, the temperature of the system must be decreased. The temperature required for $p_{CO} = p_{CO_2} = 0.5$ is calculated as follows. For reaction (i)

$$C + O_2 = CO_2 \qquad \Delta G_{(i)}^{\circ} = -394,100 - 0.84T \text{ J}$$

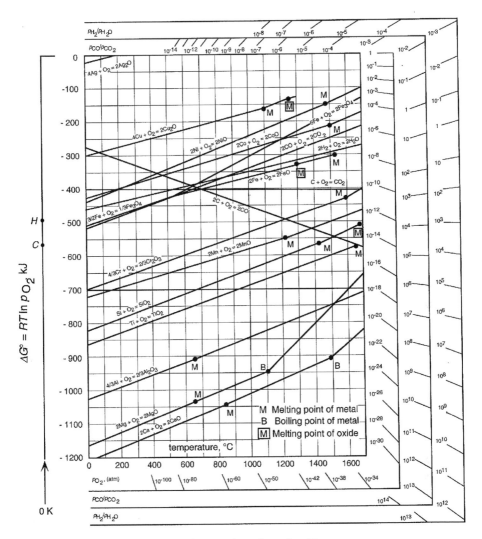

Figure 12.13 The Ellingham diagram for selected oxides.

If the pressure of the CO_2, which is produced at 1 atm, is decreased to 0.5 atm, then, for the change of state

$$CO_2(T,P = 1 \text{ atm}) \rightarrow CO_2(T,P = 0.5 \text{ atm}) \qquad \text{(v)}$$

the decrease in Gibbs free energy is $\Delta G_{(v)} = RT \ln 0.5$, and hence, for the reaction

$$C_{(gr)} + O_{2(g,P=1 \text{atm})} = CO_{2(g,P=0.5 \text{atm})}$$
$$\Delta G_{(vi)} = \Delta G_{(i)}^\circ + \Delta G_{(v)} \qquad \text{(vi)}$$
$$= -394,100 - 0.84T + RT \ln 0.5 \text{ J}$$

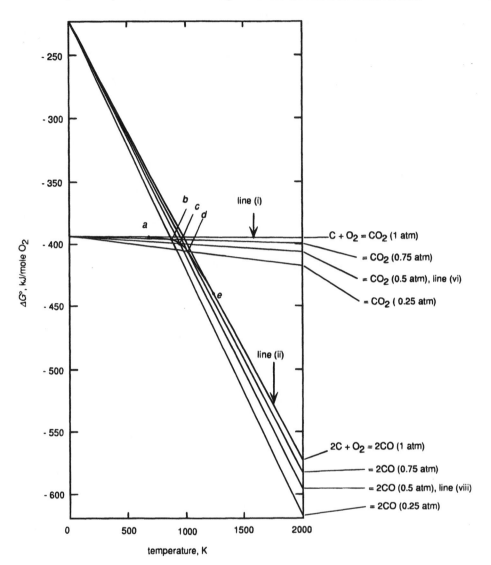

Figure 12.14 The effect of varying the pressures of the product gases of the reactions $C_{(gr)} + O_{2(g,\ p=1\ atm)} = CO_{2(g)}$ and $2C_{(gr)} + O_{2(g,\ P=1\ atm)} = 2CO_{(g)}$ on the variations of ΔG with T for the two reactions.

This line is obtained on the Ellingham diagram by rotating the line for reaction (i) clockwise about its point of intersection with the $T = 0$ axis until, at the temperature T, the vertical separation between line (i) and line (vi) is $RT \ln 0.5$. This is illustrated in Fig. 12.14. Similarly the Ellingham line for the reaction.

$$2C_{(gr)} + O_{2(g,P=1\ atm)} = 2CO_{(g,P=0.5\ atm)} \tag{vii}$$

is obtained as the sum of $\Delta G_{(ii)}^{\circ}$ and ΔG for the change of state

$$2CO(T,P = 1 \text{ atm}) \rightarrow 2CO(T,P = 0.5 \text{ atm})$$

i.e.,

$$\Delta G_{(vii)} = -223{,}400 - 175.3T + 2RT \ln 0.5 \text{ J}$$

This line is obtained by rotating the line for reaction (ii) clockwise about its point of intersection with the $T = 0$ axis until, at any temperature T, the vertical separation between line (ii) and line (i) is $2RT \ln 0.5$.

Combination of reactions (vi) and (vii) gives

$$C_{(gr)} + CO_{2(g,0.5 \text{ atm})} = 2CO_{(g,0.5 \text{ atm})} \qquad \text{(viii)}$$

for which

$$\Delta G_{(viii)} = \Delta G_{(iv)}^{\circ} + RT \ln 0.5$$

Thus, CO_2 and CO, each at 0.5 atm pressure, are in equilibrium with solid C at that temperature at which $\Delta G_{(viii)} = 0$, i.e., at the temperature of intersection of the lines (vi) and (vii) in Fig. 12.14 (the point c).

The temperature at which CO (at 0.25 atm) and CO_2 (at 0.75 atm) are in equilibrium with solid C is obtained in a similar manner as the intersection of line (i) rotated clockwise until, at T, it has been displaced a vertical distance $RT \ln 0.75$, and line (ii) rotated clockwise until, at T, it has been displaced a vertical distance $2RT$ in 0.25. This is the point b in Fig. 12.14. Similarly the point d in Fig. 12.14 is the temperature at which CO at 0.75 atm pressure and CO_2 at 0.25 atm pressure are in equilibrium with solid C. For a mixture of CO and CO_2 at 1 atm pressure in equilibrium with solid C, the variation of percent CO by volume in the gas with temperature is shown in Fig. 12.15. Fig. 12.15 includes the points a, b, c, d, and e drawn in Fig. 12.14.

Fig. 12.15 shows that, at temperatures less than 600 K, the equilibrium gas is virtually CO_2 at 1 atm pressure and, at temperatures higher than 1400 K, the equilibrium gas is virtually CO at 1 atm pressure. These points are, respectively, the points a and e in Fig. 12.14. Thus in Fig. 12.14 the variation, with temperature, of the Gibbs free energy change for oxidation of solid C to produce a CO–CO_2 mixture at 1 atm pressure which is in equilibrium with solid C, is given by line (i) up to the point a, then by the line $abcde$, and then by the line (ii) beyond the point e.

At any temperature T, the CO–CO_2 mixture in equilibrium with C exerts an equilibrium oxygen pressure via the equilibrium

$$2CO + O_2 = 2CO_2$$

for which

$$\Delta G_{(iii)}^{\circ} = -564{,}800 + 173.62T \text{ J} = -RT \ln \left(\frac{p_{CO_2}^2}{p_{CO}^2 p_{O_2}} \right)$$

$$= 2RT \ln \left(\frac{p_{CO}}{p_{CO_2}} \right)_{\text{eq. with C}} + RT \ln p_{O_2(\text{eq.})}$$

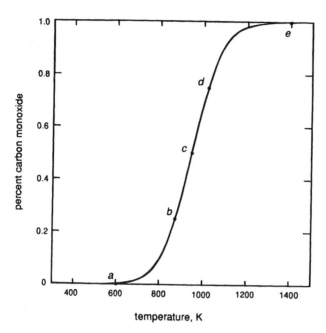

Figure 12.15   The variation, with temperature, of the composition of the CO–CO$_2$ gas mixture in equilibrium with solid graphite at $P_{total} = 1$ atm.

Thus

$$\ln p_{O_2(eq.T)} = -\frac{564{,}800}{8.3144T} + \frac{173.62}{8.3144} + 2\ln\left(\frac{p_{CO_2}}{p_{CO}}\right)_{eq.\ with\ C} \quad (12.12)$$

If it is required that solid carbon be used as a reducing agent to reduce a metal oxide MO$_2$ at the temperature T, then $p_{O_2(eq.T)}$ in Eq. (12.12) must be lower than $p_{O_2(eq.T)}$ for the equilibrium $M + O_2 = MO_2$ (see Sec. 12.7).

The Equilibrium 2CO + O$_2$ = 2CO$_2$

The Ellingham line for the above reaction is shown in Fig. 12.16 as the line *cs*. Being the variation, with temperature, of the standard Gibbs free energy of formation, $\Delta G°$, this line is for the reaction which produces CO$_2$ at 1 atm pressure from CO at 1 atm pressure and O$_2$ at 1 atm pressure. The effect of producing the CO$_2$ at any pressure, P, other the 1 atm (from CO and O$_2$ each at 1 atm) is the rotation of the Ellingham line *cs* about the point *c*, clockwise if $P < 1$ atm and anticlockwise if $P > 1$ atm. For the given value of P the rotation is such that, as before, at the temperature T, the vertical displacement of *cs* is $2RT$ in P. A series of lines, radiating from the point *c*, can thus be drawn for different pressures of CO$_2$ produced from CO and O$_2$, each at

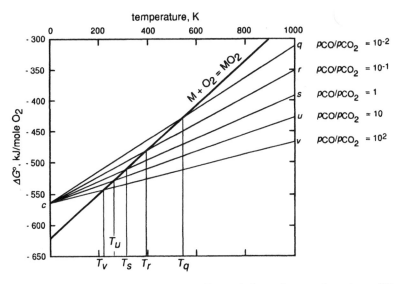

Figure 12.16 Illustration of the effect of the ratio p_{CO_2}/p_{CO} in a CO_2–CO gas mixture on the temperature at which the equilibrium $M + CO_2 = MO + CO$ is established.

1 atm. Fig. 12.16 shows four of these lines, cq for CO_2 produced at 10^2 atm, cr for CO_2 produced at 10 atm, cu for CO_2 produced at 0.1 atm, and cv for CO_2 produced at 10^{-2} atm. The significance of this series of lines, with respect to the possibility of using CO–CO_2 gas mixtures as reducing agents for the metal oxide MO_2, is illustrated as follows. The Ellingham line for the reaction $M + O_2 = MO_2$ is drawn in Fig. 12.16. This intersects the line cs at the temperature T_s, which is thus the temperature at which the standard Gibbs free energy change for reaction

$$MO_2 + 2CO = M + 2CO_2 \qquad\qquad \text{(ix)}$$

is zero, i.e.,

$$\Delta G^\circ_{(ix)} \text{ at } T_s = 0 = -RT \ln \left(\frac{p_{CO_2}}{p_{CO}} \right)^2$$

and thus

$$\frac{p_{CO_2}}{p_{CO}} = 1$$

At temperatures higher than T_s a CO–CO_2 mixture of $p_{CO}/p_{CO_2} = 1$ is reducing with respect to MO_2, and at temperatures lower than T_s it is oxidizing with respect to the metal M. If it is required that a CO–CO_2 mixture be made reducing with respect to MO_2 at temperatures lower than T_s, then the ratio p_{CO}/p_{CO_2} must be increased to a value greater than unity. The Ellingham line for the reaction $M + O_2 = MO_2$ intersects the line cu at the temperature T_u, and T_u is thus the temperature at which the reaction

$$MO_2 + 2CO(1\text{ atm}) = M + 2CO_2(0.1\text{ atm}) \tag{x}$$

is at equilibrium, i.e.,

$$\Delta G_{(x)} \text{ at } T_u = \Delta G^\circ_{(ix)} + 2RT \ln 0.1$$

But, by definition,

$$\Delta G^\circ_{(ix)} = -RT \ln \left(\frac{p_{CO_2}}{p_{CO}}\right)^2_{eq}$$

and thus

$$\Delta G_{(x)} = 0 = -2RT \ln \left(\frac{p_{CO_2}}{p_{CO}}\right)_{eq} + 2RT \ln 0.1$$

Thus $(p_{CO}/p_{CO_2})_{eq} = 10$ at T_u, and hence, by decreasing the temperature from T_s to T_u, the CO/CO_2 ratio must be increased from 1 to 10 in order to maintain reaction equilibrium.

Similarly at T_v the equilibrium CO/CO_2 ratio is 100, at T_r the equilibrium CO/CO_2 ratio is 0.1, and at T_q the equilibrium ratio is 0.01. Thus a CO/CO_2 nomographic scale can be added to the Ellingham diagram and for any reaction

$$MO_2 + 2CO = M + 2CO_2$$

the equilibrium CO/CO_2 ratio at any temperature T is read off the nomographic scale as that point which is collinear with the point C and the point $\Delta G^\circ_T, T = T$, for the reaction $M + O_2 = MO_2$. This scale is drawn in Fig. 12.13.

Fig. 12.15 is generated by reading off the equilibrium CO/CO_2 ratios for the reaction $C + O_2 = CO_2$ up to the point a in Fig. 12.14, then the CO/CO_2 ratios along the line $abcde$, and finally the CO/CO_2 ratios for the reaction $2C + O_2 = 2CO$ beyond the point e.

In an exactly similar manner the H_2/H_2O nomographic scale is added to Fig. 12.13 by considering the effect of the variation of the pressure of H_2O on the reaction equilibrium

$$2H_2 + O_2 = 2H_2O$$

The equilibrium H_2/H_2O ratio at the temperature T for the reaction

$$MO_2 + 2H_2 = M + 2H_2O$$

is read off the H_2/H_2O scale as the point which is collinear with the points H and $\Delta G^\circ_T, T = T$ for the reaction $M + O_2 = MO_2$.

12.7 GRAPHICAL REPRESENTATION OF EQUILIBRIA IN THE SYSTEM METAL–CARBON–OXYGEN

The main criteria for graphical representation of equilibria in a system are (1) the amount of information provided and (2) clarity and both of these considerations are dependent on the coordinates chosen for use in the graphical representation. As the nomographic scale for the ratio CO/CO_2 in Fig. 12.13 shows that the range of inter-est of values of p_{CO}/p_{CO_2} is 10^{-14} to 10^{14}, it is convenient to present this ratio on a logarithmic scale. Fig. 12.17, which uses the coordinates $\log (p_{CO_2}/p_{CO})$ and T, rep-resents a convenient method of clear presentation of reaction equilibrium in the car-bon–oxygen and carbon–oxygen–metal systems. From Eq. (iii), for

$$2CO_{(g)} + O_{2(g)} = 2CO_{2(g)}$$

$$\Delta G^{\circ}_{(iii)} = -546{,}800 + 173.62T \text{ J}$$

$$= -RT \ln \left(\frac{p^2_{CO_2}}{p^2_{CO}p_{O_2}} \right)$$

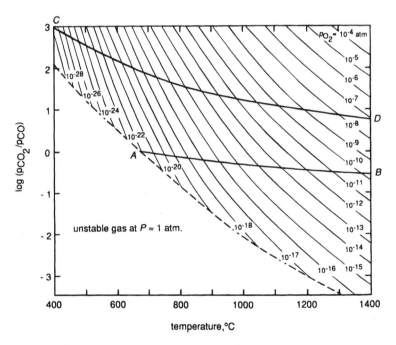

Figure 12.17   The relationship between the partial pressure of oxygen in a CO_2–CO gas mixture and temperature. The broken line is the variation, with temperature, of the composition of the gas which is in equilibrium with graphite at 1 atm pressure. The lines AB and CD represent, respectively, the equilibria $Fe + CO_2 = FeO + CO$ and $Co + CO_2 = CoO + CO$.

and thus

$$\log\left(\frac{p_{CO_2}}{p_{CO}}\right) = \frac{1}{2}\log p_{O_2} + \frac{564,800}{2 \times 2.303 \times 8.3144T} - \frac{173.62}{2 \times 2.303 \times 8.3144} \quad \text{(xi)}$$

and, for any given value of p_{O_2}, this gives the variation of the given oxygen isobar with $\log(p_{CO_2}/p_{CO})$ and temperature. The oxygen isobars in the range 10^{-29} to 10^{-4} atm are drawn as a function of $\log(p_{CO_2}/p_{CO})$ and T in Fig. 12.17.

The equilibrium

$$C_{(gr)} + CO_{2(g)} = 2CO_{(g)} \quad \text{(iv)}$$

sets a lower limit on the CO_2/CO ratio which can be obtained at any temperature,

$$\Delta G^\circ_{(iv)} = 170,700 - 174.5T \text{ J} = -RT\ln\left(\frac{p^2_{CO}}{p_{CO_2}}\right)$$

and thus, at a total pressure of 1 atm, i.e., when $p_{CO_2} = 1 - p_{CO}$,

$$\frac{p^2_{CO}}{1 - p_{CO}} = \exp\left(\frac{-170,700}{8.3144T}\right)\exp\left(\frac{174.5}{8.3144}\right) = x$$

or

$$p^2_{CO} + p_{CO}x - x = 0$$

solution of which gives

$$p_{CO} = \frac{-x + \sqrt{x^2 + 4x}}{2}$$

$$p_{CO_2} = \frac{2 + x - \sqrt{x^2 + 4x}}{2}$$

and

$$\frac{p_{CO_2}}{p_{CO}} = \frac{2 + x - \sqrt{x^2 + 4x}}{\sqrt{x^2 + 4x} - x} \quad \text{(xii)}$$

The value of p_{CO_2}/p_{CO} given by Eq. (xii) is the minimum obtainable value at the temperature T. If an attempt is made to mix CO and CO_2 (at $P_{total} = 1$ atm) in a ratio p_{CO_2}/p_{CO} lower than that given by Eq. (xii), carbon will precipitate until, thereby, the ratio is increased to its unique value required for equilibrium with carbon at the temperature T. The variation of $\log(p_{CO_2}/p_{CO})_{eq.C/CO/CO_2}$ with temperature is shown as the broken line in Fig. 12.17. The variation of the corresponding minimum values of p_{O_2} with temperature in Fig. 12.17 is given by the intersections of the oxygen isobars with the carbon deposition line.

Equilibria such as

$$MO + CO = M + CO_2$$

can readily be presented on plots such as Fig. 12.17. For example, for

$$FeO_{(s)} + CO_{(g)} = Fe_{(s)} + CO_{2(g)}$$

$\Delta G° = -22,800 + 24.26T$ J and thus the variation of the equilibrium ratio CO_2/CO with temperature is given by

$$\log \left(\frac{p_{CO_2}}{p_{CO}} \right)_{eq.FeO/Fe} = \frac{22,800}{2.303 \times 8.3144T} - \frac{24.26}{2.303 \times 8.3144}$$

This variation is drawn as the line AB in Fig. 12.17 and thus any gas, the state of which lies above the line AB, is oxidizing with respect to Fe, and states below AB are reducing with respect to FeO. The variation of $\log p_{O_2(eq.T,Fe/FeO)}$ with temperature is given by the intersections of the oxygen isobars with the line AB. The temperature at which AB intersects the carbon deposition line is the minimum temperature at which solid FeO can be reduced to solid Fe by graphite, and this is the temperature at which $Fe_{(s)}$, $FeO_{(s)}$, $C_{(s)}$, and the $CO–CO_2$ atmosphere at 1 atm pressure coexist in equilibrium, i.e., it is the temperature at which

$$p_{O_2(eq.C/CO/CO_2)} = p_{O_2(eq.Fe/FeO)}$$

The line CD in Fig. 12.17 represents the variation of $\log (p_{CO_2}/p_{CO})$ with T for the equilibrium

$$CoO_{(s)} + CO_{(g)} = Co_{(s)} + CO_{2(g)}$$

for which

$$\Delta G° = -48,500 + 14.9T \text{ J}$$

and

$$\log \left(\frac{p_{CO_2}}{p_{CO}} \right)_{eq.Co/CoO} = \frac{48,500}{2.303 \times 8.3144T} - \frac{14.9}{2.303 \times 8.3144}$$

As the equilibrium constant K for the reaction

$$MO + CO = M + CO_2$$

is given by $p_{CO_2}/p_{CO(eq.T,M/MO)}$, a plot of $\log p_{CO_2}/p_{CO(eq.T,M/MO)}$ vs. $1/T$ is a plot of $\log K$ vs. $1/T$. Fig. 12.18 shows the information given in Fig. 12.17 as such a plot. With respect to the amount of information which can be obtained from a graphical representation of equilibria in a system, Fig. 12.18 is a better representation than is Fig. 12.17. As

$$\left[\frac{\partial \ln K}{\partial \left(\frac{1}{T} \right)} \right] = -\frac{\Delta H°}{R}$$

the slope of a tangent to an equilibrium line at the temperature T gives the value of $-\Delta H°/R$. If $\Delta c_p = 0$, $\log K$ is a linear function of $1/T$. Thus the slope of the line AB in Fig. 12.18 equals $-\Delta H°/R$ for the reaction $FeO + CO = Fe + CO_2$, and the slope of the line CD equals $-\Delta H°/R$ for the reaction $CoO + CO = Co + CO_2$. Also the intercepts of the tangential slopes, or the lines themselves if they are linear, with the $1/T = 0$ axis give the corresponding values of $\Delta S°/R$ for the reactions. From Eq. (xi) the slope of any oxygen isobar equals $-\Delta H°/2R$ for the reaction $2CO + O_2 = 2CO_2$, and thus the oxygen isobars in Fig. 12.18 are parallel lines.

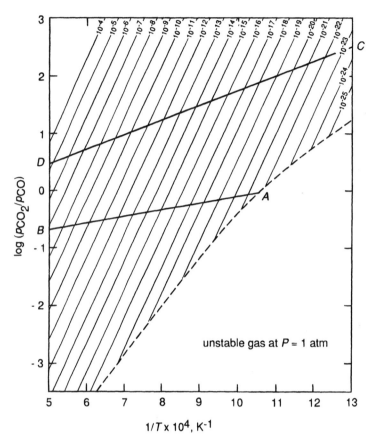

Figure 12.18 Fig. 12.17 reproduced as $\log(p_{CO_2}/p_{CO})$ vs. $1/T$.

12.8 SUMMARY

1. The facts that (1) a pure species occurring as a condensed phase exerts a unique saturated vapor pressure at the temperature T and (2) the dependence of the Gibbs free energy of a condensed phase on pressure (at low pressure) is negligibly small facilitate a convenient definition of the standard state of species occurring as condensed phases. This standard state is simply the pure species in its stable condensed state at the temperature T. Using this standard state, the equilibrium constant for a reaction involving pure condensed phases and a gas phase can be written in terms of the partial pressures of those species which occur only in the gas phase. For example, for the oxidation of a pure metal to its pure stoichiometric oxide, the equilibrium constant is given by $1/p_{O_2(eq.T)}$, where $p_{O_2(eq.T)}$ is the unique partial pressure of oxygen required for equilibrium between the metal, its oxide, and the gas phase at the temperature T. This pressure is such that

$$G_M^\circ + G_{O_2}^\circ + RT \ln p_{O_2(eq.T)} = G_{MO_2}^\circ$$

or

$$\Delta G_T^\circ = -RT \ln K = -RT \ln \left(\frac{1}{p_{O_2(eq.T)}} \right)$$

2. Determination of the equilibrium state of a chemical reaction system requires knowledge of the variation, with temperature, of the standard Gibbs free energy change for the reaction. This relationship can be obtained from thermochemical data, i.e., from a knowledge of the standard enthalpy and entropy changes at a single temperature (usually ΔH_{298}° and ΔS_{298}°) and the variations, with temperature, of the constant pressure molar heat capacities of the reactants and products, or it can be determined from knowledge of the variation, with temperature, of the equilibrium constant for the reaction. For the oxidation of a pure metal to its pure oxide, experimental measurement of the variation of $p_{O_2(eq.T)}$ with temperature gives the variation of ΔG° with temperature as

$$\ln \left(\frac{1}{p_{O_2(eq.T)}} \right) = -\frac{\Delta H^\circ}{RT} + \frac{\Delta S^\circ}{R} = -\frac{\Delta G^\circ}{RT}$$

If ΔH° and ΔS° are independent of temperature (i.e., if $\Delta c_p = 0$), ln K is a linear function of $1/T$.

3. The fitting of the variation of ΔG° with T to a line produces the so-called Ellingham line for the reaction, and a plot of ΔG° vs. T is known as an Ellingham diagram. Ellingham lines plotted on a single diagram for a series of similar reactions, e.g., for the formation of oxides, sulfides, etc., provide a convenient representation of the relative stabilities of the compounds. The addition of p_{O_2}, p_{CO}/p_{CO_2}, and p_{H_2}/p_{H_2O} nomographic scales to these diagrams facilitates geometric determination of $p_{O_2(eq.T)}$, $p_{CO}/p_{CO_2(eq.T)}$, and $p_{H_2}/p_{H_2O(eq.T)}$ for, respectively, the equilibria $2M + O_2 = 2MO$, $M + CO_2 = MO + CO$, and $M + H_2O = MO + H_2$.

12.9 NUMERICAL EXAMPLES

Example 1

Compare the relative efficiencies of H_2 and CO as reducing agents for metal oxides. For $CO + \frac{1}{2}O_2 = CO_2$,

$$\Delta G_{(i)}^\circ = -282,400 + 86.81T \text{ J} \tag{i}$$

and for $H_2 + \frac{1}{2}O_2 = H_2O$

$$\Delta G_{(ii)}^\circ = -247,500 + 55.85T \text{ J} \tag{ii}$$

The values of ΔH° and ΔS° for these two reactions cause their Ellingham lines to intersect at 1127 K, as shown in Fig. 12.13, with $\Delta G_{(ii)}^\circ$ being more negative than $\Delta G_{(i)}^\circ$ at temperatures higher than 1125 K and $\Delta G_{(i)}^\circ$ being more negative than $\Delta G_{(ii)}^\circ$ at temperatures lower than 1125 K. This indicates that H_2 is the more efficient reducing agent at higher temperatures and that CO is the more efficient reducing agent at lower temperatures. Consider the reduction of CoO by each of H_2 and CO at 1673 K and at 873 K.

For $CoO_{(s)} = Co_{(s)} + \frac{1}{2}O_{2(g)}$

$$\Delta G^\circ_{(iii)} = +233{,}900 - 71.85T \text{ J} \tag{iii}$$

Combination of $\Delta G^\circ_{(i)}$ and $\Delta G^\circ_{(iii)}$ gives

$$\Delta G^\circ_{(iv)} = -48{,}500 + 14.96T \text{ J} \tag{iv}$$

for $CoO + CO = Co + CO_2$, and combination of $\Delta G^\circ_{(iii)}$ and $\Delta G^\circ_{(ii)}$ gives

$$\Delta G^\circ_{(v)} = -12{,}500 - 17.05T \text{ J} \tag{v}$$

for $CoO + H_2 = Co + H_2O$. The positive value of $\Delta S^\circ_{(v)}$ causes $\Delta G^\circ_{(v)}$ to become more negative with increasing temperature and the negative value of $\Delta S^\circ_{(iv)}$ causes $\Delta G^\circ_{(iv)}$ to become less negative with increasing temperature. At 1673 K $\Delta G^\circ_{(v)} = -41{,}024$ J, and thus

$$K_{(v),\ 1673\ K} = \exp\left(\frac{41{,}024}{8.3144 \times 1673}\right) = 19.1 = \left(\frac{p_{H_2O}}{p_{H_2}}\right)_{eq.}$$

Thus, if H_2 at 1673 K is passed through a column of CoO, which is long enough that reaction equilibrium is achieved before the gas leaves the column, the fraction of H_2 which is consumed before equilibrium is reached is $19.1/20.1 = 0.95$, and thus 1 mole of H_2 is required to reduce 0.95 moles of CoO.

At 1673 K $\Delta G^\circ_{(iv)} = -23{,}470$ J, and thus

$$K_{(iv),\ 1673} = 5.40 = \left(\frac{p_{CO_2}}{p_{CO}}\right)_{eq.}$$

and thus the fraction of CO which is consumed by the reduction reaction at 1673 K before equilibrium is reached is $5.40/6.40 = 0.843$. Thus 1 mole of CO is required to reduce 0.843 moles of CoO.

At 873 K, $\Delta G^\circ_{(v)} = -27{,}384$ J, which gives

$$K_{(v),\ 873} = 43.5 = \left(\frac{p_{H_2O}}{p_{H_2}}\right)_{eq.}$$

and thus the fraction of H_2 consumed is $43.5/44.5 = 0.978$ and 1 mole of H_2 reduces 0.978 moles of CoO.

At 873, $\Delta G^\circ_{(iv)} = -35{,}440$ J, which gives

$$K_{(iv),\ 873} = 132 = \left(\frac{p_{CO_2}}{p_{CO}}\right)_{eq.}$$

Thus the fraction of CO consumed is $132/133 = 0.992$, and 1 mole of CO reduces 0.992 moles of CoO. Thus H_2 is the more efficient reducing agent at higher temperatures and CO is the more efficient at lower temperatures, and decreasing the temperature at which the reduction reaction is conducted increases the efficiencies of both reductants.

Example 2

Consider the reduction of solid ZnO by CO to form Zn vapor and CO_2 according to

$$ZnO_{(s)} + CO_{(g)} = Zn_{(v)} + CO_{2(g)} \tag{i}$$

For $ZnO_{(s)} = Zn_{(v)} + \frac{1}{2}O_{2(g)}$

$$\Delta G^\circ_{(ii)} = 460{,}200 - 198T \text{ J} \tag{ii}$$

and for $CO_{(g)} + \frac{1}{2}O_{2(g)} = CO_{2(g)}$

$$\Delta G^\circ_{(iii)} = -282{,}400 + 86.81T \text{ J} \tag{iii}$$

Summing $\Delta G^\circ_{(ii)}$ and $\Delta G^\circ_{(iii)}$ gives

$$\Delta G^\circ_{(i)} = 177{,}800 - 111.2T \text{ J}$$

The equilibrium involves three components (Zn, O, and C) and two phases (a solid and a gas) and thus, from the phase rule, has

$$F = C + 2 - P = 3 + 2 - 2 = 3$$

degrees of freedom. However, the stoichiometric requirement that $p_{Zn} = p_{CO_2}$ uses one of the degrees of freedom and thus the equilibrium is fixed when the temperature and total pressure are fixed. Calculate the composition of the gas phase at 950°C and $P = 1$ atm.

At 1223 K

$$\Delta G^\circ_{(i)} = 177{,}800 - (111.2 \times 1223) = 41{,}800 \text{ J}$$

and thus

$$K_{(iii)\ 1223\ K} = \exp\left(\frac{-41{,}800}{8.3144 \times 1223}\right) = 0.0164 = \frac{p_{Zn}p_{CO_2}}{p_{CO}}$$

From the stoichiometry, $p_{Zn} = p_{CO_2}$ and the total pressure P is

$$P = p_{CO} + p_{Zn} + p_{CO_2} \tag{iv}$$

Thus, with $P = 1$ atm, $p_{CO} = 1 - 2p_{Zn}$ and

$$0.0164 = \frac{p_{Zn}^2}{1 - 2p_{Zn}} \tag{v}$$

which has the solution $p_{Zn} = 0.113$ atm. Therefore $p_{CO_2} = 0.113$ atm and $P_{CO} = 1 - (2 \times 0.113) = 0.775$ atm. At $P = 1$ atm the mole fractions of the species in the gas phase are equal to their partial pressures.

To what value must the total pressure of the gas mixture be increased in order to cause condensation of the zinc vapor at 1223 K? Condensation occurs when the partial pressure of Zn reaches the value of the saturated vapor pressure of liquid zinc at 1223 K. The saturated vapor pressure of liquid zinc is given by

$$\ln p^\circ_{Zn,(l)} \text{ (atm)} = \frac{-15{,}250}{T} - 1.255 \ln T + 21.79$$

which gives $p_{Zn,(l)}^{\circ} = 1.49$ atm at 1223 K. Therefore

$$K_{(iii)\ 1223\ K} = 0.0164 = \frac{1.49^2}{P - (2 \times 1.49)}$$

which has the solution $P = 138$ atm. Thus $p_{Zn} = p_{CO} = 1.49$ atm and $p_{CO_2} = 138 - (2 \times 1.49) = 135$ atm. At the point of condensation of zinc the mole fractions of the species in the gas phase are $X_{Zn} = X_{CO_2} = 1.49/138 = 0.011$ and $X_{CO} = 135/138 = 0.978$.

Calculate the composition of the gas phase if the total pressure is increased to 150 atm at 1223 K. The system now contains three phases and the equilibrium thus has two degrees of freedom. Condensation of the zinc eliminates the stoichiometric requirement that $p_{Zn} = p_{CO_2}$, but phase equilibrium between liquid zinc and zinc vapor at 1223 K requires that the partial pressure of zinc be the saturated value of 1.49 atm. Thus

$$K_{1223\ K} = 0.0164 = \frac{1.49 \times p_{CO_2}}{p_{CO}} \tag{vi}$$

and

$$P = 150 = 1.49 + p_{CO} + p_{CO_2} \tag{vii}$$

Simultaneous solution of Eqs. (vi) and (vii) gives $p_{CO} = 146.9$ atm and $p_{CO_2} = 1.61$ atm. The mole fractions of the species in the gas phase are thus: $X_{Zn} = 1.49/150 = 0.01$, $X_{CO_2} = 1.61/150 = 0.011$, and $X_{CO} = 146.9/150 = 0.98$.

Consider, now, the reduction of ZnO by graphite to form zinc vapor, CO and CO_2 according to

$$ZnO_{(s)} + C_{(gr)} = Zn_{(v)} + CO_{(g)} \tag{viii}$$

and

$$2ZnO_{(s)} + C_{(gr)} = 2Zn_{(v)} + CO_{2(g)} \tag{ix}$$

For $2C + O_2 = 2CO$,

$$\Delta G_{(x)}^{\circ} = -223{,}400 - 175.3T \text{ J} \tag{x}$$

and for $C + O_2 = CO_2$

$$\Delta G_{(xi)}^{\circ} = -394{,}100 - 0.84T \text{ J} \tag{xi}$$

Combination of $\Delta G_{(ii)}^{\circ}$ and $\dfrac{\Delta G_{(x)}^{\circ}}{2}$ gives

$$\Delta G_{(viii)}^{\circ} = 348{,}500 - 285.7T \text{ J}$$

for $ZnO + C = Zn + CO$ and combination of $2\Delta G_{(ii)}^{\circ}$ and $\Delta G_{(xi)}^{\circ}$ gives

$$\Delta G_{(ix)}^{\circ} = 526{,}300 - 396.8T \text{ J}$$

for $2ZnO + C = 2Zn + CO_2$.

The equilibrium involves three components and three phases (ZnO, graphite, and a gas phase) and thus, according to the phase rule, has two degrees of freedom. However, as stoichiometric ZnO is the only source of oxygen and zinc in the gas phase, one of the degrees of freedom is used by the requirement that equal numbers of moles of Zn and O occur in the gas phase. Alternatively, as ZnO has a fixed composition, the system can be considered to be the quasi-binary ZnO-C, in which case the phase rule gives one degree of freedom to the equilibrium. Thus, fixing one of (1) the temperature, (2) the total pressure, (3) p_{Zn}, (4) p_{CO}, or (5) p_{CO_2} fixes the equilibrium. Determine the equilibrium state at 1223 K.

At 1223 K, $\Delta G^\circ_{(viii)1223\ K} = -850$ J and thus

$$K_{(viii)1223\ K} = \exp\left(\frac{850}{8.3144 \times 1223}\right) = 1.087 = p_{Zn}\,p_{CO} \qquad (xii)$$

and $\Delta G^\circ_{(ix)1223\ K} = 40{,}960$ J, in which case

$$K_{(ix)1223\ K} = \exp\left(\frac{-40{,}960}{8.3144 \times 1223}\right) = 0.018 = p^2_{Zn}\,p_{CO_2} \qquad (xiii)$$

The requirement that $n_{Zn}/n_O = 1$ in the gas phase leads to

$$\frac{n_{Zn}}{n_O} = 1 = \frac{n_{Zn}}{n_{CO} + 2n_{CO_2}} = \frac{p_{Zn}}{p_{CO} + 2p_{CO_2}}$$

Thus

$$p_{Zn} = p_{CO} + 2p_{CO_2} \qquad (xiv)$$

Substitution of Eq. (xiv) into Eq. (xii) gives

$$(p_{CO} + 2p_{CO_2})p_{CO} = 1.08 \qquad (xv)$$

and substitution of Eq. (xiv) into Eq. (xiii) gives

$$(p_{CO} + 2p_{CO_2})^2 p_{CO_2} = 0.018 \qquad (xvi)$$

Simultaneous solution of Eqs. (xv) and (xvi) gives $p_{CO} = 1.023$ atm and $p_{CO_2} = 0.016$ atm, and Eq. (xiv) gives $p_{Zn} = 1.023 + (2 \times 0.016) = 1.055$ atm. The total pressure at which the equilibrium exists at 1223 K is thus $1.055 + 1.023 + 0.016 = 2.094$ atm.

Consider now the temperature at which the total pressure is 1 atm. Rewriting Eqs. (xii) and (xiii) to include temperature as a variable gives

$$K_{(viii),\ T} = \exp\left(\frac{-\Delta G^\circ_{(viii)}}{RT}\right) = p_{Zn}\,p_{CO}$$

or, substituting for $\Delta G^\circ_{(viii)}$ and p_{Zn},

$$\exp\left(\frac{-348{,}500}{8.3144T}\right)\exp\left(\frac{285.7}{8.3144}\right) = (p_{CO} + 2p_{CO_2})p_{CO} \qquad (xvii)$$

and

$$K_{(ix), T} = \exp\left(\frac{-\Delta G_{(ix)}^{\circ}}{RT}\right) = p_{Zn}^2 p_{CO_2}$$

or, substituting for $\Delta G_{(ix)}^{\circ}$ and p_{Zn},

$$K_{(ix), T} = \exp\left(\frac{-526,300}{8.3144T}\right) \exp\left(\frac{396.8}{8.3144}\right) = (p_{CO} + 2p_{CO_2})^2 p_{CO_2} \quad \text{(xviii)}$$

The third equation is

$$P = 1 = p_{Zn} + p_{CO} + p_{CO_2}$$

or

$$1 = (p_{CO} + 2p_{CO_2}) + p_{CO} + p_{CO_2} \quad \text{(xix)}$$

Simultaneous solution of Eqs. (xvii), (xviii), and (xix) gives

$$T = 1172 \text{ K}$$
$$p_{CO} = 0.489 \text{ atm}$$

and

$$p_{CO_2} = 0.007 \text{ atm}$$

Thus $p_{Zn} = 0.489 + (2 \times 0.007) = 0.503$ atm, and the total pressure is $0.489 + 0.503 + 0.007 = 1$ atm.

Example 3

During the chlorination of NiO in a reactor at 900 K it is required that 90% conversion of the chlorine gas be achieved during a single pass through the reactor. Calculate the required total gas pressure.

The reaction is

$$NiO_{(s)} + Cl_{2(g)} = NiCl_{2(s)} + \frac{1}{2}O_{2(g)}$$

for which $\Delta G_{900 \text{ K}}^{\circ} = -15,490$ J. Thus

$$K_{900 \text{ K}} = \exp\left(\frac{15,490}{8.3144 \times 900}\right) = 7.925$$

From the stoichiometry of the reaction,

$$NiO_{(s)} + Cl_{2(g)} = NiCl_{2(s)} + \frac{1}{2}O_{2(g)}$$

moles before reaction	1		
moles after reaction	$1 - x$	x	$\frac{1}{2}x$

For 90% conversion of the Cl_2, $x = 0.9$, and thus $n_{Cl_2} = 0.1$, $n_{O_2} = 0.45$, and $n_T = 0.55$. In a Cl_2–O_2 mixture at the pressure P, the partial pressures of chlorine and oxygen are thus

$$p_{Cl_2} = \frac{0.1}{0.55}P = 0.182P \quad \text{and} \quad p_{O_2} = \frac{0.45}{0.55}P = 0.818P$$

Thus

$$K_{900\,K} = 7.925 = \frac{p_{O_2}^{1/2}}{p_{Cl_2}} = \frac{(0.818P)^{1/2}}{0.182P} = \frac{4.969}{P^{1/2}}$$

which has the solution $P = 0.393$ atm.

Example 4

What is the equilibrium state of a CO–CO_2–H_2–H_2O gas mixture produced by mixing CO_2 and H_2 in the molar ratio 1:1 at 1000 K and a total pressure of 1 atm?

The reaction which occurs is

$$CO_2 + H_2 = CO + H_2O \tag{i}$$

As the molar ratio of CO_2 to H_2 in the initial mixture is 1:1 and $P = 1$ atm, then, before reaction begins, $p_{CO_2} = p_{H_2} = 0.5$ atm. From the stoichiometry of the reaction, at any time during the reaction, $p_{CO_2} = p_{H_2}$ and $p_{CO} = p_{H_2O}$. At equilibrium

$$K_{p,(i)} = \frac{p_{CO}\,p_{H_2O}}{p_{CO_2}\,p_{H_2}} \tag{ii}$$

The total pressure is

$$P = p_{CO_2} + p_{CO} + p_{H_2O} + p_{H_2} = 1$$

but, as

$$p_{CO_2} = p_{H_2} \quad \text{and} \quad p_{CO} = p_{H_2O}$$

then

$$P = 1 = 2p_{H_2O} + 2p_{H_2}$$

Thus

$$p_{H_2O} = 0.5 - p_{H_2} \quad \text{and} \quad p_{CO_2} = p_{H_2}$$

Substitution of which into Eq. (ii) gives

$$K_{p,(i)} = \frac{(0.5 - p_{H_2})^2}{p_{H_2}^2}$$

The standard Gibbs free energy change for the reaction given by Eq. (i) is $\Delta G_{(i)}^\circ = 36{,}000 - 32T$ J, and thus $\Delta G_{(i)1000\,K}^\circ = 4000$ J and

$$K_{p,(i),\,1000\,K} = \exp\left(\frac{-4000}{8.3144 \times 1000}\right) = 0.618 = \frac{(0.5 - p_{H_2})^2}{p_{H_2}^2}$$

which has the solution $p_{H_2} = 0.28$ atm. Thus, at reaction equilibrium,

$$p_{H_2} = p_{CO_2} = 0.28 \text{ atm}$$

and

$$p_{H_2O} = p_{CO} = 0.22 \text{ atm}$$

Consider that this equilibrated gas is contained at 1 atm pressure and 1000 K in a rigid vessel of constant volume. What happens if some CaO is placed in the vessel? The reactions

$$CaO_{(s)} + 5CO_{(g)} = CaC_{2(s)} + 3CO_{2(g)} \tag{iii}$$

$$CaO_{(s)} + H_2O_{(g)} = Ca(OH)_{2(s)} \tag{iv}$$

and

$$CaO_{(s)} + CO_{2(g)} = CaCO_{3(s)} \tag{v}$$

are possible.

Consider the possible formation of CaC_2 according to (iii). Summing

$$\Delta G^\circ = -48{,}620 - 36.1T \text{ J} \quad \text{for } Ca + 2C = CaC_2$$
$$\Delta G^\circ = -1{,}182{,}000 - 2.4T \text{ J} \quad \text{for } 3C + O_2 = 3CO_2$$

$$\Delta G^\circ = 633{,}140 - 99T \text{ J} \quad \text{for } CaO = Ca + \frac{1}{2}O_2$$

and

$$\Delta G^\circ = 560{,}000 + 438.3T \text{ J} \quad \text{for } 5CO = 5C + \frac{5}{2}O_2$$

gives

$$\Delta G_{(iii)}^\circ = -37{,}480 + 300.7T \text{ J}$$

thus $\Delta G_{(iii),\,1000\,K}^\circ = 263{,}220$ J and

$$K_{p,(iii),\,1000\,K} = 1.78 \times 10^{-14} = \left(\frac{p_{CO_2}^3}{p_{CO}^5}\right)_{(eq)}$$

Thus, if the CaO were to react with the CO in the gas mixture (which exists at $p_{CO} = 0.22$ atm) to form CaC_2 and CO_2, the pressure of CO_2 in the gas mixture would have to be less than

$$(1.78 \times 10^{-14} \times 0.22^5)^{1/3}$$

i.e., less than 2.09×10^{-6} atm. As the partial pressure of CO_2 in the gas is 0.28 atm, reaction (iii) does not occur.

Consider the possible formation of $Ca(OH)_2$ according to reaction (iv), $\Delta G^{\circ}_{(iv)} = -117,600 + 145T$ J and thus $\Delta G^{\circ}_{(iv), 1000 K} = 27,400$ J and

$$K_{p,(iv), 1000 K} = 0.037 = \frac{1}{p_{H_2O(eq)}}$$

Thus the pressure of water vapor required for equilibrium between CaO, $Ca(OH)_2$, and water vapor at 1000 K is $1/0.037 = 20$ atm. As the actual pressure of water vapor in the vessel is 0.22 atm, reaction (iv) does not occur.

Consider the possible formation of $CaCO_3$ according to reaction (v):

$$\Delta G^{\circ}_{(v)} = -168,400 + 144T \text{ J}$$

Thus, $\Delta G^{\circ}_{(v), 1000 K} = -24,400$ J and hence

$$K_{p,(v), 1000 K} = 18.82 = \frac{1}{p_{CO_2(eq)}}$$

or, for equilibrium between CaO, $CaCO_3$, and CO_2 at 1000 K, p_{CO_2} must be $1/18.82 = 0.053$ atm. The actual partial pressure of CO_2 in the vessel is greater than 0.053 and thus the CO_2 reacts with the CaO to form $CaCO_3$. Consider that an excess of CaO is added to the vessel so that some CaO remains after the formation of $CaCO_3$ has decreased the partial pressure of CO_2 in the vessel to 0.053 atm. Now calculate the new equilibrium state of the gas.

The removal of CO_2 from the gas has two effects: (1) the pressure exerted by the gas in the constant volume is decreased and (2) the equilibrium of reaction (i) is shifted to the left. However, as all of the hydrogen in the vessel, occurring as H_2 or as H_2O, remains in the constant-volume gas phase, the sum $p_{H_2} + p_{H_2O}$ is not changed by the shift in the equilibrium. Also, from the stoichiometry of reaction (i), $p_{CO} = p_{H_2O}$ during the shift. Thus, at the new equilibrium state

$$p_{H_2} + p_{H_2O} = 0.5 \text{ atm}$$
$$p_{CO} = p_{H_2O}$$
$$p_{CO_2} = 0.053 \text{ atm}$$

and

$$K_{p,(i), 1000 K} = 0.618 = \frac{p_{H_2O}^2}{0.053 \times (0.5 - p_{H_2O})}$$

This has the solution $p_{H_2O} = 0.113$ atm, and thus the new equilibrium state is

$$p_{H_2O} = p_{CO} = 0.113 \text{ atm}$$
$$p_{H_2} = 0.387 \text{ atm}$$
$$p_{CO_2} = 0.053 \text{ atm}$$
$$P = 0.666 \text{ atm}$$

What happens now if graphite is introduced to the system? If excess graphite is added, the equilibrium

$$C_{(gr)} + CO_{2(g)} = 2CO_{(g)} \tag{vi}$$

must be established;

$$\Delta G^\circ_{(vi)} = 170{,}700 - 174.5T \text{ J}$$

Thus $\Delta G^\circ_{(vi), 1000 \text{ K}} = -3800$ J and

$$K_{p,(vi), 1000 \text{ K}} = 1.579 = \left(\frac{p^2_{CO}}{p_{CO_2}}\right)_{(eq)}$$

Thus, as $p_{CO_2} = 0.053$ atm is required for the CaO–$CaCO_3$–CO_2 equilibrium, the value of p_{CO} in the gas mixture must change from 0.113 to

$$(1.579 \times 0.053)^{1/2} = 0.289 \text{ atm}$$

to establish the C–CO–CO$_2$ equilibrium. Again, as all of the hydrogen remains in the gas phase, $p_{H_2} + p_{H_2O} = 0.5$ atm, and so

$$K_{p,(i), 1000 \text{ K}} = 0.618 = \frac{0.289 \times (0.5 - p_{H_2})}{0.053 p_{H_2}}$$

which has the solution $p_{H_2} = 0.449$ atm. Thus the newly equilibrated gas mixture, which is now in equilibrium with CaO, CaCO$_3$, and graphite, is

$$p_{H_2} = 0.449 \text{ atm}$$
$$p_{H_2O} = 0.051 \text{ atm}$$
$$p_{CO} = 0.289 \text{ atm}$$
$$p_{CO_2} = 0.053 \text{ atm}$$
$$P = 0.842 \text{ atm}$$

Consider now that the graphite is added *before* the CaO, i.e., that graphite is added to the original gas mixture in which $p_{CO} = p_{H_2O} = 0.022$ atm and $p_{H_2} = p_{CO_2} = 0.28$ atm, contained in the rigid vessel at 1000 K. The equilibrium (vi) is established which requires that the values of p_{CO} and p_{CO_2} in the mixture must change to conform with

$$K_{p,(vi), 1000 \text{ K}} = 1.579 = \frac{p^2_{CO}}{p_{CO_2}}$$

Before any reaction the partial pressure of CO in the gas mixture is 0.22 atm, which, for the C–CO–CO$_2$ equilibrium, would require $p_{CO_2} = (0.22)^2/1.579 = 0.031$ atm (which is lower than the value occurring in the gas mixture), or, for the existing p_{CO_2} of 0.28, establishment of the C–CO–CO$_2$ equilibrium would require $p_{CO} = (1.579 \times 0.28)^{1/2} = 0.665$ atm (which is higher than the value occurring in the gas mixture).

Thus reaction (vi) must proceed from left to right and the gas phase equilibrium (i) must shift from right to left until, simultaneously

$$\frac{p_{CO}^2}{p_{CO_2}} = 1.579 \quad \text{and} \quad \frac{p_{CO}p_{H_2O}}{p_{CO_2}p_{H_2}} = 0.618$$

As before, $p_{H_2} + p_{H_2O} = 0.5$ atm, and the fourth condition (required for determination of the values of the four partial pressures) is obtained from consideration of the oxygen and hydrogen mole balances. The reaction of graphite with CO_2 to form CO does not change the number of moles of oxygen in the gas phase. In the original mixture, $CO_2/H_2 = 1$, and thus equal numbers of moles of oxygen and hydrogen occur in the gas phase. The number of moles of oxygen in the gas is

$$2n_{CO_2} + n_{CO} + n_{H_2O}$$

and the number of moles of hydrogen is

$$2n_{H_2} + 2n_{H_2O}$$

Thus, in the gas mixture,

$$2n_{CO_2} + n_{CO} + n_{H_2O} = 2n_{H_2O} + 2n_{H_2}$$

or

$$n_{CO_2} + \frac{1}{2}n_{CO} = n_{H_2} + \frac{1}{2}n_{H_2O}$$

Under conditions of constant volume and temperature, $p_i \propto n_i$ and thus

$$p_{CO_2} + \frac{1}{2}p_{CO} = p_{H_2} + \frac{1}{2}p_{H_2O}$$

which, in combination with

$$p_{H_2O} + p_{H_2} = 0.5$$

gives

$$p_{H_2O} = 1 - 2p_{CO_2} - p_{CO}$$

Thus,

$$p_{CO_2} = \frac{p_{CO}^2}{1.579} = 0.633p_{CO}^2$$

$$p_{H_2O} = 1 - 2p_{CO_2} - p_{CO} = 1 - 1.266p_{CO}^2 - p_{CO}$$
$$p_{H_2} = 0.5 - p_{H_2O} = 1.266p_{CO}^2 + p_{CO} - 0.5$$

substitution of which into Eq. (ii) gives

$$K_{p,(i),\,1000\,K} = 0.618 = \frac{(1 - 1.266p_{CO}^2 - p_{CO})p_{CO}}{(1.266p_{CO}^2 + p_{CO} - 0.5) \times 0.633p_{CO}^2}$$

or

$$p_{CO}^3 + 3.346p_{CO}^2 + 1.624p_{CO} - 2.019 = 0 \qquad \text{(vii)}$$

Eq. (vii) has the solution $p_{CO} = 0.541$ atm, and thus the new equilibrium state is

$$p_{CO} = 0.541 \text{ atm}$$
$$p_{CO_2} = 0.185 \text{ atm}$$
$$p_{H_2} = 0.412 \text{ atm}$$
$$p_{H_2O} = 0.088 \text{ atm}$$
$$P = 1.226 \text{ atm}$$

Now add excess CaO to the system. The partial pressure of CO_2 in the gas ($p_{CO_2} = 0.185$ atm) is greater than the value of 0.053 atm required for equilibrium between CaO, $CaCO_3$, and CO_2 at 1000 K. Thus the CO_2 reacts with the CaO to form $CaCO_3$ until, thereby, the partial pressure of CO_2 has been decreased to the equilibrium value of 0.053 atm, and the gas phase equilibrium shifts in order to maintain the C–CO–CO_2 equilibrium. Thus, at the new equilibrium,

$$p_{CO_2} = 0.053 \text{ atm} \quad \text{and} \quad p_{CO} = (1.579 \times 0.053)^{1/2} = 0.289 \text{ atm}$$

Also

$$K_{p,(i),\ 1000\ K} = 0.618 = \frac{0.289p_{H_2O}}{0.053p_{H_2}}$$

which gives

$$\frac{p_{H_2O}}{p_{H_2}} = 0.133$$

which, with $p_{H_2} + p_{H_2O} = 0.5$ atm, gives $p_{H_2} = 0.449$ atm and $p_{H_2O} = 0.051$ atm. Thus the new equilibrium gas is

$$p_{H_2} = 0.449 \text{ atm}$$
$$p_{H_2O} = 0.051 \text{ atm}$$
$$p_{CO} = 0.289 \text{ atm}$$
$$p_{CO_2} = 0.053 \text{ atm}$$
$$P = 0.842 \text{ atm}$$

which, necessarily, is the same state as that produced by introducing the CaO before the graphite.

PROBLEMS

12.1 To what temperature must $MgCO_3$ be heated in an atmosphere containing a partial pressure of CO_2 of 10^{-2} atm to cause decomposition of the carbonate?

12.2 Using the standard Gibbs free energies of formation of NiO from solid Ni and liquid Ni, calculate the melting temperature, molar heat of melting, and the molar entropy of melting of nickel.

12.3 Calculate the temperature at which pure Ag_2O decomposes to Ag metal and O_2 gas when heated in (1) pure oxygen at 1 atm pressure, and (2) in air.

12.4 Determine the maximum pressure of water vapor in wet hydrogen at 1 atm pressure in which chromium can be heated without oxidation occurring at 1500 K. Is the oxidation of Cr by water vapor exothermic or endothermic?

12.5 A mixture of argon gas and hydrogen gas at 1 atm total pressure is passed through a reaction vessel containing a mixture of liquid Sn and liquid $SnCl_2$ at 900 K. The composition of the gas leaving the vessel is 50% H_2, 7% HCl, and 43% Ar. Has equilibrium been attained between the gas phase and the liquid phases in the vessel?

12.6 Fe and FeO are in equilibrium with a gas mixture of composition 71.8% CO– 28.2% CO_2 at 1273 K. Which of the two solid phases disappears if the composition of the gas is held constant and the temperature of the system is decreased?

12.7 Calculate the vapor pressure of Mg exerted at 1400°C by the system in which the reaction equilibrium

$$4MgO_{(s)} + Si_{(s)} = 2Mg_{(g)} + Mg_2SiO_{4(s)}$$

is established.

12.8 One gram of $CaCO_3$ is placed in an evacuated rigid vessel of volume 1 liter at room temperature, and the system is heated. Calculate (1) the highest temperature at which the $CaCO_3$ phase is present, (2) the pressure in the vessel at 1000 K, and (3) the pressure in the vessel at 1500 K. The molecular weight of $CaCO_3$ is 100.

12.9 Calculate the total pressure ($p_{SO_3} + p_{SO_2} + p_{O_2}$) exerted by equilibrated CoO and $CoSO_4$ at 1223 K.

12.10 A gas mixture initially containing 90% CO, 0.4% COS, and 9.6% inert constituents (by volume) is passed over sponge iron at 1000 K to remove sulfur by the following reaction:

$$COS_{(g)} + Fe_{(s)} = CO_{(g)} + FeS_{(s)}$$

(1) Assuming that the effluent gas is in equilibrium with Fe and FeS, calculate the percentage of sulfur removed from the gas by reaction with the sponge iron. (2) Calculate the partial pressure of S_2 in the effluent gas.

12.11 An Ar–H_2O gas mixture of $p_{H_2O} = 0.9$ atm ($P_{total} = 1$ atm) is passed over solid CaF_2, as a result of which CaO forms according to

$$CaF_{2(s)} + H_2O_{(g)} = CaO_{(s)} + 2HF_{(g)}$$

The reaction proceeds to equilibrium and solid CaO and solid CaF_2 are mutually immiscible. When the gas flow rate (measured at 298 K and 1 atm pressure) over the sample is 1 liter per minute, the measured rates of weight loss of the sample are 2.69×10^{-4} and 8.30×10^{-3} grams per hour at 900 and

1100 K, respectively. Use these data to calculate the variation of $\Delta G°$ for the above reaction with temperature. The atomic weights are

$$O = 16, F = 19, \text{ and } Ca = 40.08.$$

12.12 Magnetite (Fe_3O_4) is reduced to sponge iron (Fe) in a continuous reactor operating at 800 K using methane gas (CH_4) as the reducing agent. The gaseous reaction product leaving the reactor at a total pressure of 1 atm is a mixture of CO, CO_2, H_2, and H_2O with a negligible methane content. The gas is at equilibrium with the Fe–Fe_3O_4 mixture in the reactor. Calculate the consumption of methane as moles of methane used per mole of Fe produced.

12.13 Three equations for the oxidation of Mg according to $Mg + \frac{1}{2}O_{2(g)} = MgO_{(s)}$ are

$$\Delta G° = -604{,}000 - 5.36T \ln T + 142.0T \text{ J} \qquad \text{(i)}$$

$$\Delta G° = -759{,}800 - 13.4T \ln T + 317T \text{ J} \qquad \text{(ii)}$$

and

$$\Delta G° = -608{,}100 - 0.44T \ln T + 112.8T \text{ J} \qquad \text{(iii)}$$

One of these expressions is for the oxidation of solid Mg, one is for the oxidation of liquid Mg, and one is for the oxidation of gaseous Mg. Determine which equation is for which oxidation and calculate the melting and normal boiling temperature of Mg.

12.14 200 grams of liquid zinc are placed in a crucible at 1030 K. Two moles of air are bubbled through the liquid zinc, and the gas comes to equilibrium with the liquid before leaving the system. If the total pressure of the gas remains constant at 0.8 atm throughout the process, how many grams of *metallic* zinc are left in the crucible? The atomic weights of Zn and O are, respectively, 65.38 and 16.

12.15 Methane gas is burned with twice the amount of stoichiometric air (i.e, the initial ratio of CH_4 and O_2 in the mixture is 0.25), and the combustion gas produced, in which the concentration of methane is negligible, is used to calcine $CaCO_3$ in an isothermal furnace. The gas and the solids are at 1080 K and the pressure of the gas is maintained constant at 1 atm. How many moles of $CaCO_3$ are decomposed per mole of CH_4 burned?

12.16 Mercuric oxide (HgO) is placed in a vessel which is then evacuated, filled with nitrogen, and heated to 600 K, at which temperature it is observed that the total pressure in the vessel is 2 atm. Calculate the mole fractions of O_2 and Hg vapor in the gas phase.

12.17 In Fig. 12.17 the line *AB,* which represents the equilibrium

$$FeO_{(s)} + CO_{(g)} = Fe_{(s)} + CO_{2(g)}$$

intersects the carbon deposition line at the point A ($T = 972$ K, $p_{CO(s)} = 0.595$ atm, $p_{CO_2} = 0.405$ atm, $P = 1.000$ atm). At what total pressure does the point of intersection of the lines occur at 1000 K, and what are the values of the partial pressures of CO and CO_2 in this state?

Chapter 13

REACTION EQUILIBRIA IN SYSTEMS CONTAINING COMPONENTS IN CONDENSED SOLUTION

13.1 INTRODUCTION

Dissolving the pure component i in a condensed solution which is in contact with a vapor phase causes a decrease in the vapor pressure exerted by i from the value p_i° (exerted by pure i) to p_i (exerted by i when it occurs in solution). This decrease in the equilibrium vapor pressure corresponds, via Eq. (8.9), to a decrease of $RT \ln (p_i/p_i^\circ)$ in the value of the partial molar Gibbs free energy of i in the vapor phase. As phase equilibrium is maintained between the vapor phase and the condensed solution, the partial molar Gibbs free energy of i in the solution is $RT \ln (p_i/p_i^\circ)$ lower than the molar Gibbs free energy of pure condensed i at the temperature T. As the activity, a_i, of i in the solution with respect to pure i is defined by p_i/p_i°, then the partial molar Gibbs free energy of i in the condensed solution is lower than the molar Gibbs free energy of pure i by the amount $RT \ln a_i$. The value of p_i, and hence a_i, depends on the composition and the nature of the components of the solution and on temperature; and inasmuch that the solution of i affects the value of $\overline{G}_i$, it necessarily affects the equilibrium state of any chemical reaction system in which the component i is involved.

As an example, consider the equilibrium between silica, silicon, and oxygen gas,

$$SiO_{2(s)} = Si_{(s)} + O_{2(g)}$$

From Eq. (11.4), the equality

$$\overline{G}_{SiO_2} = \overline{G}_{Si} + \overline{G}_{O_2}$$

is the criterion for reaction equilibrium at any temperature and total pressure. If the SiO_2 and the Si present in the system are both pure, and the pure solids are chosen as the standard states, then

$$G^\circ_{SiO_2} = G^\circ_{Si} + \overline{G}_{O_2}$$

or

$$G^\circ_{SiO_2} = G^\circ_{Si} + G^\circ_{O_2} + RT \ln p_{O_2(eq.T)}$$

It has been seen in Chap. 11 that, as the values of G°_i are dependent only on temperature, then at the temperature T there exists a unique partial pressure of oxygen, $p_{O_2(eq.T)}$, at which equilibrium occurs in the system. This unique oxygen pressure is calculated as

$$p_{O_2(eq.T)} = \exp\left[\frac{1}{RT}(G^\circ_{SiO_2} - G^\circ_{Si} - G^\circ_{O_2})\right]$$

and, if it is required to reduce pure silica at the temperature T, the oxygen pressure in the system must be lower than $p_{O_2(eq.T)}$.

Suppose now that the silica occurs at the activity a_{SiO_2} in an Al_2O_3–SiO_2 solution. The criterion for equilibrium among SiO_2, Si, and O_2 is still

$$\overline{G}_{SiO_2} = \overline{G}_{Si} + \overline{G}_{O_2}$$

but now

$$\overline{G}_{SiO_2} = G^\circ_{SiO_2} + RT \ln a_{SiO_2}$$

and thus, in terms of standard Gibbs free energies,

$$G^\circ_{SiO_2} + RT \ln a_{SiO_2} = G^\circ_{Si} + G^\circ_{O_2} + RT \ln p'_{O_2(eq.T)}$$

Thus, for a given value of a_{SiO_2} there now exists a new unique equilibrium oxygen pressure, $p'_{O_2(eq.T)}$, which is given by

$$p'_{O_2(eq.T)} = p_{O_2(eq.T)}a_{SiO_2}$$

and so, if it is required to reduce SiO_2 from an Al_2O_3 solution to form pure Si, the oxygen pressure in the system must be lower than $p'_{O_2(eq.T)}$.

It is thus seen that the possibility of reducing SiO_2 from an Al_2O_3–SiO_2 solution to produce pure Si with a gas of given partial pressure of oxygen at a given temperature is determined by the solution thermodynamics of the system Al_2O_3–SiO_2. Generally, the calculation of the equilibrium state of any reaction involving components in condensed solution requires knowledge of the thermodynamic properties of the various solutions present in the system. The influence of solution thermodynamics on reaction equilibria is examined in this chapter.

13.2 THE CRITERIA FOR REACTION EQUILIBRIUM IN SYSTEMS CONTAINING COMPONENTS IN CONDENSED SOLUTION

Consider the general reaction

$$aA + bB = cC + dD$$

occurring at the temperature T and the pressure P. If none of the reactants or products of the reaction occurs in its standard state, the change in the Gibbs free energy for the reaction is

$$\Delta G = c\overline{G}_C + d\overline{G}_D - a\overline{G}_A - b\overline{G}_B \tag{13.1}$$

If, however, all of the reaction and products occur in their standard states, the change in the Gibbs free energy is the standard Gibbs free energy change, $\Delta G°$, given by

$$\Delta G° = cG_C° + dG_D° - aG_A° - bG_B° \tag{13.2}$$

Subtraction of Eq. (13.2) from Eq. (13.1) gives

$$\Delta G - \Delta G° = c(\overline{G}_C - G_C°) + d(\overline{G}_D - G_D°) - a(\overline{G}_A - G_A°) - b(\overline{G}_B - G_B°) \tag{13.3}$$

For a component i occurring in some state other than its standard state, Eq. (9.28) gives

$$\overline{G}_i = G_i° + RT \ln a_i$$

where a_i is the activity of i with respect to the standard state, and thus Eq. (13.3) can be written as

$$\Delta G - \Delta G° = c(RT \ln a_C) + d(RT \ln a_D) - a(RT \ln a_A) - b(RT \ln a_B) \tag{13.4}$$

$$= RT \ln \left(\frac{a_C^c a_D^d}{a_A^a a_B^b} \right) = RT \ln Q$$

where $Q = a_C^c a_D^d / a_A^a a_B^b$ is called the activity quotient. Reaction equilibrium is established when the reaction has proceeded to such an extent that

$$a\overline{G}_A + b\overline{G}_B = c\overline{G}_C + d\overline{G}_D$$

i.e., that the Gibbs free energy of the system at the fixed temperature and pressure has been minimized, or that ΔG for the reaction is zero. Thus, at equilibrium,

$$\Delta G° = -RT \ln Q^{eq} \tag{13.5}$$

where Q^{eq} is the value of the activity quotient at equilibrium. From Eq. (11.8),

$$\Delta G° = -RT \ln K$$

and thus

$$Q^{eq} = K$$

i.e., at reaction equilibrium the activity quotient is numerically equal to the equilibrium constant K.

Consider the oxidation of the pure solid metal M by gaseous oxygen to form the pure solid metal oxide MO_2:

$$M_{(s)} + O_{2(g)} = MO_{2(s)} \qquad (i)$$

at the temperature T and the pressure P. For this reaction

$$Q = \frac{a_{MO_2}}{a_M a_{O_2}}$$

As M and MO_2 are pure, i.e., occur in their standard states, then $a_M = a_{MO_2} = 1$, and, from the formal definition of activity, the activity of oxygen gas is given as

$$a_{O_2} = \frac{\text{the pressure of oxygen in the gas phase}}{\text{the pressure of oxygen in its standard state}}$$

As the standard state for gaseous species has been chosen as being the gas at 1 atm pressure and the temperature of interest, then the activity of oxygen in the gas phase is equal to its partial pressure (assuming ideal behavior of the gas). Thus

$$Q = \frac{1}{p_{O_2}} \quad \text{and} \quad Q^{eq} = \frac{1}{p_{O_2(eq.T)}} = K$$

Now consider that the metal in equilibrium with its pure oxide and oxygen in a gaseous atmosphere occurs at the activity a_M in a solution. In this case

$$Q^{eq} = \frac{1}{a_M p_{O_2(eq.T)}} = K$$

and, as K is dependent only on temperature and $a_M < 1$, it is seen that the oxygen pressure required to maintain equilibrium between M in solution and pure MO_2 is larger than that required for equilibrium between pure M and pure MO_2 at the same temperature. Similarly, if the pure metal M is in equilibrium with MO_2 occurring at the activity a_{MO_2} in solution and oxygen in a gas phase, then

$$Q^{eq} = \frac{a_{MO_2}}{p_{O_2(eq.T)}} = K$$

in which case the oxygen pressure required for the equilibrium between pure M and its oxide in solution is lower than that for the equilibrium between pure M and pure MO_2.

In Fig. 13.1 the line ab is drawn as the variation of the standard Gibbs free energy change, with temperature for the oxidation

$$M_{(s)} + O_{2(g,\ 1\ atm)} = MO_{2(s)} \qquad (i)$$

At the temperature T, $\Delta G° = cd$, and the oxygen pressure for equilibrium between pure solid M and pure solid MO_2 is drawn as the point e on the oxygen pressure nomographic scale. Now consider the reaction

$$M_{(in\ solid\ solution\ at\ a_M)} + O_{2(g,\ 1\ atm)} = MO_{2(s)} \qquad (ii)$$

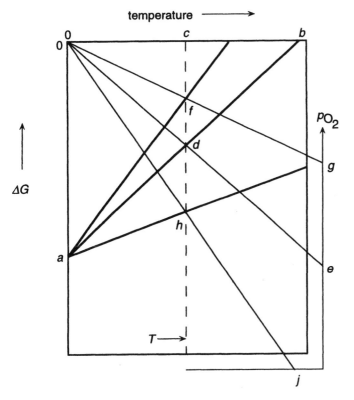

Figure 13.1 The influence of non-unit activities of the reactants and products of a reaction on the ΔG–T relationship for the reaction.

for which, at the temperature T, the Gibbs free energy change is $\Delta G_{(ii)}$. Reaction (ii) can be written as the sum of

$$M_{(\text{in solid solution at } a_M)} \rightarrow M_{(s, \text{ pure})} \tag{iii}$$

for which

$$\Delta G_{(iii)} = G^{\circ}_M - \overline{G}_M = -RT \ln a_M$$

and

$$M_{(s)} + O_{2(g, \text{ 1 atm})} = MO_{2(s)} \tag{i}$$

for which

$$\Delta G_{(i)} = \Delta G^{\circ} \text{ at the temperature } T$$

Thus

$$\Delta G_{(ii)} = \Delta G_{(i)} + \Delta G_{(iii)}$$
$$= \Delta G^{\circ} - RT \ln a_M$$

At the temperature T, $\Delta G°$ is a negative quantity, and, as $a_M < 1$, then $\Delta G_{(ii)}$ is a smaller negative quantity, drawn in Fig. 13.1 as cf. The effect of the solution of M on its oxidation is thus an anticlockwise rotation of the standard Gibbs free energy line about the point a ($\Delta G°$ at $T = 0$), with the extent of this rotation being such that, at the temperature T, the vertical separation from the standard line equals $RT \ln a_M$. The extent of the rotation is determined by the value of a_M, and the oxygen pressure required for equilibrium between M in solution and pure MO_2, is increased from e to g in accordance with

$$K_T = \frac{1}{p_{O_2(T,eq\ M_{(s)}/MO_{2(s)})}}$$

$$= \frac{1}{a_M p_{O_2(T,eq\ M_{(at\ a_M)}/MO_{2(s)})}}$$

Consider the reaction

$$M_{(s,\ pure)} + O_{2(g,\ 1\ atm)} = MO_{2(in\ solid\ solution\ at\ a_{MO_2})} \tag{iv}$$

for which, at the temperature T, the Gibbs free energy change is $\Delta G_{(iv)}$. Reaction (iv) can be written as the sum of

$$MO_{2(s,\ pure)} \rightarrow MO_{2(in\ solution\ at\ a_{MO_2})} \tag{v}$$

and reaction (i), i.e.,

$$\Delta G_{(iv)} = \Delta G° + RT \ln a_{MO_2}$$

At the temperature T, $\Delta G°$ is a negative quantity, and, as $a_{MO_2} < 1$, then $\Delta G_{(iv)}$ is a larger negative quantity, drawn in Fig. 13.1 as ch. The effect of the solution of MO_2 on the oxidation of M is thus seen to be a clockwise rotation of the Gibbs free energy line about the point a with the extent of the rotation being such that, at the temperature T, the vertical separation from the standard Gibbs free energy line equals $RT \ln a_{MO_2}$. The extent of the rotation is determined by the value of a_{MO_2}, and the oxygen pressure required for equilibrium between pure M and MO_2 in solution is decreased from e to j in accordance with

$$K_T = \frac{1}{p_{O_2(T,eq\ M_{(s)}/MO_{2(s)})}}$$

$$= \frac{a_{MO_2}}{p_{O_2[T,eq\ M_{(s)}/MO_2(at\ a_{MO_2})]}}$$

In the general case,

$$M_{(in\ solution\ at\ a_M)} + O_{2(g,\ at\ p_{O_2})} = MO_{2(in\ solution\ at\ a_{MO_2})} \tag{vi}$$

for which, at the temperature T, the free energy change is $\Delta G_{(vi)}$:

$$\Delta G_{(vi)} = \Delta G^\circ - RT \ln a_M - RT \ln p_{O_2} + RT \ln a_{MO_2}$$

$$= \Delta G^\circ + RT \ln \frac{a_{MO_2}}{a_M p_{O_2}}$$

$$= \Delta G^\circ + RT \ln Q$$

At equilibrium, the values of a_M, a_{MO_2}, and p_{O_2} are such that $\Delta G_{(vi)} = 0$, and thus

$$\Delta G^\circ = -RT \ln Q^{eq} = -RT \ln K$$

as in Eq. (13.5).

Example

Examine the conditions under which a liquid Fe–Mn alloy can be in equilibrium with an FeO–MnO liquid solution in an atmosphere containing oxygen at 1800°C. For

$$Mn_{(l)} + \frac{1}{2}O_{2(g)} = MnO_{(l)} \qquad (i)$$

$$\Delta G_{(i)}^\circ = -344,800 + 55.90T \text{ J}$$

and for

$$Fe_{(l)} + \frac{1}{2}O_{2(g)} = FeO_{(l)} \qquad (ii)$$

$$\Delta G_{(ii)}^\circ = -232,700 + 45.13T \text{ J}$$

The pertinent equilibrium is

$$FeO_{(l)} + Mn_{(l)} = MnO_{(l)} + Fe_{(l)} \qquad (iii)$$

for which

$$\Delta G_{(iii),2073 \text{ K}}^\circ = \Delta G_{(i),2073 \text{ K}}^\circ - \Delta G_{(ii),2073 \text{ K}}^\circ$$
$$= -228,900 + 138,800 = -90,100 \text{ J}$$
$$= -8.3144 \times 2073 \ln K_{(iii),2073 \text{ K}} \text{ J}$$

Therefore,

$$K_{(iii),2073 \text{ K}} = 186 = \frac{(a_{MnO})[a_{Fe}]}{(a_{FeO})[a_{Mn}]}$$

where

(a_{MnO}) = the activity of MnO in the liquid oxide phase with respect to pure MnO,
(a_{FeO}) = the activity of FeO in the liquid oxide phase with respect to iron-saturated liquid iron oxide,

$[a_{Mn}]$ = the activity of Mn in the liquid metal phase with respect to pure liquid Mn,
$[a_{Fe}]$ = the activity of Fe in the liquid metal phase with respect to pure liquid Fe.

As both the liquid metal solution and the liquid oxide solution exhibit Raoultian behavior, the condition for phase equilibrium between the two is

$$\frac{(X_{MnO})[X_{Fe}]}{(X_{FeO})[X_{Mn}]} = 186 \tag{iv}$$

or

$$\frac{[X_{Fe}]}{[X_{Mn}]} = 186 \frac{(X_{FeO})}{(X_{MnO})}$$

A series of tie lines joining the compositions of equilibrated metal and oxide solutions is shown in Fig. 13.2. Consider the metallic alloy of composition $X_{Fe} = 0.5$. Eq. (iv) gives

$$1 = 186 \frac{(X_{FeO})}{1 - (X_{FeO})}$$

or

$$(X_{FeO}) = 0.00535$$

and thus, when a metallic alloy of composition $X_{Fe} = 0.5$ is equilibrated with an oxide solution, the composition of the latter is $X_{FeO} = 0.00535$. Consider now the influence of the partial pressure of oxygen in the gaseous atmosphere:

$$\Delta G_{(i)}^{\circ} = -RT \ln K_{(i)} = -RT \ln \frac{(a_{MnO})}{[a_{Mn}]p_{O_2}^{1/2}}$$

Thus, as

$$\Delta G_{(i),2073\ K}^{\circ} = -228,900\ \text{J}$$

$$K_{(i),2073\ K} = 5.856 \times 10^5 = \frac{(a_{MnO})}{[a_{Mn}]p_{O_2}^{1/2}}$$

or, as the metallic and oxide solutions are ideal,

$$\frac{(X_{MnO})}{[X_{Mn}]} = 5.856 \times 10^5 p_{O_2}^{1/2} \tag{v}$$

Similarly,

$$\Delta G_{(ii)}^{\circ} = -RT \ln \frac{(a_{FeO})}{[a_{Fe}]p_{O_2}^{1/2}}$$

and, as

$$\Delta G_{(ii),2073\ K}^{\circ} = -138,800\ \text{J}$$

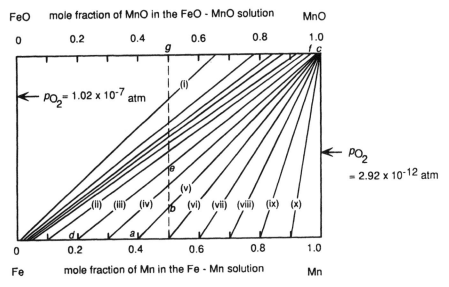

Figure 13.2 Tie lines between the compositions of equilibrated metallic and oxide alloys in the system Fe–Mn–O. The tie lines are also oxygen isobars with oxygen pressures as follows: (i) 1.24×10^{-8} atm, (ii) 2.65×10^{-10} atm, (iii) 7.0×10^{-11} atm, (iv) 3.14×10^{-11} atm, (v) 1.79×10^{-11} atm, (vi) 1.15×10^{-11} atm, (vii) 8.04×10^{-12} atm, (viii) 5.92×10^{-12} atm, (ix) 4.46×10^{-12} atm, (x) 3.58×10^{-12} atm.

then

$$K_{(ii),2073\,\mathrm{K}} = 3143 = \frac{(a_{\mathrm{FeO}})}{[a_{\mathrm{Fe}}]p_{\mathrm{O}_2}^{1/2}}$$

or

$$\frac{(a_{\mathrm{FeO}})}{[a_{\mathrm{Fe}}]} = 3143\, p_{\mathrm{O}_2}^{1/2} \tag{vi}$$

The ideal Raoultian behavior of the two solutions allows Eq. (vi) to be written as

$$\frac{(X_{\mathrm{FeO}})}{[X_{\mathrm{Fe}}]} = 3143 p_{\mathrm{O}_2}^{1/2}$$

Consider the equilibrium between the metallic alloy of $X_{\mathrm{Fe}} = 0.5$ and the oxide solution of $X_{\mathrm{FeO}} = 0.00535$. From Eq. (v)

$$p_{\mathrm{O}_2} = \left(\frac{0.99465}{0.5}\, \frac{1}{5.856 \times 10^5}\right)^2 = 1.15 \times 10^{-11}\ \text{atm}$$

and from Eq. (vi)

$$p_{O_2} = \left(\frac{0.00535}{0.5} \frac{1}{3143} \right)^2 = 1.15 \times 10^{-11} \text{ atm}$$

Thus the tie line connecting the compositions of the equilibrated metallic and oxide alloy is also the oxygen isobar in Fig. 13.2.

Thus, at any fixed oxygen pressure, the individual ratios

$$\frac{(a_{FeO})}{[a_{Fe}]} \left(= \frac{(X_{FeO})}{[X_{Fe}]} \right) \quad \text{and} \quad \frac{(a_{MnO})}{[a_{Mn}]} \left(= \frac{(X_{MnO})}{[X_{Mn}]} \right)$$

are fixed by Eqs. (v) and (vi), combination of which gives

$$\frac{(X_{MnO})[X_{Fe}]}{(X_{FeO})[X_{Mn}]} = \frac{5.856 \times 10^5}{3143} = 186$$

in accordance with Eq. (iv).

Consider the oxidation of a finite quantity of a liquid metallic alloy of composition $X_{Fe} = 0.5$ by an infinite oxygen-containing gaseous atmosphere in which the partial pressure of oxygen is slowly increased. From Fig. 13.2 the metal phase is stable when the partial pressure of oxygen is less than 1.15×10^{-11} atm. At $p_{O_2} = 1.15 \times 10^{-11}$ atm the metallic alloy is in equilibrium with an oxide solution of $X_{FeO} = 0.00535$. Increasing the partial pressure of oxygen to 1.79×10^{-11} atm moves the state of the system to the state b on the (v) isobar in Fig. 13.2. In this state a metallic alloy of $X_{Fe} = 0.6$ (at a) is in equilibrium with an oxide solution of $X_{FeO} = 0.0053$ (at c), and the relative quantities of the two phases are given by application of the lever rule to the tie line (v), i.e., the fraction of the system occurring as the metallic alloy in state a is bc/ac, and the fraction occurring as the oxide solution in state c is ab/ac. Increasing the oxygen pressure to 7.0×10^{-11} atm moves the system to the state e on the (iii) isobar where a metallic alloy of $X_{Fe} = 0.8$ (at d) is in equilibrium with an oxide solution of $X_{FeO} = 0.021$ (at f). The ratio of metallic alloy to oxide solution occurring is ef/de. Continued increase in the partial pressure of oxygen moves the state of the system upwards along the broken line in Fig. 13.2, during which the ratio of oxide to metal phase increases and the mole fraction of Fe in the metal phase and the mole fraction of FeO in the oxide phase increase. When the composition of the oxide reaches $X_{FeO} = 0.5$ (at g), the infinitesimal amount of equilibrium metal phase has the composition $X_{Mn} = 0.00535$ and the oxygen pressure is 2.55×10^{-8} atm. The oxidation of Fe–Mn alloys at 2073 K occurs between the limits of oxygen pressure 2.92×10^{-12} atm, for the equilibrium between pure Mn and pure MnO, and 1.02×10^{-7} atm for the equilibrium between pure Fe and pure FeO.

The establishment of equilibrium (iii) requires that the ΔG–T lines for the oxidation of Fe and Mn intersect at 2073 K, i.e., that $\Delta G_{(iii),2073 \text{ K}} = 0$. For any oxidation $2M + O_2 = 2MO$, clockwise rotation of the ΔG–T line (e.g., the line ab in Fig. 13.1) about its point of intersection with the $T = 0$ axis occurs when the ratio a_{MO}/a_M is decreased to a value less than unity, and, conversely, anticlockwise rotation of the line occurs when the ratio a_{MO}/a_M is increased to a value greater than unity. Also, as the equilibrium constant K is a function only of temperature, then, at any oxygen pressure p_{O_2} in the system M–MO$_2$–O$_2$ at the temperature T, the equilibrium ratio a_{MO}/a_M must be

$$\frac{a_{MO}}{a_M} = \frac{p_{O_2}^{1/2}}{p_{O_2 \text{(eq}T\text{ pure M/pure MO)}}^{1/2}}$$

where $p_{O_2\text{(eq}T\text{ pure M/pure MO)}}$ is that unique oxygen pressure at the temperature T required for equilibrium between pure M and pure MO. Thus, for any oxygen pressure within the allowed limits, equilibrium (iii) occurs when Eqs. (v) and (vi) are satisfied, and, under these conditions, the ΔG–T lines for the oxidation of Fe and Mn intersect at 1800°C.

Thus, as a consequence of the ability to vary a_M and a_{MO}, equilibrium (iii) can be established at any T and any p_{O_2} (within the above-mentioned limits). This is in contrast to the situation illustrated in Figs. 12.4 and 12.5, in which, if both metals and both oxides are present in their pure states, then an equilibrium such as (iii) can only be achieved at the single unique state (unique T and unique p_{O_2}) at which the Ellingham lines for the two oxidation reactions intersect with one another. The restrictions on general multicomponent multiphase equilibria are discussed in Sec. 13.4.

13.3 ALTERNATIVE STANDARD STATES

Up to this point the standard state of a component of a system has been chosen as being the pure component in its stable state of existence at the temperature of interest. This is called the Raoultian standard state; in Fig. 13.3 the Raoultian standard state for the component B is located at point r.

In situations in which the pure component exists in a physical state which differs from that of the solution the Henrian standard state may be more convenient than the Raoultian standard state. Such situations include the solution of a gas in a solid or liquid solvent and the solution of a solid in a liquid solvent. The Henrian standard state is obtained from consideration of Henry's law, which, strictly being a limiting law obeyed by the solute B at infinite dilution, is expressed as

$$\frac{a_B}{X_B} \to k_B \quad \text{as } X_B \to 0$$

where a_B is the activity of B in the solution with respect to the Raoultian standard state, and k_B is the Henry's law constant at the temperature T. Alternatively, Henry's law can be written as

$$\frac{a_B}{X_B} \to \gamma_B^\circ \quad \text{as } X_B \to 0 \tag{13.6}$$

where γ_B° $(= k_B)$ is the constant activity coefficient which quantifies the difference between Raoultian solution behavior of B and Henrian solution behavior of B. If the solute obeys Henry's law over a finite range of composition, then, over this range,

$$a_B = \gamma_B^\circ X_B$$

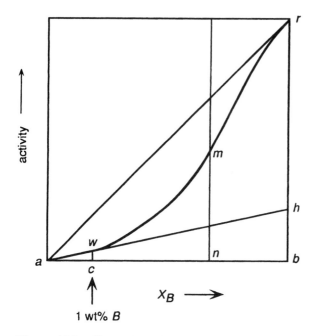

Figure 13.3 Illustration of the Raoultian, Henrian, and 1 weight percent standard states for component B in a binary A–B system.

The Henrian standard state is obtained by extrapolating the Henry's law line to $X_B = 1$. This state (the point h in Fig. 13.3) represents pure B in the hypothetical nonphysical state in which it would exist as a pure component, if it behaved as it does in dilute solution. The activity of B in the Henrian standard state with respect to the Raoultian standard state having unit activity is given by Eq. (13.6) as

$$a_B = \gamma_B^\circ$$

Thus, in Fig. 13.3, if the length rb is unity, then $hb = \gamma_B^\circ$.

Having defined the Henrian standard state thus, the activity of B in a solution, with respect to the Henrian standard state having unit activity, is given by

$$h_B = f_B X_B \tag{13.7}$$

where h_B is the Henrian activity and f_B is the Henrian activity coefficient. In the range of composition in which the solute B obeys Henry's law, $f_B = 1$, and the solute exhibits Henrian ideality.

The mole fraction of B in an A–B solution is related to the weight percent of B by

$$X_B = \frac{\dfrac{\text{wt\%}B}{MW_B}}{\dfrac{\text{wt\%}B}{MW_B} + \dfrac{(100 - \text{wt\%}B)}{MW_A}}$$

where MW_A and MW_B are, respectively, the molecular weights of A and B. Thus, in dilute solution, as the mole fraction of B is virtually proportional to the weight percentage of B, i.e.,

$$X_B \sim \frac{\text{wt\%}B \times MW_A}{100 \times MW_B}$$

a third standard state can be introduced. This is the 1 weight percent standard state, which is defined as

$$\frac{h_{B(1\ \text{wt\%})}}{\text{wt\%}B} \to 1 \text{ as wt\%}B \to 0$$

and is located at the point on the Henry's law line which corresponds to a concentration of 1 weight percent B (the point w in Fig. 13.3). With respect to the 1 weight percent standard state having unit activity, the activity of B, $h_{B(1\ \text{wt\%})}$ is given by

$$h_{B(1\ \text{wt\%})} = f_{B(1\ \text{wt\%})}\text{wt\%}B \tag{13.8}$$

where $f_{B(1\ \text{wt\%})}$ is the 1 weight percent activity coefficient, and in the range of composition in which the solute B obeys Henry's law, $f_{B(1\ \text{wt\%})} = 1$ and hence

$$h_{B(1\ \text{wt\%})} = \text{wt\%}B$$

which is of considerable practical convenience.

From consideration of the similar triangles awc and ahb in Fig. 13.3, the activity of B in the 1 wt% standard state with respect to the Henrian standard state having unity activity is

$$\frac{wc}{hb} = \frac{ac}{ab} = \frac{MW_A}{100MW_B}$$

and, with respect to the Raoultian standard state having unit activity, is

$$\frac{\gamma_B^\circ MW_A}{100MW_B}$$

The value of the equilibrium constant for any reaction, being equal to the quotient of the activities of the reactants and products at reaction equilibrium, necessarily depends on the choice of standard states for the components. Similarly the magnitude of ΔG° for the reaction depends on the choice of standard states and thus, in order to convert from the use of one standard state to another, it is necessary that the differences between the Gibbs free energies of the standard states be known.

For the change of standard state,

$$B_{(\text{in the Raoultian standard state})} \to B_{(\text{in the Henrian standard state})}$$

$$G_B^\circ (R \to H) = G_{B(H)}^\circ - G_{B(R)}^\circ = RT \ln \frac{a_{B(\text{in the Henrian standard state})}}{a_{B(\text{in the Raoultian standard state})}}$$

where both activities are measured on the same activity scale. On either the Raoultian or Henrian scales

$$\frac{a_{B(\text{in the Henrian standard state})}}{a_{B(\text{in the Raoultian standard state})}} = \frac{hb}{rb} = \gamma_B^\circ$$

and thus

$$\Delta G_{B(R \to H)}^\circ = RT \ln \gamma_B^\circ \tag{13.9}$$

where γ_B° is the Henrian activity coefficient *at the temperature T.*

For the change of standard state,

$$B_{(\text{in the Henrian standard state})} \to B_{(\text{in the 1 wt\% standard state})}$$

$$\Delta G_B^\circ(H \to 1\ \text{wt\%}) = G_{B(1\ \text{wt\%})}^\circ - G_{B(H)}^\circ = RT \ln \frac{a_{B(\text{in the 1 wt\% standard state})}}{a_{B(\text{in the Henrian standard state})}}$$

where, again, both activities are measured on the same scale:

$$\frac{a_{B(\text{in the 1 wt\% standard state})}}{a_{B(\text{in the Henrian standard state})}} = \frac{wc}{hb} = \frac{ac}{ab} = \frac{MW_A}{100 MW_B}$$

and thus

$$\Delta G_{B(H \to 1\text{wt\%})}^\circ = RT \ln \left(\frac{MW_A}{100 MW_B} \right) \tag{13.10}$$

Combination of Eqs. (13.9) and (13.10) gives

$$\Delta G_{B(R \to 1\text{wt\%})}^\circ = RT \ln \left(\frac{\gamma_B^\circ MW_A}{100 MW_B} \right) \tag{13.11}$$

for the change of standard state, Raoultian $\to$ 1 wt%. Using the subscript (R) to denote the Raoultian standard state, the subscript (H) to denote the Henrian standard state, and the subscript (wt%) to denote the 1 weight percent standard state, consider again the oxidation of metal M to form the oxide MO_2 at the temperature T:

$$M_{(R)} + O_{2(g)} = MO_{2(R)}$$

For this equilibrium

$$\Delta G_{(R)}^\circ = -RT \ln K_{(R)} = -RT \ln \frac{a_{MO_2}}{a_M p_{O_2}}$$

If M occurs in dilute solution, in which case it may be more convenient to use the Henrian standard state for M, then

$$M_{(H)} + O_{2(g)} = MO_{2(R)}$$

$$\Delta G_{(H)}^\circ = \Delta G_{(R)}^\circ - \Delta G_{M(R \to H)}^\circ$$

i.e.,

$$-RT \ln K_{(H)} = -RT \ln K_{(R)} - RT \ln \gamma_M^\circ$$

or

$$RT \ln \frac{a_{MO_2}}{h_M p_{O_2}} = RT \ln \frac{a_{MO_2}}{a_M p_{O_2}} + RT \ln \gamma_M^\circ$$

Thus

$$a_M = h_M \gamma_M^\circ \tag{13.12}$$

which relates the activity of M in solution with respect to the Raoultian standard state to the activity of M in solution with respect to the Henrian standard state, e.g., in the case of composition m in Fig. 13.3,

$$a_B = \frac{mn}{rb} = \frac{mn}{hb} \frac{hb}{rb} = h_B \gamma_B^\circ$$

Similarly, if it is convenient to use the 1 wt% standard state for M, then

$$M_{(1 \text{ wt\%})} + O_{2(g)} = MO_{2(R)}$$

$$\Delta G_{(1 \text{ wt\%})}^\circ = \Delta G_{(R)}^\circ - \Delta G_{M(R \to 1 \text{ wt\%})}^\circ$$

or

$$-RT \ln K_{(1 \text{ wt\%})} = -RT \ln K_{(R)} - RT \ln \frac{\gamma_M^\circ MW_{\text{solvent}}}{100 MW_M}$$

or

$$-RT \ln \frac{a_{MO_2}}{f_{M(1 \text{ wt\%})} \text{wt\% } M \cdot p_{O_2}} = -RT \ln \frac{a_{MO_2}}{a_M p_{O_2}} - RT \ln \frac{\gamma_M^\circ MW_{\text{solvent}}}{100 MW_M}$$

or

$$a_M = f_{M(1 \text{ wt\%})} \cdot \text{wt\% } M \cdot \gamma_M^\circ \cdot \frac{MW_{\text{solvent}}}{100 MW_M} \tag{13.13}$$

Example

The activity of silicon in binary Fe–Si liquid alloys, a_{Si}, is shown in Fig. 13.4 at two temperatures. As is seen, Si exhibits considerable negative deviation from Raoult's law, e.g., at $X_{Si} = 0.1$ and 1420°C, $a_{Si} = 0.00005$. Thus, in considering dilute solutions of Si in Fe, there is an advantage to using either the Henrian standard state or the 1 weight percent standard state.

For the change of standard state from Raoultian to Henrian at the temperature T,

$$\Delta G_{Si(R \to H)}^\circ = RT \ln \gamma_{Si}^\circ$$

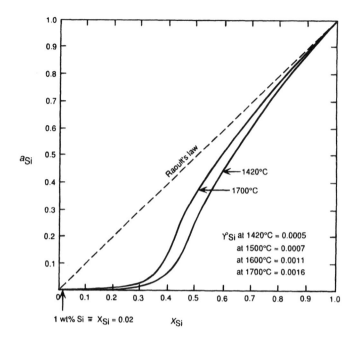

Figure 13.4   The activity of Si in Fe–Si melts at 1420°C and 1700°C.

and the experimentally measured variation of log γ_{Si}° with temperature is

$$\log \gamma_{Si}^{\circ} = -\frac{6230}{T} + 0.37$$

and thus

$$\Delta G_{Si(R \rightarrow H)}^{\circ} = 8.3144T \times 2.303 \log \gamma_{Si}^{\circ}$$
$$= -119{,}300 + 7.08T \text{ J}$$

Also, for the change of standard state from Henrian to 1 wt% in Fe at the temperature T

$$\Delta G_{Si(H \rightarrow 1 \text{ wt%})}^{\circ} = RT \ln \frac{MW_{Fe}}{100 MW_{Si}}$$
$$= RT \ln \left(\frac{55.85}{100 \times 28.09} \right)$$
$$= -32.6T \text{ J}$$

so, for the change $Si_{(R)} \rightarrow Si_{(1 \text{ wt% in Fe})}$,

$$\Delta G_{(R \rightarrow 1 \text{ wt%})}^{\circ} = \Delta G_{(R \rightarrow H)}^{\circ} + \Delta G_{(H \rightarrow 1 \text{ wt%})}^{\circ} \qquad (i)$$
$$= -119{,}300 - 25.5T \text{ J}$$

Now, given that a liquid Fe–Si alloy is in equilibrium with an SiO_2-saturated FeO–SiO_2 melt (in which $a_{SiO_2} = 1$) and an atmosphere containing oxygen, calculate the relationship between the equilibrium weight percentage of Si in the Fe–Si alloy and the oxygen pressure in the gaseous atmosphere. For the reaction

$$Si_{(l)} + O_{2(g)} = SiO_{2(s)} \tag{ii}$$

$\Delta G^\circ_{(ii)} = -952,700 + 204T$ joules in the temperature range 1700 to 2000 K. Thus the standard Gibbs free energy for the reaction

$$Si_{(1 \text{ wt\%})} + O_{2(g)} = SiO_{2(s)} \tag{iii}$$

is obtained as $\Delta G^\circ_{(ii)} - \Delta G^\circ_{(i)}$, namely

$$\Delta G^\circ_{(iii)} = -833,400 + 229.5T \text{ J}$$

$$= -RT \ln \frac{a_{SiO_2}}{h_{Si(1 \text{ wt\%})} p_{O_2}}$$

As $a_{SiO_2} = 1$, then

$$\ln h_{Si(1 \text{ wt\%})} = -\frac{833,400}{8.3144T} + \frac{229.5}{8.3144} - \ln p_{O_2}$$

If it can be assumed that Si in Fe obeys Henry's law over some initial range of composition, then, in this range, $h_{Si(1 \text{ wt\%})} = \text{wt\% Si}$, and thus

$$\ln \text{wt\% Si} = -\frac{100,200}{T} + 27.60 - \ln p_{O_2}$$

Thus, to produce an equilibrium melt containing 1 wt% Si at 1600°C, the partial pressure of oxygen must be 5.57×10^{-12} atm, and the oxygen pressure for any other weight percentage of Si at 1600°C is calculated from

$$p_{O_2} = \frac{5.57 \times 10^{-12}}{\text{wt\% Si}}$$

The error in this calculation caused by the assumption of Henrian behavior in some initial range of composition is demonstrated when the calculation is considered again in Sec. 13.9.

13.4 THE GIBBS PHASE RULE

In Chap. 7 it was found that the number of degrees of freedom available an equilibrium in a one-component system is related to the number of phases present by means of a simple rule. This rule, the Gibbs phase rule, was easily derived because of the simplicity of graphical representation of phase equilibria in a one-component system. However, phase relations in a multicomponent system can be complicated, and in such systems the Gibbs phase rule is a powerful tool in the determination of possible equilibria and the restrictions on these equilibria. The general derivation of the phase rule is as follows.

Consider a system containing C chemical species, $i, j, k, \ldots$ (none of which enters into chemical reaction with any other), which occur in P phases, $\alpha, \beta, \gamma, \ldots$. As the thermodynamic state of each of the P phases is determined by specification of its temperature, pressure, and composition (where composition is expressed in terms of $C - 1$ composition variables such as mole fractions or weight percentages), then the state of the entire system is specified when its $P(C + 1)$ variables are fixed. The conditions that the entire system be at complete equilibrium are

$$T_\alpha = T_\beta = T_\gamma = \cdots (P - 1) \text{ equalities of temperature}$$
$$P_\alpha = P_\beta = P_\gamma = \cdots (P - 1) \text{ equalities of pressure}$$
$$a_{i(\alpha)} = a_{i(\beta)} = a_{i(\gamma)} = \cdots (P - 1) \text{ equalities of the activity of the species } i$$
$$a_{j(\alpha)} = a_{j(\beta)} = a_{j(\gamma)} = \cdots (P - 1) \text{ equalities of the activity of the species } j$$

and so on for each of the C chemical species. Thus the total number of equilibrium conditions, given as the number of required equations among the variables of the system, is

$$(P - 1)(C + 2)$$

The number of degrees of freedom, F, which an equilibrium in the system may have is defined as the maximum number of variables which may be independently altered in value without disturbing the equilibrium. This number F is obtained as the difference between the total number of variables available to the system and the minimum number of equations among these variables that is required to maintain the equilibrium, i.e.,

$$F = P(C + 1) - (P - 1)(C + 2)$$
$$= C + 2 - P \tag{13.14}$$

In a system of nonreacting species, the number of species C equals the number of components in the system. However, if some of the species enter into reaction with one another, such that the equilibrium of the system includes a number of reaction equilibria—in addition to the phase, temperature, and pressure equilibria—then the number of equations among the variables which must be specified is increased by R, the number of such independent reaction equilibria occurring in the system. For example, if the species i and j react to form the species k, then the establishment of the reaction equilibrium requires that, in each of the P phases,

$$\overline{G}_i + \overline{G}_j = \overline{G}_k$$

which increases the number of equations among the variables by one. Thus if the system contains N species, among which there are R independent reaction equilibria, then

$$F = P(N + 1) - (P - 1)(N + 2) - R$$
$$= (N - R) + 2 - P$$

In order that the phase rule, as given by Eq. (13.14), be generally applicable to both reactive and nonreactive systems, the number of components in the former is defined as

$$C = N - R$$

C can be determined as either the minimum number of chemical species required to produce the system at equilibrium, or the number of species in the system minus the number of independent reaction equilibria among these species.

Example 1

Consider, again, the example discussed in Sec. 13.2, in which an examination was made of the conditions under which a liquid Fe–Mn solution and a liquid FeO–MnO solution can be in equilibrium with an oxygen-containing atmosphere. This is a three-component system (Fe–Mn–O) existing in three phases (metal-oxide-gas) and thus, from the phase rule, the equilibrium has two degrees of freedom, which can be selected from the variables T, p_{O_2}, $[X_{Fe}]$, $[X_{Mn}]$, (X_{FeO}), and (X_{MnO}). With five species (O_2, Fe, Mn, FeO, MnO) and three components, there are two independent reaction equilibria, which can be selected as

$$\text{Fe} + \frac{1}{2}O_2 = \text{FeO}$$

for which

$$K_{(i),T} = \frac{a_{FeO}}{a_{Fe}\,p_{O_2}^{1/2}} = \frac{(X_{FeO})}{[X_{Fe}]p_{O_2}^{1/2}} \tag{i}$$

and

$$\text{Mn} + \frac{1}{2}O_2 = \text{MnO}$$

for which

$$K_{(ii),T} = \frac{a_{MnO}}{a_{Mn}\,p_{O_2}^{1/2}} = \frac{(X_{MnO})}{[X_{Mn}]p_{O_2}^{1/2}} \tag{ii}$$

(a) If T and p_{O_2} are chosen as the independent variables,

$$\frac{(X_{FeO})}{[X_{Fe}]} = K_{(i),T} \cdot p_{O_2}^{1/2}$$

is fixed by Eq. (i), and

$$\frac{(X_{MnO})}{[X_{Mn}]} = \frac{1 - (X_{FeO})}{1 - [X_{Fe}]} = K_{(ii),T} \cdot p_{O_2}^{1/2}$$

is fixed by Eq. (ii). Thus (X_{FeO}) [and hence (X_{MnO})] and $[X_{Fe}]$ (and hence $[X_{Mn}]$) are fixed.

(b) If T and $[X_{Fe}]$ are chosen as the independent variables, $[X_{Mn}] = 1 - [X_{Fe}]$ is automatically fixed:

$$\frac{(X_{MnO})}{X_{FeO}} = \frac{1 - (X_{FeO})}{(X_{FeO})} = \frac{K_{(ii),T}\,[X_{Mn}]}{K_{(i),T}\,[X_{Fe}]}$$

is fixed by Eqs. (i) and (ii), which fixes (X_{FeO}) [and hence (X_{MnO})], and

$$p_{O_2}^{1/2} = \frac{(X_{FeO})}{[X_{Fe}]K_{(i),T}}$$

is fixed by Eq. (i).

(c) If p_{O_2} and $[X_{Fe}]$ are chosen as the independent variables, from Eq. (i),

$$(X_{FeO}) = K_{(i),T} \cdot [X_{Fe}] \cdot p_{O_2}^{1/2} = \exp\left(\frac{-\Delta H_{(i)}^{\circ}}{RT}\right)\exp\left(\frac{\Delta S_{(i)}^{\circ}}{R}\right) \cdot [X_{Fe}] \cdot p_{O_2}^{1/2}$$

and from Eq. (ii),

$$(X_{MnO}) = 1 - (X_{FeO}) = K_{(ii),T} \cdot [X_{Mn}] \cdot p_{O_2}^{1/2} = \exp\left(\frac{-\Delta H_{(ii)}^{\circ}}{RT}\right)\exp\left(\frac{\Delta S_{(ii)}^{\circ}}{R}\right) \cdot [X_{Mn}] \cdot p_{O_2}^{1/2}$$

simultaneous solution of which fixes T and (X_{FeO}). Thus the fixing of any two of the variables fixes the values of all of the others. In the previous discussion of this example, $T = 1800°C$ and p_{O_2} were selected as the independent variables.

Example 2

Consider the system M–MO–O$_2$ in which the reaction equilibrium

$$M_{(s)} + \frac{1}{2}O_{2(g)} = MO_{(s)}$$

is established. This system has three phases (the condensed phases M and MO and the gaseous oxygen phase) and two components (M and O). The available variables are the temperature T and the total pressure P. As, at equilibrium, the species M and MO occur in fixed states, i.e., M saturated with oxygen and MO saturated with M, the activities of these two species are fixed. Thus, the total pressure P is the sum of the oxygen pressure and the saturated vapor pressures of the solid phases M and MO, and as the latter two are fixed at any given temperature, then the value of P can be varied only by varying p_{O_2}. From the phase rule, $F = C + 2 - P = 2 + 2 - 3 = 1$, and thus the equilibrium has only one degree of freedom. Thus either T can be arbitrarily fixed, in which case the equilibrium constant K_T, and hence $p_{O_2(eqT)}$ is fixed; or $P = p_{O_2} +$ (the vapor pressures of M and MO) can be arbitrarily fixed, in which case the value of K_T, and hence T, is fixed. If an inert gas is added to the system, then $P = p_{O_2} + p_{inert\,gas} +$ (the vapor pressures of M and MO), and so the values of p_{O_2} and $p_{inert\,gas}$ may be independently varied. The addition of the inert gas as the third component increases the number of degrees of freedom to two, but the additional degree of freedom is restricted to variation of the value of $p_{inert\,gas}$, i.e., in addition to either T or p_{O_2} being independently variable, $p_{\text{inert gas}}$ may be independently varied.

Consider the equilibrium

$$M_{(s)} + CO_{2(g)} = MO_{(s)} + CO_{(g)}$$

This three-component, three-phase equilibrium has two degrees of freedom which may be selected from T, P, p_{CO}, and p_{CO_2}. For example, fixing T and P uniquely fixes p_{CO} and p_{CO_2} via

$$K_T = \frac{p_{CO}}{p_{CO_2}} \quad \text{and} \quad P = p_{CO} + p_{CO_2}$$

If the system contains the solid carbide MC, then the three-component four-phase system (M + MO + MC + gas) has one degree of freedom which again can be selected from T, P, p_{CO}, and p_{CO_2}. As $R = N - C = 5 - 3 = 2$, the two independent reaction equilibria can be selected as

$$M_{(s)} + CO_{2(g)} = MO_{(s)} + CO_{(g)} \qquad \text{(i)}$$

and

$$M_{(s)} + 2CO_{(g)} = MC_{(s)} + CO_{2(g)} \qquad \text{(ii)}$$

Fixing T fixes

$$K_{T(i)} = \frac{p_{CO}}{p_{CO_2}}$$

and

$$K_{T(ii)} = \frac{p_{CO_2}}{p_{CO}^2}$$

which uniquely fixes the values of p_{CO} and p_{CO_2}, and hence $P = p_{CO} + p_{CO_2}$.

If solid carbon is also present, in which case the system contains the phases M, MO, MC, C, and gaseous CO and CO_2, then the number of independent reaction equilibria is increased by one, e.g., the independent equilibria

$$M_{(s)} + CO_{2(g)} = MO_{(s)} + CO_{(g)} \qquad \text{(i)}$$
$$M_{(s)} + 2CO_{(g)} = MC_{(s)} + CO_{2(g)} \qquad \text{(ii)}$$

and

$$C_{(s)} + CO_{2(g)} = 2CO_{(g)} \qquad \text{(iii)}$$

occur, and the number of phases present is increased by one. In this case $F = 0$ and the system is invariant, occurring at a unique T and at unique values of p_{CO} and p_{CO_2}.

In a multiphase, multicomponent system in which several independent reaction equilibria occur, the number of such equilibria can be calculated as follows. First write a chemical reaction equation for the formation of each species present from its constituents. For example, in the above example

$$M + \frac{1}{2}O_2 = MO \tag{a}$$

$$M + C = MC \tag{b}$$

$$C + \frac{1}{2}O_2 = CO \tag{c}$$

$$C + O_2 = CO_2 \tag{d}$$

Then combine these reaction equations in such a manner that those elements not considered to be present in the system are eliminated. The resulting number of reaction equations is then the number of independent reaction equilibria, R. In the above system the species present are M, MO, MC, C, CO, and CO_2. Thus, from Eq. (a)

$$\frac{1}{2}O_2 = MO - M$$

and thus, in Eq. (c)

$$C + MO - M = CO \quad \text{or} \quad C + MO = CO + M \tag{iv}$$

in Eq. (d)

$$C + 2MO - 2M = CO_2 \quad \text{or} \quad C + 2MO = CO_2 + 2M \tag{v}$$

and in Eq. (b)

$$M + C = MC \tag{vi}$$

Thus three independent equilibria occur, combination of which produces other equilibria which occur in the system, e.g.,

$$MC + 2MO = CO_2 + 3M \tag{vii}$$

$$MC + MO = CO + 2M \tag{viii}$$

$$M + CO_2 = MO + CO \tag{i}$$

$$M + 2CO = CO_2 + MC \tag{ii}$$

$$C + CO_2 = 2CO \tag{iii}$$

When any three of equilibria (i) to (viii) are established, then the other five are established.

Example 3

Consider equilibrium in the ternary system Si–C–O at 1000°C. The solid phases which can exist in this system are Si, SiO_2, SiC, C, and the gas phase is a mixture of CO and CO_2. Just as fixing the activity of one component in a binary system fixes the activity of the other component, the fixing of the activities of two of the components in a ternary system fixes the activity of the third component. Thus when the activities of C and O_2 are fixed in the system Si–C–O, the activity of Si is fixed and a definite equilibrium state exists. Thus two-dimensional representation of the phase stability can be considered, (1) at constant temperature with a_C and p_{O_2} as the variables, or (2) at constant a_C (or constant p_{O_2}) with T and p_{O_2} (or a_C) as the variables.

Application of the phase rule to a three-component system indicates that an equilibrium among five phases has no degrees of freedom. Thus, if a gas phase is always present,

1. Four condensed phases can be in equilibrium with one another and a gas phase at an invariant state.
2. Three condensed phases can be in equilibrium with one another and a gas phase at an arbitrarily chosen temperature.
3. Two condensed phases can be in equilibrium with one another and a gas phase at an arbitrarily chosen temperature and an arbitrarily chosen value of a_C or p_{O_2}.
4. One condensed phase can be in equilibrium with the gas phase at an arbitrarily chosen temperature and arbitrarily chosen values of a_C and p_{O_2}.

As four solid phases can exist, there are $(4 \times 3 \times 2)/(3 \times 2) = 4$ possible equilibria involving three condensed phases and a gas phase, $(4 \times 3)/2 = 6$ possible equilibria involving two condensed phases and a gas phase, and four possible equilibria involving one condensed phase and the gas phase. Consider construction of the phase stability diagram for the system Si–C–O at 1000°C using log a_C and log p_{O_2} as the variables. The six possible equilibria involving two condensed phases and a gas phase are

(1) Si–SiO_2–gas
(2) Si–SiC–gas
(3) SiC–SiO_2–gas
(4) SiC–C–gas
(5) SiO_2–C–gas
(6) Si–C–gas

and the four possible equilibria involving three condensed phases and a gas phase are

Si–SiO_2–SiC–gas
Si–SiC–C–gas
SiO_2–SiC–C–gas
Si–SiO_2–C–gas

1. The equilibrium, Si–SiO_2–gas phase. For the reaction

$$Si_{(s)} + O_{2(g)} = SiO_{2(s)}$$

$$\Delta G^\circ_{(i)1273\,K} = -683,400 \text{ J} \tag{i}$$

$$= -RT \ln \frac{1}{p_{O_2}}$$

$$= 8.3144 \times 1273 \times 2.303 \log p_{O_2}$$

or $\log p_{O_2} = -28.04$. Thus, at 1273 K, the equilibrium between Si and SiO_2 requires $\log p_{O_2} = -28.04$ and this equilibrium, which is drawn as line AB in Fig. 13.5a, is independent of a_C. At lower values of p_{O_2} Si is stable relative to SiO_2 and, at higher values, SiO_2 is stable relative to Si.

2. The equilibrium, Si–SiC–gas phase. For the reaction

$$Si_{(s)} + C_{(s)} = SiC_{(s)}$$

$$\Delta G^\circ_{(iii)1273\ K} = -63,300\ J \tag{ii}$$

$$= -RT \ln \frac{1}{a_C}$$

$$= 8.3144 \times 1273 \times 2.303 \log a_C$$

or $\log a_c = -2.60$, Thus, at 1273 K the equilibrium between Si and SiC requires $\log a_C = -2.60$, and this equilibrium, which is drawn as line CD in Fig. 13.5a, is independent of p_{O_2}. At lower values of a_C, Si is stable relative to SiC and, at higher values, SiC is stable relative to Si.

3. The equilibrium, SiC–SiO$_2$–gas phase. Lines AB and CD intersect at the point P ($\log p_{O_2} = -28.04$, $\log a_C = -2.60$), which is the unique state at which Si, SiC, and SiO_2 are in equilibrium with one another and a gas phase at 1273 K, and the variation of $\log a_C$ with $\log p_{O_2}$ required for the equilibrium SiC–SiO$_2$ must pass through this point. Combination of Eqs. (i) and (ii) gives

$$SiC_{(s)} + O_{2(g)} = SiO_{2(s)} + C_{(s)} \tag{iii}$$

for which

$$\Delta G^\circ_{(iii)1273\ K} = -620,100\ J$$

$$= -8.3144 \times 1273 \times 2.303 \log \frac{a_C}{p_{O_2}}$$

or

$$\log a_C = \log p_{O_2} + 25.44$$

which is drawn as line EF in Fig. 13.5a. In states above this line SiC is stable relative to SiO_2 and, below the line, SiO_2 is stable relative to SiC.

4. The equilibrium SiC–C–gas. The equilibrium between solid SiC and solid C is a phase equilibrium which exists only at $a_C = 1$, or $\log a_C = 0$. Thus, in Fig. 13.5a, the equilibrium between SiC and C exists along the $\log a_C = 0$ line at values of $\log p_{O_2}$ less than -25.44, the point of intersection of line EF with the $\log a_C = 0$ line.

5. The equilibrium SiO$_2$–C–gas. As with the equilibrium between SiC and C, the phase equilibrium between SiO_2 and C requires $\log a_C = 1$, and thus occurs along the $\log a_C = 0$ line at values of $\log p_{O_2}$ greater than -25.44.

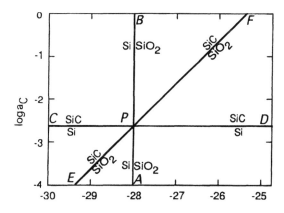

Figure 13.5 (*a*) Construction of the phase stability diagram for the system Si–C–O at 1273 K.

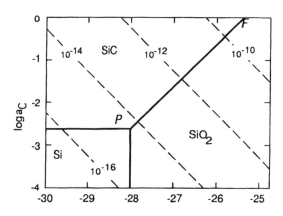

Figure 13.5 (*b*) CO_2 isobars in the phase stability diagram for Si–C–O at 1273 K.

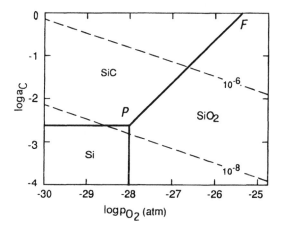

Figure 13.5 (*c*) CO isobars in the phase stability diagram for Si–C–O at 1273 K.

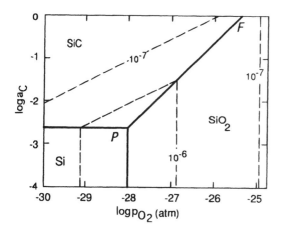

Figure 13.5 (d) SiO isobars in the phase stability diagram for Si–C–O at 1273 K.

6. **The equilibrium Si–C–gas.** The standard Gibbs free energy change at 1273 K for the reaction

$$Si_{(s)} + C_{(s)} = SiC_{(s)}$$

is $\Delta G^\circ_{(ii)1273\,K} = -63,300$ J, which, being negative, indicates that Si and C spontaneously react with one another to form SiC until either the C or the Si is consumed. Thus Si and C cannot be in equilibrium with one another. If the molar ratio Si/C in the system is greater than 1 the C is consumed by the reaction and equilibrium between the product SiC and the remaining Si is attained, and if the ratio is less than 1 the Si is consumed by the reaction and an equilibrium is attained between the SiC produced and the remaining C.

The solid carbon phase exists only along the log $a_C = 1$ line, and thus Fig. 13.5a contains fields of stability of the single phases Si, SiO_2, and SiC. Consequently, of the six lines in the diagram radiating from point P, three represent stable equilibria involving two condensed phases and a gas phase, and three represent metastable equilibria involving two condensed phases and a gas phase, and the problem is how to distinguish between the two types of equilibria. It is a property of such diagrams that the lines of metastable and stable equilibria radiate alternatively from a point such as P (see, e.g., Figs. 7.11 and 7.12). Thus, one set of lines is *PA-PC-PF* and the other is *PE-PB-PD*. In Fig. 13.5a Si is stable relative to SiO_2 in states to the left of *PA* and is stable relative to SiC in states below *PC*. Thus the line *PE* represents the metastable equilibrium between SiC and SiO_2. This identifies the stable equilibrium lines as being *PA-PC-PF* and, as shown in Fig. 13.5b, defines the fields of stability of a single condensed phase with a gas phase, which, at constant temperature, have two degrees of freedom. These fields meet at lines which represent the equilibrium between two condensed phases and a gas phase (which, at constant temperature, have 1 degree of freedom), and the lines meet at points representing equilibrium among three condensed phases and a gas phase. The point P is the state of equilibrium of Si, SiC, and SiO_2 and a gas phase, and the point F is the state of equilibrium of SiC, SiO_2, C, and a gas phase.

The partial pressures of CO and CO_2 are determined by the activities of carbon and oxygen in the system, and the iso-p_{CO_2} and iso-p_{CO} lines can be placed on the stability diagram as follows. For the reaction

$$C_{(s)} + O_{2(g)} = CO_{2(g)} \tag{iv}$$

$$\Delta G^{\circ}_{(iv)1273 \text{ K}} = -395,200 \text{ J} = -8.3144 \times 1273 \times 2.303 \log \frac{p_{CO_2}}{a_C p_{O_2}}$$

Therefore,

$$16.21 = \log p_{CO_2} - \log a_C - \log p_{O_2}$$

or

$$\log a_C = -\log p_{O_2} - 16.21 + \log p_{CO_2}$$

Thus, as shown in Fig. 13.5b, the CO_2 isobars in the phase stability diagram are straight lines with slopes of -1. Similarly, for

$$C_{(s)} + \frac{1}{2} O_{2(g)} = CO_{(g)} \tag{v}$$

$$\Delta G^{\circ}_{(v)1273 \text{ K}} = -223,300 \text{ J} = -8.3144 \times 1273 \times 2.303 \log \frac{p_{CO}}{a_C p_{O_2}^{1/2}}$$

or

$$\log a_C = -\frac{1}{2} \log p_{O_2} - 9.16 + \log p_{CO}$$

and, as shown in Fig. 13.5c, the CO isobars are straight lines with slopes of minus one-half.

The gaseous species SiO also occurs in the system Si–C–O, and SiO isobars can be drawn on the stability diagram. As the partial pressure of SiO is determined by the activity of Si and the partial pressure of oxygen, the individual single condensed phase stability fields have to be considered separately, and each equilibrium must involve the condensed phase of interest, SiO gas and C and/or O_2.

In the field of stability of Si the equilibrium is

$$Si_{(s)} + \frac{1}{2} O_{2(g)} = SiO_{(g)} \tag{vi}$$

for which

$$\Delta G^{\circ}_{(vi)1273 \text{ K}} = -209,200 \text{ J} = -8.3144 \times 1273 \times 2.303 \log \frac{p_{SiO}}{p_{O_2}}$$

or

$$\log p_{SiO} = \frac{1}{2} \log p_{O_2} + 8.59$$

Thus, as shown in Fig. 13.5d, the SiO isobars in the stability field of Si are vertical lines, and p_{SiO} increases with increasing p_{O_2}.

In the SiO_2 field, the equilibrium is

$$SiO_{2(s)} = SiO_{(g)} + \frac{1}{2} O_{2(g)} \tag{vii}$$

for which combination of $\Delta G^\circ_{(vii)1273\,K}$ and $\Delta G^\circ_{(i)1273\,K}$ gives

$$\Delta G^\circ_{(vii)1273\,K} = -474{,}200\ J = -8.3144 \times 1273 \times 2.303 \log \left(p_{SiO} p_{O_2}^{1/2}\right)$$

which gives

$$\log p_{SiO} = -\frac{1}{2} \log p_{O_2} - 19.45$$

Thus, as shown in Fig. 13.5d, the SiO isobars in the field of stability of SiO_2 are vertical lines and p_{SiO} decreases with increasing p_{O_2}.

In the field of stability of SiC the equilibrium is

$$SiC_{(s)} + \frac{1}{2} O_{2(g)} = SiO_{(g)} + C_{(s)} \tag{viii}$$

for which combination of $\Delta G^\circ_{(vi)1273\,K}$ and $\Delta G^\circ_{(ii)1273\,K}$ gives

$$\Delta G^\circ_{(viii)1273\,K} = -145{,}900\ J = -8.3144 \times 1273 \times 2.303 \log \frac{p_{SiO} \cdot a_C}{p_{O_2}^{1/2}}$$

or

$$\log a_C = \frac{1}{2} \log p_{O_2} - \log p_{SiO} + 5.99$$

Thus, as shown in Fig. 13.5d, the SiO isobars in the field of stability of SiC are straight lines with slopes of one-half. At any temperature, the maximum value of p_{SiO} occurs in states in which Si and SiO_2 are in equilibrium with a gas phase.

The activities of carbon and oxygen used as the variables in the construction of Fig. 13.5 are determined by the individual values of p_{CO} and p_{CO_2}, which establish the equilibria

$$CO + \frac{1}{2} O_2 = CO_2$$

and

$$C + CO_2 = 2CO$$

and thus the phase stability diagram can be constructed using p_{CO} and p_{CO_2} as the variables. The equilibrium corresponding to Eq. (i) is

$$Si_{(s)} + 2CO_{2(g)} = SiO_{2(s)} + 2CO_{(g)} \tag{ix}$$

Combination of the Gibbs free energy changes for the reactions given by Eqs. (i), (iv), and (v) gives

$$\Delta G^{\circ}_{(ix)1273\,K} = -339,600\ J = -8.3144 \times 1273 \times 2.303 \log \frac{p^2_{CO}}{p^2_{CO_2}}$$

which gives

$$\log p_{CO} = \log p_{CO_2} + 6.97$$

This is drawn as the line AB in Fig. 13.6a (and corresponds to the line AB in Fig. 13.5a). Above the line Si is stable relative to SiO_2, and below the line SiO_2 is stable relative to Si.

The equivalent of the equilibrium given by Eq. (ii) is

$$Si_{(s)} + 2CO_{(g)} = SiC_{(s)} + CO_{2(g)} \tag{x}$$

for which

$$\Delta G^{\circ}_{(x)1273\,K} = -11,900\ J = -8.3144 \times 1273 \times 2.303 \log \frac{p_{CO_2}}{p^2_{CO}}$$

or

$$\log p_{CO} = \frac{1}{2} \log p_{CO_2} - 0.24$$

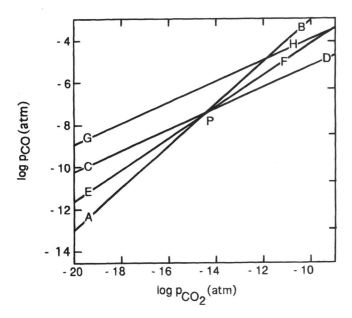

Figure 13.6 (a) Construction of the phase stability diagram for the system Si–C–O at 1273 K.

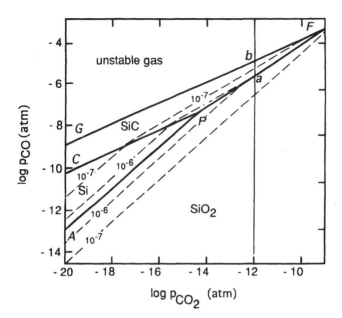

Figure 13.6 (b) The phase stability diagram for the system Si–C–O at 1273 K showing the 10^{-6} and 10^{-7} atm SiO isobars.

This is drawn as line CD in Fig. 13.6a. Above the line SiC is stable relative to Si, and below the line Si is stable relative to SiC. The lines AB and CD intersect at P, which is thus the invariant point at which Si, SiC, SiO_2, and a gas phase are in equilibrium.

The equivalent of the equilibrium given by Eq. (iii) is

$$SiC_{(s)} + 3CO_{2(g)} = SiO_{2(s)} + 4CO_{(g)} \tag{xi}$$

for which

$$\Delta G^\circ_{(xi)1273\,K} = -327{,}700\ J = -8.3144 \times 1273 \times 2.303 \log \frac{p^4_{CO}}{p^3_{CO_2}}$$

or

$$\log p_{CO} = 0.75 \log p_{CO_2} + 3.36$$

This is drawn as the line EF in Fig. 13.6a. SiC is stable relative to SiO_2 above the line, and SiO_2 is stable relative to SiC below the line.

The "carbon line" which is equivalent to the log $a_C = 0$ line in Fig. 13.5 is obtained from

$$C_{(s)} + CO_{2(g)} = 2CO_{(g)} \tag{xii}$$

for which

$$\Delta G^\circ_{(xii)1273\,K} = -51{,}400\ J = -8.3144 \times 1273 \times 2.303 \log \frac{p^2_{CO}}{p_{CO_2}}$$

which gives

$$\log p_{CO} = \frac{1}{2} \log p_{CO_2} - 1.05$$

This is drawn as the line GH in Fig. 13.6a. The phase stability fields, identified in the same manner as was used in Fig. 13.5a, are shown in Fig. 13.6b. The Si phase field is bounded by APC, the SiC phase field is bounded by $CPFG$, and the SiO_2 phase field lies below the line APF. The region above GF is an unstable gas. Any gas in this region precipitates carbon according to

$$2CO_{(g)} \rightarrow CO_{2(g)} + C_{(s)}$$

until, thereby, the ratio p_{CO_2}/p_{CO} is that required for equilibrium with C at $a_C = 1$ at 1273 K. Isobars for SiO gas are calculated as follows. In the field of stability of Si the equivalent of the equilibrium given by Eq. (vi) is

$$Si_{(s)} + CO_{2(g)} = SiO_{(g)} + CO_{(g)} \tag{xiii}$$

for which

$$\Delta G^\circ_{(xiii)1273 \text{ K}} = -37,300 \text{ J} = -8.3144 \times 1273 \times 2.303 \log \frac{p_{SiO}p_{CO}}{p_{CO_2}}$$

which gives

$$\log p_{CO} = \log p_{CO_2} + 1.53 - \log p_{SiO}$$

In the field of stability of SiO_2 the equivalent of the equilibrium given by Eq. (vii) is

$$SiO_{2(s)} + CO_{(g)} = SiO_{(g)} + CO_{2(g)} \tag{xiv}$$

for which

$$\Delta G^\circ_{(xiv)1273 \text{ K}} = 302,300 \text{ J} = -8.3144 \times 1273 \times 2.303 \log \frac{p_{CO_2}p_{SiO}}{p_{CO}}$$

which gives

$$\log p_{CO} = \log p_{CO_2} + 12.41 + \log p_{SiO}$$

In the field of stability of SiC the equilibrium equivalent to that given by Eq. (viii) is

$$SiC_{(s)} + 2CO_{2(g)} = SiO_{(g)} + 3CO_{(g)} \tag{xv}$$

for which

$$\Delta G^\circ_{(xv)1273 \text{ K}} = -25,400 \text{ J} = -8.3144 \times 1273 \times 2.303 \log \frac{p^3_{CO}p_{SiO}}{p^2_{CO_2}}$$

which gives

$$\log p_{CO} = \frac{2}{3} \log p_{CO_2} - 0.35 - \frac{1}{3} \log p_{SiO}$$

The 10^{-7} and 10^{-6} atm SiO isobars are shown in Fig. 13.6b, and comparison of Figs. 13.5 and 13.6 shows that the latter is produced by distorting the former such that the CO isobars are horizontal lines and the CO_2 isobars are vertical lines. The points P and F in Fig. 13.6b correspond to the points P and F in Fig. 13.5.

The full phase stability diagram is three-dimensional with the activities of two components (a_C and p_{O_2}) or the partial pressures of CO and CO_2 plus temperature as the axes. Two-dimensional phase stability diagrams which show the influence of temperature can be drawn if the activity of one component is held constant. Consider the phase stability diagram using p_{CO} and $1/T$ as coordinates and drawn for a constant partial pressure of CO_2 of 10^{-12} atm. The diagram will be drawn for temperatures in the range 1250–1990 K, and thus, as the melting temperature of Si is 1683 K, equilibria involving both solid and liquid Si will have to be considered. In Fig. 13.7a the vertical line at $1/T = 5.942 \times 10^{-4}\,T^{-1}$ (1683 K) represents the melting temperature of Si, and thus liquid Si occurs to the left of the line, and solid Si occurs to the right.

1. The equilibrium, $Si_{(s)}$–SiO_2–gas phase. The equilibrium is

$$Si_{(s)} + 2CO_{2(g)} = SiO_{2(s)} + 2CO_{(g)} \qquad (xvi)$$

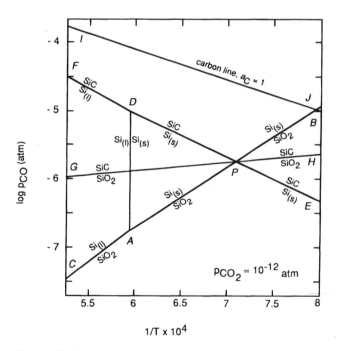

Figure 13.7 (a) Construction of the phase stability diagram for the system Si–C–O using log p_{CO} and $1/T$ as coordinates at a constant partial pressure of CO_2 of 10^{-12} atm.

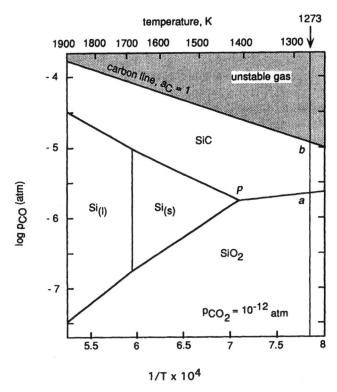

Figure 13.7 (*b*) The phase stability diagram for the system Si–C–O at a constant partial pressure of $CO_2 = 10^{-12}$ atm.

for which

$$\Delta G^{\circ}_{(xvi)} = -337,300 - 0.02T \text{ J}$$

Thus

$$\frac{-337,300}{8.3144 \times 2.303T} - \frac{0.02}{8.3144 \times 2.303} = -\log \frac{p^2_{CO}}{p^2_{CO_2}}$$

which can be rearranged to give

$$\log p_{CO} = \frac{8807}{T} + 5.22 \times 10^{-4} + \log p_{CO_2}$$

This line, with $\log p_{CO_2} = -12$, is drawn as AB in Fig. 13.7*a*. Silicon is stable relative to SiO_2 above the line, and SiO_2 is stable relative to Si below the line.

The equilibrium, $Si_{(l)}$–SiO_2–gas phase: The equilibrium is

$$Si_{(l)} + 2CO_{2(g)} = SiO_{2(s)} + 2CO_{(g)} \qquad \text{(xvii)}$$

for which

$$\Delta G^{\circ}_{(xvii)} = -387,900 + 30.18T \text{ J}$$

Thus

$$\frac{-387,900}{8.3144 \times 2.303T} + \frac{30.18}{8.3144 \times 2.303} = -\log \frac{p_{CO}^2}{p_{CO_2}^2}$$

which gives

$$\log p_{CO} = \frac{10,130}{T} - 0.788 + \log p_{CO_2}$$

With $p_{CO_2} = 10^{-12}$ atm, this gives line CA in Fig. 13.7a. Liquid Si is stable relative to SiO_2 above the line, and SiO_2 is stable relative to liquid Si below the line.

2. The equilibrium $Si_{(s)}$–SiC–gas phase. The equilibrium is

$$Si_{(s)} + 2CO_{(g)} = SiC_{(s)} + CO_{2(g)} \tag{xviii}$$

for which

$$\Delta G_{(xviii)}^\circ = -243,750 + 182.11T \text{ J}$$

which yields

$$\log p_{CO} = -\frac{6364}{T} + 4.76 + \frac{1}{2} \log p_{CO_2}$$

With $\log p_{CO_2} = -12$, this gives line DE in Fig. 13.7a. SiC is stable relative to solid Si above the line and Si is stable relative to SiC below the line.
    The equilibrium, $Si_{(l)}$–SiC–gas phase: The equilibrium is

$$Si_{(l)} + 2CO_{(g)} = SiC_{(s)} + CO_{2(g)} \tag{xix}$$

for which

$$\Delta G_{(xix)}^\circ = -293,300 + 211.5T \text{ J}$$

This gives

$$\log p_{CO} = -\frac{7659}{T} + 5.52 + \frac{1}{2} \log p_{CO_2}$$

which, with $\log p_{CO_2} = -12$, gives line FD in Fig. 13.7a. Liquid Si is stable with respect to SiC below the line and SiC is stable with respect to Si above the line.

3. The equilibrium SiC–SiO_2–gas phase. The equilibrium is

$$SiC_{(s)} + 3CO_{2(g)} = SiO_{2(s)} + 4CO_{(g)} \tag{xx}$$

which gives

$$\log p_{CO} = \frac{1221}{T} + 2.38 + 0.75 \log p_{CO_2}$$

This is drawn, with $\log p_{CO_2} = -12$, as line GH in Fig. 13.7a. SiC is stable relative to SiO_2 above the line and SiO_2 is stable relative to SiC below the line.

The carbon line, at which $a_C = 1$, is determined by the equilibrium

$$C_{(s)} + CO_{2(g)} = 2CO_{(g)}$$

for which

$$\Delta G° = 170,700 - 174.5T \text{ J}$$

which gives

$$\log p_{CO} = -\frac{4457}{T} + 4.55 + \frac{1}{2} \log p_{CO_2}$$

With $p_{CO_2} = 10^{-12}$ atm, this gives line IJ in Fig. 17.7a. States above this line represent unstable gas. The stability fields are identified as follows:

1. In the area APD solid silicon is stable with respect to SiO_2 and SiC, and thus this area is the field of stability of solid Si.
2. In the areas $FDAC$ liquid silicon is stable with respect to SiC and SiO_2, and thus this area is the field of stability of liquid Si.
3. In the area $FDPHJI$ SiC is stable with respect to Si and SiO_2, and thus this area is the field of stability of SiC.
4. Below the line $CAPH$ SiO_2 is stable with respect to Si and SiC, and thus this is the field of stability of SiO_2.

These fields of stability are shown in Fig. 13.7b. The phase stability diagram shows that, with $p_{CO_2} = 10^{-12}$ atm, 1410 K (the point P) is the minimum temperature at which silicon is stable, and that, with increasing temperature, the field of stability of silicon widens at the expense of SiC and SiO_2. The slopes of the lines in Fig. 13.7 are related to the standard enthalpy changes for the equilibrium reactions, e.g., the slope of the line dividing the solid Si and the SiO_2 fields (the line AB in Fig. 13.7a) is 8807, which obtained as $-\Delta H°/2R$ for the reaction given by Eq. (xvi). The points a and b on the 10^{-12} atm CO_2 isobar in Fig. 13.6b correspond with the points a and b on the 1273 K isotherm in Fig. 13.7b.

13.5 BINARY SYSTEMS CONTAINING COMPOUNDS

The phase relationships in a two-component system can be represented on an isobaric phase diagram using temperature and composition as coordinates, and these are the phase diagrams normally encountered in materials science. If the two components react with one another to form compounds then, in such systems, chemical reaction equilibria and phase equilibria are synonymous.

Consider the binary system $A–B$, the phase diagram for which is shown in Fig. 13.8. The negative departures from ideality in the solid state are sufficiently large that compounds are formed, with there being negligible solubility of A in B, or B in A, and negligible range of nonstoichiometry in the compounds AB_3, AB, and A_3B. The system contains the equilibria

$$3A + B = A_3B, \quad A + B = AB, \quad A + 3B = AB_3$$

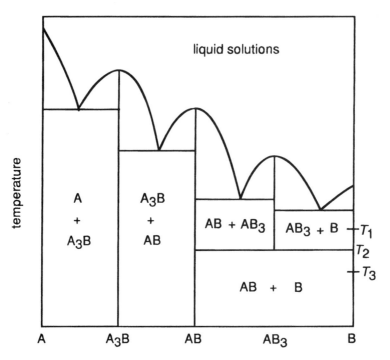

Figure 13.8 The phase diagram for the system $A–B$ in which three stoichiometric compounds are formed.

and, if one of the components (B) is appreciably volatile and the other (A) is not, then the thermodynamics of the system can be determined from a knowledge of the variation of p_B with composition. The variation of p_B with composition at the temperature T_1 is shown in Fig. 13.9. In the range of composition between B and AB_3 virtually pure B exists in equilibrium with AB_3 (saturated with B), and, as pure B exists, the pressure exerted by the system is p_B°, the saturated vapor pressure of B at the temperature T_1. In the range of composition between AB (saturated with B) and AB_3 (saturated with A) the constant pressure exerted by the system is p_B', in the range between A_3B (saturated with B) and AB (saturated with A) it is p_B'', and in the range between A (saturated with B) and A_3B (saturated with A) it is p_B'''. In each of these ranges of composition the two-component, three-phase equilibrium has one degree of freedom, which is used when T_1 is specified, which requires that $P = p_B$ within these ranges at fixed temperature. The activity of B in the system, defined as p_B/p_B°, is thus

$$\frac{p_B^\circ}{p_B^\circ} = 1 \text{ in the range } AB_3–B$$

$$\frac{p_B'}{p_B^\circ} \text{ in the range } AB–AB_3$$

$$\frac{p_B''}{p_B^\circ} \text{ in the range } A_3B–AB$$

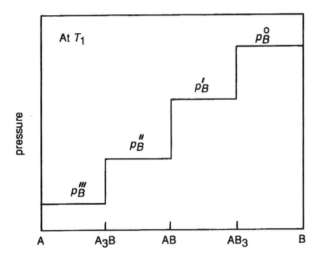

At T_1

p_B^o

p_B'

p_B''

p_B'''

pressure

A A_3B AB AB_3 B

Figure 13.9 The variation, with composition, of the vapor pressure of component B in the system shown in Fig. 13.8 at the temperature T_1.

and

$$\frac{p_B'''}{p_B^o} \text{ in the range } A\text{–}A_3B$$

Thus, as $\Delta \overline{G}_B^M = \overline{G}_B - G_B^o = RT \ln p_B/p_B^o$, the Gibbs free energy-composition diagram at the temperature T_1 is as shown in Fig. 13.10. As Fig. 13.10 is drawn for 1 mole of the system, then

$$hb = \Delta G_{(i)} = \Delta G \text{ for the reaction}$$

$$0.75 \text{ mole } A + 0.25 \text{ mole } B = A_{0.75}B_{0.25} \text{ or } 0.25 \text{ moles of } A_3B$$

$$gc = \Delta G_{(ii)} = \Delta G \text{ for the reaction}$$

$$0.5 \text{ mole } A + 0.5 \text{ mole } B = A_{0.5}B_{0.5} \text{ or } 0.5 \text{ moles of } AB$$

$$fd = \Delta G_{(iii)} = \Delta G \text{ for the reaction}$$

$$0.25 \text{ mole } A + 0.75 \text{ mole } B = A_{0.25}B_{0.75} \text{ or } 0.25 \text{ moles of } AB_3$$

These three changes in Gibbs free energy can be determined geometrically as follows:

$$ek = \Delta \overline{G}_B'^M = RT \ln \frac{p_B'}{p_B^o}$$

$$em = \Delta \overline{G}_B''^M = RT \ln \frac{p_B''}{p_B^o}$$

and

$$en = \Delta \overline{G}_B'''^M = RT \ln \frac{p_B'''}{p_B^o}$$

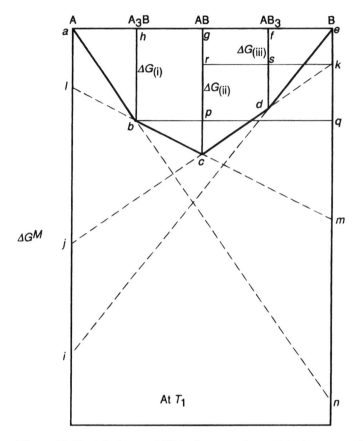

Figure 13.10 The integral Gibbs free energies at T_1 in the system shown in Fig. 13.8.

Thus, from consideration of the similar triangles *ahb* and *aen*,

$$\frac{\Delta G_{(i)}}{\Delta\overline{G}_B^{mM}} = \frac{1}{4}$$

and hence

$$\Delta G_{(i)} = \frac{1}{4}\Delta\overline{G}_B^{mM}$$

Consideration of the similar triangles *bpc* and *bqm* gives

$$\frac{pc}{qm} = \frac{bp}{bq} = \frac{1}{3}$$

But

$$pc = gc - gp = \Delta G_{(ii)} - \Delta G_{(i)}$$

and

$$qm = em - eq = \Delta\overline{G}_B^{''M} - \Delta G_{(i)}$$

Thus

$$3(\Delta G_{(ii)} - \Delta G_{(i)}) = (\Delta\overline{G}_B^{''M} - \Delta G_{(i)})$$

or

$$\Delta G_{(ii)} = \frac{1}{3}\Delta\overline{G}_B^{''M} + \frac{2}{3}\Delta G_{(i)}$$

Consideration of the similar triangles rck and sdk gives

$$\frac{rc}{sd} = \frac{rk}{sk} = 2$$

But

$$rc = gc - gr = \Delta G_{(ii)} - \Delta\overline{G}_B^{'M}$$

and

$$sd = fd - fs = \Delta G_{(iii)} - \Delta\overline{G}_B^{'M}$$

such that

$$\Delta G_{(iii)} = \frac{1}{2}\Delta G_{(ii)} + \frac{1}{2}\Delta\overline{G}_B^{'M}$$

Thus,

$$\Delta G_{(i)} = \frac{1}{4}RT \ln \frac{p_B'''}{p_B^{\circ}}$$

$$\Delta G_{(ii)} = \frac{1}{3}RT \ln \frac{p_B'''}{p_B^{\circ}} + \frac{1}{6}RT \ln \frac{p_B'''}{p_B^{\circ}}$$

and

$$\Delta G_{(iii)} = \frac{1}{6}RT \ln \frac{p_B''}{p_B^{\circ}} + \frac{1}{12}RT \ln \frac{p_B'''}{p_B^{\circ}} + \frac{1}{2}RT \ln \frac{p_B'}{p_B^{\circ}}$$

and hence for

$$3A + B = A_3B \quad \Delta G_{(i)}^{\circ} = 4\Delta G_{(i)}$$

$$A + B = AB \quad \Delta G_{(ii)}^{\circ} = 2\Delta G_{(ii)} = \frac{2}{3}\Delta\overline{G}_B^{''M} + \frac{4}{3}\Delta G_{(i)}$$

$$A + 3B = AB_3 \quad \Delta G_{(iii)}^{\circ} = 4\Delta G_{(iii)} = 2\Delta G_{(ii)} + 2\Delta\overline{G}_B^{'M}$$

Fig. 13.8 shows that, below the temperature T_2, the compound AB_3 is unstable with respect to AB and B, with the invariant equilibrium

$$AB_{(s)} + 2B_{(s)} = AB_{3(s)}$$

occurring at the temperature T_2. The standard Gibbs free energy change for the above reaction is calculated as $\Delta G_{(iii)}^\circ - \Delta G_{(ii)}^\circ$, which equals $2RT \ln p_B'/p_B^\circ$. Thus at $T > T_2$

$$\Delta G_{(iii)}^\circ - \Delta G_{(ii)}^\circ < 0 \quad \text{and} \quad p_B' < p_B^\circ$$

at T_2

$$\Delta G_{(iii)}^\circ - \Delta G_{(ii)}^\circ = 0 \quad \text{and} \quad p_B' = p_B^\circ$$

and at $T < T_2$

$$\Delta G_{(iii)}^\circ - \Delta G_{(ii)}^\circ > 0 \quad \text{and} \quad p_B' > p_B^\circ$$

The variations of the Gibbs free energy with temperature at T_2 and T_3 are shown in Figs. 13.11a and b, respectively, which illustrate graphically that, at T_2, $\Delta G_{(iii)} = \frac{1}{2}\Delta G_{(ii)}$ and, at $T_3 |\Delta G_{(iii)}| < |\frac{1}{2}\Delta G_{(ii)}|$, such that, in the range of composition B–AB, at temperatures below T_2, the system occurring as either $AB + AB_3$ or $AB_3 + B$ is metastable with respect to its occurrence as $AB + B$.

In considering the thermodynamic properties of a system such as is shown in Fig. 13.8, two approaches can be made, namely:

1. The consideration that the compounds are ordered solid solutions
2. The consideration that the compounds are formed by the chemical reaction of A with B

1. Consider the compound AB_3 to be an ordered solid solution of A and B in the molar ratio $\frac{1}{3}$. Then, in Fig. 13.10

$$fd = \Delta G^M = RT (X_A \ln a_A + X_B \ln a_B) = RT (0.25 \ln a_A + 0.75 \ln a_B) \quad \text{(i)}$$
$$= RT \ln a_A^{0.25} a_B^{0.75}$$

2. Consider the compound AB_3 to form as the product of the reaction

$$A + 3B = AB_3$$

for which the change in the standard Gibbs free energy is $\Delta G_{(iii)}^\circ$. Then

$$fd = 0.25\Delta G_{(iii)}^\circ = -0.25RT \ln K_{(iii)} = -RT \ln \left(\frac{a_{AB_3}}{a_A a_B^3}\right)^{0.25} \quad \text{(ii)}$$

$$= RT \ln \left(\frac{a_A^{0.25} a_B^{0.75}}{a_{AB_3}^{0.25}}\right)$$

As AB_3 is a "line compound," i.e., has a negligible range of nonstoichiometry, it exists at a fixed composition and, hence, exists in a fixed state. If this fixed state is chosen as being the standard state, in which $a_{AB_3} = 1$, then Eq. (ii) becomes

$$fd = 0.25\Delta G_{(iii)}^\circ = RT \ln a_A^{0.25} a_B^{0.75}$$

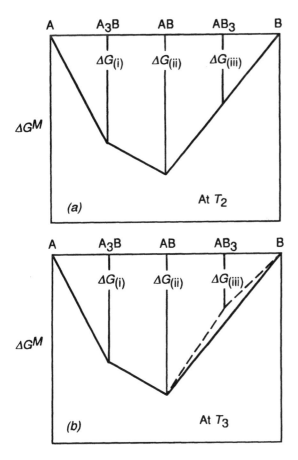

Figure 13.11 (*a*) The integral Gibbs free energies in the system shown in Fig. 13.8 at the temperature T_2. (*b*) The integral Gibbs free energies in the system shown in Fig. 13.8 at the temperature T_3.

which is identical with Eq. (i). In both Eqs. (i) and (ii), the standard states of A and B are the pure solid elements at the temperature T.

The variations of the activities of A and B in the compound AB_3 are limited by the separation of B and AB, e.g., when AB_3 is in equilibrium with B, $a_B = 1$ and thus $a_A = \exp(4\Delta G_{(iii)}/RT) = \exp(4\Delta G^\circ_{(iii)}/RT)$ ($RT \ln a_A = ai$ in Fig. 13.10). If the activity of B is decreased to a value less than unity, then AB_3 is no longer saturated with B, and the activity of A in the compound increases in accordance with Eq. (iv). The minimum activity of B in AB_3 is determined by saturation of AB_3 with A, at which point the compound AB appears. The minimum activity which B may have is obtained from Fig. 13.10 as $RT \ln a_B = ek$, and the corresponding maximum activity of A is obtained from $RT \ln a_A = aj$. Nonsaturation of AB_3 with either A or B occurs when the partial pressure of B exerted by the compound lies between the limits p'_B and p°_B.

Similar consideration can be made with respect to the compounds AB and A_3B, e.g., in Fig. 13.10:

$$gc = \Delta G^M = \frac{1}{2}\Delta G^\circ_{(ii)} = RT \ln a_A^{0.5} A_B^{0.5}$$

and

$$hb = \Delta G^M = 0.25\Delta G^\circ_{(i)} = RT \ln a_A^{0.75} A_B^{0.25}$$

If A and B are partially soluble in one another and the compounds A_2B and AB_2 (identified as phases β and γ, respectively) have measurable ranges of nonstoichiometry, the phase diagram is as shown in Fig. 13.12a. Again, if B is appreciably volatile and A is not, then the variation of vapor pressure with composition at the temperature T_1 is as shown in Fig. 13.12b, and the corresponding variation of the Gibbs free energy is as shown in Fig. 13.12c.

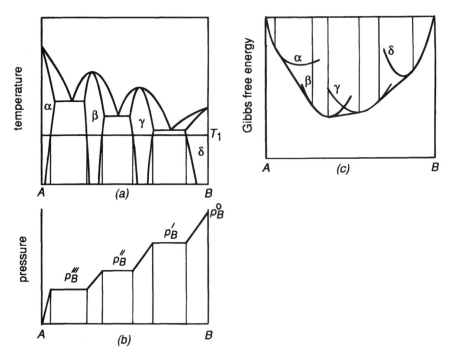

Figure 13.12 (a) The phase diagram for the system A–B. (b) The partial vapor pressure of B at the temperature T_1. (c) The molar Gibbs free energies at the temperature T_1.

Example 1

The phase diagram for the system Ga–GaP is shown in Fig. 13.13. Calculate the partial pressure of phosphorus vapor, p_{P_2}, exerted by the GaP liquidus melt at 1273 K. The standard Gibbs free energy change for the reaction

$$Ga_{(l)} + \frac{1}{2} P_{2(g)} = GaP_{(s)}$$

is

$$\Delta G^\circ = -178{,}800 + 96.2T + 3.1T \ln T - 3.61 \times 10^{-3}T^2 - \frac{1.035 \times 10^5}{T}$$

Thus, at 1273 K,

$$\Delta G^\circ_{1273\,K} = -3.968 \times 10^4 \, J = -8.3144 \times 1273 \ln K_{1273\,K}$$

which gives

$$K_{1273\,K} = 24.97 = \frac{a_{GaP}}{a_{Ga} p_{P_2}^{1/2}}$$

In the above expression, a_{Ga} is the activity of Ga in the liquidus melt with respect to liquid Ga as the standard state, and, as the liquidus melt is in equilibrium with pure solid GaP, the activity of GaP, a_{GaP}, is unity. The variation of the liquidus composition with temperature in the range 1173–1373 K can be expressed as

$$\ln X_P = -\frac{16{,}550}{T} + 9.902$$

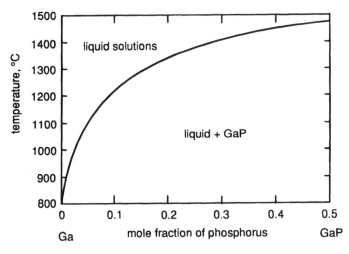

Figure 13.13 The phase diagram for the system Ga–GaP.

which gives the liquidus composition at 1273 K as $X_P = 0.045$. In view of this low solubility of the solute P, it can be assumed that the solvent Ga obeys Raoult's law, in which case the activity of Ga in the liquidus melt is 0.955, and hence the partial pressure of P_2 exerted by the liquidus melt is

$$p_{P_2} = \left(\frac{1}{24.97 \times 0.955} \right)^2$$
$$= 1.76 \times 10^{-3} \text{ atm}$$

Example 2

The phase diagram for the system Mg–Si is shown in Fig. 13.14. Determine the extent to which the phase diagram can be calculated assuming that the liquid solutions exhibit regular solution behavior. Magnesium melts at 921 K and has a Gibbs free energy change on melting of $\Delta G^\circ_{m,\text{Mg}} = 8790 - 9.54T$ J, silicon melts at 1688 K and has a Gibbs free energy change on melting of $\Delta G^\circ_{m,\text{Si}} = 50{,}630 - 30.0T$ J, and the intermetallic compound Mg_2Si melts at 1358 K and has a Gibbs free energy change on melting of $\Delta G_{m,\text{Mg}_2\text{Si}} = 85{,}770 - 63.2T$ J. The standard Gibbs free energy change for the reaction

$$2\text{Mg}_{(l)} + \text{Si}_{(s)} = \text{Mg}_2\text{Si}_{(s)} \tag{i}$$

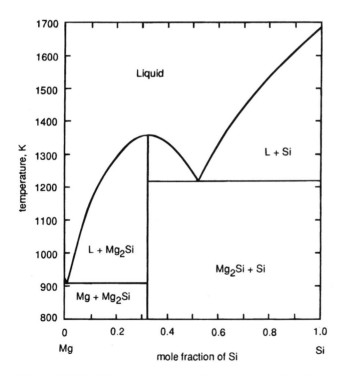

Figure 13.14 The phase diagram for the system Mg–Si.

is

$$\Delta G^\circ_{(i)} = -100{,}400 + 39.3T \text{ J}$$

The Gibbs free energy diagram at 1358 K for the system, using liquid as the standard state for Mg and solid as the standard state for Si, is shown in Fig. 13.15. $\Delta G^\circ_{(i)} = -47{,}030$ J at the melting temperature of Mg_2Si (1358 K), and thus the Gibbs free energy of formation of $Mg_{2/3}Si_{1/3} = -47{,}030/3 = -15{,}676$ J, and this is the length of the line de in Fig. 13.15. At 1358 K $\Delta G^\circ_{m,Si} = 9890$ J, and thus the point b in Fig. 13.15, which represents the free energy of liquid Si relative to solid Si, lies 9890 J above the point a. Consequently the length of the line cd is $9890/3 = 3297$ J, and the length of the line ce is $3297 + 15{,}676 = 18{,}973$ J. Thus the Gibbs free energy of formation of solid $Mg_{2/3}Si_{1/3}$ from liquid Mg and liquid Si is $-18{,}919$ J. However, at the melting temperature of 1358 K $G^\circ_{Mg_2Si,(s)} = G^\circ_{Mg_2Si,(l)}$, and thus the Gibbs free energy of formation of liquid $Mg_{2/3}Si_{1/3}$ from liquid Mg and liquid Si at 1358 K is also 973 J. Thus the line representing the molar Gibbs free energy of formation of melts in the system at 1358 K passes through the point e and from the general expression for the formation of a regular solution

$$\Delta G^M = RT \left(X_{Mg} \ln X_{Mg} + X_{Si} \ln X_{Si} \right) + \Omega X_{Mg} X_{Si}$$

at $X_{Si} = \frac{1}{3}$,

$$-18{,}973 = 8.3144 \times 1358 \left(\frac{2}{3} \ln \frac{2}{3} + \frac{1}{3} \ln \frac{1}{3} \right) + \Omega \frac{1}{3} \cdot \frac{2}{3}$$

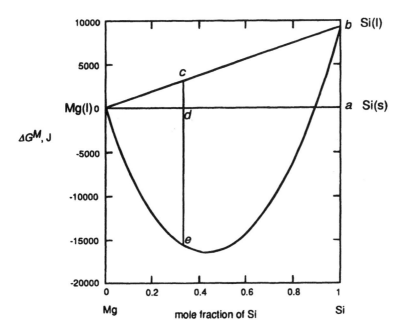

Figure 13.15   Molar Gibbs free energies in the system Mg–Si at 1358 K.

which gives $\Omega = -53,040$ J.

Combination of $\Delta G^\circ_{(i)}$ and the Gibbs free energy change for the melting of Si gives

$$\Delta G^\circ_{(ii)} = -151,030 + 69.3T \text{ J}$$

for the reaction

$$2Mg_{(l)} + Si_{(l)} = Mg_2Si_{(s)} \tag{ii}$$

Thus

$$-151,030 + 69.3T = -RT \ln K$$

$$= -RT \ln \frac{a_{Mg_2Si,(s)}}{a^2_{Mg}a_{Si}}$$

As melts on the Mg_2Si liquidus line are saturated with Mg_2Si, $a_{Mg_2Si,(s)} = 1$, and thus the variations of the activities of Mg and Si with temperature along the Mg_2Si liquidus line are given by

$$-15,030 + 69.3T = 2RT \ln a_{Mg} + RT \ln a_{Si} \tag{iii}$$

In a regular solution

$$RT \ln a_i = RT \ln X_i + \Omega(1 - X_i)^2$$

and thus Eq. (iii) becomes

$$-15,030 + 69.3T = 2RT \ln (1 - X_{Si}) - 2 \times 53,040X^2_{Si} \tag{iv}$$
$$+ RT \ln X_{Si} - 53,040(1 - X_{Si})^2$$

Eq. (iv), which is the equation of the Mg_2Si liquidus, is quadratic and gives two values of X_{Si} at each temperature, with one being the liquidus composition in the Mg–Mg_2Si sub-binary and the other being the liquidus composition in the Mg_2Si–Si sub-binary. Eq. (iv) is drawn as the broken line *abc* in Fig. 13.16.

The Si liquidus line is obtained from Eq. 10.22 as

$$\Delta \overline{G}^M_{Si(l)} = -\Delta G^\circ_{m,Si}$$

i.e.,

$$RT \ln X_{Si} + \Omega(1 - X_{Si})^2 = -50,630 + 30.0T$$

which gives

$$T = \frac{50,630 - 53,040(1 - X_{Si})^2}{30.0 - 8.3144 \ln X_{Si}} \tag{v}$$

Eq. (v) is drawn as the broken line *cd* in Fig. 13.16.

Similarly, the Mg liquidus line is given by

$$RT \ln (1 - X_{Si}) + \Omega X^2_{Si} = -8790 + 9.52T$$

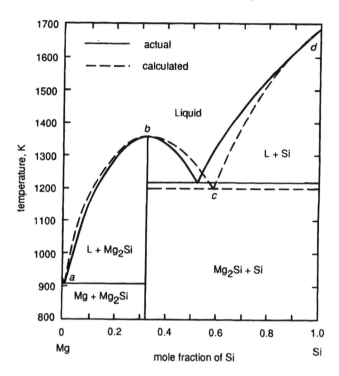

Figure 13.16 Comparison between the calculated and the actual phase diagram for the system Mg–Si.

which gives

$$T = \frac{8790 - 53{,}040X_{Si}^2}{9.52 - 8.3144 \ln (1 - X_{Si})^2} \tag{vi}$$

The calculated diagram shows good agreement with the actual diagram; the calculated eutectic temperature and eutectic composition in the Mg_2Si–Si sub-binary are, respectively, 1200 K and $X_{Si} = 0.58$, which are close to the actual values of 1218 K and $X_{Si} = 0.53$, and the eutectic composition and temperature in the Mg–Mg_2Si sub-binary coincide with the actual values.

13.6 GRAPHICAL REPRESENTATION OF PHASE EQUILIBRIA

13.6.1 Phase Equilibria in the System Mg–Al–O

Consider the phase equilibria in the system Mg–Al–O at 1073 K. At 1073 K liquid Mg and Al are completely miscible in one another, and MgO, Al_2O_3, and the spinel $MgAl_2O_4$ occur as the products of oxidation of the metallic alloys. The stabilities of the oxides are determined by the activities of Al and Mg in the liquid metallic alloys and by the standard Gibbs free energies of formation of the oxides. For

$$2Al_{(l)} + \frac{3}{2}O_{2(g)} = Al_2O_{3(s)} \tag{i}$$

$$\Delta G^{\circ}_{(i),1073 \text{ K}} = -1,323,000 \text{ J} = -RT \ln \frac{a_{Al_2O_3}}{a^2_{Al}p^{3/2}_{O_2}}$$

which gives

$$2 \log a_{Al} + 1.5 \log p_{O_2} - \log a_{Al_2O_3} = -64.39 \tag{ia}$$

For

$$Mg_{(l)} + \frac{1}{2}O_{2(g)} = MgO_{(s)} \tag{ii}$$

$$\Delta G^{\circ}_{(ii),1073 \text{ K}} = -484,300 \text{ J} = -RT \ln \frac{a_{MgO}}{a_{Mg}p^{1/2}_{O_2}}$$

which gives

$$\log p_{O_2} = -2 \log a_{Mg} + \log a_{MgO} - 47.14 \tag{iia}$$

For

$$Mg_{(l)} + 2Al_{(l)} + 2O_{2(g)} = MgAl_2O_{4(s)} \tag{iii}$$

$$\Delta G^{\circ}_{(iii),1073 \text{ K}} = -1,854,000 \text{ J} = -RT \ln \frac{a_{MgAl_2O_4}}{a_{Mg}a^2_{Al}p^2_{O_2}}$$

which gives

$$\log p_{O_2} = -\frac{1}{2} \log a_{Mg} - \log a_{Al} + \log a_{MgAl_2O_4} - 90.24 \tag{iiia}$$

Combination of the reactions given by Eqs. (i), (ii), and (iii) gives

$$MgO_{(s)} + Al_2O_{3(s)} = MgAl_2O_{4(s)} \tag{iv}$$

for which

$$\Delta G^{\circ}_{(iv),1073 \text{ K}} = -46,700 \text{ J} = -RT \ln \frac{a_{MgAl_2O_4}}{a_{MgO}a_{Al_2O_3}}$$

or

$$\log a_{MgO} + \log a_{Al_2O_3} - \log a_{MgAl_2O_4} = -2.273 \tag{iva}$$

The Gibbs free energy diagram for the system MgO–Al_2O_3 is shown in Fig. 13.17. From $\Delta G^{\circ}_{(iv),1073 \text{ K}}$ the Gibbs free energy of formation of $(MgO)_{1/2}(Al_2O_3)_{1/2}$ is $-23,350 \text{ J}$, and the existence of $MgAl_2O_4$ as a stable phase requires that the activities of both MgO and Al_2O_3 have values between $\log a = -2.273$ and $\log a = 0$, the values of which are determined by Eq. (iva). From Eq. (iva) and Fig. 13.17 the logarithm of the activity of Al_2O_3 in $MgAl_2O_4$ (at $a_{MgAl_2O_4} = 1$) which is saturated with MgO

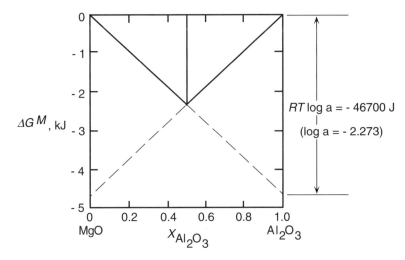

Figure 13.17   Molar Gibbs free energies in the system $MgO–Al_2O_3$ at 1073 K.

(at $a_{MgO} = 1$) is -2.273, and, from symmetry, the logarithm of the activity of MgO in $MgAl_2O_4$ (at $a_{MgAl_2O_4} = 1$) which is saturated with Al_2O_3 (at $a_{Al_2O_3} = 1$) is -2.273. Thus, if the activity of either MgO or Al_2O_3 in the system is less than antilog (-2.273), the spinel of $MgAl_2O_4$ is not stable.

The measured activities of Mg and Al at 1073 K in the system Mg–Al have been fitted by the equations*

$$\log a_{Mg} = \log X_{Mg} - 0.68(1 - X_{Mg})^3$$

and

$$\log a_{Al} = \log(1 - X_{Mg}) - 1.02X_{Mg}^2 + 0.68X_{Mg}^3$$

The phase stability diagram for the system at 1073 K, using $\log p_{O_2}$ and $\log a_{Mg}$ as coordinates, is shown in Fig. 13.18a (the fixing of the activities of O_2 and Mg in the ternary system Mg–Al–O at constant temperature fixes the activity of Al). From Eq. (ii$a$), equilibrium between pure liquid Mg and MgO occurs at $\log p_{O_2} = -47.14$, which is shown as the point a in Fig. 13.18a. The three-phase equilibrium involving the Mg–Al melt, MgO, and the gas phase is determined by Eq. (ii$a$) and is shown as the line ab in Fig. 13.18a. The addition of Al to the liquid alloy decreases a_{Mg} and hence increases the value of the pressure of oxygen required to maintain $a_{MgO} = 1$. Also, in moving along ab from a toward b the activity of $a_{Al_2O_3}$, given by Eq. (i), increases, and at the point b it reaches the value of antilog (-2.273), which with $a_{MgO} = 1$ makes $a_{MgAl_2O_4} = 1$. Thus, the four-phase equilibrium melt–MgO–$MgAl_2O_4$–gas equilibrium occurs at the point b. The three-phase equilibrium involving the

*G. R. Belton and Y. K. Rao, "A Galvanic Study of Activities in Mg–Al Liquid Alloys," *Trans. Met. Soc.* (1969), vol. 245, p. 2189.

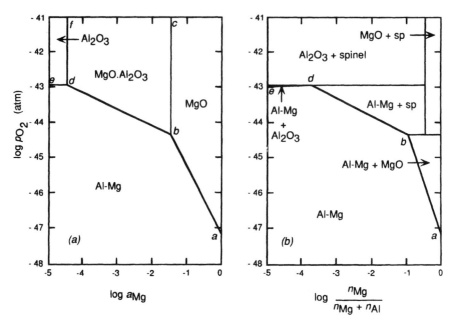

Figure 13.18 (*a*) The phase stability diagram for the system Al–Mg–O at 1073 K. (*b*) The phase diagram for the system Al–Mg–O at 1073 K.

Mg–Al melt, $MgAl_2O_4$, and the gas phase, determined by Eq. (iii*a*), occurs along the line *bd*. In moving along the line from *b* to *d* the dilution of Mg in the melt causes a_{MgO} to decrease from unity at *b* and $a_{Al_2O_3}$ to increase and, at the point *d*, $a_{Al_2O_3} = 1$ and log $a_{MgO} = -2.273$. Thus *d* represents the four-phase equilibrium melt–Al_2O_3–$MgAl_2O_4$–gas. At activities of Mg less than that at *d* the activity of MgO is less than antilog (-2.273), and thus the spinel $MgAl_2O_4$ is not stable. The line *ed* represents the equilibrium involving a melt, Al_2O_3, and a gas phase, given by Eq. (i*a*). However, as $a_{Mg} = 5.75 \times 10^{-5}$ at the point *d*, a_{Al} is virtually unity, and thus the line *ed* is virtually horizontal at log p_{O_2} [obtained from Eq. (i*a*)] $= -42.93$. The lines *df* and *bc* represent, respectively, the three-phase equilibria Al_2O_3–$MgAl_2O_4$–gas and $MgAl_2O_4$–MgO–gas equilibria. The lines in Fig. 13.18*a* identify the fields of stability of Al–Mg liquid alloys, MgO, $MgAl_2O_4$, and Al_2O_3 at 1073 K.

The activity-composition relationships given by Eqs. (v) and (vi) allow the phase stability diagram presented as Fig. 13.18*a* to be converted to the phase diagram shown in Fig. 13.18*b*, in which log p_{O_2} and log $n_{Mg}/(n_{Mg} + n_{Al})$ (where n_{Mg} and n_{Al} are, respectively, the numbers of moles of Mg and Al in the system) are used as coordinates. The line *ab* gives the compositions of the metallic melts saturated with MgO, the line *bd* gives the compositions of the melts saturated with $MgAl_2O_4$, and the line *ed* gives the compositions of the melts saturated with Al_2O_3.

Consider the sequence of oxidation of Al–Mg alloys of $X_{Mg} = 0.333, 0.2$, and 0.01 at 1073 K. With $X_{Mg} = 0.333$ and log $p_{O_2} = -48$, the system exists at the state *a* in Fig. 13.19. When the oxygen pressure is increased to log $p_{O_2} = -45.6$ the sys-

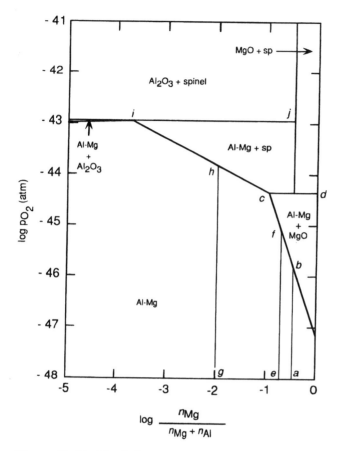

Figure 13.19 The influence of composition and oxygen pressure on the equilibrium states of existence in the system Mg-Al-O.

tem exists at b, in which state the alloy is in equilibrium with MgO. Further increase in the oxygen pressure causes the precipitation of MgO from the melt, which decreases the mole fraction of X_{Mg} in the melt and causes its composition to move along the MgO-saturation line from b toward c. At log $p_{O_2} = -44.3$ the melt of composition c is saturated with MgO, and further increase in the oxygen pressure causes all of the melt of composition c to react with all of the MgO at d to form the spinel $MgAl_2O_4$. A melt of $X_{Mg} = 0.2$ equilibrated with oxygen at log $p_{O_2} = -48$ exists at the state e in Fig. 13.19, and increasing the oxygen pressure to log $p_{O_2} = -45.1$ brings the alloy into equilibrium with MgO at the state f. Further increase in the oxygen pressure to log p_{O_2} to -44.3 causes MgO to precipitate and moves the composition of the melt along the MgO-saturation line from f to c. Further increase in the oxygen pressure causes all of the MgO to react with some of the melt to produce $MgAl_2O_4$ and a melt saturated with $MgAl_2O_4$. With further oxidation $MgAl_2O_4$ is precipitated from the melt, and the composition of the melt is moved along the $MgAl_2O_4$-saturation line from c toward i. Although the precipitation of the spinel from the melt removes Al and Mg atoms from the melt in the ratio $Al/Mg = 2$, the values of X_{Mg} in the melts are low enough that the removal of Al atoms makes a negligible change in X_{Al} but causes a significant decrease in X_{Mg}. At log $p_{O_2} = -43.8$ the melt at the state i is doubly saturated with Al_2O_3 and $MgAl_2O_4$, and further increase in the oxygen pressure causes the disappearance of the melt. Increasing the oxygen pressure exerted on a melt of $X_{Mg} = 0.01$ causes saturation of the melt with $MgAl_2O_4$ at log $p_{O_2} = -43.8$ (the state h), and further increase in the oxygen pressure causes oxidation to proceed as described above.

13.6.2 Phase Equilibria in the System Al–C–O–N Saturated with Carbon

Consider phase stability in the system Al–C–O–N, saturated with carbon at 2000 K, and identify the conditions under which AlN can be contained in a graphite crucible at 2000 K without the formation of Al_4C_3. The solid phases which occur in the quaternary system are AlN, Al_4C_3, Al_4O_4C, and Al_2O_3, and, as the system is saturated with graphite, the minimum number of phases which can co-exist in equilibrium with one another is three (graphite, a gas phase, and a second condensed phase). The number of degrees of freedom available to this three-phase equilibrium is

$$F = C + 2 - P = 4 + 2 - 3 = 3$$

which can be selected as T, p_{O_2}, and p_{N_2}. An isothermal phase stability diagram can thus be constructed using log p_{N_2}, and log p_{O_2} as coordinates. For

$$4Al_{(l)} + 3C_{(s)} = Al_4C_{3(s)} \tag{i}$$

$$\Delta G^{\circ}_{(i),2000\ K} = -74{,}060\ J$$

and thus an equilibrium involving pure liquid Al, graphite, and solid Al_4C_3 does not exist at 2000 K. The activity of Al in carbon-saturated Al_4C_3 is obtained from

$$K_{(i),2000\ K} = \exp\left(\frac{74{,}060}{8.3144 \times 2000}\right) = 85.95 = \frac{a_{Al_4C_3}}{a_{Al}^4 a_C^3} = \frac{1}{a_{Al}^4}$$

which gives $a_{Al} = 0.327$, relative to pure liquid Al. For

$$4AlN_{(s)} + 3C_{(s)} = Al_4C_{3(s)} + 2N_{2(g)} \tag{ii}$$

$$\Delta G^{\circ}_{(ii),2000\ K} = 917{,}900\ J = -2 \times 8.3144 \times 2000 \times 2.303\ \log p_{N_2}$$

which gives log $p_{N_2} = -11.98$ for the equilibrium involving solid graphite, solid AlN, solid Al_4C_3, and a gas phase. The line representing this equilibrium is drawn as ab in Fig. 13.20a. Graphite and AlN are stable relative to graphite and Al_4C_3 in states above the line, and graphite and Al_4C_3 are stable relative to graphite and AlN in states below the line.

For

$$Al_4O_4C_{(s)} + 2C_{(s)} = Al_4C_{3(s)} + 2O_{2(g)} \tag{iii}$$

$$\Delta G^{\circ}_{(iii),2000\ K} = 1{,}333{,}000\ J = -2 \times 8.3144 \times 2000 \times 2.303\ \log p_{O_2}$$

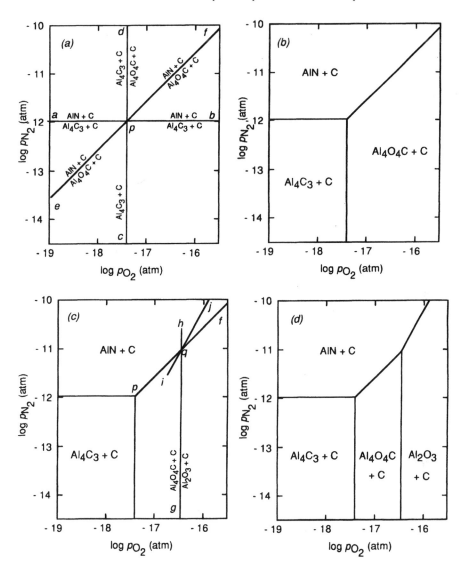

Figure 13.20 Construction of the phase stability diagram for the system Al–C–O–N saturated with carbon at 2000 K.

which gives $\log p_{O_2} = -17.40$ for the equilibrium involving solid graphite, solid Al_4C_3, solid Al_4O_4C, and a gas phase. This is drawn as line cd in Fig. 13.20a. To the left of this line graphite and Al_4C_3 are stable relative to graphite and Al_4O_4C, and to the right of the line the reverse is the case.

For

$$4Al_{(l)} + 2O_{2(g)} + C_{(s)} = Al_4O_4C_{(s)} \tag{iv}$$

$$\Delta G^\circ_{(iv),2000\ K} = -1,407,000\ J$$

and thus, for carbon-saturated Al_4O_4C

$$K_{(iv),2000\,K} = \exp\left(\frac{1,407,000}{8.3144 \times 2000}\right) = 5.58 \times 10^{36} = \frac{1}{a_{Al}^4 p_{O_2}^2}$$

which, with $\log p_{O_2} = -17.40$, gives $a_{Al} = 0.327$. Thus the activity of Al in C-saturated Al_4O_4C has the same value as that in C-saturated AlN. The point of intersection of lines ab and cd is the state in which the five-phase equilibrium involving four solid phases and a gas phase occurs.

For

$$Al_4O_4C_{(s)} + 2N_{2(g)} = C_{(s)} + 2O_{2(g)} + 4AlN_{(s)} \tag{v}$$

$$\Delta G^\circ_{(v),2000\,K} = 415,000\,J = -2 \times 8.3144 \times 2000 \times 2.303 \log \frac{p_{O_2}}{p_{N_2}}$$

which gives

$$\log p_{N_2} = \log p_{O_2} + 5.42$$

for the equilibrium among graphite, solid AlN, solid Al_4O_4C, and a gas phase. This is drawn as line ef in Fig. 13.20a. Carbon-saturated AlN is stable relative to carbon-saturated Al_4O_4C above the line, with the reverse being the case below the line. Inspection of Fig. 13.20a shows the following:

1. In the area below ap and to the left of pc, Al_4C_3 is stable with respect to AlN and Al_4O_4C.
2. In the area above ap and above pf, AlN is stable with respect to Al_4C_3 and Al_4O_4C.
3. In the area below pf and to the right of pc, Al_4O_4C is stable with respect to AlN and Al_4C_3.

Thus, without consideration of the stability of Al_2O_3, the fields of stability of AlN, Al_4C_3, and Al_4O_4C are as shown in Fig. 13.20b.

For

$$2Al_2O_{3(s)} + C_{(s)} = O_{2(g)} + Al_4O_4C_{(s)} \tag{vi}$$

$$\Delta G^\circ_{(vi),2000\,K} = 630,800\,J = -8.3144 \times 2000 \times 2.303 \log p_{O_2}$$

which gives $\log p_{O_2} = -16.47$ for the equilibrium among graphite Al_4O_4C, Al_2O_3, and a gas phase. This is drawn as the line gh in Fig. 20c, and its point of intersection with pf is the state of the five-phase equilibrium among graphite, AlN, Al_4O_4C, Al_2O_3, and a gas phase.

For

$$Al_2O_{3(s)} + N_{2(g)} = 2AlN_{(s)} + 1.5O_{2(g)} \tag{vii}$$

$$\Delta G^\circ_{(vii),2000\,K} = 522,900\,J = -8.3144 \times 2000 \times 2.303 \log \frac{p_{O_2}^{1.5}}{p_{N_2}}$$

which gives

$$\log p_{N_2} = 1.5 \log p_{O_2} + 13.65$$

for the equilibrium given by Eq. (vii). This line, which is independent of the activity of C in the system, is drawn as *ij* in Fig. 13.20c. As the line *pq* represents a stable equilibrium, the other two lines radiating from the point *q* which represent stable equilibria are *qg* and *gj*, and the full phase stability diagram is as shown in Fig. 13.20d. The diagram shows that AlN can be heated in a graphite crucible without forming Al_4C_3 if the pressures of oxygen and nitrogen are such that the thermodynamic state lies in the field of stability of AlN.

13.7 THE FORMATION OF OXIDE PHASES OF VARIABLE COMPOSITION

Fig. 13.21 shows the variation, with composition at some temperature T, of the molar Gibbs free energy for the metal M-oxygen system in which measurable solubility of oxygen in metallic M occurs and the oxides "MO" and "M_3O_4" have variable compositions. Starting with pure M, increasing the pressure of oxygen causes the molar Gibbs free energy to move from *f* along the line *fi* until, at $p_{O_2} = p_{O_2(M/MO)}$, the metal is saturated with oxygen and the metal-saturated "MO" phase of composition M_bO_a appears. If pure metal M and oxygen gas at 1 atm pressure at the temperature T are chosen as the standard states, then

$$\Delta G_{\text{(metal-saturated MO)}} = jk = RT(b \ln a_M + a \ln p_{O_2}^{1/2})$$
$$= RT \ln a_M^b \, p_{O_2}^{1/2a}$$

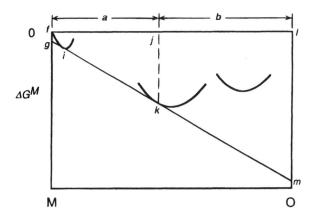

Figure 13.21 The integral Gibbs free energies of the system M–O which forms oxide phases of variable composition and which shows a significant solubility of oxygen in metallic M.

in which, in Fig. 13.21, $fg = RT \ln a_M$ and $lm = RT \ln p_{O_{2(M/MO)}}$. Or, for the reaction

$$bM_{(s)} + \frac{1}{2} aO_{2(g)} = M_bO_{a(s)}$$

$$\Delta G^\circ = jk = RT \ln \left(\frac{a_M^b p_{O_2}^{1/2a}}{a_{M_bOa}} \right)$$

which, if metal-saturated oxide is chosen as the standard state, is identical with the preceding expression.

It is convenient to write the oxidation such that it involves the consumption of an integral number of gram-atoms of oxygen. For example, for the consumption of 1 gram-atom of oxygen,

$$yM + \frac{1}{2} O_2 = M_yO \quad \text{where } y = \frac{b}{a}$$

$$\Delta G^\circ = RT \ln \left(\frac{a_M^y p_{O_2}^{1/2}}{a_{M_yO}} \right)$$

or, for the consumption of 1 gram-mole of oxygen (2 gram-atoms),

$$\Delta G^\circ = RT \ln \left(\frac{a_M^{2y} p_{O_2}}{a_{M_yO}^2} \right)$$

If the solubility of oxygen in the metal is virtually zero, then fg in Fig. 13.21 shrinks to a point, and Fig. 13.21 is redrawn as Fig. 13.22. In this case choosing pure M, oxygen gas at 1 atm pressure and the temperature T and the oxide of composition M_yO as the standard states gives, for the oxidation,

$$yM + \frac{1}{2} O_2 = M_yO$$

$$\Delta G^\circ = RT \ln p_{O_2(M/MO)}^{1/2}$$

where $p_{O_2(M/MO)}$ is the pressure of oxygen required for equilibrium between the metal M and the metal-saturated oxide "MO," given by $lm = RT \ln p_{O_2(M/MO)}$.

If the oxygen pressure is increased to a value greater than $p_{O_2(M/MO)}$, the metal phase disappears, the oxygen content of the "MO" phase increases, the molar Gibbs free energy of the system moves along the line kn, and the activities of M and "MO" vary accordingly. In a classic investigation Darken and Gurry* determined the phase relationships occurring in the system Fe–O by varying the oxygen pressure and temperature and observing the consequential changes in phase and phase composition. Their diagram, drawn for the components FeO and Fe$_2$O$_3$, is shown in Fig. 13.23. Consider the wustite ("FeO") phase field which, at

---

*L. S. Darken and R. W. Gurry, "The System Iron-Oxygen, I: The Wustite Field and Related Equilibria," *J. Am. Chem. Soc.* (1945), vol. 67, p. 1398: "The System Iron-Oxygen, II: Equilibria and Thermodynamics of Liquid Oxide and Other Phases," *J. Am. Chem. Soc.* (1946), vol. 68, p. 798.

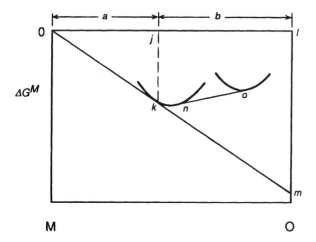

Figure 13.22 The integral Gibbs free energies of the system M–O which forms oxide phases of variable composition and which shows a negligible solubility of oxygen in metallic M.

1100°C, extends from the composition m to the composition n. The variation of a_{Fe} in the wustite phase can be calculated from the experimentally determined variation of the composition of wustite with oxygen pressure using the Gibbs-Duhem equation:

$$X_{Fe}\, d \ln a_{Fe} + X_O\, d \ln a_O = 0$$

i.e.,

$$\log a_{Fe} = -\int \frac{X_O}{X_{Fe}}\, d \log a_O = -\int \frac{X_O}{X_{Fe}}\, d \log p_{O_2}^{1/2}$$

where the upper limit of integration is the oxygen pressure in equilibrium with the wustite composition of interest, and the lower limit is $p_{O_2(Fe/"FeO")}$ (the oxygen pressure at which wustite of composition m is in equilibrium with oxygen-saturated metallic iron), at which composition $a_{Fe} = 1$. Having thus determined the variations of a_{Fe} with composition, the corresponding variations of $a_{"FeO"}$ are determined as follows. If the standard state for oxygen gas is selected as being $p_{O_2(Fe/"FeO")}$ at the temperature of interest, then for

$$y Fe_{(s)} + \frac{1}{2} O_{2(g,\ at\ p_{O_2(Fe/"FeO")})} = Fe_y O_{(s)}$$

as the standard states are in equilibrium with one another, $\Delta G° = 0$, and thus $K = 1$, i.e.,

$$a_{"FeO"} = a_{Fe} a_O \quad \text{where } a_O = \left(\frac{p_{O_2}}{p_{O_2(Fe/"FeO")}} \right)^{1/2}$$

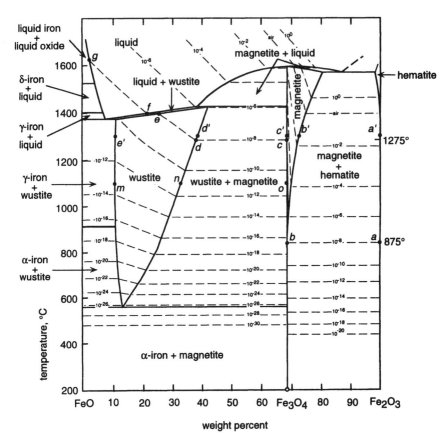

Figure 13.23 The phase diagram for the system FeO–Fe_2O_3 showing the positions of the oxygen (atm) isobars.

or

$$\log a_{\text{"FeO"}} = \log a_{Fe} + \log a_O$$

$$= -\int \frac{X_O}{X_{Fe}} d\log a_O + \int d\log a_O$$

$$= -\int \left(\frac{X_O}{X_{Fe}} - 1 \right) d\log p_{O_2}^{1/2}$$

in which the integration limits are the same as before.

The variations of a_{Fe}, $a_{\text{"FeO"}}$, and a_O across the wustite field at 1100°, 1200°, and 1300°C are shown in Fig. 13.24, and from these variations, the molar Gibbs free energy curve for wustite, kn in Fig. 13.22, can be determined.

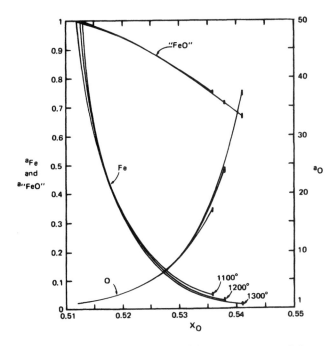

Figure 13.24 The activities of iron, oxygen, and iron-saturated wustite in the wustite phase field at several temperatures.

For a fixed composition, the partial molar heats of solution of metal and oxygen in the wustite can be obtained from the Gibbs-Helmholtz relationship as

$$\Delta \overline{H}_O^M = R \frac{\partial \ln p_{O_2}^{1/2}}{\partial (1/T)}$$

and

$$\Delta \overline{H}_{Fe}^M = R \frac{\partial \ln a_{Fe}}{\partial (1/T)}$$

The variations of $\Delta \overline{H}_O^M$ and $\Delta \overline{H}_{Fe}^M$ with composition are shown in Fig. 13.25. In Fig. 13.24, at the composition of wustite $X_{Fe} = 0.5185$, the activity of Fe is independent of temperature and thus, as shown in Fig. 13.25, the partial molar heat of mixing of Fe at this composition is zero.

At the temperature T, the limit of increase of p_{O_2} above homogeneous stable wustite is $p_{O_2("FeO"/Fe_3O_4)}$, the oxygen pressure at which wustite of composition n is in

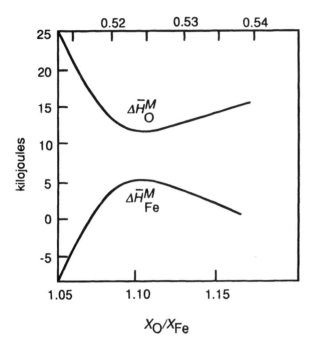

Figure 13.25 The partial molar heats of solution of iron and oxygen in wustite.

equilibrium with magnetite of composition o (in Fig. 13.23), and, if these compositions are chosen as the standard states, then for

$$3\text{"FeO"}_{(\text{saturated with oxygen})} + \frac{1}{2} O_{2(g)} = Fe_3O_{4(\text{saturated with Fe})}$$

$$\Delta G^\circ = RT \ln p_{O_2(\text{"FeO"}/Fe_3O_4)}^{1/2}$$

Fig. 13.23 shows that the composition of wustite in equilibrium with magnetite varies significantly with temperature. Thus the heat of formation of magnetite from wustite cannot be calculated by application of the Gibbs-Helmholtz equation to the variation of $p_{O_2(\text{"FeO"}/Fe_3O_4)}$ with temperature (the Gibbs-Helmholtz partial differential is for constant total pressure and constant composition). However, as the composition of magnetite in equilibrium with wustite is independent of temperature, the change in enthalpy for the reaction

$$3Fe_{(s)} + 2O_{2(g)} = Fe_3O_{4(\text{saturated with Fe})}$$

can be obtained using the Gibbs-Helmholtz relationship, i.e.,

$$K = \frac{1}{a_{Fe}^3 p_{O_2}^2}$$

and thus

$$\Delta H^\circ = -R \frac{d \ln K}{d(1/T)} = R\left[3\frac{d \ln a_{\text{Fe}}}{d(1/T)} + 2\frac{d \ln p_{O_2}}{d(1/T)} \right]$$

In this expression, a_{Fe} and p_{O_2} are the respective values for equilibrium between wustite and magnetite obtained from the data in Fig. 13.24.

Below 550°C, homogeneous wustite is metastable with respect to iron and magnetite. This situation corresponds directly to Fig. 13.11b, in that, below 550°C, the double tangent drawn from pure Fe to the curve for the Gibbs free energy of magnetite lies below the curve for wustite. At 550°C this double tangent becomes a triple tangent and the two-component, four-phase equilibrium is invariant.

The phase diagram at 1 atm total pressure shown in Fig. 13.23 has, superimposed on it, oxygen isobars which trace the loci of variation of equilibrium composition with temperature under a fixed oxygen pressure in the system. For example, consider a small quantity of hematite at room temperature held in a gas reservoir of $p_{O_2} = 10^{-8}$ atm, the volume of which is sufficiently large that any oxygen gas produced by the reduction of the oxide has an insignificant effect on the pressure of oxygen in the gas reservoir. Let the oxide be heated slowly enough that equilibrium with the gas phase is maintained. From Fig. 13.23 it is seen that the oxide remains as homogeneous hematite until 875°C is reached, at which temperature 10^{-8} atm is the invariant partial pressure of oxygen required for the equilibrium

$$2Fe_3O_4 + \frac{1}{2}O_2 = 3Fe_2O_3$$

At 875°C magnetite of composition b is in equilibrium with hematite of composition a, and any increase in temperature upsets the equilibrium toward the magnetite side, with the consequent disappearance of the hematite phase. Further increase in temperature moves the composition of the oxide along the 10^{-8} atm isobar in the magnetite phase field until 1275°C is reached, at which temperature 10^{-8} atm is the invariant partial pressure of oxygen required for the equilibrium

$$3\text{``FeO''} + \frac{1}{2}O_2 = Fe_3O_4$$

At 1275°C wustite of composition d is in equilibrium with magnetite of composition c. Further increase in temperature causes the disappearance of the magnetite phase, and the composition of the solid homogeneous wustite moves along the 10^{-8} atm oxygen isobar until the solidus temperature of 1400°C is reached, in which state solid wustite of composition e melts to form a liquid oxide of composition f at $p_{O_2} = 10^{-8}$ atm. Continued increase in temperature moves the composition of the liquid oxide along the 10^{-8} atm isobar to saturation with iron at the temperature 1635°C, where the liquid oxide has the composition g, and oxygen-saturated liquid iron appears. In this state the equilibrium

$$Fe_{(l)} + \frac{1}{2}O_{2(g)} = \text{``FeO''}_{(l)}$$

is established. An increase in temperature beyond 1635°C causes the disappearance of the liquid oxide phase and a decrease in the dissolved oxygen content of the liquid iron.

Similarly, isothermal reduction of hematite is achieved by decreasing the partial pressure of oxygen in the system. For example, from Fig. 13.23, as 1300°C hematite is the stable phase until the partial pressure of oxygen has been decreased to 1.34×10^{-2} atm, in which state magnetite of composition b' is in equilibrium with hematite of composition a'. Magnetite is then stable until the partial pressure of oxygen has been decreased to 2.15×10^{-8} atm, where wustite of composition d' is in equilibrium with magnetite of composition c'. Wustite is then stable until the partial pressure of oxygen has been decreased to 1.95×10^{-11} atm where solid iron appears in equilibrium with wustite of composition e'. Further decrease in the pressure of oxygen causes the disappearance of the oxide phase.

Fig. 13.26 shows the phase relationships in a plot of log p_{O_2} vs. temperature T and the paths a–g and a'–e' correspond to those in Fig. 13.23. In that Fig. 13.26 does

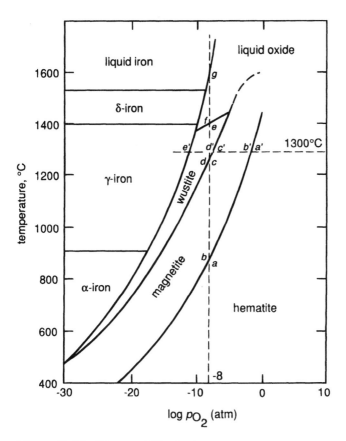

Figure 13.26 Phase stability in the system Fe–Fe$_2$O$_3$ as a function of temperature and log p_{O_2}.

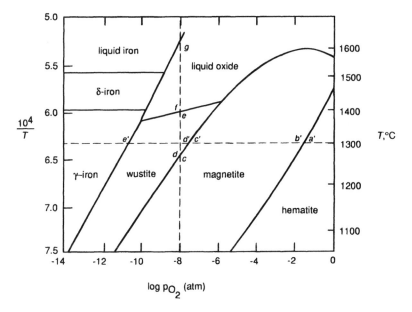

Figure 13.27 Phase stability in the system Fe–Fe$_2$O$_3$ as a function of log p_{O_2} and 1/T.

not contain the compositions of the coexisting oxide phases, it is less useful than the normal composition-temperature phase diagram containing oxygen isobars. Fig. 13.27 shows the phase equilibria on a plot of log p_{O_2} vs. 1/T. In this figure the slope of any invariant three-phase equilibrium line at any temperature, i.e., d log $p_{O_2}/d(1/T)$, equals $\Delta H/(8.3144 \times 2.303)$, where ΔH is the change in enthalpy per mole of oxygen consumed during the oxidation phase change. A linear variation of log p_{O_2} with 1/T occurs in ranges of temperature over which the compositions of the equilibrated phase are constant. Multiplying the abscissa in Fig. 13.27 by 2.303RT and plotting the result against T produces the Ellingham diagram shown in Fig. 13.28, in which, again, the paths a–g and a'–e' correspond to those in Fig. 13.23. Except for the Fe$_3$O$_4$–Fe$_2$O$_3$ line in Fig. 13.28, the lines are drawn for oxidation reactions involving the consumption of 1 mole of oxygen, i.e., of the type

$$Fe_xO_y + O_2 = Fe_xO_{(2+y)}$$

in which the lower oxide of composition Fe$_x$O$_y$ is in equilibrium with the higher oxide of composition Fe$_x$O$_{(2+y)}$. The Fe$_3$O$_4$–Fe$_2$O$_3$ line is hypothetical and applies to the stoichiometric compounds (stoichiometric Fe$_3$O$_4$ contains Fe at a higher activity than does the composition in equilibrium with hematite). In the Ellingham diagram, lines which radiate from the origin ($\Delta G° = 0$, $T = 0$ K) are oxygen isobars. The distinct advantage of Ellingham-type representation of reaction and phase equilibria is its ability to indicate, at a glance, the relative stabilities of a large number of metal-oxygen systems, as was seen in Fig. 12.13.

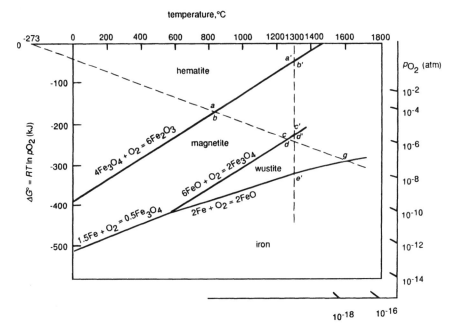

Figure 13.28 Phase stability in the system Fe–Fe_2O_3 as a function of $\Delta G°$ and temperature.

13.8 THE SOLUBILITY OF GASES IN METALS

It is invariably found that molecular gases dissolve in metals as atoms. For example, if pure liquid Ag is brought into contact with oxygen gas at a relatively low pressure the following series of events occur: (1) molecules of O_2 striking the surface of the liquid Ag become adsorbed on the surface, (2) the adsorbed molecules dissociate to form O atoms adsorbed on the surface, and (3) the adsorbed O atoms diffuse from the surface into the bulk melt. The overall reaction can be written as

$$\frac{1}{2}O_{2(g)} = [O]_{\text{in Ag}}$$

and equilibrium is attained when the partial molar Gibbs free energy of oxygen in solution in the liquid Ag, $\overline{G}_O$, is equal to the molar Gibbs free energy of oxygen in the gas phase, $\frac{1}{2}G_{O_2}$. The standard state for a gas dissolved in a metal can be chosen as the 1 wt% standard state, discussed in Sec. 13.3, or as the 1 atom percent (1 at%) standard state, which is the point on the Henry's law line at a mole fraction of solute of 0.01. Thus the standard change in the molar Gibbs free energy which occurs when a gaseous species A_2, at 1 atm pressure and the temperature T, is dissolved in a metal at the concentration $X_A = 0.01$ (1 at%), at the temperature T, according to

$$\frac{1}{2} A_{2(g,\ P=1\text{atm})} = [A]_{(1\ \text{at}\%)} \tag{13.15}$$

is

$$\Delta G^{\circ}_{1\ \text{at}\%} = -RT \ln K_{1\ \text{at}\%} = -RT \ln \frac{[h_A]_{(1\ \text{at}\%)}}{p_{A_2}^{1/2}} \tag{13.16}$$

in which $[h_A]_{(1\ \text{at}\%)}$ is the activity of A in solution in the metal relative to the 1 at% standard state. If the solute obeys Henry's law,

$$K_{1\ \text{at}\%} = \frac{[\text{at}\%\ A]}{p_{A_2}^{1/2}} \tag{13.17}$$

If, however, the 1 wt% standard state is chosen for the solute, then for

$$\frac{1}{2} A_{2(g,P=1\ \text{atm})} = [A]_{(1\ \text{wt}\%)} \tag{13.18}$$

$$\Delta G^{\circ}_{1\ \text{wt}\%} = -RT \ln K_{1\ \text{wt}\%} = -RT \ln \frac{[h_A]_{(1\ \text{wt}\%)}}{p_{A_2}^{1/2}} \tag{13.19}$$

in which $[h_A]_{(1\ \text{wt}\%)}$ is the activity of A in solution relative to the 1 wt% standard state. If Henry's law is obeyed,

$$K_{1\ \text{wt}\%} = \frac{[\text{wt}\%\ A]}{p_{A_2}^{1/2}} \tag{13.20}$$

For the solution of oxygen in liquid Ag in the range 1213–1573 K

$$\Delta G^{\circ}_{1\ \text{at}\%} = -14,310 + 5.44T\ \text{J} \tag{i}$$

A concentration of 1 wt% O in Ag corresponds to a mole fraction of

$$\frac{\dfrac{1}{16}}{\dfrac{1}{16} + \dfrac{99}{107.9}} = 0.0638$$

and thus the change in the molar Gibbs free energy for

$$[O]_{(\text{in Ag. 1 at}\%)} = [O]_{(\text{in Ag. 1 wt}\%)}$$

is

$$\Delta G^{\circ}_{(ii)} = 8.3144T \ln \frac{0.0638}{0.01} = 15.40T\ \text{J} \tag{ii}$$

The sum of $\Delta G^\circ_{1\ at\%}$ and $\Delta G^\circ_{(ii)}$ gives

$$\Delta G^\circ_{1\ wt\%} = -14{,}310 + 20.84T \text{ J} \qquad (iii)$$

For the solution of oxygen in solid Ag in the range 573–1173 K

$$\Delta G^\circ_{1\ at\%} = 49{,}620 - 15.77T \text{ J} \qquad (iv)$$

Oxygen obeys Henry's law in liquid Ag, and thus, from Eqs. (i), (13.16), and (13.17), the solubility of oxygen in liquid silver is

$$\text{at\% O} = p_{O_2}^{1/2} \exp\left(\frac{1721}{T} - 0.654\right) \qquad (v)$$

and, from Eqs. (iv), (13.16), and (13.17), the solubility of oxygen in solid Ag is

$$\text{at\% O} = p_{O_2}^{1/2} \exp\left(-\frac{5967}{T} + 1.90\right) \qquad (vi)$$

The phase equilibria in the system Ag–O at an oxygen pressure of 1 atm are shown in Fig. 13.29. Eq. (i) gives $\Delta H^\circ_{1\ at\%}$ for the change of state

$$\frac{1}{2}O_{2(g,P=1\ atm)} = [O]_{(1\ at\%\ in\ liquid\ Ag)}$$

as $-14{,}310$ J. Thus, as the enthalpy change is negative, decreasing the temperature causes the equilibrium to shift to the right, with the consequence that, at constant oxygen pressure, the solubility of oxygen in liquid Ag increases with decreasing temperature. From Eq. (v) the maximum solubility of O in liquid Ag in Fig. 13.29 is 2.14 at% at 940°C (the state b). Eq. (vi) gives the solubility of O in solid Ag at 940°C as 0.049 at% (the state d), and thus the transformation

$$Ag_{(l,2.14\ at\%\ O)} \rightarrow Ag_{(s,0.049\ at\%\ O)} + O_{2(g,P=1\ atm)}$$

occurs at 940°C. The evolution of oxygen during the freezing of oxygen-containing liquid Ag causes the phenomenon of "spitting" during which droplets of liquid silver are ejected from the freezing mass. Eq. (iv) gives $\Delta H^\circ_{1\ at\%}$ for the change of state

$$\frac{1}{2}O_{2(g,P=1\ atm)} = O_{1\ at\%\ in\ solid\ Ag}$$

as 49,620 J, which being positive, requires that the solubility of O in solid Ag decreases with decreasing temperature. From Eq. (vi) the solubility of O in solid Ag at an oxygen pressure of 1 atm decreases from 0.049 at% at 940°C to 1.7×10^{-5} at% at 190°C (the point e in Fig. 13.29). For the reaction

$$2Ag_{(s)} + \frac{1}{2}O_{2(g)} = Ag_2O_{(s)}$$

$$\Delta G^\circ = -30{,}540 + 66.11T \text{ J}$$

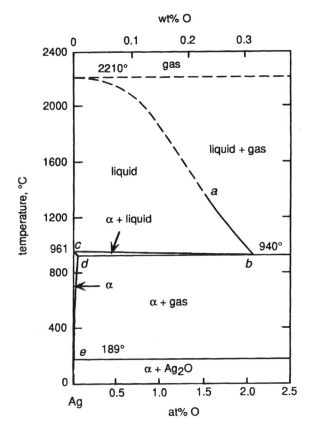

Figure 13.29 Phase equilibria in the system Ag–O at an oxygen pressure of 1 atm.

The temperature at which $p_{O_2,eq} = 1$ atm is thus $30,540/66.11 = 462$ K (189°C), and the three-phase invariant equilibrium is shown at 189°C in Fig. 13.29. The solubility of oxygen in Ag is shown as a function of temperature and oxygen pressure in Fig. 13.30. On a plot of log (at% O) versus inverse temperature the oxygen isobars in a single phase field are parallel lines, the slopes of which are determined by the molar heats of solution of oxygen. From Eq. (v), the slope of the lines in the phase field of liquid silver is 1721, and from Eq. (vi), the slope in the phase field of solid silver is −5967. The points b and d on the 1 atm isobar at 940°C correspond with the points b and d in Fig. 13.29. Increasing the oxygen pressure decreases the temperature at which the equilibrium involving liquid Ag, solid Ag, and O_2 gas occurs and increases the temperature at which the equilibrium involving solid Ag, Ag_2O, and O_2 gas occurs. The two temperatures coincide at 508°C when the oxygen pressure is 414 atm (the state A in Fig. 13.30). This four-phase equilibrium in a binary system has zero degrees of freedom.

Recasting Eq. (13.17) as

$$[\text{at\% A}] = k(T)p_{A_2}^{1/2} \tag{13.21}$$

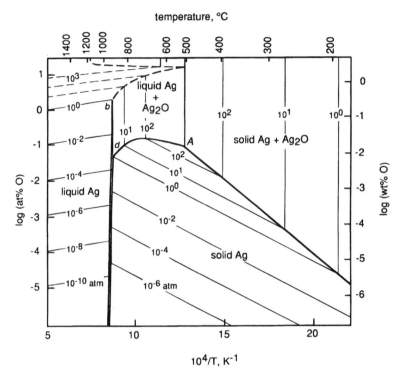

Figure 13.30   The solubility of oxygen in silver as a function of oxygen pressure and temperature.

gives an equation known as Sieverts's law, and the temperature-dependent constant in Eq. (13.21), $k(T)$, which is known as Sieverts's constant, is evaluated as the concentration of A in the metal equilibrated, at the temperature T, with gaseous A_2 at 1 atm pressure. Sieverts measured the solubility of oxygen in liquid Ag in 1907.*

13.9 SOLUTIONS CONTAINING SEVERAL DILUTE SOLUTES

The behavior of a dilute solute in a binary solution is determined by the nature and magnitude of the interactions between the solute and solvent atoms. However, when a second dilute solute is added, three types of interaction occur, namely solvent-solute I, solvent-solute II, and solute I-solute II, and the thermodynamic behavior of the system is determined by the relative magnitudes of the three types of interaction.

---

*A. Sieverts and J. Hagenacker, "Über die Loslichkeit von Wasserstoff und Sauerstoff in festem und geschmolzenem Silber," *Z. phys. Chem.* (1907), vol. 68, p. 115.

Consider the exposure of liquid iron to a gaseous mixture of hydrogen and oxygen. The equilibrium

$$H_{2(g)} + \frac{1}{2}O_{2(g)} = H_2O_{(g)} \tag{i}$$

is established in the gas phase, and thus

$$p_{H_2}p_{O_2}^{1/2} = \frac{p_{H_2O}}{K_{(i)}}$$

As both hydrogen and oxygen have some limited solubility in liquid iron, both gases will dissolve atomically until their respective activities in the iron, with respect to the 1 atm pressure standard state, equal the respective partial pressures in the gas phase. Alternatively, with respect to the 1 wt% in Fe standard state,

$$\frac{1}{2}O_{2(g)} = [O]_{(1 \text{ wt\% in Fe})} \tag{ii}$$

and

$$\frac{1}{2}H_{2(g)} = [H]_{(1 \text{ wt\% in Fe})} \tag{iii}$$

for which

$$h_{O(1 \text{ wt\%})} = K_{(ii)}p_{O_2}^{1/2}$$

and

$$h_{H(1 \text{ wt\%})} = K_{(iii)}p_{H_2}^{1/2}$$

equilibrium in the metal phase is given as

$$h_{H(1 \text{ wt\%})}^2 h_{O(1 \text{ wt\%})} = \frac{K_{(iii)}^2 K_{(ii)}}{K_{(i)}} p_{H_2O}$$

or

$$f_{H(1 \text{ wt\%})}^2 f_{O(1 \text{ wt\%})}[\text{wt\% H}]^2[\text{wt\% O}] = \frac{K_{(iii)}^2 K_{(ii)}}{K_{(i)}} p_{H_2O}$$

The solubilities of H and O (expressed as weight percentages) are thus determined by the values of the activity coefficients of H and O, and the questions to be answered are

1. How is the activity coefficient of O in Fe influenced by the presence of H?
2. How is the activity coefficient of H in Fe influenced by the presence of O?

This problem is dealt with by the introduction of interaction coefficients and interaction parameters.

In the binary A–B, the activity of B in dilute solution with respect to the Henrian standard state is given by

$$h_B = f_B^B X_B$$

If, holding the concentration of B constant, the addition of a small amount of C changes the value of the activity coefficient of B to f_B, then the difference between f_B and f_B^B is quantified by the expression

$$f_B = f_B^B f_B^C \qquad (13.22)$$

where f_B^C is called the interaction coefficient of C on B and is a measure of the effect, on the behavior of B, of the presence of a specific concentration of C, at the same concentration of B. Similarly, if a small amount of D is added to the A–B solution, as a result of which the value of the activity coefficient of B changes from f_B^B to f_B, then

$$f_B = f_B^B f_B^D$$

Now consider the system A–B–C–D. Mathematical analysis of such a system is possible only if f_B^D is independent of the concentration of C and if f_B^C is independent of the concentration of D. Consider that the interaction coefficient of the solute i on the solute j is independent of the other solutes present, in which case the interaction coefficients may be combined by means of a Taylor's expansion of $\ln f_i$ as a function of the concentrations of the solutes, e.g., for the binary system A–B in which A is the solvent,

$$\ln f_B = \text{some function of the mole fraction of B}$$

$$= \ln f_B^\circ + \left(\frac{\partial \ln f_B}{\partial X_B} \right) X_B + \frac{1}{2} \left(\frac{\partial^2 \ln f_B}{\partial X_B^2} \right) X_B^2 + \cdots$$

In this expression the partial derivatives are the limiting values reached as $X_B \to 0$ For the multicomponent system A–B–C–D

$$\ln f_B = \text{some function of the mole fractions of B, C, and D}$$

$$= \ln f_B^\circ + \left(\frac{\partial \ln f_B}{\partial X_B} X_B + \frac{\partial \ln f_B}{\partial X_C} X_C + \frac{\partial \ln f_B}{\partial X_D} X_D \right)$$

$$+ \left(\frac{1}{2} \frac{\partial^2 \ln f_B}{\partial X_B^2} X_B^2 + \frac{\partial^2 \ln f_B}{\partial X_B \partial X_C} X_B X_C + \cdots \right)$$

in which, again, the partial derivatives are the limiting values as the mole fractions of the solutes approach zero. At very low concentration the terms containing the products of mole fractions are small enough to be ignored, and also, choice of the Henrian standard state makes $f_B^\circ = 1$. Thus

$$\ln f_B = \frac{\partial \ln f_B}{\partial X_B} X_B + \frac{\partial \ln f_B}{\partial X_C} X_C + \frac{\partial \ln f_B}{\partial X_D} X_D$$

$$= \varepsilon_B^B X_B + \varepsilon_B^C X_C + \varepsilon_B^D X_D \qquad (13.23)$$

where

$$\varepsilon_j^i = \frac{\partial \ln f_j}{\partial X_i}\bigg|_{X_i \to 0}$$

is called the interaction parameter of i on j and is obtained as the limiting slope of a plot of $\ln f_j$ against X_i at constant X_j. ε_j^i and ε_i^j are related as follows. For the general system

$$\frac{\partial^2 G}{\partial n_i \partial n_j} = \frac{\partial \overline{G}_i}{\partial n_j} = \frac{\partial \overline{G}_j}{\partial n_i}$$

and as $\partial \overline{G}_i = RT \, \partial \ln a_i = RT \, \partial \ln f_i$, then

$$\frac{\partial \ln f_j}{\partial n_i} = \frac{\partial \ln f_i}{\partial n_j}$$

and thus

$$\varepsilon_i^j = \varepsilon_j^i \tag{13.24}$$

It is often more convenient to consider the concentrations of the solutes in terms of weight percentages and to use logarithms to the base 10, in which case Eq. (13.23) becomes

$$\log f_B = \frac{\partial \log f_B}{\partial \text{ wt\%B}} \text{wt\% B} + \frac{\partial \log f_B}{\partial \text{ wt\%C}} \text{wt\% C} + \frac{\partial \log f_B}{\partial \text{ wt\%D}} \text{wt\%D} \tag{13.25}$$

$$= e_B^B \text{ wt\%B} + e_B^C \text{ wt\%C} + e_B^D \text{ wt\%D}$$

Multiplying Eq. (13.25) by 2.303 and comparing, term by term, with Eq. (13.23) gives

$$\varepsilon_B^i X_i = 2.303 \, e_B^i \text{ wt\%}i$$

and as, at small concentrations of B and i,

$$X_i \sim \frac{\text{wt\%}i \cdot \text{MW}_A}{100 \text{MW}_i}$$

then

$$e_B^i = \frac{1}{230.0} \frac{\text{MW}_A}{\text{MW}_i} \varepsilon_B^i$$

and

$$e_B^i = \frac{\text{MW}_B}{\text{MW}_i} e_i^B$$

Pehlke and Elliott* have determined that nitrogen, dissolved in liquid iron at 1600°C, obeys Sieverts's law according to

$$[\text{wt\% N}] = k p_{\text{N}_2}^{1/2} \tag{iv}$$

---

*R. Pehlke and J. F. Elliott, "Solubility of Nitrogen in Liquid Iron Alloys, I: Thermodynamics," *Trans. Met. Soc. AIME* (1960), vol. 218, p. 1088.

where $k = 0.045$ at 1873 K, and they have measured the effects of the presence of a second dilute solute on the thermodynamics of nitrogen dissolved in liquid iron. These systems are particularly amenable to experimental study because of the ease with which the activity of nitrogen can be controlled by the gas phase. The interaction parameters, e_N^i, are determined by maintaining the nitrogen in the melt at constant activity and measuring the variation in the solubility of nitrogen with concentration of the second solute. As nitrogen in liquid iron obeys Henry's (Sieverts's) law, $f_N^N = 1$, and thus e_N^N (and ε_N^N) are zero. Consequently, the first terms in Eqs. (13.23) and (13.25) are zero. If the addition of a second solute X to the Fe–N binary (equilibrated with a fixed p_{N_2}) causes a change in the dissolved nitrogen content from [wt% N]$_{Fe-N}$ to [wt% N]$_{Fe-N-X}$. Then, from Eq. (iv),

$$k = \frac{[wt\% \text{ N}]_{\text{in Fe-N}}}{p_{N_2}^{1/2}} = \frac{f_N^X[wt\% \text{ N}]_{\text{in Fe-N-X}}}{p_{N_2}^{1/2}}$$

and thus f_N^X is obtained experimentally as

$$f_N^X = \left(\frac{[wt\% \text{ N}]_{\text{in Fe-N}}}{[wt\% \text{ N}]_{\text{in Fe-N-X}}} \right)_{T, p_{N_2}}$$

The variation of [wt% N] with [wt% X] is shown for several second solutes in Fig. 13.31, and the corresponding variation of log f_N^X with [wt% X] is shown in Fig. 13.32. The values of e_N^X are obtained as the slopes of the linear portions of the lines in Fig. 13.32.

Thus in a multicomponent liquid iron alloy containing several solutes including nitrogen, if the effect of any one solute on f_N is independent of the presence of any other solute, then the total effect of the solutes on f_N is the sum of their individual effects, and if log f_N^X is a linear function of [wt% X], then f_N is given by Eq. (13.25). If, however, the concentrations of X are higher than the limits of linear variation of log f_N^X with [wt% X], then a graphical solution is required. In these cases the value of log f_N^X for each value of [wt% X] is read from the graph (Fig. 13.32), and log f_N is obtained as

$$\log f_N = \sum_X \log f_N^X$$

or

$$f_N = \prod_X f_N^X$$

Fig. 13.32 indicates that, as a general rule, e_N^X is a negative quantity when X forms a nitride which is more stable than iron nitride, and that the order of increasing magnitude of $|e_N^X|$ follows the order of increasing magnitude of the Gibbs free energy of formation of the nitride of X. Similarly, e_N^X is a positive quantity when X has a greater affinity for iron than either X has for N or iron has for nitrogen. The values of the interaction coefficients for several elements in dilute solution in iron at 1600°C are listed in Table 13.1.

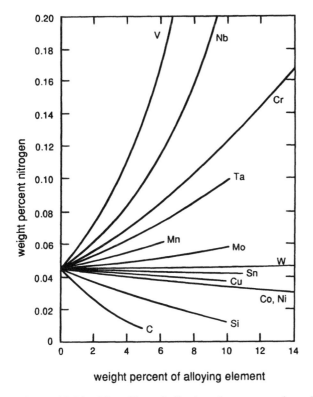

Figure 13.31 The effect of alloying elements on the solubility of nitrogen at 1 atm pressure in liquid binary iron alloys at 1600°C.

Example 1

In view of the introduction of interaction parameters, the example of the Si–O equilibrium in liquid Fe, discussed in Sec. 13.3, can now be reexamined. In this example it was determined that, for

$$Si_{(1 \text{ wt\% in Fe})} + O_{2(g)} = SiO_{2(s)}$$

$$\Delta G° = -833,400 + 229.5T \text{ J}$$

For $\frac{1}{2}O_{2(g)} = [O]_{(1 \text{ wt\% in Fe})}$

$$\Delta G° = -111,300 - 6.41T \text{ J} \tag{i}$$

and thus, for

$$Si_{(1 \text{ wt\% in Fe})} + 2O_{(1 \text{ wt\% in Fe})} = SiO_{2(s)}$$

$$\Delta G° = -610,800 + 242.32T \text{ J} \tag{ii}$$

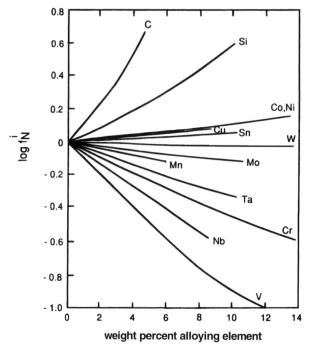

Figure 13.32 The activity coefficients of nitrogen in binary iron alloys at 1600°C.

Table 13.1 Interaction coefficients for dilute solutions of elements dissolved in liquid iron at 1600°C

		Element (j)										
		Al	C	Co	Cr	H	Mn	N	Ni	O	P	S	Si
Element (i)	Al	4.8	11	—	—	(34)	—	(0.5)	—	−160	—	4.9	6
	C	(4.8)	22	1.2	−2.4	(72)	—	(11.1)	1.2	(−9.7)	—	9	10
	Co	—	(6)	—	—	(11)	—	(4.7)	—	(2.6)	—	—	—
	Cr	—	(−10)	—	—	(−11)	—	(−16.6)	—	(−13)	—	(−3.55)	—
	H	1.3	6.0	0.18	−0.22	0	−0.14	—	0	—	1.1	0.8	2.7
	Mn	—	—	—	—	(−7.7)	—	(−7.8)	—	(0)	—	(−4.3)	(0)
	N	0.3	13	1.1	−4.5	—	−2	0	1	5.0	5.1	1.3	4.7
	Ni	—	(5.9)	—	—	(0)	—	(4.2)	0	(2.1)	—	(0)	(1.0)
	O	−94	−13	0.7	−4.1	—	0	(5.7)	0.6	−20	7.0	−9.1	−14
	P	—	—	—	—	(34)	—	(11.3)	—	(13.5)	—	(4.3)	(9.5)
	S	5.8	(24)	—	−2.2	(26)	−2.5	(3.0)	0	(−18)	4.5	−2.8	6.6
	Si	(6.3)	24	—	—	(76)	0	(9.3)	0.5	(−25)	8.6	(5.7)	32

Some interaction coefficients $e_i^j \times 10^2$ for dilute solutions of elements dissolved in liquid iron at 1600°C. Values in parentheses are calculated from $e_i^j = (MW_i/MW_j)e_j^i$. (From J. F. Elliott, M. Gleiser, and V. Ramakrishna, *Thermochemistry for Steelmaking*, vol. 2, Addison-Wesley, Reading, MA, 1963.)

From Eq. (i) at 1600°C,

$$\frac{h_{O(1 \text{ wt\%})}}{p_{O_2}^{1/2}} = 2.746 \times 10^3 \tag{iii}$$

and, from Eq. (ii) at 1600°C,

$$\frac{a_{SiO_2}}{h_{Si(1 \text{ wt\%})}h_{O(1 \text{ wt\%})}^2} = 2.380 \times 10^4 \tag{iv}$$

Thus with $p_{O_2} = 5.57 \times 10^{-12}$ atm and $a_{SiO_2} = 1$, Eq. (iii) gives

$$h_{O(1 \text{ wt\%})} = 6.48 \times 10^{-3} \tag{v}$$

and Eq. (iv) gives

$$h_{O(1 \text{ wt\%})}^2 h_{Si(1 \text{ wt\%})} = 4.0 \times 10^{-5} \tag{vi}$$

Division of Eq. (vi) by $h_{O(1 \text{ wt\%})}^2$ from Eq. (v) gives

$$h_{Si(1 \text{ wt\%})} = 1 \tag{vii}$$

In the previous treatment the assumption that Si obeys Henry's law leads to the conclusion that

$$h_{Si(1 \text{ wt\%})} = \text{wt\% Si} = 1$$

At 1600°C, from Table 13.1,

$$e_O^{Si} = -0.14 \quad e_O^O = -0.2$$
$$e_{Si}^O = -0.25 \quad e_{Si}^{Si} = 0.32$$

Thus, from Eq. (v)

$$\log f_O + \log[\text{wt\% O}] = \log (6.48 \times 10^{-3})$$

or

$$-0.2 \times [\text{wt\% O}] - 0.14 \times [\text{wt\% Si}] + \log[\text{wt\% O}] = -2.188 \tag{viii}$$

and from Eq. (vii)

$$\log f_{Si} + \log[\text{wt\% Si}] = \log (1)$$

or

$$0.32 [\text{wt\% Si}] - 0.25[\text{wt\% O}] + \log[\text{wt\% Si}] = 0 \tag{ix}$$

Computer solution of Eqs. (viii) and (ix) gives

$$[\text{wt\% Si}] = 0.631 \quad \text{and} \quad [\text{wt\% O}] = 0.00798$$

In the example in Sec. 13.3, in which the effect of dissolved oxygen was ignored and it was assumed that $f_{Si} = 1$, the equilibrium weight percentage of Si in iron when $a_{SiO_2} = 1$ and $p_{O_2} = 5.57 \times 10^{-12}$ atm was 1.0. It is of interest to determine which of the two initial assumptions, (1) that $e_{Si}^{Si} = 0$, and (2) $e_{Si}^O = e_O^{Si} = 0$, contributes more to the error in the initial calculation. Use $e_{Si}^{Si} = 0.32$ and assume that e_{Si}^{O} and e_{O}^{Si} are zero. From Eq. (ix)

$$0.32[wt\% \text{ Si}] + \log [wt\% \text{ Si}] = 0$$

which gives [wt% Si] = 0.629, and from Eq. (viii)

$$-0.20 \times [wt\% \text{ O}] - 0.24 \times 0.629 + \log [wt\% \text{ O}] = -2.188$$

which gives [wt% O] = 0.00797. The error introduced by ignoring the interaction between Si and O in solution in Fe is thus seen to be negligible in comparison with that introduced by assuming that Si obeys Henry's law over some initial range of composition.

Example 2

Calculate the equilibrium oxygen content of an Fe–C–O alloy which, at 1600°C, contains 1 wt% C and is under a pressure of 1 atm of CO:

For $C_{(gr)} + \frac{1}{2}O_{2(g)} = CO_{(g)}$, $\Delta G° = -111,700 - 87.65T$ J
For $C_{(gr)} = C_{(1 \text{ wt\% in Fe})}$, $\Delta G° = 22,600 - 42.26T$ J
For $\frac{1}{2} O_{2(g)} = O_{(1 \text{ wt\% in Fe})}$, $\Delta G° = -111,300 - 6.41T$ J
Thus, for

$$C_{(1 \text{ wt\%})} + O_{(1 \text{ wt\%})} = CO_{(g)}, \quad \Delta G° = -23,000 - 38.98T \text{ J}$$

Therefore, $\Delta G°_{1873 \text{ K}} = -96,010$ J and

$$\frac{p_{CO}}{h_C h_O} = 476$$

Thus

$$h_C h_O = f_C[wt\% \text{ C}] f_O[wt\% \text{ O}] = 2.1 \times 10^{-3} p_{CO}$$

At 1600°C,

$$e_C^C = 0.22$$
$$e_O^O = -0.2$$
$$e_C^O = -0.097$$
$$e_O^C = -0.13$$

Thus for 1 wt% C and $p_{CO} = 1$ atm,

$$\log [wt\% \text{ O}] - 0.297 [wt\% \text{ O}] = -2.768$$

solution of which gives [wt% O] = 0.00171. If all of the interaction parameters had been ignored, the weight percentage of O would have been calculated as 0.00210.

Example 3

The partial pressure of hydrogen in the atmosphere is such that an Fe–C–Ti melt containing 1 wt% C and 3 wt% Ti contains 5 parts per million (by weight) of hydrogen at 1600°C. Calculate the vacuum which is required to decrease the hydrogen content of the melt to 1 ppm, given that $e_H^{Ti} = -0.08$, that $e_H^C = 0.06$, and that hydrogen in pure iron obeys Henry's law up to a solubility of 0.0027 wt% under a pressure of 1 atm of hydrogen at 1600°C.

For the equilibrium between gaseous hydrogen and dissolved H, given as $\frac{1}{2}H_{2(g)} = [H]_{(1\ wt\%\ in\ Fe)}$,

$$K = \frac{f_{H(1\ wt\%)}[wt\%\ H]}{p_{H_2}^{1/2}}$$

In pure iron, as H obeys Henry's law, $f_{H(1\ wt\%)} = 1$, and thus

$$K_{1873\ K} = 0.0027$$

Thus

$$\log f_{H(1\ wt\%)} + \log[wt\%\ H] - \frac{1}{2}\log p_{H_2} = \log 0.0027$$

But

$$\log f_{H(1\ wt\%)} = e_H^H[wt\%\ H] + e_H^{Ti}[wt\%\ Ti] + e_H^C[wt\%\ C]$$

As $f_{H(1\ wt\%)} = 1$, $e_H^H = 0$, and hence, at 1600°C,

$$e_H^{Ti}[wt\%\ Ti] + e_H^C[wt\%\ C] + \log[wt\%\ H] - \frac{1}{2}\log p_{H_2} = \log 0.0027$$

When $[wt\%\ H] = 5 \times 10^{-4}$,

$$\log p_{H_2} = 2 \times [(-0.08 \times 3) + (0.06 \times 1) + \log(5 \times 10^{-4}) - \log 0.0027]$$
$$= -1.825$$

which gives $p_{H_2} = 0.015$ atm. Similarly, when $[wt\%\ H] = 1 \times 10^{-4}$, $p_{H_2} = 6 \times 10^{-4}$ atm. Thus

$$[wt\%\ H] = 5\ ppm \quad when\ p_{H_2} = 0.015\ atm \quad and\ P_{(total)} = 1\ atm$$

and so

$$[wt\%\ H] = 1\ ppm \quad when\ p_{H_2} = 0.0006\ atm$$

and

$$P_{total} = \frac{0.0006}{0.015} = 0.04\ atm$$

Thus, in order to achieve the desired decrease in the content of dissolved H, the total pressure must be decreased from 1 to 0.04 atm.

13.10 SUMMARY

1. Reaction equilibrium in the reaction

    $$aA + bB = cC + dD$$

    is established when the reaction has proceeded to the extent that

    $$a\overline{G}_A + b\overline{G}_B = c\overline{G}_C + d\overline{G}_D$$

    i.e., ΔG for the reaction is zero.

2. The state of reaction equilibrium is determined by the standard Gibbs free energy change for the reaction, $\Delta G°$, via

    $$\Delta G° = -RT \ln K$$

    where K is the equilibrium constant for the reaction given by the quotient of the activities of the reactants and products at reaction equilibrium, i.e.,

    $$K = \frac{a_C^c a_D^d}{a_A^a a_B^b}$$

3. The Raoultian standard state of a thermodynamic component is the pure component in its stable state of existence at the temperature of interest. The Henrian standard state is obtained from consideration of Henry's law, which, strictly being a limiting law obeyed by the solute B at infinite dilution, is expressed as

    $$\frac{a_B}{X_B} \to k_B \quad \text{as } X_B \to 0$$

    where a_B is the activity of B in the solution with respect to the Raoultian standard state and k is the Henry's law constant at the temperature T. Alternatively,

    $$\frac{a_B}{X_B} \to \gamma_B° \quad \text{as } X_B \to 0$$

    where $\gamma_B°(= k_B)$ is the constant activity coefficient which quantifies the difference between Raoultian and Henrian solution behavior of B. If the solute obeys Henry's law over a finite range of composition, then, over this range

    $$a_B = \gamma_B° X_B$$

    The Henrian standard state is obtained by extrapolating the Henry's law line to $X_B = 1$, and the activity of B in the Henrian standard state with respect to the Raoultian standard state having unit activity is

    $$a_B = \gamma_B°$$

    The activity of B in a solution with respect to the Henrian standard state having unit activity is given by

    $$h_B = f_B X_B$$

where h_B is the Henrian activity and f_B is the Henrian activity coefficient. In the range of composition in which B obeys Henry's law, $f_B = 1$.

The 1 weight percent standard state is defined as

$$\frac{h_{B(1\ wt\%)}}{wt\%B} \to 1 \quad \text{as } wt\%B \to 0$$

and is located at that point on the Henry's law line which corresponds to a concentration of 1 weight percent B. With respect to the 1 weight percent standard state having unit activity, the activity of B, $h_{B(1\ wt\%)}$, is given by

$$h_{B(1\ wt\%)} = f_{B(1\ wt\%)} wt\%B$$

where $f_{B(1\ wt\%)}$ is the 1 weight percent activity coefficient. The activities are related via

$$a_B = h_B \gamma_B^\circ$$

and

$$a_B = f_{B(1\ wt\%)} \cdot wt\%B \cdot \gamma_B^\circ \cdot \frac{MW_{solvent}}{100MW_B}$$

4. The Gibbs phase rule is

$$F = C + 2 - P$$

in which C is the number of components in the system and F is the number of degrees of freedom available to the equilibrium involving P phases. With R independent reaction equilibria involving N species, the Gibbs phase rule is

$$F = (N - R) + 2 - P$$

where $C = N - R$

5. In a solution of solvent A and several dilutes solutes B, C, and D,

$$\ln f_B = \varepsilon_B^B X_B + \varepsilon_B^C X_C + \varepsilon_B^D X_D$$

where

$$\varepsilon_j^i = \left. \frac{\partial \ln f_j}{\partial X_i} \right|_{X_i \to 0}$$

is the interaction parameter of i on j. The interaction parameters are related to one another by

$$\varepsilon_i^j = \varepsilon_j^i$$

If the concentrations of the dilute solutes are expressed in weight percent, then

$$\log f_B = e_B^B\, wt\%B + e_B^C\, wt\%C + e_B^D\, wt\%D$$

where

$$e_i^j = \frac{\partial \log f_j}{\partial \text{ wt\% } i}\bigg|_{\text{wt\% } i \to 0}$$

The two interaction parameters of i and j are related by

$$e_j^i = \frac{\text{MW}_j}{\text{MW}_i} e_i^j$$

Also

$$\varepsilon_j^i X_i = 2.303 e_j^i \text{wt\% } i$$

13.11 NUMERICAL EXAMPLES

Example 1

100 grams of silica and 100 grams of graphite are placed in a rigid vessel of volume 20 liters, which is evacuated at room temperature and then heated to 1500°C, at which temperature the quartz and graphite react to form SiC. Calculate

1. The equilibrium partial pressures of CO and SiO in the vessel at 1500°C
2. The mass of SiC formed
3. The mass of graphite consumed to form CO and SiC

The equilibrium attained in the vessel is best seen by constructing the phase stability diagram at 1773 K. In Ex. 2 in Sec. 13.4 an isothermal phase stability diagram was constructed using p_{CO} and p_{CO_2} as the independent variables. However, in the present problem the equilibrium values of p_{CO} and p_{SiO} are required, and thus the phase stability diagram at 1773 K will be constructed using p_{CO} and p_{SiO} as the independent variables. When CO and SiO exist at equilibrium, the activity of O in the CO equals the activity of O in the SiO. Thus, at a given partial pressure of CO, the activity of C in the CO is fixed, and at a given partial pressure of SiO the activity of Si is fixed. The condensed phases liquid silicon, solid SiC, solid SiO_2, and graphite can exist in the system, and thus the number of possible equilibria involving two condensed phases and a gas phase is $(4 \times 3)/2 = 6$. However, as was seen in Ex. 2 in Sec. 13.4, Si and C cannot exist in equilibrium with one another. The standard molar Gibbs free energies of formation of the four compounds of interest at 1773 K are

Compound	$\Delta G^\circ_{1773 \text{ K}}$, J
$SiO_{2(s)}$	$-595,900$
$SiO_{(g)}$	$-246,100$
$SiC_{(s)}$	$-56,990$
$CO_{(g)}$	$-266,900$

Each of the five equilibria involving two condensed phases and a gas phase must include CO and SiO.

1. Equilibrium among Si, SiO$_2$, CO, and SiO. The equilibrium is

$$SiO_2 + C = SiO + CO \qquad (i)$$

for which $\Delta G^\circ_{1773\,K} = 82,900$ J. Thus, for this equilibrium

$$\log p_{SiO} = -\log p_{CO} - 2.44$$

which is drawn as line 1 in Fig. 13.33. Note that, as both gases occur on the same side of the equation describing the equilibrium, determination cannot be made as to which condensed phase is stable above line 1 and which condensed phase is stable below the line.

2. Equilibrium among SiO$_2$, SiC, SiO, and CO. The equilibrium is

$$2SiO_2 + SiC = 3SiO + CO \qquad (ii)$$

for which $\Delta G^\circ_{1773\,K} = 243,590$ J. This gives

$$\log p_{SiO} = -\frac{1}{3}\log p_{CO} - 2.39$$

which is drawn as line 2 in Fig. 13.33. Again, a determination cannot be made as to which condensed phase is stable above the line and which is stable below the line.

3. Equilibrium among SiO$_2$, Si, and SiO. This equilibrium is independent of the pressure of CO and is written as

$$SiO_2 + Si = 2SiO \qquad (iii)$$

for which $\Delta G^\circ_{1773\,K} = 103,700$ J. Thus

$$\log p_{SiO} = -1.53$$

which is drawn as line 3 in Fig. 13.33. Again, indication of the stability of the condensed phases is not given.

4. Equilibrium among Si, SiC, SiO, and CO. This is

$$2Si + CO = SiC + SiO \qquad (iv)$$

for which $\Delta G^\circ_{1773\,K} = 36,190$ J. Thus, for the equilibrium

$$\log p_{SiO} = \log p_{CO} + 1.06$$

which is drawn as line 4. In this equilibrium, Si is stable relative to SiC above the line, and SiC is stable relative to Si below the line.

5. Equilibrium among SiC, C, SiO, and CO. The equilibrium is

$$SiC + CO = 2C + SiO \qquad (v)$$

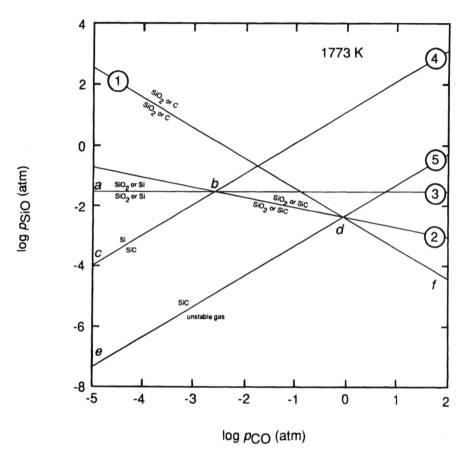

Figure 13.33 Construction of the phase stability diagram for the system Si–C–O at 1773 K.

for which $\Delta G^\circ_{1773 \text{ K}} = 77,790$ J. Thus

$$\log p_{SiO} = \log p_{CO} - 2.29$$

which is drawn as line 5 in Fig. 13.33. As carbon exists at unit activity along this line, SiC is stable relative to graphite above the line and an unstable gas occurs below the line.

Inspection of Fig. 13.33 shows that

1. SiC is stable relative to Si below the line bc
2. SiC is stable relative to graphite above the line ed and
3. SiC or SiO_2 are stable below the line bd

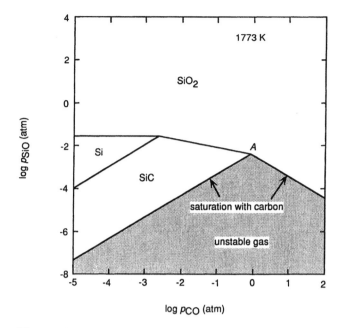

Figure 13.34 The phase stability diagram for the system Si–C–O at 1773 K.

This identifies the area *cbde* as the field of stability of SiC. It is then seen that (1) Si is stable relative to SiC above the line *cb,* and (2) Si or SiO_2 is stable below the line *ab.* This identifies the area *abc* as the field of stability of liquid Si. Thus (1) SiO_2 is stable relative to Si above the line *ab,* (2) SiO_2 is stable relative to SiC above the line *bd,* and (3) SiO_2 or graphite is stable above the line *df.*

Thus the field of stability of SiO_2 lies above the line *abdf,* and the phase stability diagram is as shown in Fig. 13.34. The phase stability diagram shows that graphite and quartz react with one another to produce SiC until the SiO_2–graphite–SiC equilibrium is reached at the state *A,* which is the intersection of lines 1, 2, and 5 in Fig. 13.33. Simultaneous solution of the equations of any two of these lines gives the state *A* as

$$\log p_{CO} = -0.075 \quad (p_{CO} = 0.844 \text{ atm})$$

and

$$\log p_{SiO} = -0.2365 \quad (p_{SiO} = 4.32 \times 10^{-3} \text{ atm})$$

The mass of SiC produced and the mass of graphite consumed are obtained by conducting a mass balance on Si, C, and O. The atomic weights of C, O, and Si are, respectively, 12, 16, and 28.09. Thus, before any reactions begin, the vessel contains $100/60.09 = 1.6642$ moles of SiO_2 and $100/12 = 8.3333$ moles of C. Thus the vessel contains 1.6642 moles of Si, 3.3283 moles of O, and 8.3333 moles of C. When reaction equilibrium is attained at 1773 K, the number of moles of CO in the gas phase is calculated as

$$n_{CO} = \frac{p_{CO}V}{RT} = \frac{0.844 \times 20}{0.082057 \times 1773} = 0.1160 \text{ moles}$$

and the number of moles of SiO in the gas phase is

$$n_{SiO} = \frac{4.315 \times 10^{-3} \times 20}{0.082057 \times 1773} = 5.9318 \times 10^{-4} \text{ moles}$$

Thus the gas phase contains 0.1160 moles of C, 0.1166 moles of O, and 5.9318×10^{-4} moles of Si. Consequently, at equilibrium the solid phases contain

$$1.6642 - 5.9318 \times 10^{-4} = 1.6636 \text{ moles of Si}$$

$$3.3286 - 0.1166 = 3.2120 \text{ moles of O}$$

and

$$8.3333 - 0.1160 = 8.2173 \text{ moles of C}$$

All of the oxygen in the solids occurs in the silica, and thus $3.2120/2 = 1.6060$ moles of silicon in the solids occurs in the silica. The remaining $1.6636 - 1.6060 = 0.0576$ moles of silicon in the solids occurs in the SiC. Thus 0.0576 moles, or $0.0576 \times 40.09 = 2.31$ g of SiC are formed. The number of moles of graphite consumed equals the number of moles of SiC formed plus the number of moles of CO produced, i.e., $0.0576 + 1160 = 0.1736$ moles, or $0.1736 \times 12 = 2.08$ grams.

The equilibrium partial pressures of CO_2 and O_2, which are, respectively, 5.9×10^{-5} and 1.3×10^{-16} atm, are small enough that the CO_2 and O_2 produced in the gas phase do not need to be included in the mass balance.

Example 2

Determine the conditions under which an Fe–Cr–O melt is in equilibrium with (1) solid Cr_2O_3 and (2) solid $FeO \cdot Cr_2O_3$ at 1600°C. For

$$2Cr_{(s)} + \frac{3}{2}O_{2(g)} = Cr_2O_{3(s)} \quad \Delta G^\circ_{(i)} = -1,120,300 + 259.8T \text{ J} \qquad \text{(i)}$$

and for

$$\frac{1}{2}O_{2(g)} = [O]_{(1 \text{ wt\% in Fe})} \quad \Delta G^\circ_{(ii)} = -111,070 - 5.87T \text{ J} \qquad \text{(ii)}$$

At 1600°C, Fe–Cr melts exhibit Raoultian ideality, and the molar heat of melting of Cr, at its equilibrium melting temperature of 2173 K, is 21,000 J. Thus for $Cr_{(s)} = Cr_{(l)}$,

$$\Delta G^\circ_m = \Delta H^\circ_m - T\frac{\Delta H^\circ_m}{T_m} = 21,000 - 9.66T \text{ J}$$

and for $Cr_{(l)} = [Cr]_{(1 \text{ wt\% in Fe})}$

$$\Delta G = RT \ln \frac{55.85}{100 \times 52.01} = -37.70 \, T \, J$$

Therefore, for $Cr_{(s)} = [Cr]_{(1 \text{ wt\% in Fe})}$,

$$\Delta G^\circ_{(iii)} = 21,000 - 47.36T \, J \tag{iii}$$

The standard Gibbs free energy change for the reaction

$$2[Cr]_{(1 \text{ wt\%})} + 3[O]_{(1 \text{ wt\%})} = Cr_2O_{3(s)} \tag{iv}$$

is thus

$$\Delta G^\circ_{(iv)} = \Delta G^\circ_{(i)} - 3\Delta G^\circ_{(ii)} - 2\Delta G^\circ_{(iii)}$$
$$= -829,090 + 372.13T \, J$$
$$= -RT \ln \frac{a_{Cr_2O_3}}{h^2_{Cr(1 \text{ wt\%})} \cdot h^3_{O(1 \text{ wt\%})}}$$

or, at 1873 K,

$$\log \frac{h^2_{Cr(1 \text{ wt\%})} \cdot h^3_{O(1 \text{ wt\%})}}{a_{Cr_2O_3}} = -3.68 \tag{v}$$

Saturation of the melt with solid Cr_2O_3 occurs at $a_{Cr_2O_3} = 1$, and, if the interactions between Cr and O in solution are ignored, and it is assumed that oxygen obeys Henry's law, Eq. (v) can be written as

$$\log [\text{wt\% Cr}] = -1.5 \log [\text{wt\% O}] - 1.84 \tag{vi}$$

which is the variation of [wt% Cr] with [wt% O] in liquid iron required for equilibrium with solid Cr_2O_3 at 1600°C. Eq. (vi) is drawn as line (vi) in Fig. 13.35.
For

$$Fe_{(l)} + 2Cr_{(s)} + 2O_{2(g)} = FeO \cdot Cr_2O_{3(s)}$$
$$\Delta G^\circ_{(vii)} = -1,409,420 + 318.07T \, J \tag{vii}$$

and thus, for the reaction

$$Fe_{(l)} + 22[Cr]_{(1 \text{ wt\%})} + 4[O]_{(1 \text{ wt\%})} = FeO \cdot Cr_2O_{3(s)} \tag{viii}$$
$$\Delta G^\circ_{(viii)} = \Delta G^\circ_{(vii)} - 2\Delta G^\circ_{(iii)} - 4\Delta G^\circ_{(ii)}$$
$$= -1,007,140 + 436.27T \, J$$
$$= -RT \ln \frac{a_{FeO \cdot Cr_2O_3}}{a_{Fe} \cdot h^2_{Cr(1 \text{ wt\%})} \cdot h^4_{O(1 \text{ wt\%})}}$$

or, at 1873 K,

$$\log \frac{a_{Fe} \cdot h^2_{Cr(1 \text{ wt\%})} \cdot h^4_{O(1 \text{ wt\%})}}{a_{FeO \cdot Cr_2O_3}} = -5.30 \tag{ix}$$

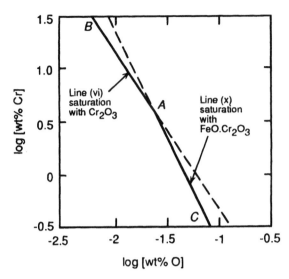

Figure 13.35 The variations of the concentrations of Cr with the concentrations of O in Fe saturated with Cr_2O_3 and $FeO \cdot Cr_2O_3$ at 1600°C.

Saturation of the melt with $FeO \cdot Cr_2O_3$ occurs at $a_{FeO \cdot Cr_2O_3} = 1$ and, with the same assumptions as before, and $a_{Fe} = X_{Fe} = 1 - X_{Cr}$, the variation of [wt% Cr] with [wt% O] required for equilibrium with solid $FeO \cdot Cr_2O_3$ at 1600°C is

$$\log (1 - X_{Cr}) + 2 \log[\text{wt\% Cr}] + 4 \log [\text{wt\% O}] = -5.30. \tag{x}$$

In solutions sufficiently dilute that $X_{Fe} \sim 1$, Eq. (x) can be simplified as

$$\log [\text{wt\% Cr}] = -2 \log [\text{wt\% O}] - 2.65 \tag{xi}$$

Eq. (xi) is drawn as line (x) in Fig. 13.35. Lines (vi) and (x) intersect at the point A, $\log [\text{wt\% Cr}] = 0.59$, $\log [\text{wt\% O}] = -1.62$ (wt% O = 0.024, wt% Cr = 3.89), which is the composition of the melt which is simultaneously saturated with solid Cr_2O_3 and $FeO \cdot Cr_2O_3$. From the phase rule, equilibrium in a three-component system (Fe–Cr–O) among four phases (liquid Fe–Cr–O, solid Cr_2O_3, solid $FeO \cdot Cr_2O_3$, and a gas phase) has one degree of freedom, which, in the present case, has been used by specifying the temperature to be 1873 K. Thus the activities of Fe, Cr, and O are uniquely fixed, and hence [wt% Cr] and [wt% O] are uniquely fixed. The equilibrium oxygen pressure in the gas phase is obtained from Eq. (ii) as

$$\Delta G^{\circ}_{(ii),1873 \text{ K}} = -122,065 \text{ J} = -8.3144 \times 1873 \ln \frac{[\text{wt\% O}]}{p_{O_2}^{1/2}}$$

which, with [wt% O] = 0.024, gives $p_{O_2(eq)} = 8.96 \times 10^{-11}$ atm. The positions of the lines in Fig. 13.35 are such that, in melts of [wt% Cr] > 3.89, Cr_2O_3 is the stable phase in equilibrium with saturated melts along the line AB and, in melts in which [wt% Cr] < 3.89, $FeO \cdot Cr_2O_3$ is the stable phase in equilibrium with saturated melts along the line AC. Alternatively, Cr_2O_3 is the stable phase in equilibrium with saturated melts of [wt% O] < 0.024, and FeO·Cr$_2$O$_3$ is the stable phase in equilibrium with saturated melts of [wt% O] > 0.024. Consider a melt in which log [wt% Cr] = 1.5. From Fig. 13.35, or Eq. (vi), the oxygen content at this chromium level required for equilibrium with Cr$_2$O$_3$ (at the point B in Fig. 13.35) is 5.93×10^{-3} wt%, or log [wt% O] = -2.25. From Eq. (v), the activity of Cr$_2$O$_3$ in this melt with respect to solid Cr$_2$O$_3$ is unity, and hence, the melt is saturated with respect to solid Cr$_2$O$_3$. However, from Eq. (ix), in the same melt, i.e., X_{Fe} = 0.668, [wt% Cr] = 31.6, [wt% O] = 0.00593, the activity of FeO·Cr$_2$O$_3$ with respect to solid FeO·Cr$_2$O$_3$ is only 0.2. Thus the melt is saturated with respect to Cr$_2$O$_3$ and is undersaturated with respect to FeO·Cr$_2$O$_3$. Moving along the line BA from B toward A, $a_{Cr_2O_3}$ = 1, and $a_{FeO \cdot Cr_2O_3}$ = 1 increases from 0.2 at B to unity at A in the doubly saturated melt. Consider a melt in which log [wt% Cr] = -0.5. From Fig. 13.35 the oxygen content required for saturation with FeO·Cr$_2$O$_3$ is 0.084 wt% (log [wt% O] = -1.075 at the point C in Fig. 13.35). From Eq. (ix), the activity of FeO·Cr$_2$O$_3$ in this melt is unity. However, from Eq. (v), the activity of Cr$_2$O$_3$ in the melt, with respect to solid Cr$_2$O$_3$, is only 0.285. Thus, this melt is saturated with FeO·Cr$_2$O$_3$ and is undersaturated with Cr$_2$O$_3$. On moving along the line CA from C toward A, $a_{FeO \cdot Cr_2O_3}$ is unity and $a_{Cr_2O_3}$ increases from 0.285 at C to unity at A.

If the various solute-solute interactions had been considered, Eq. (v), with $a_{Cr_2O_3}$ = 1 would be written as

$$2 \log h_{Cr(1 \text{ wt\%})} + 3 \log h_{O(1 \text{ wt\%})} = -3.68$$

or

$$2 \log f_{Cr(1 \text{ wt\%})} + 2 \log [\text{wt\% Cr}] + 3 \log f_{O(1 \text{ wt\%})} + 3 \log [\text{wt\% O}] = -3.68$$

or

$$2e_{Cr}^{Cr} \cdot [\text{wt\% Cr}] + 2e_{Cr}^{O} \cdot [\text{wt\% O}] + 2 \log [\text{wt\% Cr}] + 3e_{O}^{O} \cdot [\text{wt\% O}]$$
$$+ 3e_{O}^{Cr} \cdot [\text{wt\% Cr}] + 3 \log [\text{wt\% O}] = -3.68$$

With

$$e_{Cr}^{O} = 0 \quad e_{O}^{O} = -0.2 \quad e_{O}^{Cr} = -0.041 \quad \text{and} \quad e_{Cr}^{O} = -0.13$$

this gives

$$-0.43[\text{wt\% O}] + 0.0615 [\text{wt\% Cr}] + \log [\text{wt\% Cr}]$$
$$+ 1.5 \log [\text{wt\% O}] = -1.84 \tag{xii}$$

which is drawn as line (xii) in Fig. 13.36.

Similarly, with $a_{FeO \cdot Cr_2O_3}$ = 1, Eq. (ix) would be written as

$$\log X_{Fe} + 2 \log h_{Cr(1 \text{ wt\%})} + 4 \log h_{O(1 \text{ wt\%})} = -5.30$$

or

$$\log X_{Fe} + 2e_{Cr}^{Cr} \cdot [\text{wt\% Cr}] + 2e_{Cr}^{O} \cdot [\text{wt\% O}] + 2 \log [\text{wt\% Cr}] + 4e_{O}^{O} \cdot [\text{wt\% O}]$$
$$+ 4e_{O}^{Cr} \cdot [\text{wt\% Cr}] + 4 \log [\text{wt\% Cr}] = -5.30$$

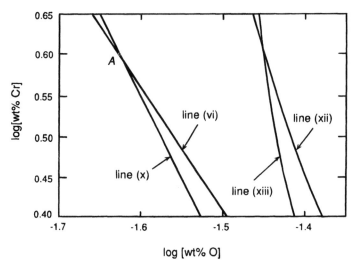

Figure 13.36 Fig. 13.35 amended to take into consideration the interactions between the solutes, Cr and O, in liquid Fe.

or

$$\log X_{Fe} - 1.06 \,[wt\% \,O] - 0.164 \,[wt\% \,Cr] + 2 \log \,[wt\% \,Cr]$$
$$+ \, 4 \log \,[wt\% \,O] = -5.30$$

which is drawn as line (xiii) in Fig. 13.36. Lines (xii) and (xiii) intersect at log [wt% Cr] = 0.615, log [wt% O] = -1.455 ([wt% Cr] = 4.12, [wt% O] = 0.035). When the interactions among the solute were ignored the point of intersection, *A*, was obtained as [wt% Cr] = 3.89, [wt% O] = 0.024.

PROBLEMS

13.1 Air at atmospheric pressure is blown over a Cu-rich copper-gold liquid solution at 1500 K. If only the copper is oxidized (to form pure solid Cu_2O), calculate the minimum activity of Cu which can be obtained in the solution.

13.2 Magnesium can be removed from Mg–Al liquid solution by selectively forming the chloride $MgCl_2$. Calculate the activity of Mg in the liquid Mg–Al system which can be achieved at 800°C by reacting the solution with an H_2–HCl gas mixture containing hydrogen at essentially 1 atm pressure and $p_{HCl} = 10^{-5}$ atm to form pure liquid $MgCl_2$.

13.3 The partial pressure of oxygen in equilibrium with pure liquid lead and pure liquid lead oxide at 1200 K is 2.16×10^{-9} atm. When SiO_2 is added to the liquid PbO to form a lead silicate melt the oxygen pressure in equilibrium with pure liquid lead and the silicate melt is decreased to 5.41×10^{-10} atm. Calculate the activity of PbO in the lead silicate melt.

13.4 Copper, present as an impurity in liquid Pb, can be removed by adding PbS to the Cu–Pb alloy and allowing the exchange reaction

$$2Cu_{(s)} + PbS_{(s)} = Cu_2S_{(s)} + Pb_{(l)}$$

to come to equilibrium.

The solid sulfides are mutually immiscible, Pb is insoluble in solid Cu, and the Cu liquidus, below 850°C, can be represented by

$$\log X_{Cu} = -\frac{3500}{T} + 2.261$$

where X_{Cu} is the solubility of Cu in liquid Pb. If Cu obeys Henry's law in liquid Pb, calculate the extent to which Cu can be removed from liquid Pb by this process at 800°C. Would the extent of purification of the lead be increased by increasing or by decreasing the temperature?

13.5 A CH_4–H_2 gas mixture at 1 atm total pressure, in which $p_{H_2} = 0.957$ atm, is equilibrated with an Fe–C alloy at 1000 K. Calculate the activity of C with respect to graphite in the alloy. What would the value of p_{H_2} in the gas mixture (at $P_{total} = 1$ atm) have to be in order to saturate the Fe with graphite at 1000 K?

13.6 Calculate the activity of FeO in an FeO–Al_2O_3–SiO_2 melt below which the FeO cannot be reduced to pure liquid iron by a CO–CO_2 mixture of $p_{CO}/p_{CO_2} = 10^5$ at 1600°C.

13.7 A piece of iron is to be heat-treated at 1000 K in a CO–CO_2–H_2O–H_2 gas mixture at 1 atm pressure. The gas mixture is produced by mixing CO_2 and H_2 and allowing the equilibrium $CO_2 + H_2 = CO + H_2O$ to establish. Calculate (*a*) the minimum H_2/CO_2 ratio in the inlet gas which can be admitted to the furnace without oxidizing the iron, (*b*) the activity of carbon (with respect to graphite) in the equilibrated gas of this initial minimum H_2/CO_2 ratio, (*c*) the total pressure to which the equilibrated gas would have to be raised to saturate the iron with graphite at 1000 K, and (*d*) the effect, on the partial pressure of oxygen in the equilibrated gas, of this increase in total pressure.

13.8 An Fe–Mn solid solution containing $X_{Mn} = 0.001$ is in equilibrium with an FeO–MnO solid solution and an oxygen-containing gaseous atmosphere at 1000 K. How many degrees of freedom does the equilibrium have? What is the composition of the equilibrium oxide solution, and what is the partial pressure of oxygen in the gas phase? Assume that both solid solutions are Raoultian in their behavior.

13.9 The elements A and B, which are both solid at 1000°C, form two stoichiometric compounds A_2B and AB_2, which are also both solid at 1000°C. The system A–B does not contain any solid solutions. A has an immeasurably small vapor pressure at 1000°C, and, for the change of state, $B_{(s)} = B_{(v)}$

$$\Delta G° = 187,220 - 108.8T \text{ J}$$

The vapor pressure exerted by an equilibrated AB_2–A_2B mixture is given by

$$\log p(atm) = -\frac{11,242}{T} + 6.53$$

and the vapor pressure exerted by an equilibrated $A–A_2B$ mixture is given by

$$\log p(\text{atm}) = -\frac{12,603}{T} + 6.9$$

From these data, calculate the standard Gibbs free energies of formation of A_2B and AB_2.

13.10 For the change of standard state $V_{(s)} = V_{(1 \text{ wt\% in Fe})}$

$$\Delta G° = -15,480 - 45.61T \text{ J}$$

Calculate the value of $\gamma_V°$ at 1600°C. If a liquid Fe–V solution is equilibrated with pure solid VO and a gas containing $p_{O_2} = 4.72 \times 10^{-10}$ atm, calculate the activity of V in the liquid solution (a) with respect to solid V as the standard state, (b) with respect to liquid V as the standard state, (c) with respect to the Henrian standard state, and (d) with respect to the 1 wt% in iron standard state.

13.11 When an Fe–P liquid solution is equilibrated at 1900 K with solid CaO, solid $3CaO \cdot P_2O_5$ and a gas phase containing $p_{O_2} = 10^{-10}$ atm, the activity of P in the iron, with respect to the 1 wt% in Fe standard state, is 20. Given that $\Delta G°_{1900 \text{ K}} = -564,600$ J for

$$3CaO_{(s)} + P_2O_{5(g)} = 3CaO \cdot P_2O_{5(s)}$$

and $\Delta G° = -122,200 - 19.22T$ J for

$$\frac{1}{2} P_{2(g)} = P_{(1 \text{ wt\% in Fe})}$$

calculate $\Delta G°_{1900 \text{ K}}$ for the reaction

$$P_{2(g)} + \frac{5}{2} O_{2(g)} = P_2O_{5(g)}$$

13.12 Liquid iron, contained in an Al_2O_3 crucible under a gaseous atmosphere of $p_{O_2} = 3 \times 10^{-12}$ atm at 1600°C, contains its equilibrium contents of dissolved oxygen and aluminum. To what value must be p_{O_2} be raised in order that solid hercynite ($FeO \cdot Al_2O_3$) appears in equilibrium with the melt and with solid Al_2O_3? What is the activity of Al (with respect to the 1 wt% in Fe standard state) in this state? How many degrees of freedom does this equilibrium have at 1600°C? Given

$$\frac{1}{2} O_{2(g)} = O_{(1 \text{ wt\% in Fe})} \quad \Delta G° = -111,070 - 5.87T \text{ J}$$

$$Al_{(l)} = Al_{(1 \text{ wt\% in Fe})} \quad \Delta G° = -43,100 - 32.26T \text{ J}$$

$$FeO \cdot Al_2O_{3(s)} = Fe_{(l)} + O_{(1 \text{ wt\% in Fe})} + Al_2O_{3(s)}$$

$$\Delta G° = 146,230 - 54.35T \text{ J}$$

13.13 UC_2 can be equilibrated with UC and C at high temperature and can be equilibrated with U_2C_3 and C at lower temperatures. Calculate the maximum and minimum temperatures at which UC_2 can exist.

13.14 In the Pigeon process for the production of magnesium, dolomite (CaO·MgO) is reduced by silicon to form magnesium vapor and $2CaO·SiO_2$. Calculate the equilibrium pressure of magnesium vapor produced by this reaction at 1200°C. The Gibbs free energy of formation of dolomite from CaO and MgO is small enough that it can be ignored.

13.15 What is the minimum value that the activity of MgO can have in $MgO·Al_2O_3$ at 1000°C?

13.16 A mixture of ZnO and graphite is placed in an evacuated vessel and heated to 1200 K. Calculate the partial pressures of Zn, CO, and CO_2 that are developed.

13.17 An assemblage of solid CaO, MgO, $3CaO·Al_2O_3$, and liquid Al exerts an equilibrium vapor pressure of Mg of 0.035 atm at 1300 K. Write the equation for the appropriate reaction equilibrium. Calculate the standard Gibbs free energy of formation of $3CaO·Al_2O_3$ from CaO and Al_2O_3 and the activity of Al_2O_3 in CaO-saturated $3CaO·Al_2O_3$ at 1300 K.

13.18 An iron-carbon melt containing 0.5 wt% C is prepared in an alumina crucible under an atmosphere of $p_{CO} = 1$ atm at 1600°C. Calculate the equilibrium concentrations of O and Al in the melt (*a*) ignoring all solute-solute interactions and (*b*) considering the solute-solute interactions. The interaction coefficients are listed in Table 13.1.

13.19 It is required that PbO be eliminated from an ore containing PbO, PbS, and $PbSO_4$ by converting it to PbS or $PbSO_4$ by reaction with an SO_2–O_2 gas. Although the pressure of O_2 in the gas can vary within wide limits, the partial pressure of SO_2 may not be higher than 0.5 atm. Calculate the maximum temperature at which it can be guaranteed that the PbO phase will be eliminated.

13.20 Cementite, Fe_3C, is metastable with respect to carbon-saturated α-iron and graphite at 950 K and 1 atm pressure. Given that the molar volumes of α-Fe, graphite, and Fe_3C at 950 K are, respectively, 7.32, 5.40, and 23.92 cm^3/mole, calculate the pressure, at 950 K, at which Fe_3C is in equilibrium with carbon-saturated α-Fe and graphite. At what temperature, at 1 atm pressure, is carbon-saturated γ-Fe and graphite in equilibrium with cementite?

13.21 An experiment is being conducted on an equilibrated mixture of CaO and $CaCO_3$ contained in a closed vessel at 1200 K. The mixture is contaminated by iron in the form of hematite (Fe_2O_3). The contaminant would not be harmful to the experiment if it occurred as either wustite (FeO) or as cementite (Fe_3C). The necessary changes in the chemical form of the contaminant can be effected by admitting CO gas to the vessel. Calculate the allowable limits of p_{CO} in the vessel for the occurrence of the contaminant (*a*) as wustite and (*b*) as cementite.

13.22 A Cu–Au alloy of $X_{Cu} = 0.5$ is being annealed at 600°C in deoxidized argon. The argon is deoxidized by being passed over heated pure copper turnings prior to its admission to the annealing furnace. The solid Cu–Au system is virtually regular in its solution behavior, with a molar Gibbs excess free energy of mixing given by

$$G^{XS} = -28,280X_{Cu}X_{Au} \text{ J}$$

Assuming that equilibrium is attained in the deoxidizing furnace, calculate the maximum temperature at which the deoxidizing furnace can be operated without causing oxidation of the copper in the Cu–Au alloy being annealed.

13.23 In a dew-point experiment a Cu–Zn alloy is placed in one end of an evacuated and closed tube, and is heated to 900°C. When the other end of the tube is cooled to 740°C Zn vapor begins to condense. Calculate the activity of Zn in the alloy relative to pure zinc.

13.24 A crucible containing 100.0 g of silver at 1000°C is placed in the reaction chamber of a Sieverts's apparatus. The chamber is evacuated and filled with 50 cm^3 (STP) of argon, which measures the dead volume of the chamber. An external manometer reads the pressure of argon as 0.9 atm. The chamber is re-evacuated and filled with oxygen, and it is found that 251.5 cm^3 (STP) are required to produce a gas pressure of 0.9 atm in the chamber. Calculate the solubility of O in the Ag (as atom percent) and calculate the value of the Sieverts's law constant at 1000°C.

13.25 Silicon and manganese are commonly used together as deoxidizers for liquid steel. At 1600°C,

$$[Mn]_{(1\ wt\%\ in\ Fe)} + [O]_{(1\ wt\%\ in\ Fe)} = MnO_{(s)} \quad K = 23.5$$

$$[Si]_{(1\ wt\%\ in\ Fe)} + 2[O]_{(1\ wt\%\ in\ Fe)} = SiO_{2(s)} \quad K = 27,840$$

The values of the equilibrium constants show that SiO_2 is considerably more stable than MnO. Why, then, is a mixture of Mn and Si more effective as a de-oxidizing agent than Si alone? The activities of MnO and SiO_2, with respect to solids as the standard states in MnO–SiO_2 melts at 1600°C, are shown in Fig. 13.37.

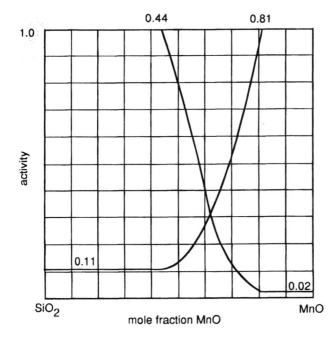

Figure 13.37 Activities in the system MnO–SiO_2 at 1600°C.

Chapter 14

PHASE DIAGRAMS FOR BINARY SYSTEMS IN PRESSURE-TEMPERATURE-COMPOSITION SPACE

14.1 INTRODUCTION

The development of fabricating processes conducted in evacuated systems or at high pressure requires knowledge of the influence of pressure on phase equilibria in materials systems. Vacuum arc melting, electron beam melting, and pulsed laser deposition are examples of processes that proceed at low pressures and high temperatures. In pulsed laser deposition the flux of laser irradiation focused on the target material is high enough to cause melting and vaporization. Calculation of the transient temperature profiles that are developed during the process requires the assumption of local thermal and phase equilibrium between the liquid and the vapor plume. Control of the composition of the plume produced by irradiation of a multi-component target requires knowledge of the influence of pressure on the phase equilibria.

14.2 A BINARY SYSTEM EXHIBITING COMPLETE MUTUAL SOLUBILITY OF THE COMPONENTS IN THE SOLID AND LIQUID STATES

A schematic phase diagram for the binary system A–B in pressure-temperature-composition space is shown in Fig. 14.1. The system exhibits complete mutual solubility of A and B in the solid and liquid states. The front and back faces of the figure are, respectively, the one-component phase diagrams for the components B

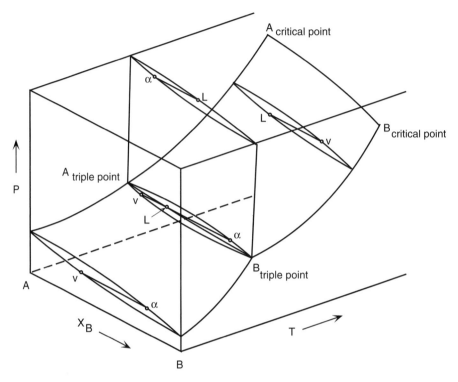

Figure 14.1 Phase diagram for a binary system A–B in pressure-temperature-composition space, in which there is complete mutual solubility in the solid and liquid states.

and A. As discussed in Chapter 7, these one-component phase diagrams contain areas in which the equilibrium existence of a single phase (solid, liquid, and vapor) has two degrees of freedom. These areas meet at lines that represent phase equilibrium between two phases (the variation, with pressure, of the melting temperature of the solid and the variations, with temperature, of the saturated vapor pressures of the solid and liquid phases). These equilibria have one degree of freedom. The three lines for the 2-phase equilibrium meet at the triple point where the 3-phase equilibrium has no degree of freedom. The line representing the saturated vapor pressure of the liquid phase ends at the critical point, where the molar volumes of the liquid and vapor phases have the same value and the enthalpy of boiling is zero.

Increasing the number of components in the system from one to two increases the number of degrees of freedom for each phase equilibrium by one and thus the binary phase diagram occurs in pressure-temperature-composition space. The equilibria of the single phases, α-solid solutions, liquid solutions, and mixtures of vapors, have three degrees of freedom and their state spaces are volumes. The 2-phase equilibria have two degrees of freedom and thus the compositions of equilibrated phases occur on the surfaces of envelopes that pass through the diagram. The 2-phase α–v equilibrium begins on the line representing the saturated vapor pressure of solid B, enters the diagram to form two surfaces of an envelope, and ends on the line representing the saturated vapor pressure of pure A. The upper surface of this envelope contains the compositions of α-solid solutions in equilibrium with a mixture of vapors and the lower surface contains the compositions of the mixtures of vapors in equilibrium with α-solid solutions. These equilibrated compositions lie at the extremities of tie-lines drawn through the envelope and all of the tie-lines are isobaric and isothermal. A tie-line between the compositions of an equilibrated α–v phase pair is shown in Fig. 14.1. The upper surface of the envelope gives the minimum pressure at which an α-solid solution can exist at any temperature. The 2-phase α–L begins on the melting line of B, enters the diagram as the two surfaces of an envelope, and terminates on the melting line of A. The compositions of the equilibrated α–L phase lie at the extremities of tie-lines and a tie-line between an equilibrated α–L pair is shown in Fig. 14.1. The surface for the α-solid gives the maximum temperature at which an α-solid solution can exist at any pressure and thus the α-surfaces in the α–v equilibria and the α–L equilibria are the limits of the volume where α-solid solutions are stable. Similarly, an envelope between the saturated vapor pressures of liquid B and liquid A gives the compositions of 2-phase L–v equilibria. This envelope ends at a single line connecting the critical points of the two pure components. A tie-line between the composition of an equilibrated L–v pair is shown in Fig. 1. The surfaces giving the compositions of the liquid phase in the 2-phase α–L and L–v equilibria are the limits of the volume state space in the diagram when the liquid solutions are stable. The surface giving the composition of the vapor phases in the 2-phase α–v and L–v equilibria gives the limits of the vapor volume.

The three envelopes meet at three lines originating at the triple points of the pure components, each of which gives the composition of a phase in the 3-phase equilibrium. The 3-phase equilibrium in a binary system has one degree of freedom and a tie-line joining the compositions of equilibrated α–L–v phases is shown in Fig. 14.1.

The projection of the phase diagram on the $P–T$ plane is shown in Fig. 14.2. The line joining the triple points of A and B through the diagram is a continuum of isobaric and isothermal tie-lines joining the compositions of the solid, liquid, and vapor phases in the 3-phase equilibrium. These tie-lines are normal to the $P–T$ plane. Isobaric sections of Fig. 14.2a at P_1 to P_6 are shown in Fig. 14.3. At P_1 the 2-phase region between the phase field of α-solid solutions and the liquid phase field opens at a, the melting temperature of A, and closes at c, the melting temperature of B. Similarly, the 2-phase region between the vapor and liquid phase fields opens at b, the boiling temperature of A, and closes at d, the boiling temperature of B. The compositions of the vapor and liquid phases in equilibrium with each other and the compositions of the liquids and α-solid solutions in equilibrium with each other lie at the extremities of tie-lines drawn through the 2-phase regions. The pressure P_2 is the triple point pressure for A and thus the isobaric section shows the two 2-phase regions coinciding with each other to produce a 3-phase equilibrium at pure A. The

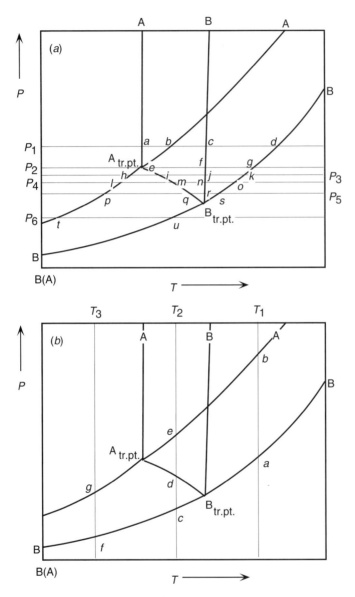

Figure 14.2 *P–T* projection of the phase diagram shown in Fig. 14.1.

pressures P_3, P_4, and P_5 are between the triple point pressures of A and B and thus the isobaric sections show 3-phase equilibria at *i*, *m*, and *q*. With decreasing pressure the temperature at which the 3-phase equilibrium occurs increases and the position of the tie-line moves from the A-rich side of the section to the B-rich side. The pressure P_6 is lower than the triple point pressure of B and thus the liquid state does not

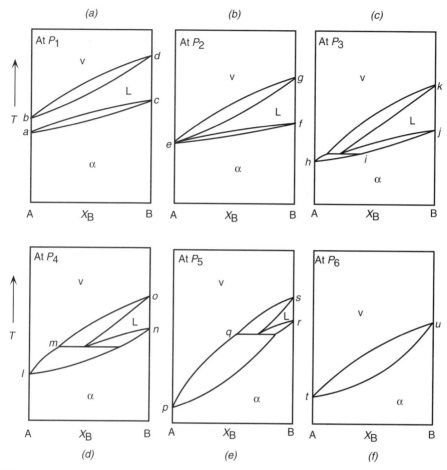

Figure 14.3 Isobaric sections of the *P–T* phase diagram shown in Fig. 14.2*a*.

appear on the isobaric section at P_6. The 2-phase region between the phase fields of vapor and α opens at *t*, the sublimation point of A, and closes at *u*, the sublimation point of B.

Three isothermal sections of Fig. 14.2*b* are shown in Fig. 14.4. At T_1 the 2-phase region between the fields of stability of vapor and liquid enters the section at *a*, the melting point of B, and terminates at *b*, the melting point of A. Within the 2-phase region the compositions of the phases in equilibrium with each other occur at the extremities of tie-lines drawn through the 2-phase region. At T_2 the 2-phase region between the fields of stability of the α-solid solution and vapor originates at *c*, the sublimation point of B, and extends to *d* in which state the 3-phase equilibrium exists. The composition of the liquid in this 3-phase equilibrium lies between those of the vapor and the α-solid solution. The 2-phase region between

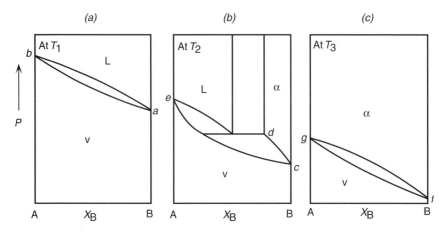

Figure 14.4 Isothermal sections of the *P–T* projection shown in Fig. 14.2*b*.

the fields of stability of vapor and liquid originates at *d* and terminates at *e*, the boiling point of A. At pressures less than 100 atm the compositions of the condensed phases are virtually independent of pressure and thus the 2-phase region between the fields of stability of liquid and α-solid solutions is defined by virtually vertical lines. At T_3 the 2-phase region between the fields of stability of vapor and α-solid solution originates at *f*, the sublimation point of B, and terminates at *g*, the sublimation point of A.

14.3 A BINARY SYSTEM EXHIBITING COMPLETE MUTUAL SOLUBILITY IN THE SOLID AND LIQUID STATES AND SHOWING MINIMA ON THE MELTING, BOILING, AND SUBLIMATION CURVES

Fig. 14.5 shows the phase diagram, in pressure-temperature-composition space, of a binary system that exhibits complete mutual solubility in the solid and liquid states and shows minima on the melting, boiling, and sublimation curves. At the minima the phase transformations do not involve a change in composition and thus these compositions behave as quasi-unary systems, in that the phase transition involves a thermal arrest at the temperature of the minimum during the phase transformation. The phase diagram is shown as a projection on the *P–T* plane in Fig. 14.6. The points 1 and 2 are, respectively, the triple points of A and B. The line 3–4 is the projection of the position of the minimum on the melting curves. This is the full line drawn from the minimum on the melting curve to the position on the 3-phase equilibrium in Fig. 14.5 Lines 5–6 and 6–7 are the corresponding minima on, respectively, the sublimation curves and the boiling curves. This is the broken line in Fig. 14.5 originating at the minimum on the front face, passing through the

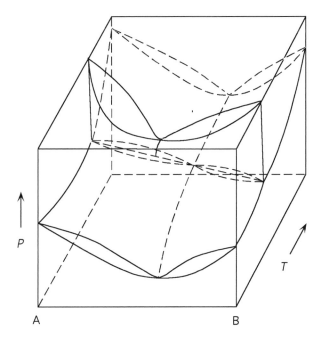

Figure 14.5 Minima on the melting, boiling, and sublimation curves of the binary system A–B which exhibits complete mutual solubility in the solid and liquid states.

point of 3-phase equilibrium, and terminating at the minimum on the rear face. The position of the line of 3-phase equilibrium in Fig. 14.6 is such that isobaric sections show two 3-phase equilibria in the range of pressure between slightly below point 1 and slightly above point 6 and isothermal sections show two 3-phase equilibria in the range of temperature between slightly below point 2 and slightly above point 4.

The isobaric sections at P_1 to P_4, indicated in Fig. 14.6a, are shown in Fig. 14.7. At P_1 the 2-phase α–L region originates at b, the melting temperature of B, passes through a minimum at a, where the compositions of the α and the liquid are the same, and terminates at d, the melting temperature of A. The 2-phase region v–L originates at c, the boiling temperature of B, and terminates at the upper edge of the diagram. The isobaric section at P_2 contains a 3-phase equilibrium at g. The region of 2-phase α–L equilibrium originates at h, the melting temperature of A, passes through a minimum at e, and terminates on the tie-line of the 3-phase equilibrium at g. The region of 2-phase v–L equilibrium originates at f, the sublimation point of B, and terminates on the 3-phase tie-line at g. The section requires that the composition of α in the 3-phase equilibrium be between those of the vapor and the liquid. The 2-phase v–L region originates on the 3-phase tie-line and terminates at the upper edge of the diagram. The section at P_3 contains two 3-phase equilibria—one at the temperature of l and the other at the lower temperature of j. The 2-phase α–v equilibrium originates at k and terminates on the tie-line of the 3-phase equilibrium at l. The

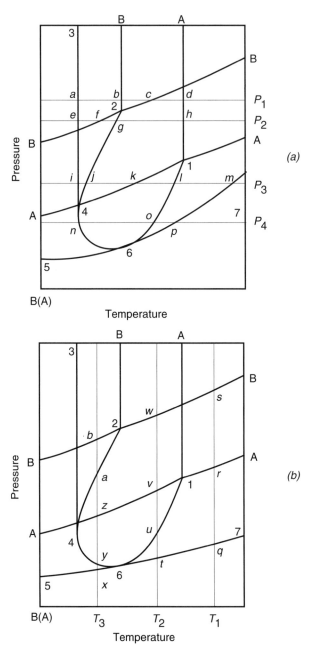

Figure 14.6   The *P–T* projection for the binary system shown in Fig. 14.5.

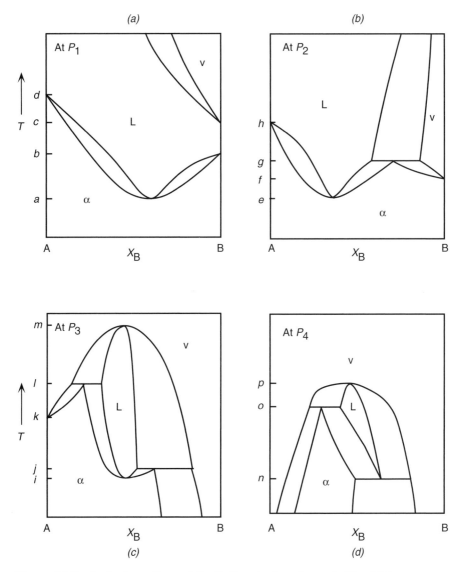

Figure 14.7 Isobaric sections of the *P–T* projection shown in Fig. 14.6*a*.

2-phase v–L equilibrium originates on the 3-phase tie-line at *l*, passes through a maximum at *m*, and terminates on the tie-line of the 3-phase equilibrium at *j*. The 2-phase region of the α–L equilibrium originates on the tie-line at *l*, passes through a minimum at *i*, and terminates on the tie-line of the 3-phase equilibrium at *j*. The region of the 2-phase α–v equilibrium originates on the tie-line at *j* and terminates at the lower edge of the diagram. The occurrence of two 3-phase equilibria requires that the stability field of the liquid phase be confined to an area in the center of the diagram. The section at P_4 contains the same two 3-phase equilibria as occur at P_3 at o and n, but does not show a minimum on the melting curve. This requires that the positions of the α and L phases on the tie-line at n on the P_4 isobar be the reverse of those of the α and liquid on the tie-line at j on the isobar at P_3.

The isothermal sections at the temperatures T_1 to T_3, indicated in Fig. 14.6b are shown in Fig. 14.8. At T_1 the region of 2-phase v–L equilibrium origi-

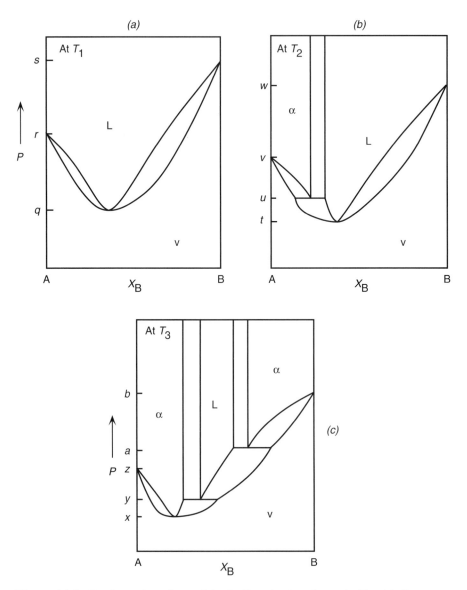

Figure 14.8 Isothermal sections of the P–T projection shown in Fig. 14.6b.

nates at r, the boiling point of A, passes through a minimum at q, and terminates at s, the boiling point of B. The section at T_2 contains a 3-phase equilibrium at u and a minimum on the boiling curve at t. The regions of 2-phase equilibria α–v and L–v originate, respectively, at v and w and terminate on the 3-phase tie-line. The limits of the α and liquid phase fields are virtually vertical lines. The section at T_3 contains two 3-phase equilibria, at y and a, and a minimum on the sublimation curve at x. The occurrence of the two 3-phase equilibria again requires that the field of stability of the liquid phase be confined to an area in the center of the diagram.

14.4 A BINARY SYSTEM CONTAINING A EUTECTIC EQUILIBRIUM AND HAVING COMPLETE MUTUAL SOLUBILITY IN THE LIQUID

The phase diagram, in pressure-temperature-composition space, for a binary system A–B containing a eutectic equilibrium and having complete mutual solubility in the liquid state is shown in Fig. 14.9. The points p and q are, respectively, the triple points of A and B and the system can exist in four phases: α-solid solutions, β-solid solutions, liquid, and as a mixture of vapors, each of which occurs in a discrete volume in the phase diagram. The invariant 4-phase equilibrium occurs at a(α)–b(L)–c(v)–d(β). The state volume occupied by α-solid solutions is shown in Fig. 14.10. Compositions on the surface a–g–j–n are equilibrium with β-solid solutions, compositions on the surface h–g–a–p are in equilibrium with mixtures of vapors, and compositions on the surface a–n–o–p are in equilibrium with liquid solutions. The state volume occupied by β-solid solutions is at the B-rich side of the diagram and is similar to that of the α-solid solutions. The state volume occupied by liquid solutions is shown in Fig. 14.11. Compositions on the surface s–p–b–q–t are in equilibrium with mixtures of vapors, compositions on the surface b–p–o–m are in equilibrium with α-solid solutions, and compositions on the surface q–b–m–r are in equilibrium with β-solid solutions. The state volume occupied by mixtures of vapors is shown in Fig. 14.12. Compositions on the surface t–s–p–c–q are in equilibrium with liquid solutions, compositions on the surface c–p–h–f are in equilibrium with α-solid solutions, and compositions on the surface c–f–i–q are in equilibrium with β-solid solutions.

In Fig. 14.9 the 3-phase α–β–v equilibrium exists on a–d–e–g with the α-solid solutions lying on a–g, mixtures of vapor lying on c–f, and the β-solid solutions lying on d–e. The 3-phase α–L–v equilibrium lies on a–b–p, with α lying on a–p, vapor lying on c–p, and liquid lying on b–p. The 3-phase β–L–v equilibrium lies on c–d–q with liquid on b–q, β on d–q, vapor on c–q. Finally, the 3-phase α–β–L equilibrium lies on a–d–l–n, with vapor on a–n, β on d–l, and liquid on b–m.

Fig. 14.13 shows the P–T projection for the system shown in Fig. 14.9. States 1 and 2 are, respectively, the triple points of the components B and A. Point 3 is the invariant state at which the 4-phase equilibrium (α–β–liquid–vapor) occurs in the system and the four lines (3–5, 3–2, 3–4, and 3–1) originating at the point 3

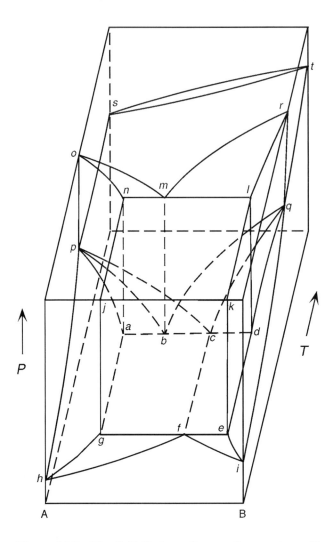

Figure 14.9 The *P–T–X* phase diagram for a system with a eutectic and complete solubility in the liquid state.

represent the four 3-phase equilibria occurring in the system. The occurrence of the 3-phase equilibrium liquid–α–β at a temperature lower than the melting temperatures of A and B indicates that the binary system contains a eutectic equilibrium.

Consider the isobaric section of Fig. 14.13*a* at the pressure P_1, which contains phase equilibria at the temperatures *a, b, c, d,* and *e;* the 3-phase α–β–L equilibrium at *a*, the melting of A at *b*, the boiling of A at *c*, the melting of B at *d*, and the boiling of B at *e*. The isobaric section is shown in Fig. 14.14*a*. Pure liquid B and pure vapor B coexist in equilibrium at *e*. Decreasing the temperature to values lower than *e* gives

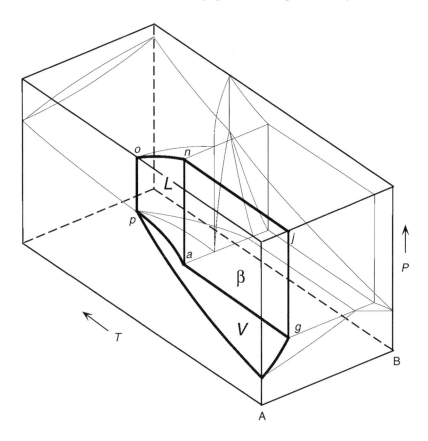

Figure 14.10   The α phase field in the *P–T–X* phase diagram for a system with a eutectic and complete solubility in the liquid state.

a 2-phase equilibrium between liquid solutions and mixtures of vapors, with compositions given by two lines originating on the pure B axis at *e* and moving into the binary. Phases in equilibrium with each other exist at the extremities of horizontal tie-lines drawn through the 2-phase region at any temperature and the lines coincide with each other and terminate on the pure A axis at *c*, the state in which pure liquid A coexists in equilibrium with pure vapor A. Liquidus and β-solidus lines originate at *d*, the melting temperature of B, and the liquidus and α-solidus lines originate at *b*, the melting temperature of A. These lines move down into the binary until the compositions of the liquidus melts coincide at *a*, the temperature of the 3-phase α–β–L eutectic equilibrium. The solvus lines begin at the compositions of α and β at *a* and move down into the phase diagram.

 Consider the isobaric section at P_2, which contains phase equilibria at *f, g, h, i,* and *j*: the 3-phase α–β–L equilibrium at *f*, the 3-phase α–L–v equilibrium at *g*, boiling of A at *h*, the 3-phase β–L–v equilibrium at *i*, and boiling of B at *j*. The procedure

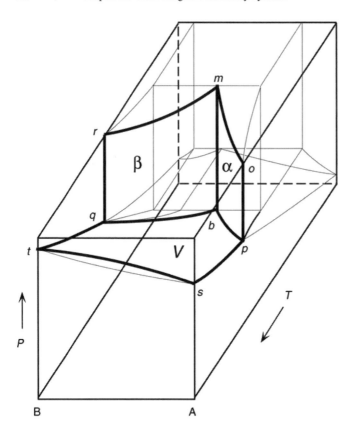

Figure 14.11 The liquid phase field in the *P–T–X* phase diagram for a system with a eutectic and complete solubility in the liquid state.

for constructing the isobaric section is shown in Fig. 14.14*b*. Lines representing the compositions of equilibrated β-solid solutions and mixtures of vapors originate at the boiling point of B (at the temperature *j*) and move down into the diagram. The lines terminate at *i*, the temperature at which the 3-phase v–β–L equilibrium occurs. The composition of the liquid solution at *a* lies between those of the β-solid solution and the mixture of vapors. Similarly, lines representing the compositions of equilibrated α-solid solutions and mixtures of vapors originate at *h* (the boiling temperature of A) and move downward into the diagram. These lines terminate at *g*, the temperature of the 3-phase L–v–α equilibrium where the composition of the vapor phase lies between those of the liquid solution and the α-solid solution. The 3-phase α–β–L equilibrium occurs at *f*, with the composition of the eutectic liquid lying between those of the α- and β-solid solutions. The phase diagram is completed by drawing a line connecting the β-phase at *i* to the β-phase at *f*, a line connecting the α-phase at *g* to the α-phase at *f*, a line connecting the vapor phase at *i* to the vapor phase at *g*, and three lines joining the compositions of the liquid solutions at *i*, *g*, and *f*.

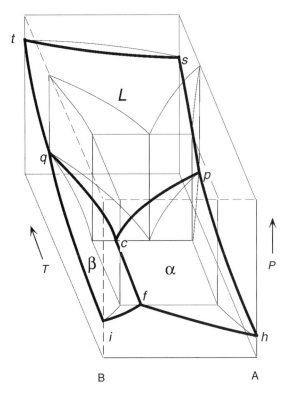

Figure 14.12   The vapor phase field in the $P-T-X$ diagram for a system with a eutectic and complete solubility in the liquid state.

The addition of the α- and β-solvus lines gives the complete phase diagram as shown in Fig. 14.14c. In the isobaric section at P_2 the compositions of the liquids are restricted to a small area in the middle of the phase diagram.

The isobaric section at P_3 is shown in Fig. 14.14d. The pressure P_3 is lower than the lowest pressure at which a liquid phase exists (the pressure at the state 3 in Fig. 14.13a). Therefore the section at P_3 does not contain a liquid phase. The section contains the 3-phase $\alpha-\beta-v$ equilibrium at k, and the boiling points of B and A at, respectively, m and l. Construction of the section is similar to that of the section at P_1. Lines representing the compositions of equilibrated solid solutions and mixtures of vapors originate at l and m and move down into the diagram. The lines terminate at k, where the compositions of the mixtures of vapors coincide and the 3-phase equilibrium $\alpha-\beta-v$ is established.

Consider the isothermal section of Fig. 14.13b at the temperature T_1 which contains phase equilibria at n, o, p, and q; boiling of B at n, boiling of A at o, the 3-phase $\beta-L-v$ equilibrium at p, and the 3-phase $\alpha-L-v$ equilibrium at q. The construction of this isothermal section is illustrated in Fig. 14.15a. Lines representing the compositions of equilibrated β-solid solutions and mixtures of vapor originate at the boil-

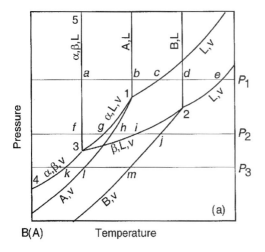

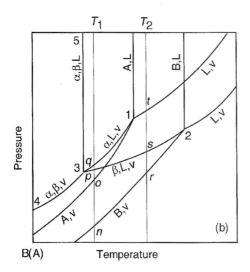

Figure 14.13 *P–T* projection of the system shown in Fig. 14.9.

ing point of B at *n* and move up into the diagram. These lines terminate at *p*, at which state the 3-phase β–L–v equilibrium exists, with the composition of the liquid lying between those of the β-phase and the mixture of vapors. Similarly, lines representing the compositions of equilibrated α-solid solutions and mixtures of vapors originating at the boiling point of A, at *o*, move up into the diagram and terminate at *q*, the state of the 3-phase L–v–α equilibrium. The phase diagram is completed by drawing two lines from *o* to *p*, one connecting the liquid phases and one connecting the vapor phases. As the compositions of the condensed phase are

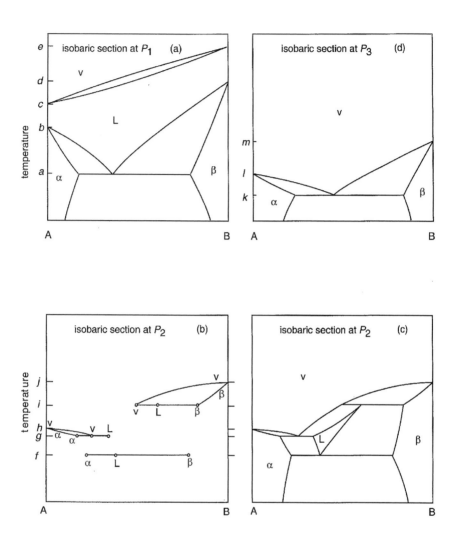

Figure 14.14 Isobaric sections of Fig. 14.13*a*.

not noticeably dependent on temperature, the limits of existence of the α and β phases and the liquid phase are virtually vertical lines originating at *p* and *q* as shown in Fig. 14.15*b*.

The isothermal section at T_2 is shown in Fig. 14.15*c*. This section contains phase equilibria at *r*, *s*, and *t;* boiling of B at *r*, the 3-phase β–L–v equilibrium at *s*, and boiling of A at *t*. Lines representing the compositions of equilibrated β-solid solutions and mixtures of vapors originate at *r* and move up into the diagram, terminating at *s*, where the 3-phase β–L–v equilibrium exists. Lines representing the compositions of equilibrated liquid solutions and mixtures of vapors originate at *t* and

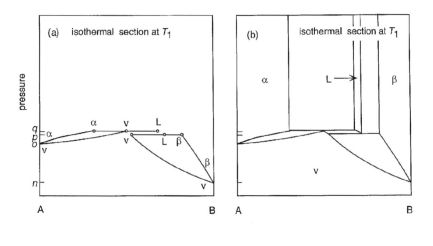

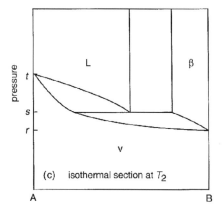

Figure 14.15 Isothermal sections of Fig. 14.13b.

move down in the diagram terminating at s and the limits of the compositions of the β- and liquid phases are again virtually vertical lines originating at s.

The isobaric section through state 3 in Fig. 14.13a is shown in Fig. 14.16a. At this pressure the three 3-phase equilibria shown in Fig. 14.14c coincide to produce a 4-phase equilibrium in which state the liquid occurs at a single composition at the temperature of the 4-phase equilibrium. The isothermal section through the state 3 in Fig. 14.13b is shown in Fig. 14.16b. At this temperature the two 3-phase equilibria shown in Fig. 14.15b coincide and the liquid phase, which exists only at pressures equal to or higher than that of the 4-phase equilibrium, has a fixed composition.

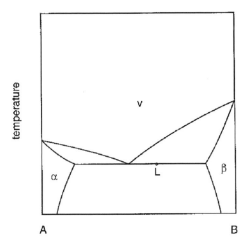

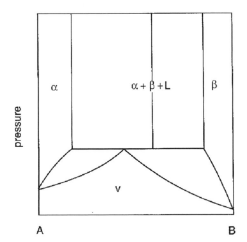

Figure 14.16 (*a*) Isobaric section of Fig. 14.9 at the pressure at which the 4-phase equilibrium occurs. (*b*) Isothermal section of Fig. 14.9 at the temperature at which the 4-phase equilibrium occurs.

14.5 A BINARY SYSTEM CONTAINING A PERITECTIC EQUILIBRIUM AND HAVING COMPLETE MUTUAL SOLUBILITY IN THE LIQUID STATE

The phase diagram, in pressure-temperature-composition space, for a binary system A–B containing a peritectic equilibrium and having complete mutual solubility in the liquid state is shown in Fig. 14.17. The points p and q are, respectively, the

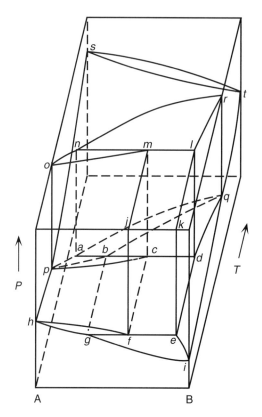

Figure 14.17 The P–T–X phase diagram for a system with a peritectic and complete solubility on the liquid state.

triple points of A and B and the system can exist in four phases: α-solid solutions, β-solid solutions, liquid, and as a mixture of vapors, each of which has a discrete volume in the phase diagram. The invariant 4-phase equilibrium occurs at $a(L)$–$b(v)$–$c(\alpha)$–$d(\beta)$. The state volume occupied by the β-solid solutions is shown in Fig. 14.18. Compositions on the surface e–k–l–d are in equilibrium with α-solid solutions, compositions on the surface l–r–q–d are in equilibrium with liquid solutions, and compositions on the surface e–d–q–i are in equilibrium with mixtures of vapor. The state volume occupied by liquid solutions is shown in Fig. 14.19. Compositions on the surface a–p–o–n are in equilibrium with α-solid solutions, compositions on the surface a–n–r–q are in equilibrium with β-solid solutions and compositions on the surface t–s–p–a are in equilibrium with mixtures of vapors. The state volume occupied by mixtures of vapor is shown in Fig. 14.20. The v–L–α equilibrium occurs in a–c–p, with the composition of the vapor phase lying on b–p, the composition of the liquid solution lying on a–p, and the composition of the α phase lying on c–p.

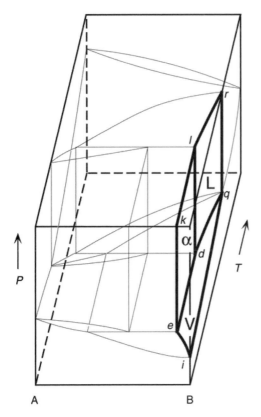

Figure 14.18 The β phase field in the P–T–X phase diagram for a system with a peritectic and complete solubility in the liquid state.

The β–L–v equilibrium occurs in a–q–d, with the compositions of the liquid solutions lying on a–q, the compositions of the vapors lying on b–q, and the compositions of the β-solid solutions lying on the line d–q. The v–α–β equilibrium occurs in b–d–e–g with the composition of the vapor lying on b–g, the composition of the α-solid solution lying on c–f, and the composition of the β-solid solution lying on d–e.

Fig. 14.21 shows the P–T projection for the binary system A–B containing a peritectic equilibrium. States 1 and 2 are, respectively, the triple points of the components A and B. Point 3 is the invariant state where the 4-phase equilibrium (α–β–liquid–vapor) occurs and the four lines (3–5, 3–2, 3–4, and 3–1) originating at the point 3 are for the four 3-phase equilibria. The occurrence of the 3-phase equilibrium liquid–α–β at a temperature between the melting temperatures of A and B indicates that the binary system contains a peritectic equilibrium.

Consider the isobaric section of Fig. 14.21a at the pressure P_1, which contains phase equilibria at the temperatures a, b, c, and d: transformation of solid A to A vapor a, the 3-phase α–L–v equilibrium at b, the 3-phase α–β–L equilibrium at c, and the melting of pure B at d. The procedure for constructing the isobaric section is shown in Fig. 14.22a. Pure solid A and A vapor coexist in equilibrium at a.

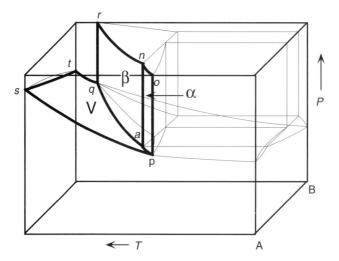

Figure 14.19 The liquid phase field in the P–T–X phase diagram for a system with a peritectic and complete solubility in the liquid state.

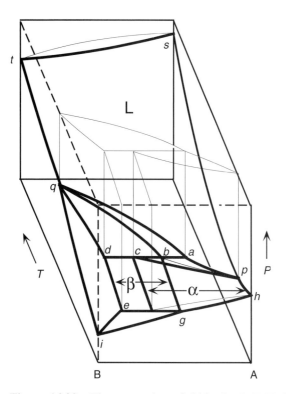

Figure 14.20 The vapor phase field in the P–T–X phase diagram for a system with a peritectic and complete solubility in the liquid state.

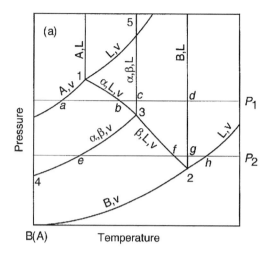

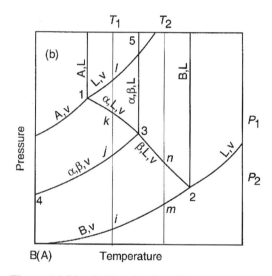

Figure 14.21 *P–T* projection of the phase diagram shown in Fig. 14.17.

Increasing the temperatures to values higher than *a* gives a 2-phase equilibrium between α-solid solutions and a mixture of A and B vapor, with compositions given by two lines originating on the pure B-axis at *a* and moving into the binary. The upper line gives the compositions of the mixtures of vapors and the lower line gives the compositions of the α-solid solutions. Equilibrated compositions exist at the extremities of horizontal tie-lines drawn through the two-phase region at any temperature.

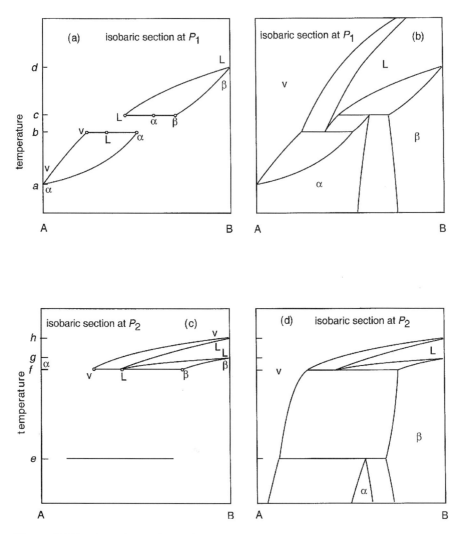

Figure 14.22 Isobaric sections of Fig. 14.21 at P_1 (a) and (b); at P_2 (c) and (d).

The 3-phase equilibrium α–L–v exists at b, with the composition of the liquid being between that of α and the mixture of vapors. This 3-phase equilibrium at b is shown in Fig. 14.22a. At the temperature d, pure liquid B is in equilibrium with pure solid B and, as the temperature is decreased to values lower than d, lines giving the compositions of equilibrated β-solid solutions and liquid solutions originate on the B-axis at d and move into the binary. This 2-phase equilibrium exists until the temperature reaches c, where the 3-phase equilibrium α–β–L occurs. The α phase has a composition between those of the β and the liquid and this equilibrium is shown at *c* in Fig. 14.22*a*. The isobaric section, shown in Fig. 14.22*b*, is obtained by drawing a line from the composition of α at *c* to the composition of α at *b* and drawing a similar line between L at *c* and L at *b*. These lines define the region of equilibrium between α-solid solutions and liquid solutions. Fig. 14.21 at P_1 shows that liquid B boils at some temperature higher than the maximum shown in Fig. 14.21, in which case the region of 2-phase equilibrium between liquid solutions and mixtures of vapors is as shown in Fig. 14.22*b*. Finally, the region of 2-phase equilibrium between α- and β-solid solutions is defined by lines starting at the α and β compositions at temperature *c* and moving downwards into the diagram as shown. The solubilities of B in α and A in β decrease with decreasing temperature.

Consider the isobaric section of Fig. 14.21*a* at the pressure P_2 which contains phase transformations at the temperatures *e*, *f*, *g*, and *h*, involving the 3-phase equilibrium α–β–v at *e*, the 3-phase β–L–v equilibrium at *f*, and the melting and boiling of pure B at, respectively, *g* and *h*. Construction of the isobaric section is outlined in Fig. 14.22*c*. Lines representing the compositions of equilibrated liquid solutions and mixtures of vapor originate at *h* and move into the diagram with decreasing temperature. Similarly, lines representing the compositions of equilibrated β-solid solutions and liquid solutions originate at *g* and move into the binary system with decreasing temperature. The compositions of the liquid solutions coincide at the temperature *f*, where the 3-phase β–L–v equilibrium is established. In order that lines on the phase diagram do not intersect with one another the left extremity on the line at *e* represents the composition of v in the 3-phase α–β–v equilibrium, the right extremity represents the composition of the β, and the composition of the α-solid solution lies between those of the β and the vapor. As shown in Fig. 14.22*d*, the 2-phase β–v region is defined by drawing a line from the composition of β at *f* to the composition of β at *e*, and by drawing a similar line between the compositions of the vapor phases at *f* and *e*. The β-phase field is defined by the two lines originating from the composition of α at *e* and the complete isobaric section is as shown in Fig. 14.22*d*.

Construction of the isothermal section at T_1 in Fig. 14.21*b* is illustrated in Fig. 14.23*a*. The section contains four phase equilibria: sublimation of B at *i*, the 3-phase α–β–v equilibrium at *j*, the 3-phase α–L–v equilibrium at *k*, and melting of A at *l*. The 2-phase β–v region originates on the pure B-axis and moves up in the diagram to terminate at *j*, the state of the 3-phase α–β–v equilibrium. The 2-phase L–v region originates on the pure A-axis at *l* and moves down in the diagram to terminate at *k*, the state of the 3-phase α–L–v equilibrium. At this equilibrium the composition of the liquid phase lies between those of the α and the vapor phases. The diagram is completed by joining the α phases and the vapor phases at *j* and *k* and drawing virtually vertical lines to delineate the 2-phase α–β and α–L regions. The complete diagram is shown in Fig. 14.23*b*. Construction of the isothermal section at T_2 is illustrated in Fig. 14.23*c* and the section is shown in Fig. 14.23*d*.

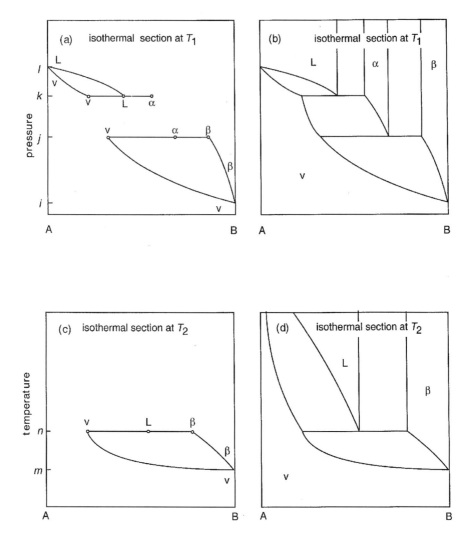

Figure 14.23 Isothermal sections of Fig. 14.21 at T_1 (a) and (b); at T_2 (c) and (d).

Fig. 14.24a shows the isobaric section of the phase diagram at the pressure at which the 4-phase α–β–L–v occurs. This equilibrium occurs at the unique pressure at which the two 3-phase equilibria in Fig. 14.22b and Fig. 14.22d merge with each other. The isothermal section of the phase diagram at the temperature at which the 4-phase equilibrium occurs is shown in Fig. 14.24b.

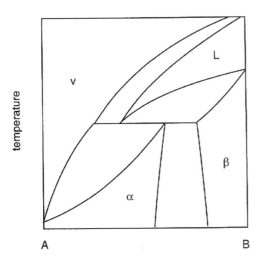

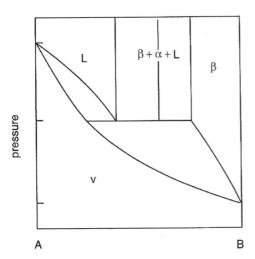

Figure 14.24 (*a*) Isobaric section at the pressure at which the 4-phase equilibrium exists. (*b*) Isothermal section at the temperature at which the 4-phase equilibrium exists.

14.6 PHASE EQUILIBRIUM IN A BINARY SYSTEM CONTAINING AN INTERMEDIATE γ PHASE THAT MELTS, SUBLIMES, AND BOILS CONGRUENTLY

Fig. 14.25*a* shows the *P–T* projection for a binary A–γ system in which the composition of the γ phase sublimes, melts, and boils congruently. States 1 and 3 are, respectively, the triple points of A and γ and the invariant 4-phase equilibrium occurs

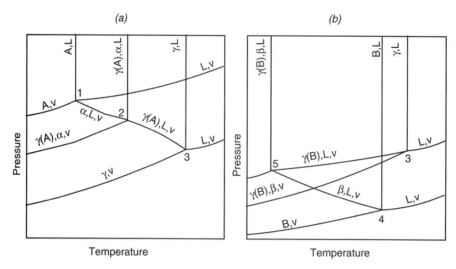

Figure 14.25 (*a*) A binary system A–γ containing a peritectic equilibrium. (*b*) a binary system B–γ containing a eutectic equilibrium.

at state 2. The designation γ(A) indicates the composition of the γ phase which is saturated with A. The temperature where the 3-phase equilibrium γ(A)–α–L occurs is between the melting temperatures of A and congruently melting γ and thus the 3-phase equilibrium is a peritectic. Fig. 14.25*b* shows the corresponding *P–T* projection for the binary system B–γ. The states 3 and 4 are, respectively, the triple points of γ and B and the 4-phase equilibrium occurs at state 5. The designation γ(B) indicates the composition of the γ phase that is saturated with B and the occurrence of the 3-phase equilibrium γ(B)–β–L at a temperature lower than the melting temperatures of B and γ indicates that the 3-phase equilibrium is a eutectic.

Figs. 14.25*a* and *b* are combined in Fig. 14.25*c* which is the *P–T* projection of the binary system A–B that contains an intermediate γ phase which sublimes, melts, and boils congruently. The isobaric section of the diagram at P_1 is shown in Fig. 14.26*a*. The tie-lines giving the compositions in the two 3-phase equilibria occur at *a*, the eutectic equilibrium in which the composition of the L lies between γ(B) and β, and at *c*, the peritectic equilibrium where the composition of α lies between those of L and γ(A). The region of 2-phase equilibrium β–L originates at *d*, the melting temperature of B, and terminates at *a*, and the region of 2-phase equilibrium α–L originates at *b* and terminates at *c*. The solidus and liquidus lines of γ originate at *c*, pass through the congruent melting temperature at *e*, where the γ phase and the liquid phase have the same composition and terminate at *a*. The addition of four solvus lines completes the section.

The isobaric section of the diagram at P_2 is shown in Fig. 14.26*b*. This section contains three 3-phase equilibria, α–L–v at *g*, γ(A)–L–α at *h*, and γ(B)–β–L at *f*. The region of 2-phase equilibrium v–α originates at *f*, the boiling temperature of A,

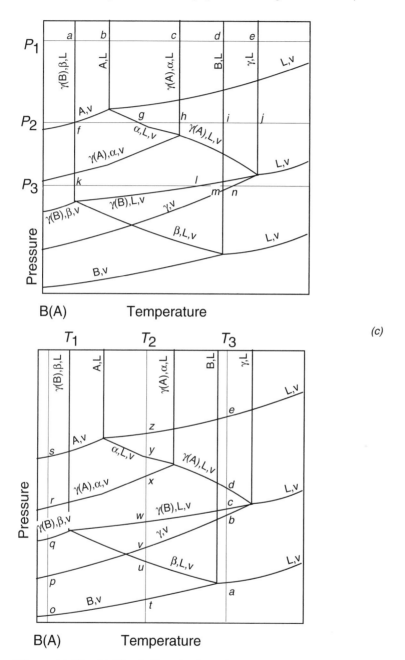

Figure 14.25 (*c*) The P–T projection of a binary system A–B which contains an intermediate phase γ that melts, sublimes, and boils congruently.

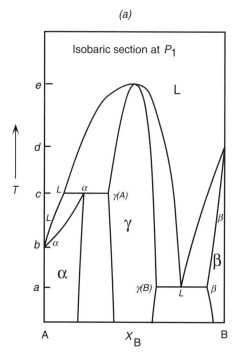

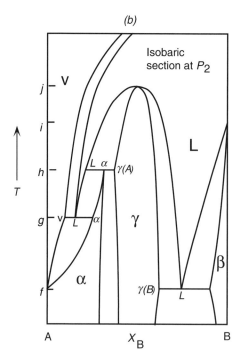

Figure 14.26 (*a*) Isobaric section of Fig. 14.25*c* at P_1; (*b*) at P_2.

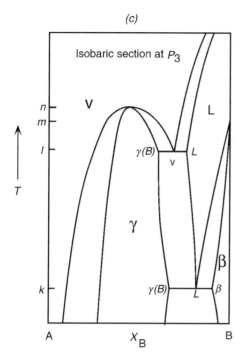

Figure 14.26   (c) Isobaric section of Fig. 14.25c at P_3.

and terminates on the 3-phase tie-line at g. In this equilibrium the composition of the liquid phase lies between those of v and α. The region of the 2-phase L–α equilibrium is defined by drawing two lines, one joining the liquid compositions at g and h and one joining the α compositions at g and h. The eutectic equilibrium at f is similar to that at a in Fig. 26a. The liquidus and solidus lines for the γ phase are also drawn from h through the congruent melting temperature at j to f in a manner similar to that shown in Fig. 26a. The region of the 2-phase equilibrium v–L originates at g and runs out of the diagram at its upper edge, which is at a temperature lower than the boiling temperature of B. The addition of the four solvus lines completes the diagram.

The isobaric section of the P–T projection at P_3, shown in Fig. 14.26c, contains the 3-phase equilibrium γ(B)–β–L at k and the 3-phase equilibrium γ(B)–L–v at l. The region of the 2-phase v–γ equilibrium enters the diagram at its lower edge, goes through the point of congruent sublimation at n, and terminates at l. The region of 2-phase equilibrium v–L originates at l and terminates at the upper edge of the diagram. The eutectic equilibrium at k is the same as occurs in Fig. 14.26a and Fig. 14.26b. The region of 2-phase γ(B)–L equilibrium originates at k and terminates at l. Adding the solvus lines completes the diagram.

The isothermal section of the P–T projection in Fig. 14.25d at T_1, which is shown in Fig. 14.27a, contains two 3-phase equilibria: γ(A)–α–v at r and γ(B)–

(a)

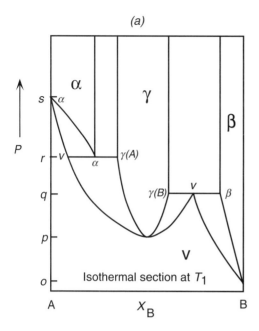

(b)

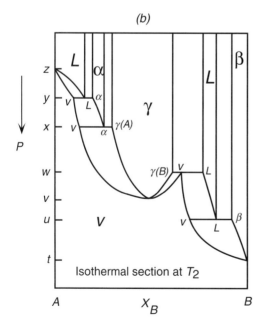

Figure 14.27 *(a)* Isothermal section of Fig. 14.25c at T_1; *(b)* at T_2.

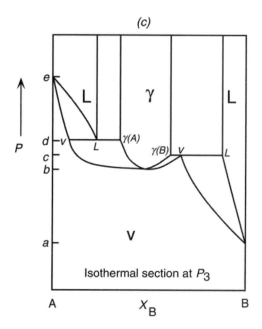

Figure 14.27 (c) Isothermal section of Fig. 14.25c at T_3.

β–v at q. The region of 2-phase equilibrium v–α originates at s, the sublimation point of A, and terminates at r, and the region of the 2-phase α–γ equilibrium is defined by vertical lines rising from the compositions α and γ(A) on the tie-line at r. The region of the 2-phase v–β equilibrium originates at o, the sublimation state of B, and terminates at q, and the region of the 2-phase equilibrium γ–β is defined by vertical lines rising from the compositions γ(B) and β on the tie-line at q. The region of 2-phase equilibrium v–γ begins at r, passes through the point of congruent sublimation of γ at p and ends at q.

The isothermal section at T_2, shown in Fig. 14.27b, evolves from Fig. 14.27a. The v–α–γ(A) equilibrium at r in Fig. 14.27a is replaced by two 3-phase equilibria v–L–γ(A) at x and v–α–L at the higher pressure y. Similarly the 3-phase equilibrium γ(B)–v–β at q in Fig. 14.27a is replaced by two 3-phase equilibria, v–L–β at u and γ(B)–v–L at the higher pressure w. Thus regions of the 2-phase equilibrium v–L can be identified between x and y and between u and w. The regions of 2-phase equilibria α–L and L–v are defined by vertical lines rising from, respectively, y and x and the regions of 2-phase equilibria β–L and L–γ are defined by vertical lines rising from u and w.

In the isothermal section at P_3, shown in Fig. 14.27c, the 3-phase equilibria at y and u in Fig. 14.27b have disappeared and Fig. 14.27c is geometrically similar to Fig. 27a, except that the fields of stability of the α- and β-solid solutions in Fig. 14.27a are replaced by fields of stability of the liquid phase.

14.7 PHASE EQUILIBRIUM IN A BINARY SYSTEM CONTAINING AN INTERMEDIATE γ PHASE THAT MELTS AND SUBLIMES CONGRUENTLY AND BOILS INCONGRUENTLY

Figs. 14.28*a* and *b* show, respectively, the *P–T* projections of binaries in the systems A–γ and B–γ that contain eutectic equilibria and in which melting and sublimation of the γ phase are congruent and boiling of the γ phase is incongruent. Inasmuch as the triple point is defined as being the point of intersection of the three lines of 2-phase equilibrium, the absence of a boiling line in Fig. 14.28*a* means that state 3 cannot be identified as being a triple point. It is simply the state at which the 3-phase equilibrium occurs. The two sections in Fig. 14.28*a* and Fig. 14.28*b* are combined in Fig. 14.28*c* and Fig. 14.28*d*.

The isobaric section at P_1 is easily discerned and is shown in Fig. 14.29*a*. It contains the eutectic equilibrium β–L–γ at *a*, the eutectic equilibrium α–L–γ at *b*, and the melting temperatures of B, A, and γ at, respectively, *c*, *d*, and *e*.

The isobaric section at P_2 is shown in Fig. 14.29*b*. The eutectic equilibrium α–L–γ occurs at *h* and the incongruent decomposition of γ to liquid and vapor begins at *k*. The occurrence of three 3-phase equilibria on the B side of the section accommodates the isolation of fields of 1-phase stability for liquid and β. At *i* and *f* the phase at the left extremity of the tie-line must be γ(B) and in the equilibria at *i* and *g* the phase at the right extremity of the tie-line must be vapor. This fixes the position of L at *i*, and hence at *g* and *f*, which, in turn, fixes the positions of β at *g* and *f*.

The isobaric section at P_3 is shown in Fig. 14.29*c*. The 3-phase equilibrium at *l* is the point at which incongruent decomposition of γ to α and vapor begins on heat-

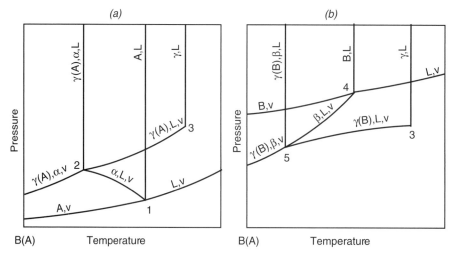

Figure 14.28 (*a*) and (*b*) *P–T* projections of, respectively, binaries in the systems A–γ and B–γ that contain eutectic equilibria and in which melting and sublimation of the γ phase are congruent and boiling of the γ phase is incongruent.

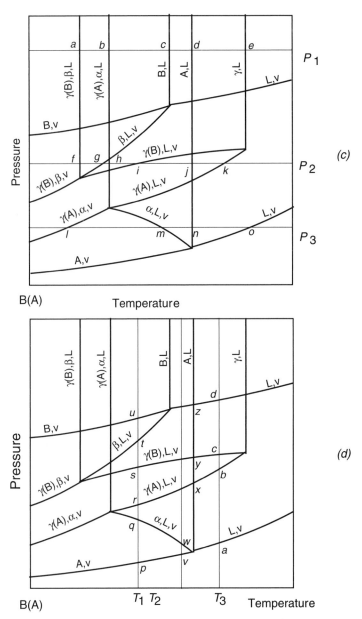

Figure 14.28 (c) and (d) *P–T* projections of a binary system A–B that contains an intermediate phase γ which melts and sublimes congruently and boils incongruently.

(a)

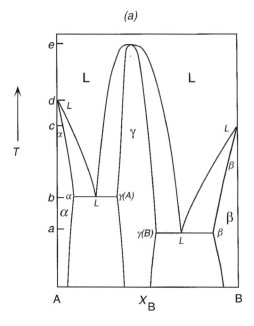

(b)

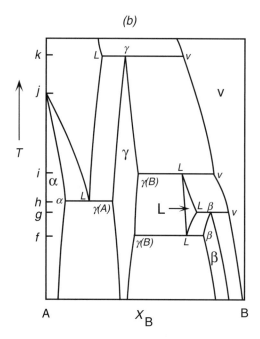

Figure 14.29 *(a)* Isobaric section of Fig. 14.28c at P_1; *(b)* at P_2.

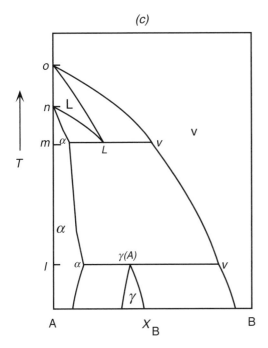

Figure 14.29 (*c*) Isobaric section of Fig. 14.28*c* at P_3.

ing and the 3-phase equilibrium α–L–v occurs at *m*. Lines drawn from *o* and *n*, respectively, the boiling and melting temperatures of A, define the fields of stability of α, liquid, and vapor.

The isothermal section of the *P–T* diagram at T_1, shown in Fig. 14.30*a*, contains four 3-phase equilibria each one of which involves a vapor phase and a liquid phase. The 3-phase equilibrium at the lowest pressure *q* involves an α-solid solution, the equilibria at *r* and *s* involve a γ-solid solution, and the equilibrium at *t* involves a β-solid solution. The region of 2-phase α–v equilibrium originates at *p* and terminates at *q* and a vertical line rising from the composition of the α-phase at *q* defines the field of stability of α. The region of 2-phase L–v equilibrium exists between *q* and *r* and vertical lines rising from the compositions of the liquids at *q* and *r* define the field of stability of a liquid. The region of 2-phase γ–v equilibrium exists between *r* and *s* and vertical lines drawn from the compositions of γ(A) at *r* and γ(B) at *s* define the field of stability of γ. A second region of 2-phase L–v equilibrium exists between *s* and *t* and vertical lines rising from the compositions of the liquids on the tie-lines at *s* and *t* define the field of stability of a second liquid. The region of 2-phase stability β–v originates at *u* and terminates at *t*, and this, along with a vertical line rising from the composition of β at *t*, defines the field of stability of β.

As shown in Fig. 14.30*b*, increasing the temperature from T_1 to T_2 causes the elimination of the field of stability of β from the isothermal section. The

(a)

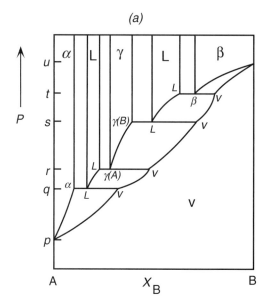

(b)

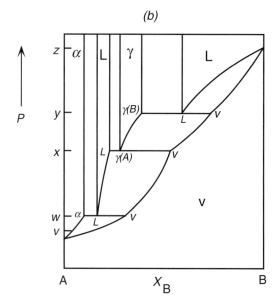

Figure 14.30 *(a)* Isothermal section of Fig. 14.28*d* at T_1; *(b)* at T_2.

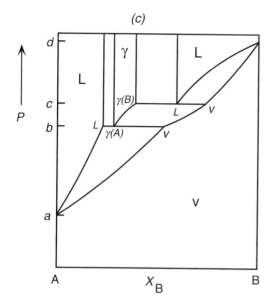

Figure 14.30 (*c*) Isothermal section of Fig. 14.28*d* at T_3.

approach to the construction of Fig. 14.30*b* is the same as that used for the construction of Fig. 14.30*a*.

Finally, as shown in Fig. 14.30*c*, increasing the temperature from T_1 to T_3 eliminates the two fields of stability of liquid occurring in the isothermal section at T_1.

14.8 PHASE EQUILIBRIUM IN A BINARY SYSTEM WITH A EUTECTIC AND ONE COMPONENT THAT EXHIBITS ALLOTROPY

The phase diagram, in pressure-temperature-composition space for a binary system with a eutectic and one component that exhibits allotropy, is shown in Fig. 14.31. The upper face of the figure shows a set of phase equilibria among condensed phases. The eutectic occurs at *p–i–q*. The temperature of transition of the low-temperature α allotrope of A to the high-temperature γ allotrope of A is given by *e*, and A and B melt at, respectively, *f* and *g*. Solid B exists in the α form and a continuous series of α-solid solutions exists between A and B. The 4-phase equilibrium occurs at *a*(v)–*b*(γ)–*c*(L)–*d*(α) and the four 3-phase equilibria occur at v–L–α (*a–d–k*), v–L–α (*a–d–j*), v–γ–L (*a–c–h*), and γ–L–α (*b–d–q–p*).

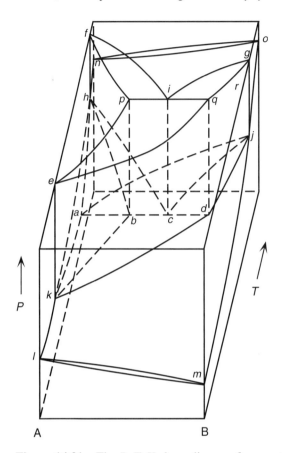

Figure 14.31 The P–T–X phase diagram for a system with a eutectic and one component that exhibits allotropy. The low-temperature allotrope of A forms a continuous series of solid solutions with B.

The state space occupied by the α-solid solutions is shown in Fig. 14.32. Solid solutions in equilibrium with γ-solid solutions exist on the surface e–q–d–k, α-solid solutions in equilibrium with the vapor phase exist on the surface k–d–j–m–l, and α-solid solutions in equilibrium with liquid solutions exist on the surface q–g–j–d. The corresponding state spaces occupied by the γ-solid solutions and the mixtures of vapors are shown, respectively, in Fig. 14.33 and Fig. 14.34. The state space occupied by the liquid solutions is the same as that shown in Fig. 14.11.

The P–T projection of the phase diagram is shown in Fig. 14.35. The invariant state of 4-phase equilibrium occurs at state 1 and the three triple points occur at states 2, 3, and 4.

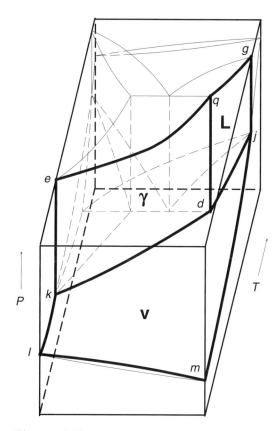

Figure 14.32 The α phase field in the *P–T–X* phase diagram for the system shown in Fig. 14.31.

The isobaric section of Fig. 14.35a at P_1 and shown in Fig. 14.36a contains two 3-phase equilibria; γ–L–α at *b* and v–γ–L at *e*. Transformation of A from the α-form to the γ-form occurs at *a* and the region of 2-phase α–γ equilibrium moves into the diagram from *a* and terminates at *b*. The region of 2-phase γ–v equilibrium originates at *c* and terminates at *e*. Phase equilibrium requires that the composition of γ lie between those of v and L at *e*, and that the composition of L lie between γ and α at *b*. The region of 2-phase equilibria γ–L, which is restricted to temperatures between *b* and *e*, is defined by drawing two lines between *b* and *e*, one connecting the compositions of γ and one connecting the compositions of liquid. The region of 2-phase equilibrium L–α originates at *d*, the melting temperature of B, and terminates at *b* and the region of 2-phase equilibrium v–L moves from *e* and exits at the upper edge of the diagram.

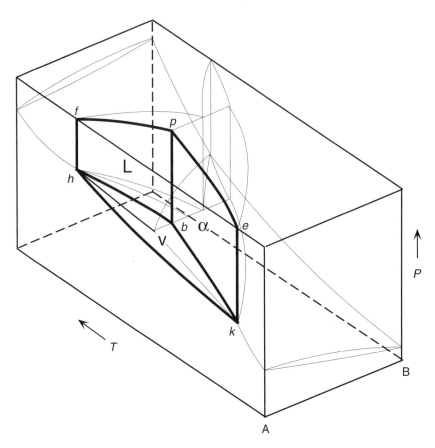

Figure 14.33 The γ phase field in the *P–T–X* phase diagram shown in Fig. 14.31.

The isobaric section of Fig. 14.35*a* at P_2 is shown in Fig. 14.36*b*. This section contains the two 3-phase equilibria shown in Fig. 14.36*a*: v–γ–L at *h*, γ–L–α at *g*, and a third 3-phase equilibrium v–γ–α at *f*. The section thus contains an isolated field of phase stability of γ and contains the regions of 2-phase equilibria: γ–α between *f* and *g*, and v–γ between *g* and *h*.

The isothermal section of Fig. 14.35*b* at T_1 is shown in Fig. 14.37*a*. The 3-phase v–γ–α equilibrium at *k* is the state at which the 2-phase equilibria v–γ, γ–α, and v–α meet. In the isothermal section at T_2, shown in Fig. 14.37*b* the 3-phase equilibrium v–γ–α, shown in Fig. 14.37*a*, has been replaced by two 3-phase equilibria v–γ–L at *o* and v–L–α at *n*, which gives a field of stability of liquid solutions between those of γ and α.

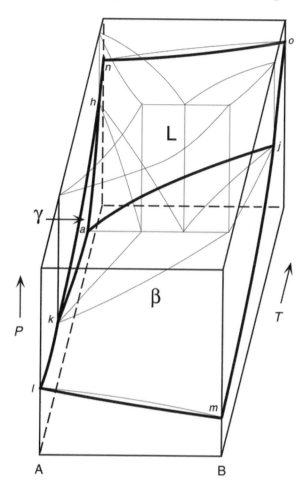

Figure 14.34 The vapor phase field in the P–T–X phase diagram for the system shown in Fig. 14.31.

14.9 A BINARY EUTECTIC SYSTEM IN WHICH BOTH COMPONENTS EXHIBIT ALLOTROPY

The P–T projection of a binary system in which both components exhibit allotropy is shown in Fig. 14.38. In this system A exists as low-temperature α and high-temperature γ, B exists as low-temperature β and high-temperature δ, the low-temperature allotropes are completely miscible in each other and the high-temperature allotropes are completely miscible in each other. Both components contain two triple points: α–γ–v at state 1 and γ–L–v at state 2 for component A, and β–δ–v at state 3 and δ–L–v at state 4 for component B. The isobaric section at P_1, shown in Fig. 14.39,

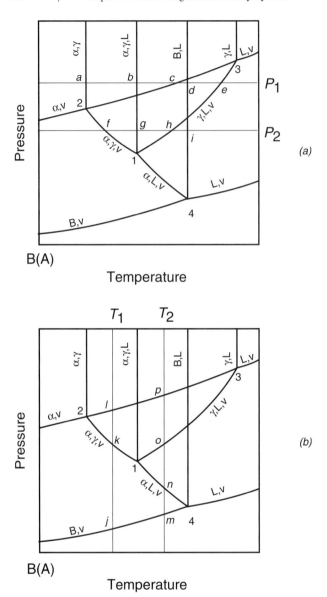

Figure 14.35 *P–T* section of the phase diagram shown in Fig. 14.31.

contains two regions of 2-phase equilibrium: low temperature solid solutions in equilibrium with high temperature solid solutions and high temperature solid solutions in equilibrium with liquid solutions.

The *P–T* projection of a binary system in which B is partially soluble in α, A is partially soluble in β, and the high-temperature allotropes are completely soluble in each other is shown in Fig. 14.40, and the isobaric section at P_1 is shown in Fig. 14.41. The section shows a eutectoid equilibrium at *a*.

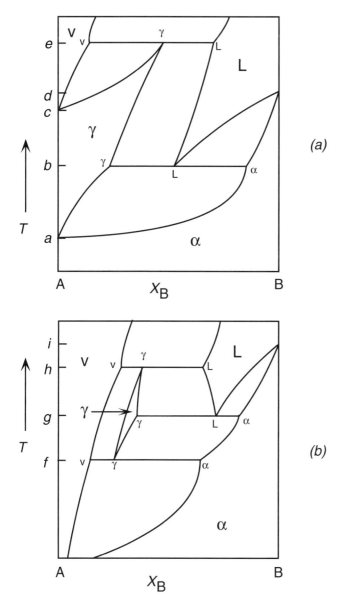

Figure 14.36 (*a*) Isobaric section of Fig. 14.35*a* at P_1; (*b*) of Fig. 14.35*a* at P_2.

The *P–T* projection of a binary system in which both the low-temperature and the high-temperature allotropes are partially soluble in each other is shown in Fig. 14.42 and the isobaric section at P_1 is shown in Fig. 14.43. The existence of five fields of 1-phase equilibrium requires that three 3-phase equilibria occur. The eutectoid equilibrium occurs at *a*, with the δ phase having a composition between

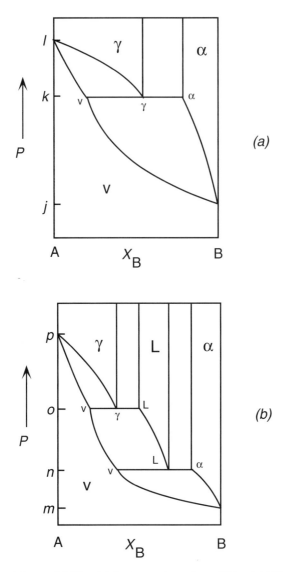

Figure 14.37   (*a*) Isothermal section of Fig. 14.35b at T_1; (*b*) at T_2.

those of α and β. The lines defining the field of 2-phase δ–β equilibrium start at δ and β at *a* and terminate at *c*, the state at which the low- and high-temperature iso-topes of B are in equilibrium with each other. The lines defining the field of 2-phase α–δ equilibrium begin at the α and δ compositions at *a* and terminate at the compositions α and δ at the 3-phase equilibrium at *d*. The region of 2-phase γ–α equilibrium originates at *b* and terminates at the γ and α compositions at *d*. The re-gion of 2-phase equilibrium γ–δ begins at *d* and terminates at *e*, the eutectic equi-

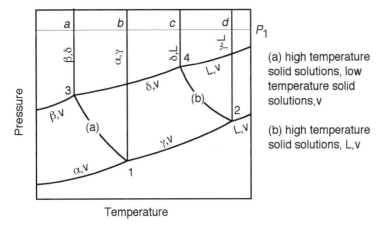

Temperature

Figure 14.38 The *P–T* projection of a binary system in which both components exhibit allotropy. A exists as low-temperature α and high-temperature γ, and B exists as low-temperature β and as high-temperature δ. The low-temperature allotropes are completely miscible in each other, as are the high-temperature allotropes.

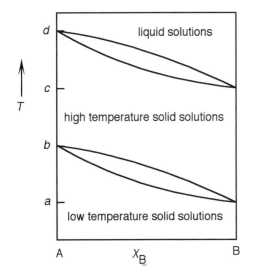

Figure 14.39 The isothermal section of Fig. 14.38 at P_1.

librium γ–L–δ. The liquidus and solidus lines for γ and δ originate at the respective melting temperatures and terminate at the eutectic, where the melt is doubly saturated with A and B.

The isobaric section of Fig. 14.42 at P_2, shown in Fig. 14.44, contains five 3-phase equilibria. The eutectoid equilibrium, occurring at *a* in Fig. 14.40 occurs

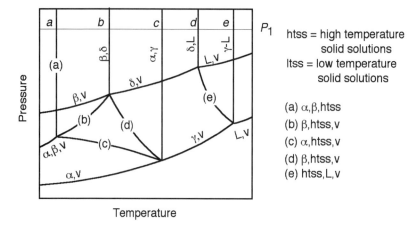

Figure 14.40 The *P–T* projection of a binary system in which both components exhibit allotropy. A exists as low-temperature α and high-temperature γ, and B exists as low-temperature β and as high-temperature δ. The low-temperature allotropes are partially soluble in one another, and the high-temperature allotropes are completely soluble in one another.

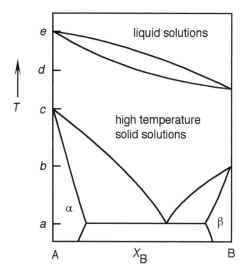

Figure 14.41 The isobaric section of Fig. 14.40 at P_1.

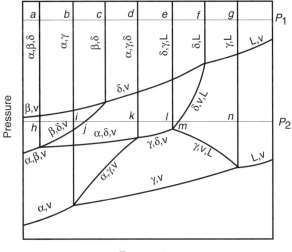

Figure 14.42 The *P–T* projection of a binary system in which both components exhibit allotropy. A exists as low-temperature α and as high-temperature γ, and B exists as low-temperature β and high-temperature δ. The low-temperature allotropes are partially soluble in one another, and the high-temperature allotropes are partially soluble in one another.

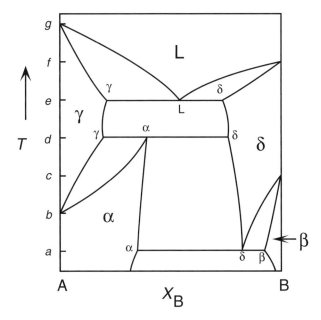

Figure 14.43 The isobaric section of Fig. 14.42 at P_1.

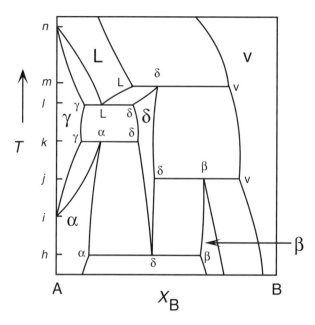

Figure 14.44 The isobaric section of Fig. 14.42 at P_2.

at h in Fig. 14.44. The region of 2-phase β–δ equilibrium exists between the eutectoid at h and the 3-phase δ–β–v equilibrium at j. The region of 2-phase equilibrium β–v originates at j and exits at the lower edge of the figure. This defines the field of stability of the β phase. The region of 2-phase α–δ equilibrium exists between the eutectoid at h and the peritectoid at k, in which state the 3-phase γ–α–δ equilibrium exists. The region of 2-phase γ–α equilibrium begins at the transition temperature i and terminates at the tie-line of the 3-phase peritectoid equilibrium γ–α–δ at k. The γ liquidus and solidus run from n to l and the δ liquidus and solidus run from the eutectic at l to the 3-phase equilibrium L–δ–v at m. The region of 2-phase equilibrium δ–v exists between j and m, and this defines the field of stability of the δ phase. The region of 2-phase equilibrium l–v begins at m and moves out of the figure at its upper edge. The section thus contains fields of stability of the α, β, γ, δ, liquid, and vapor phases.

14.10 PHASE EQUILIBRIUM AT LOW PRESSURE: THE CADMIUM–ZINC SYSTEM

Cadmium and zinc are relatively volatile metals having normal boiling temperatures of, respectively, 765 and 907°C. The phase diagram for the system at 1 atm pressure, presented in Fig. 14.45, shows a simple eutectic with a eutectic melt of $X_{Zn} =$ 0.266 at 266°C. The maximum solubilities of Zn in Cd (α) and Cd in Zn (β) are, re-

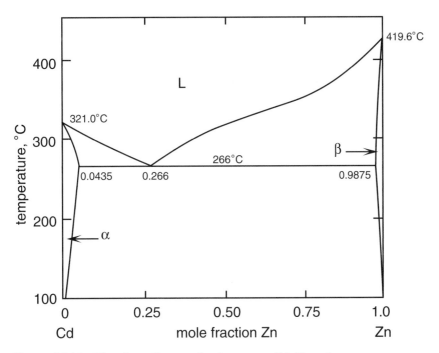

Figure 14.45 The phase diagram for the system Cd–Zn at 1 atm pressure.

spectively, X_{Zn} = 0.0435 and X_{Cd} = 0.0125. The P–T projection of the system is shown in Fig. 14.46; states 1 and 2 are, respectively, the triple points of Cd and Zn and the invariant 4-phase equilibrium α–β–L–v occurs at state 3. The line representing the states of existence of the 3-phase β–L–v equilibrium has a maximum between 380 and 390°C at approximately 4 × 10^{-4} atm.

The isobaric section at P = 10^{-4} atm, shown in Fig. 14.47, contains the states of sublimation of Cd and Zn and three 3-phase equilibria: the eutectic α–L–β, v–α–L, and v–L–b. The isothermal section at 265°C, which is shown in Fig. 14.48, contains the states of sublimation of Cd and Zn and the 3-phase equilibrium v–α–β. The isothermal section at 298°C (see Fig. 14.49), which is a higher temperature than state 3, contains a restricted field of stability of liquid bounded by the two 3-phase equilibria v–α–L and v–L–β. The isothermal section at 370°C, shown in Fig. 14.50, does not contain a field of stability of α but does contain the 3-phase equilibrium v–L–β.

14.11 PHASE EQUILIBRIUM AT HIGH PRESSURE: THE Na$_2$O·Al$_2$O$_3$·2SiO$_2$–SiO$_2$ SYSTEM

The system Na$_2$O·Al$_2$O$_3$·2SiO$_2$–SiO$_2$ is a quasi-binary in the ternary system Na$_2$O–Al$_2$O$_3$–SiO$_2$ (a quasi-binary is a system in which the composition of the liquid phase

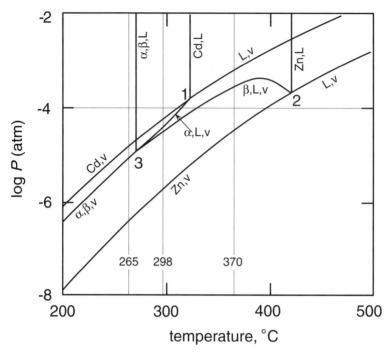

Figure 14.46 The *P–T* projection for the system Cd–Zn.

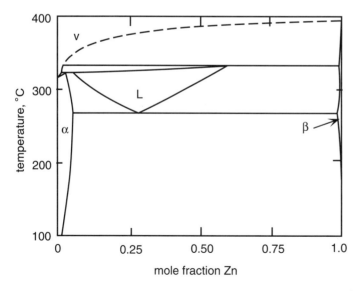

Figure 14.47 Isobaric section of the binary Cd–Zn at $P = 10^{-4}$ atm.

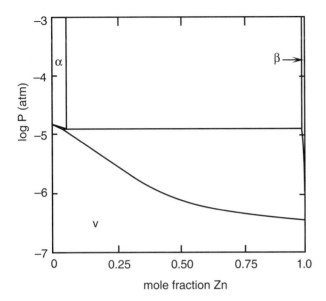

Figure 14.48 The isothermal section of the system Cd–Zn at 265°C. [A. G. Guy and S. Morozuma, "Phase Diagrams. Covering Pressure, Temperature and Composition. Use of Thermodynamic Data to Elucidate the Cadmium–Zinc Binary System," *Metal Treat. and Drop Forg.* (1959), vol. 26, pp. 21–26.]

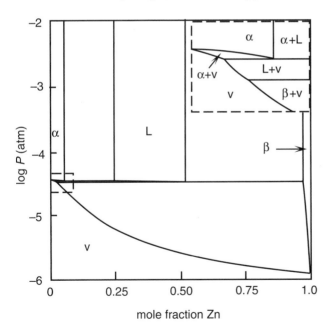

Figure 14.49 Isothermal section of the system Cd–Zn at 298°C. (A. G. Guy and S. Morozuma.)

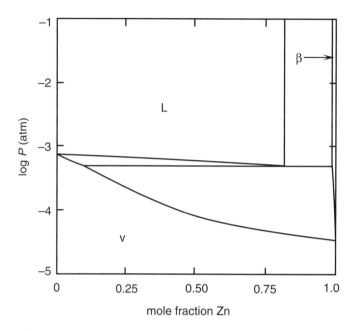

Figure 14.50 Isothermal section of the system Cd–Zn at 370°C. (A. G. Guy and S. Morozuma.)

does not deviate from the straight line drawn between the compositions of the quasi-binary components in the ternary phase diagram). Silica has six polymorphs and, within the range of states $1050°C < T < 1550°C$ and 1 atm $< P < 3.4 \times 10^4$ atm, silica can exist as tridymite, cristobalite, high quartz, and coesite.

The names, formulae, and abbreviations for the compounds that can exist in the system $Na_2O \cdot Al_2O_3 \cdot 2SiO_2 - SiO_2$ are listed in Table 14.1.

The phase diagram for the system $Na_2O \cdot Al_2O_3 \cdot 2SiO_2 - SiO_2$ at 1 atm pressure is shown in Fig. 14.51a. The figure shows the transformation of nepheline to

Table 14.1 Compounds that can exist in the system $Na_2O \cdot Al_2O_3 \cdot 2SiO_2 - SiO_2$.

Formula	Name	Abbreviated name
SiO_2	Tridymite	Tr
	Cristobalite	Cr
	Quartz	Q
	Coesite	C
$Na_2O \cdot Al_2O_3 \cdot 6SiO_2$	Albite	Ab
$Na_2O \cdot Al_2O_3 \cdot 4SiO_2$	Jadeite	Jd
$Na_2O \cdot Al_2O_3 \cdot 2SiO_2$	Carnegieite	Ca
	Nepheline	Ne

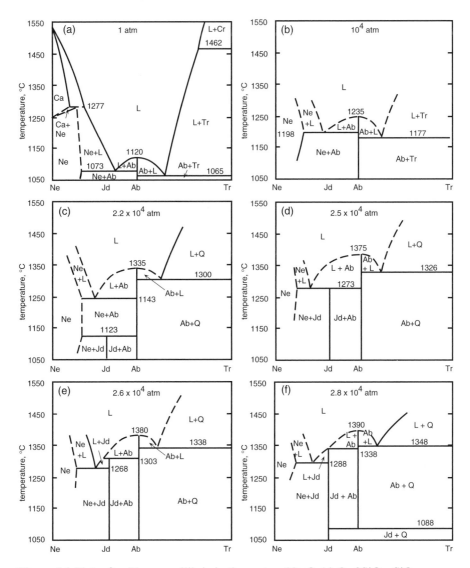

Figure 14.51 (*a–f*)   Phase equilibria in the system Na$_2$O·Al$_2$O$_3$·2SiO$_2$–SiO$_2$.

carnegieite at 1250°C, the incongruent melting of nepheline at 1277°C, the transformation of tridymite to cristobalite at 1462°C, and the congruent melting of albite at 1120°C. Fig. 14.51*b* shows that increasing the pressure to 10^4 atm increases the congruent melting temperature of albite which, consequently, increases the temperatures at which the two eutectic equilibria exist in the system. The rate of change, with pressure, of the temperature at which a phase equilibrium occurs is given by Eq. (7.5). At this pressure silica exists as tridymite. At 2.2 × 10^4 atm, Fig. 14.51*c*

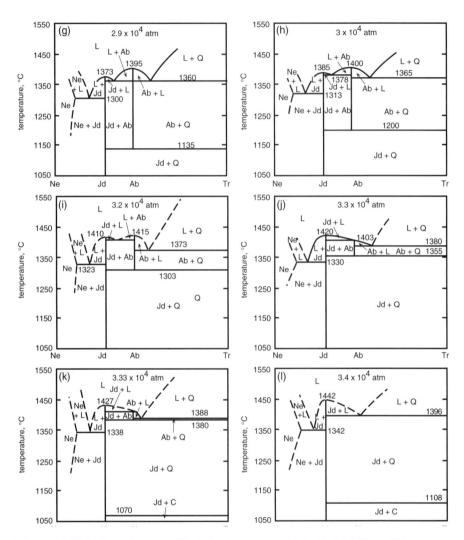

Figure 14.51 (g–l) Phase equilibria in the system $Na_2O \cdot Al_2O_3 \cdot 2SiO_2 - SiO_2$.

shows that jadeite exists as a stable phase at temperatures lower than 1123°C and that the eutectic temperature in the albite–tridymite system occurs at 157 centigrade degrees higher than the eutectic in the system nepheline–albite. Silica now occurs as quartz. Fig. 14.51*d* shows that, at 2.5×10^4 atm, an invariant 4-phase equilibrium involving nepheline, liquid, jadeite, and albite occurs at 1273°C. The equilibrium is invariant because it occurs in a quasi-binary system in which the ratio of Na_2O to Al_2O_3 is unity. At 2.6×10^4 atm, Fig. 14.51*e* shows that the stability of jadeite has increased and that it melts incongruently at 1295°C which gives rise to a eutectic equilibrium in the nepheline–jadeite system. Fig. 14.51*f* shows that, at 2.8×10^4

atm, jadite begins to melt congruently at 1338°C and that albite does not exist as a stable phase at temperatures below 1088°C. At 2.9×10^4 atm, Fig. 14.51g shows that the congruent melting temperature of albite has increased by 1 centigrade degree from its value at 2.8×10^4 atm, whereas the corresponding increase in the temperature at which albite decomposes to jadite and quartz has increased by 47 centigrade degrees. The quasi-binary system now contains three eutectic equilibria. Figs. 14.51i and j show that, with increasing pressure, the decomposition temperature of albite increases and that the limit of congruent melting of albite is reached at 3.3×10^4 atm. Fig. 14.51k shows that albite exists in the range of temperature between its decomposition temperature of 1380°C and its incongruent melting temperature of 1407°C. The phase equilibrium between quartz and coesite occurs at 1070°C. Fig. 14.51l shows that, at 3.4×10^4 atm, albite does not exist as a stable phase and the quasi-binary contains two binary eutectic equilibria.

14.12 SUMMARY

Following an introduction to the representation of phase equilibria, in temperature-pressure-composition space, of a binary system exhibiting complete mutual solubility of the components in the solid and liquid states, a similar system that exhibits minima on the melting, boiling, and sublimation curves is discussed. This is followed by a presentation of phase equilibria in a binary system containing a eutectic equilibrium and having complete mutual solubility of the components in the liquid state, a similar system containing a peritectic equilibrium, a system containing an intermediate phase that melts, boils, and sublimes congruently. A similar system in which the intermediate phase melts and sublimes congruently but boils incongruently is then discussed. This is followed by presentation of phase equilibria in a system with a eutectic and one component that exhibits allotropy and a similar system in which both components exhibit allotropy. Phase equilibria in the systems Cd–Zn and $Na_2O \cdot Al_2O_3 \cdot 2SiO_2 - SiO_2$ are examined.

REFERENCE

Y. Levinsky, "Pressure Dependent Phase Diagrams of Binary Systems," ed. G. Effenberg. Vols. 1 and 2, ASM, 1997.

Chapter 15

ELECTROCHEMISTRY

15.1 INTRODUCTION

All chemical reactions, in which products form from reactants, involve changes in the state of oxidation, or "valence state" of some or all of the participating atoms. By convention, the valence state of an atom in a compound is determined by the number of electrons which surround the nucleus of an atom. The assignment of a valence state is not influenced by the nature of the bonding between the constituent atoms, e.g., although the bonding in the HF molecule is considered to be 50% ionic in character and 50% covalent in character, where "ionic character" implies complete electron transfer from the H atom to the F atom to form the ions H^+ and F^-, and "covalent character" implies complete electron sharing to give a normal covalent HF molecule, which has zero electric dipole moment, the valence states of H and F in HF are, respectively, $+1$ and -1.

Changes in the valence state of an element are caused by the addition or removal of electrons, and thus the thermodynamic driving force of any reaction must, in some way, be related to the ease with which the required changes of valence of the participating atoms can occur, i.e., to the ease with which the necessary transfer of electrons can occur. For example, the reaction

$$AO + B = BO + A$$

involves a decrease in the valence state of A from $+2$ to zero and an increase in the valence state of B from zero to $+2$, i.e., the reaction involves the transfer of two electrons from B to A and, hence, can be written as

$$A^{2+} + B = B^{2+} + A$$

The change in free energy is thus a manifestation of the energetics of electron transfer. The reaction equation, written as the sum of

$$A^{2+} + 2e^- = A$$

and

$$B - 2e^- = B^{2+}$$

suggests the possibility of conducting the reaction as shown in Fig. 15.1. A mixture of A + AO and a mixture of B + BO are joined by two connections (a) and (b), where (a) is an electronic conductor through which only electrons can pass and (b) is an ionic conductor through which only oxygen ions can pass. The spontaneous reaction thus occurs in the following manner. Two electrons leave a B atom, travel from right to left along (a) and, on arrival at the A + AO mixture, convert an A^{2+} ion to an A atom. Simultaneously an O^{2-} ion leaves the A + AO mixture and passes through (b) to the B + BO mixture. Charge neutrality in the overall system is thus maintained, and the overall reaction can be written as

$$A^{2+}O^{2-} + B = B^{2+}O^{2-} + A$$

As a result of its being conducted in an electrochemical manner, this reaction is called an electrochemical reaction. The driving force for the transport of electrons along (a) is manifested as an electric voltage (or difference in electric potential) which can be measured by placing an external opposing voltage in the circuit (a) and adjusting this voltage until no electric current flows, at which point the electrochemical reaction ceases. At this point the external voltage exactly balances the voltage generated by the electrochemical system, i.e., the thermodynamic driving force for the chemical reaction is exactly balanced by the externally applied electric driving force. Knowledge of the mathematical relationship between these two types of force allows the former (ΔG for the reaction) to be measured. Furthermore, whereas in Chap. 12 it was seen that pure A, pure AO, pure B, and pure AO are in thermodynamic equilibrium only at the unique temperature at which the two Ellingham lines intersect (if, indeed, they do intersect) it is now seen that the four phases can be

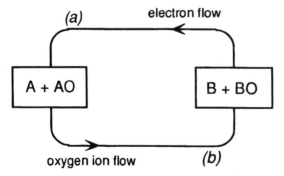

Figure 15.1 A schematic representation of an electrolytic reaction.

brought into electrochemical equilibrium at any temperature by balancing the chemical driving force with an opposing electric driving force. The properties of systems such as this are examined in this chapter.

15.2 THE RELATIONSHIP BETWEEN CHEMICAL AND ELECTRICAL DRIVING FORCES

Equation (5.8) showed that when a system undergoes a reversible process at constant temperature and pressure, the decrease in the Gibbs free energy of the system equals w'_{max}, the work (other than work of expansion) done by the system. For an increment of such a process,

$$-dG' = \delta w'_{max}$$

Consider a system which performs electrical work by transporting an electric charge across a voltage difference, i.e., from one electric potential to another. The work performed is obtained as the product of the charge transported, q (coulombs), and the electric potential difference, $\Delta\phi$ (volts), and the unit of such work is the joule ($=$ volts $\times$ coulombs). A system which is capable of performing electrical work as the result of the occurrence of a chemical reaction is called a galvanic cell, and the overall chemical reaction is represented by an equation called the cell reaction. The charge carried by 1 gram ion (i.e., Avogadro's number of ions) of unit positive charge is 96,487 coulombs and is Faraday's constant, denoted as $\mathfrak{f}$. Thus if dn gram ions of valence z are transported through a voltage difference $\Delta\phi$ maintained between the electrodes of a cell, then

$$\delta w' = z\mathfrak{f}\Delta\phi dn$$

If the transportation is conducted reversibly, in which case the electric potential difference between the electrodes of the cell is called the electromotive force or EMF, $\mathscr{E}$, of the cell, then

$$\delta w'_{max} = z\mathfrak{f}\mathscr{E}dn = -dG' \tag{15.1}$$

For the transportation of 1 mole of ions, Eq. (15.1) becomes

$$\Delta G = -z\mathfrak{f}\mathscr{E} \tag{15.2}$$

which is known as the Nernst equation.

Consider the familiar Daniell cell shown in Fig. 15.2. This cell consists of a zinc electrode dipping into an acidified aqueous solution of $ZnSO_4$ and a copper electrode dipping into an acidified aqueous solution of $CuSO_4$. The two aqueous solutions, which constitute the electrolyte of the galvanic cell (i.e., the medium through which ionic current flows), are prevented from mixing by the insertion between them of a porous diaphragm. Consider the processes which occur when a metal is immersed in an electrolyte. The asymmetric forces at the interface between the metal and the electrolyte cause rearrangement of the solvent dipoles and the charged species in such a manner that the electrolyte side of the interface becomes electrically charged. The

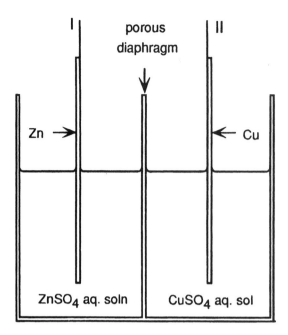

Figure 15.2 The Daniell cell.

electric field produced operates across the interface, and, depending on the direction of the field, free electrons in the metal either move toward or away from the interface until a charge is induced in the metal which is equal and opposite to that on the electrolyte side of the interface. This separation of charge causes the development of a difference in electric potential between the interface and the bulk of the electrolyte far removed from the interface. Fig. 15.3a shows the nature of the variation of electric potential with distance away from the interface into the electrolyte when the metal acquires a negative charge, and Fig. 15.3b shows the corresponding variation when the metal acquires a positive charge. The difference between the potential at the metal, ϕ^M, and the potential in the bulk of the electrolyte, ϕ^e, is called the "absolute potential" of the electrode. In the Daniell cell, before external electrical contact is made between the electrodes, the zinc electrode acquires an excess of electrons, and the copper acquires a deficit of electrons; consequently, the variation of electric potential from the zinc electrode through the electrolyte to the copper electrode is as shown schematically in Fig. 15.4. The equilibrium

$$Zn^{2+} + 2e^- = Zn$$

is established at the zinc-electrolyte interface when

$$\mu_{Zn^{2+}}(\text{in the electrolyte}) = \mu_{Zn^{2+}}(\text{in the electrode})$$

and

$$\mu_{Zn^{2+}} + 2\mu_{e^-}^{Zn} = \mu_{Zn}$$

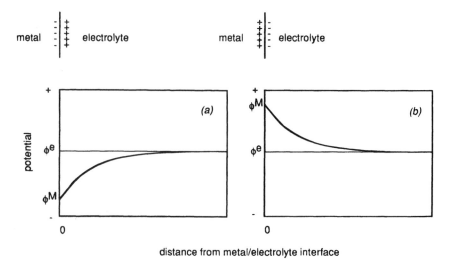

Figure 15.3 A schematic representation of the variation, with distance from a metal/electrolyte interface, of the potential when (*a*) the metal electrode acquires a negative charge, and (*b*) when the electrode acquires a positive charge.

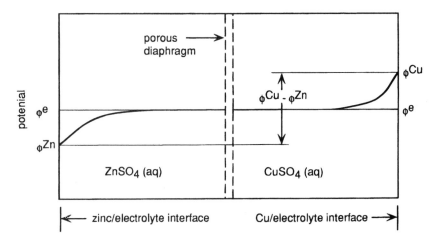

Figure 15.4 A schematic representation of the variation of potential with position in a Daniell cell. The open cell EMF of the cell is given by $\phi^{Cu} - \phi^{Zn}$.

where

μ_{Zn} is the chemical potential of the zinc atoms in the electrode,
$\mu_{Zn^{2+}}$ is the chemical potential of the zinc ions in the solution and in the electrode, and
μ_e^{Zn} is the chemical potential of the electrons in the zinc electrode.

It is seen that the value of $\mu_{Zn^{2+}(\text{in solution})}$, which is determined by the concentration of $ZnSO_4$ in the solution, determines the equilibrium value of $\mu_{e^-}^{Zn}$. Similarly the equilibrium

$$Cu^{2+} + 2e^- = Cu$$

is established at the copper-electrolyte interface when

$$\mu_{Cu^{2+}} \text{ (in the solution)} = \mu_{Cu^{2+}} \text{ (in the copper electrode)}$$

and

$$\mu_{Cu^{2+}} + 2\mu_{e^-}^{Cu} = \mu_{Cu}$$

Again it seen that the equilibrium value of $\mu_{e^-}^{Cu}$ is determined by the concentration of $CuSO_4$ in the electrolyte. If identical pieces of metal wire are joined to each of the electrodes, in Fig. 15.2, as extensions, then, as both the wire-Zn electrode and the wire-Cu electrode are electrical conductors,

$$\mu^I = \mu_{e^-}^{Zn} \quad \text{and} \quad \mu_{II} = \mu_{e^-}^{Cu}$$

The reversible transfer of dn moles of electrons from the electric potential ϕ^{Zn} to the electric potential ϕ^{Cu} involves the performance of work $\delta w'_{max}$ given by

$$\delta w'_{max} = z\mathfrak{f}(\phi^{Cu} - \phi^{Zn})dn$$

and, also, the reversible transfer of dn moles of electrons from the chemical potential μ_I to μ_{II} at constant temperature and pressure involves the performance of work $\delta w'_{max}$ and a decrease in the Gibbs free energy, dG', of the system according to

$$\delta w'_{max} = -dG' = -(\mu^{II} - \mu^I)dn$$

Thus

$$(\mu^{II} - \mu^I) = z\mathfrak{f}(\phi^{Cu} - \phi^{Zn}) \tag{15.3}$$

where, for electrons, z has the value of minus unity. Eq. (15.3) relates the chemical potential difference and the electric potential difference for electron transfer. The difference $\phi^{Cu} - \phi^{Zn}$ is the open circuit EMF, $\mathscr{E}$ of the Daniell cell.

When an opposing electric potential of magnitude $\mathscr{E}$ is applied externally between the electrodes I and II, as the chemical driving force of the cell is exactly balanced by the external opposing voltage, the entire system is at equilibrium. If the magnitude of the external opposing voltage is decreased, equilibrium no longer exists and electronic current flows through the external circuit from I to II, with the equivalent ionic current flowing through the cell. In the Daniell cell the ionic current through the cell involves the transport of SO_4^{2-} ions from the solution in the $CuSO_4$ compartment (the catholyte) to the $ZnSO_4$ compartment (the anolyte) at a rate equal to that of the addition of Zn^{2+} ions to the anolyte and, hence, equal to the rate of removal of Cu^{2+} ions from the catholyte. As the cell reaction proceeds, the concentration of $CuSO_4$ in the catholyte decreases, and the concentration of $ZnSO_4$ in the anolyte increases. As has been seen, the equilibrium values of μ_I and

μ_{II}, and hence the value of $\mathscr{E}$, depend on the concentrations of $ZnSO_4$ and $CuSO_4$ in the electrolyte, and thus, eventually, in the absence of unfavorable kinetic factors, the EMF of the cell decreases to the value of the externally applied opposing voltage, at which point the passage of electric current ceases and a new equilibrium is established. If, however, by appropriate addition and removal, the concentrations of $ZnSO_4$ and $CuSO_4$ in their respective compartments are maintained constant, the cell reaction continues indefinitely. When the externally applied voltage is finitely less than the EMF of the cell, a finite current flows and the cell reaction proceeds irreversibly. In such a situation, less than maximum work is obtained as the electrons in the external circuit are being transported through a smaller difference in voltage. In the limit of decrease of the external voltage, i.e., when the external voltage is zero and the cell is short-circuited, the degree of irreversibility of the reaction is maximized, no work is done, and the decrease in the Gibbs free energy of the system appears entirely as heat. This system corresponds to that of placing a piece of zinc in an aqueous solution of $CuSO_4$. For the production of maximum work, the cell must be operated reversibly, in which case the externally applied voltage must be only infinitesimally smaller than the EMF of the cell, giving an infinitesimal flow of current in the forward direction. If the cell can be operated reversibly, then an infinitesimal increase in the magnitude of the external voltage reverses the direction of the current flow and the direction of the cell reaction. The cell thus becomes current consuming rather than current producing, i.e., it becomes an electrolysis cell instead of a galvanic cell. This scheme is illustrated in Fig. 15.5.

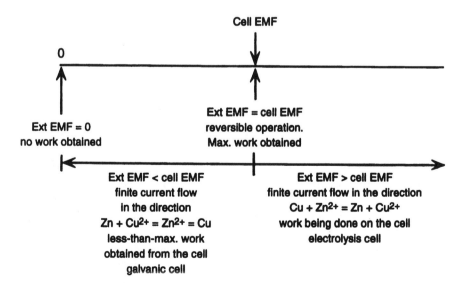

Figure 15.5 The relationship of a current-producing cell (a galvanic cell) to a current-consuming cell (an electrolysis cell).

When operating as a galvanic cell the anodic oxidation reaction

$$Zn = Zn^{2+} + 2e^-$$

occurs at the Zn anode and the cathodic reduction reaction

$$Cu^{2+} + 2e^- = Cu$$

occurs at the Cu cathode. The overall cell reaction is the sum of these, which gives

$$Zn + CuSO_3 = Cu + ZnSO_4$$

In shorthand notation this is written as

$$Zn \mid Zn^{2+}(\text{aqueous solution}) \mid Cu^{2+}(\text{aqueous solution}) \mid Cu$$

where the full vertical line indicates phase boundaries in the cell and the dashed vertical line represents the porous diaphragm separating the two aqueous solutions.

15.3 THE EFFECT OF CONCENTRATION ON EMF

In the preceding section it was seen that the EMF of a Daniell cell is dependent on the concentrations of $CuSO_4$ and $ZnSO_4$ in the catholyte and anolyte, respectively. The quantitative relationship between the concentration, or correctly, the activity, and the EMF can be introduced as follows. Consider the reaction

$$Zn + CuSO_4 = Cu + ZnSO_4$$

Although it was assumed that the reaction, as written, proceeds spontaneously from left to right, the direction depends on the states of the reactants and products. Consider the reactants and products to occur in their standard states at 298 K. Then

$$Zn_{(s)} + CuSO_4 (\text{saturated aqueous solution})$$
$$= Cu_{(s)} + ZnSO_4 (\text{saturated aqueous solution})$$

which is thermodynamically equivalent to

$$Zn_{(s)} + CuSO_{4(s)} = Cu_{(s)} + ZnSO_{4(s)}$$

At 298 K, $\Delta G = \Delta G^\circ_{298\ K} = -213,040$ J, and thus, from Eq. (15.2),

$$\mathscr{E}^\circ = -\frac{\Delta G^\circ}{z\mathsf{f}} = \frac{213,040}{2 \times 96,487} = 1.104 \text{ volts}$$

where $\mathscr{E}^\circ$, being the EMF of the cell when the reactants and products occur in their designated standard states, is termed the standard EMF of the cell. Thus, when Zn dips into a saturated aqueous solution of $ZnSO_4$ and Cu dips into a saturated aqueous solution of $CuSO_4$, the externally applied voltage which is required to balance the chemical driving force of the cell reaction at 298 K is 1.104 volts.

For the general reaction

$$aA + bB = cC + dD$$

when the reactants and products do not occur in their standard states,

$$\Delta G = \Delta G^\circ + RT \ln \frac{a_C^c a_D^d}{a_A^a a_B^b} \tag{13.4}$$

and, from Eq. (15.2), the EMF of the cell in which the above reaction is occurring electrochemically is

$$\mathscr{E} = \mathscr{E}^\circ - \frac{RT}{z\mathfrak{f}} \ln \frac{a_C^c a_D^d}{a_A^a a_B^b} \tag{15.4}$$

Thus, with pure Zn and Cu in the Daniell cell, the EMF is given as

$$\mathscr{E} = \mathscr{E}^\circ - \frac{RT}{2\mathfrak{f}} \ln \frac{a_{ZnSO_4}}{a_{CuSO_4}}$$

which, at 298 K, gives

$$\mathscr{E} = 1.104 - 0.0296 \log \frac{a_{ZnSO_4}}{a_{CuSO_4}}$$

In order that the EMF of the cell be zero, the activity quotient must be

$$10^{1.104/0.0296} = 1.98 \times 10^{37}$$

Thus, if the $ZnSO_4$ solution is saturated, i.e., $a_{ZnSO_4} = 1$, then it is required that the activity of $CuSO_4$ in the catholyte, with respect to the saturated solution as the standard state, be 5×10^{-38} in order that the occurrence of equilibrium does not require a backing EMF in the external circuit between the electrodes. The concentration gradient which exists across the porous diaphragm causes the diffusion of ions from one compartment to the other, and as diffusion is an irreversible process, the concentration gradient gives rise to a potential known as the liquid junction potential. This liquid junction potential must be minimized by such means as the use of a salt bridge between the anolyte and the catholyte.

15.4 FORMATION CELLS

An example of a cell without a liquid junction is the cell

$$Pb_{(l)} | PbO_{(l)} | O_{2(g)},(Pt)$$

in which the anode is liquid lead, the cathode is oxygen gas bubbled over a platinum (inert) wire dipping into the electrolyte, and the electrolyte is liquid lead oxide. In this cell the driving force of the cell reaction

$$Pb_{(l)} + \frac{1}{2}O_{2(g)} = PbO_{(l)}$$

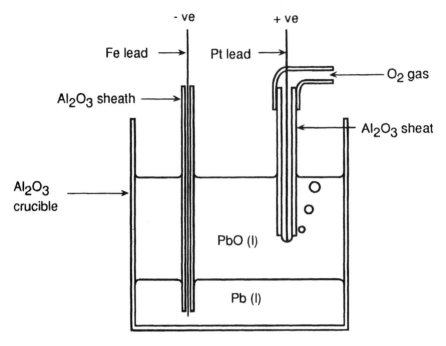

Figure 15.6 A lead oxide formation cell.

can be balanced by application of an opposing voltage between the electrodes. Such a cell, which is an example of a formation cell, is illustrated in Fig. 15.6.*

At the liquid lead anode,

$$Pb = Pb^{2+} + 2e^-$$

and at the oxygen cathode,

$$\frac{1}{2}O_2 + 2e^- = O^{2-}$$

With a pure liquid lead anode, pure liquid PbO electrolyte and oxygen gas at 1 atm pressure at the cathode, the standard Gibbs free energy of formation of lead oxide is obtained as

$$\Delta G° = -2\mathcal{F}\mathcal{E}°$$

Maintaining the oxygen pressure constant at 1 atm and varying the temperature of the cell facilitate determination of the variation of $\Delta G°$ with T. The addition to the

---

*R. Sridhar and J. H. E. Jeffes. "Thermodynamics of PbO and PbO–SiO$_2$ Melts," *Trans. Inst. Mining Met.* (1967), vol. 76, p. C44.

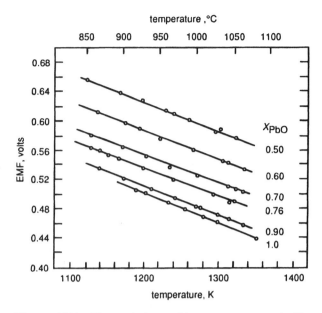

Figure 15.7 The variation, with temperature and silica of the electrolyte, of the EMF of the cell $PbO_{(l)}|PbO-SiO_{2(l)}|O_{2(1\text{ atm})}$, Pt. [From R. Sridhar and J. H. E. Jeffes, "The Thermodynamics of PbO and PbO–SiO$_2$ Melts," *Trans. Inst. Mining Met.* (1967), vol. 76, p. C44.]

electrolyte of a second oxide which (1) must be chemically more stable than PbO and (2) must not introduce any electronic conductivity (for example, SiO$_2$) influences the EMF of the cell according to

$$\mathscr{E} = \mathscr{E}° - \frac{RT}{2\mathfrak{f}} \ln \frac{a_{\text{PbO(in PbO–SiO}_2)}}{p_{O_2}^{1/2}}$$

Thus measurement of the EMF of the cell as a function of the concentration of PbO in the lead silicate from pure PbO to saturation with SiO$_2$ allows determination of the variation of a_{PbO} with composition, and hence, via the Gibbs-Duhem relationship, the variation of a_{SiO_2} with composition. The variation of $\mathscr{E}$ with temperature and composition of the lead silicate melts, obtained by Sridhar and Jeffes, are shown in Fig. 15.7. The variation of $\mathscr{E}°$ with temperature (the line in Fig. 15.7 for $X_{\text{PbO}} = 1$) gives

$$\Delta G° = -191,600 + 79.08T \text{ J}$$

for the cell reaction

$$Pb_{(l)} + \frac{1}{2}O_{2(g)} = PbO_{(l)}$$

15.5 CONCENTRATION CELLS

A cell which has identical electrodes inserted into solutions differing only in concentration is called a concentration cell. Consider the cell

$$Cu\,|\,CuSO_4(aq.\ sol'n,\ low\ concentration)\,|\,CuSO_4\ (aq.\ sol'n,\ high\ concentration)\,|\,Cu$$

The cell reaction is

$$CuSO_4\ (at\ high\ concentration) \rightarrow CuSO_4\ (at\ low\ concentration)$$

i.e., the spontaneous process is the dilution of $CuSO_4$. The standard EMF of such a cell is zero, and thus the EMF is

$$\mathscr{E} = -\frac{RT}{2\mathsf{f}}\ln\frac{a_{CuSO_4(low\ concentration)}}{a_{CuSO_4(high\ concentration)}}$$

If one of the aqueous solutions (say the high-concentration solution) contains $CuSO_4$ in its standard state then

$$\mathscr{E} = -\frac{RT}{2\mathsf{f}}\ln a_{CuSO_4(low\ concentration)} = \frac{\Delta\bar{G}^M_{CuSO_4}}{2\mathsf{f}}$$

Thus an electrochemical measurement allows the determination of the partial molar Gibbs free energy of solution of $CuSO_4$ in water. Such a cell suffers from the disadvantage of having a liquid junction potential.

A form of concentration cell of considerable importance in materials applications is the oxygen concentration cell which uses lime-stabilized zirconia as a solid electrolyte. The phase diagram for the system ZrO_2–CaO, presented as Fig. 15.8, shows that the substitution of Ca^{2+} ions for Zr^{4+} ions stabilizes the high-temperature cubic polymorph of ZrO_2, and solid solutions containing up to 20 mole percent CaO are formed. Electroneutrality in the solid solution requires that an oxygen vacancy be formed in the crystal lattice for every substitution of Ca^{2+} for Zr^{4+}, and thus, when the stabilized cubic zirconia contains x mole percent of CaO, $0.5x$ percent of the oxygen lattice sites are vacant, e.g., with 20 mole percent CaO, 10% of the oxygen sites in the crystal lattice are vacant. This causes the diffusivity of oxygen in the solid solution to be high enough that, within certain ranges of oxygen pressure and temperature, the cubic solid solution is an ionic conductor in which the oxygen ion is the only mobile species. Thus a cell constructed as follows:

$$O_2(g,\ lower\ pressure\ I),\ Pt\,|\,CaO\text{–}ZrO_2\,|\,Pt,\ O_2(g,\ higher\ pressure\ II)$$

in which the cell reaction is

$$O_2(g,\ pressure\ II) \rightarrow O_2(g,\ pressure\ I)$$

has the EMF

$$\mathscr{E} = -\frac{RT}{4\mathsf{f}}\ln\frac{p_{O_2(I)}}{p_{O_2(II)}}$$

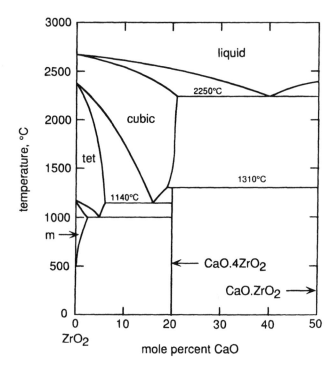

Figure 15.8 The phase diagram for the system ZrO_2–CaO (m = monoclinic polymorph, tet = tetragonal polymorph).

In the operation of the cell, oxygen ions are transported through the electrolyte from the cathode to the anode (or oxygen vacancies diffuse in the reverse direction), and electrons are transported through the external circuit from the anode to the cathode. The oxygen pressures at the electrodes can be fixed by using metal-metal oxide couples; e.g., using the couples X–XO and Y–YO, the cell becomes

$$X \,|\, XO \,|\, CaO\text{--}ZrO_2 \,|\, YO \,|\, Y$$

At the temperature T the lower anode oxygen pressure $p_{O_2(X/XO)}$ is fixed by the establishment of the chemical equilibrium

$$X_{(s)} + \frac{1}{2}O_{2(g)} = XO_{(s)}$$

and the higher cathode oxygen pressure $p_{O_2(Y/YO)}$ is fixed by the establishment of the chemical equilibrium

$$Y_{(s)} + \frac{1}{2}O_{2(g)} = YO_{(s)}$$

Thus the anode half-cell reaction can equivalently be regarded as being

$$X + \frac{1}{2}O_2(\text{at } p_{O_2(\text{eq. } T,X/XO)}) = XO$$

or

$$O^{2-} - 2e^- = \frac{1}{2}O_2 \text{ (at } p_{O_2(\text{eq. } T,X/XO)})$$

and, similarly, the cathode half-cell reaction can equivalently be regarded as being

$$YO = Y + \frac{1}{2}O_2 \text{ (at } p_{O_2(\text{eq. } T,Y/YO)})$$

or

$$\frac{1}{2}O_2 \text{ (at } p_{O_2(\text{eq. } T,Y/YO)}) + 2e^- = O^{2-}$$

The cell reaction is thus

$$YO + X = XO + Y \tag{i}$$

or

$$\frac{1}{2}O_2 \text{ (at } p_{O_2(\text{eq. } T,Y/YO)}) = \frac{1}{2}O_2 \text{ (at } p_{O_2(\text{eq. } T,X/XO)}) \tag{ii}$$

Thus, whereas the chemical equilibrium between the pure reactants and products

$$YO + X = XO + Y$$

could be obtained only at the single invariant temperature (for example, T_E in Fig. 12.4) in which state

$$p_{O_2(\text{eq. } T,X/XO)} = p_{O_2(\text{eq. } T,Y/YO)}$$

the electrochemical equilibrium

$$YO + X = XO + Y$$

can be obtained at any temperature (within the limits imposed by the performance of the ZrO_2–CaO electrolyte) by placing an external voltage in opposition to the chemical driving force of the cell. For either Eq. (i) or (ii) the chemical driving force is

$$\Delta G = RT \ln \frac{p_{O_2(\text{eq. } T,X/XO)}^{1/2}}{p_{O_2(\text{eq. } T,Y/YO)}^{1/2}}$$

and hence the cell EMF is

$$\mathcal{E} = -\frac{RT}{2\mathfrak{f}} \ln \frac{p_{O_2(eq.\ T,X/XO)}^{1/2}}{p_{O_2(eq.\ T,Y/YO)}^{1/2}}$$

or

$$\mathcal{E} = -\frac{RT}{4\mathfrak{f}} \ln \frac{p_{O_2(eq.\ T,X/XO)}}{p_{O_2(eq.\ T,Y/YO)}} \tag{iii}$$

If one of the metals, say, X, is dissolved in an inert solvent, where the requirement for "inertness" is that the equilibrium oxygen pressure for the solvent metal-solvent metal oxide equilibrium is considerably higher than $p_{O_2(eq.\ T,X/XO)}$, then the activity of X in the alloy can be obtained as follows. If X in solution is denoted $\underline{X}$, then the EMF of the cell

$$\underline{X}\,|\,XO\,|\,CaO\text{--}ZrO_2\,|\,Y\,|\,YO$$

is

$$\mathcal{E} = -\frac{RT}{4\mathfrak{f}} \ln \frac{p_{O_2(eq.\ T,\underline{X}/XO)}}{p_{O_2(eq.\ T,Y/YO)}} \tag{iv}$$

As, at the temperature T,

$$a_X p_{O_2(\underline{X},XO)}^{1/2} = p_{O_2(X,XO)}^{1/2}$$

combination of Eqs. (iii) and (iv) gives

$$\mathcal{E} - \mathcal{E}' = -\frac{RT}{4\mathfrak{f}} \ln a_X$$

and thus measurement of the variation of $\mathcal{E}'$ with composition of the alloy allows determination of the activity-composition relationship of X in the alloy. More simply, if the cell is

$$X\,|\,XO\,|\,CaO\text{--}ZrO_2\,|\,\underline{X}\,|\,XO$$

then

$$\mathcal{E} = \frac{RT}{2\mathfrak{f}} \ln \frac{p_{O_2(eq.\ T,\underline{X}/XO)}^{1/2}}{p_{O_2(eq.\ T,X/XO)}^{1/2}} = -\frac{RT}{2\mathfrak{f}} \ln a_X = -\frac{\Delta \overline{G}_X^M}{2\mathfrak{f}}$$

Similarly, if the metal oxide XO is dissolved in an inert oxide solvent, then, denoting dissolved XO as $\underline{XO}$, for the cell

$$X\,|\,\underline{XO}\,|\,CaO\text{--}ZrO_2\,|\,X\,|\,XO$$

$$\mathcal{E} = -\frac{RT}{2\mathfrak{f}} \ln \frac{p_{O_2(eq.\ T,X/\underline{XO})}^{1/2}}{p_{O_2(eq.\ T,X/XO)}^{1/2}}$$

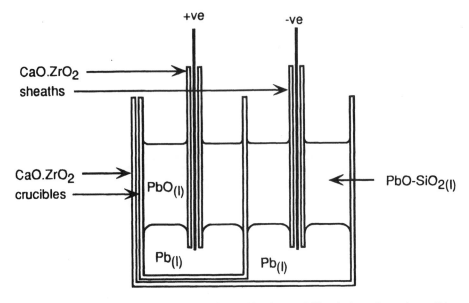

Figure 15.9 A lead oxide concentration cell using stabilized zirconia as the solid-state electrolyte.

and, as

$$a_{XO} = \frac{p_{O_2(eq.\ T,X/\underline{XO})}^{1/2}}{p_{O_2(eq.\ T,X/XO)}^{1/2}}$$

then

$$\mathscr{E} = -\frac{RT}{2\mathfrak{f}} \ln a_{XO}$$

This technique has been used by Kozuka and Samis* to measure the activities of PbO in the melt in the system PbO–SiO$_2$ using the cell

$$Pb_{(l)} \,|\, PbO_{(l,\text{in PbO–SiO}_2)} \,|\, CaO\text{–}ZrO_2 \,|\, Pb_{(l)} \,|\, PbO_{(l)}$$

Their experimental cell is illustrated in Fig. 15.9, and their results, at 1000°C, are shown, in comparison with those obtained by Sridhar and Jeffes, in Fig. 15.10.

---

*Z. Kozuka and C. S. Samis, "Thermodynamic Properties of Molten PbO–SiO$_2$ Systems," *Met. Trans. AIME* (1970), vol. 1, p. 871.

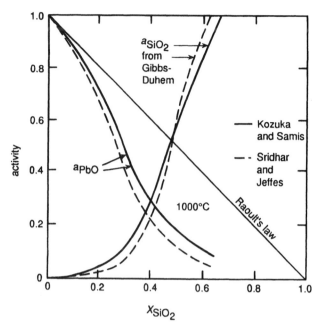

Figure 15.10 The activities of PbO in liquid PbO–SiO$_2$ determined from a PbO concentration cell [from Z. Kozuka and C. S. Samis, "Thermodynamic Properties of Molten PbO–SiO$_2$ Systems," *Met. Trans.* AIME (1970), vol. 1, p. 871], and a PbO formation cell [from R. Sridhar and J. H. E. Jeffes, "Thermodynamics of PbO and PbO–SiO$_2$ Melts," *Trans. Inst. Mining Met.* (1967), vol. 76, p. C44].

15.6 THE TEMPERATURE COEFFICIENT OF THE EMF

For any cell reaction at constant temperature and pressure,

$$\Delta G = -z \mathfrak{f} \mathscr{E}$$

Differentiation with respect to temperature at constant pressure gives

$$\left(\frac{\partial \Delta G}{\partial T} \right)_P = -z \mathfrak{f} \left(\frac{\partial \mathscr{E}}{\partial T} \right)_P = -\Delta S$$

Thus, for the cell reaction,

$$\Delta S = z \mathfrak{f} \left(\frac{\partial \mathscr{E}}{\partial T} \right)_P \tag{15.5}$$

and

$$\Delta H = -z \mathfrak{f} \mathscr{E} + z \mathfrak{f} T \left(\frac{\partial \mathscr{E}}{\partial T} \right)_P \tag{15.6}$$

Thus, from Eq. (15.5), the slopes of the lines in Fig. 15.7 are equal to $\Delta S/2\mathfrak{f}$ for the cell reaction in the formation cell $Pb + \frac{1}{2}O_2 = PbO$. The slope of the line for $X_{PbO} = 1$ gives

$$\Delta S_{(a)} = S_{PbO(l)}^\circ - S_{Pb(l)}^\circ - \frac{1}{2}S_{O_2(g)}^\circ$$

and the slopes of the lines for melts of $X_{PbO} < 1$ give

$$\Delta S_{(b)} = \overline{S}_{PbO \text{ at } X_{PbO}} - S_{Pb(l)}^\circ - \frac{1}{2}S_{O_2(g)}^\circ$$

Thus the partial molar entropy of mixing of PbO in a melt of composition X_{PbO}, $\Delta\overline{S}_{PbO}^M$, is given by $\Delta S_{(b)} - \Delta S_{(a)}$.

In concentration cells such as

$$A\,|\,A^{z+}X^{z-}\,|\,A(\text{in an A–B alloy})$$

as

$$\mathscr{E} = -\frac{RT}{z\mathfrak{f}} \ln a_A$$

the partial molar properties of A in the A–B alloy are obtained as

$$\Delta\overline{S}_A^M = z\mathfrak{f}\frac{\partial\mathscr{E}}{\partial T}$$

and

$$\Delta\overline{H}_A^M = -z\mathfrak{f}\mathscr{E} + z\mathfrak{f}T\frac{\partial\mathscr{E}}{dT}$$

Alternatively, from measurements on formation cells of the type

$$A\,|\,A^{2+}X^{2-}\,|\,X_{2(g,1\text{ atm})}$$

for which

$$\mathscr{E}_{(i)} = \mathscr{E}^\circ$$

and measurements on cells of the type

$$A(\text{in A–B alloy})\,|\,A^{2+}X^{2-}\,|\,X_{2(g,1\text{ atm})}$$

for which

$$\mathscr{E}_{(ii)} = \mathscr{E}^\circ - \frac{RT}{2\mathfrak{f}}\ln\left(\frac{1}{a_A}\right)$$

the value of a_A is obtained from

$$\mathscr{E}_{(ii)} - \mathscr{E}_{(i)} = \frac{RT}{2\mathfrak{f}}\ln a_A$$

Variation with temperature then gives

$$\Delta \bar{S}_A^M = 2\mathfrak{f}\frac{\partial}{\partial T}(\mathscr{E}_{(i)} - \mathscr{E}_{(ii)})$$

and

$$\Delta \bar{H}_A^M = 2\mathfrak{f}(\mathscr{E}_{(ii)} - \mathscr{E}_{(i)}) - 2\mathfrak{f}T\frac{\partial}{\partial T}(\mathscr{E}_{(ii)} - \mathscr{E}_{(i)})$$

Belton and Rao* measured the EMF's of the cells

$$Mg_{(l)} | MgCl_{2(l)} | Cl_{2(g,1\ atm)}$$

and

$$Mg(in\ Mg-Al)_{(l)} | MgCl_{2(l)} | Cl_{2(g,1\ atm)}$$

in the range of temperature 700°–1000°C with Mg–Al alloys in the range $X_{Mg} =$ 0.096–0.969. With pure Mg as the anode they obtained

$$\mathscr{E}° = 3.135 - 6.5 \times 10^{-4}T\ (\text{volts})$$

which gives, for the reaction

$$Mg_{(l)} + Cl_{2(g)} = MgCl_{2(l)}$$

$$\Delta G° = -604{,}970 + 125.4T\ \text{J}$$

They fitted their activities, measured at 1073 K, to Eqs. (v) and (vi) in Sec. 13.6.

15.7 HEAT EFFECTS

In examining the properties of enthalpy, H, in Chap. 5, it was noted that the change in the enthalpy of a system equals the heat entering or leaving the system during a constant pressure process only if the work of volume change is the sole form of work performed on or by the system. If an electrochemical reaction is conducted in a galvanic cell, as a result of which electrical work is performed, $\Delta H \neq q_P$.

For a change of state at constant temperature and pressure, Eq. (5.6) gave

$$\Delta G = q - w + P\Delta V - T\Delta S$$
$$= q - w' - T\Delta S$$

If $w' = 0$, then, $q = \Delta G + T\Delta S = \Delta H$. But, if the process, which involves the performance of work w' is conducted reversibly, in which case $-w' = -w'_{max} = \Delta G$, then

$$q = T\Delta S$$

---

*G. R. Belton and Y. K. Rao, "A Galvanic Cell Study of Activities in Mg–Al Liquid Alloys," *Trans. Met. Soc. AIME* (1969), vol. 245, p. 2189.

Consider the Daniel cell reaction $Zn + CuSO_4 = Cu + ZnSO_4$. When the reactants and products are in their standard states (pure metals and saturated aqueous solutions) the change in the Gibbs free energy for the cell reaction is

$$\Delta G° = -208,800 - 13.9T \text{ J}$$

If the reaction occurs as a result of placing pure solid zinc into saturated copper sulfate solution at 25°C, in which case the reaction proceeds spontaneously and w' is zero, then, for the formation of saturated zinc sulfate solid and copper, per mole of the reaction,

$$\Delta H° = -208,800 \text{ J}$$

is the heat which *flows from* the system into the thermostating heat reservoir. However, if the reaction is conducted reversibly in a Daniell cell, in which case

$$w' = -\Delta G° = 208,800 + 13.9 \times 298$$

then $q = T\Delta S = 13.9 \times 298 = +4140 \text{ J}$ is the heat that *flows into* the system from the thermostatting reservoir.

15.8 THE THERMODYNAMICS OF AQUEOUS SOLUTIONS

The composition of an aqueous solution is usually expressed in terms of the molality, m, or the molarity, M, where the molality is the number of moles of solute present per 1000 g of water, and the molarity is the number of moles of solute present in 1 liter of solution. Mole fraction, molality, and molarity are related as follows. Consider an aqueous solution of m_i moles of solute i in 1000 grams of H_2O, such that the solution is m_i molal. As the molecular weight of H_2O is 18, 1000 grams of H_2O contain 1000/18 gram-moles and hence

$$X_i = \frac{n_i}{n_i + n_{H_2O}} = \frac{m_i}{m_i + 1000/18}$$

Consider an M_i molar solution which contains M_i moles of solute i per liter of solution, i.e., in 1000ρ grams of the solution, where ρ is the density of the solution in grams/cm^3. The number of gram-moles of H_2O in the liter of solutions is $(1000\rho - M_i MW_i)/18$, where MW_i is the molecular weight of i, and thus

$$X_i = \frac{n_i}{n_i + n_{H_2O}} = \frac{M_i}{M_i + (1000\rho - M_i MW_i)/18}$$

As the solution tends toward infinite dilution,

$$m_i \to \frac{1000X_i}{18} \quad \text{and} \quad M_i \to \frac{1000\rho X_i}{18}$$

In dilute solutions, molality and molarity are essentially equal to one another e.g., an aqueous solution of NaCl of $X_{NaCl} = 10^{-3}$ is 0.0556 molal and 0.0554 molar.

Just as, in the case of dilute solutes in liquid metals, it was convenient to define the 1 weight percent standard state, and hence the 1 weight percent activity scale, as

$$h_{i(1\ wt\%)} \rightarrow [wt\%\ i] \quad \text{as} \quad [wt\%\ i] \rightarrow 0$$

with the 1 weight percent standard state located on the Henry's law line at 1 weight percent, it is convenient, in aqueous solutions, to define the analogous unit molality standard state, and hence unit molality activity scale, as

$$a_{i(m)} \rightarrow m_i \quad \text{as} \quad m_i \rightarrow 0$$

where $a_{i(m)}$ is the activity of the solute with respect to the unit molality standard state, and the unit molality standard state is located on the Henry's law line at $m_i = 1$. As before, deviations from ideality are accommodated by introducing an activity coefficient defined as

$$\gamma_{i(m)} = \frac{a_{i(m)}}{m_i}$$

Consider the electrolyte, or salt, A_aY_y which, when dissolved in water, dissociates to form A^{z+} cations and Y^{z-} cations according to

$$A_aY_y = aA^{z+} + yY^{z-}$$

When m moles of A_aY_y are dissolved in n moles of H_2O, the solution formed can be considered either as

1. A solution containing m moles of the component A_aY_y and n moles of H_2O, or
2. A solution of am moles of A^{z+} and ym moles of Y^{z-} in n moles of H_2O.

In case (1), the variation of the Gibbs free energy of the solution with composition at constant T and P is given by Eq. (9.16) as

$$dG' = \overline{G}_{A_aY_y}dm + \overline{G}_{H_2O}\,dn \tag{i}$$

In case (2), the stoichiometry of the dissociation is such that the number of moles of A^{z+}, $m_{A^{z+}}$, is am and the number of moles of Y^{z-}, $m_{Y^{z-}}$, is ym. Thus

$$dm_{A^{z+}} = a\,dm \quad \text{and} \quad dm_{Y^{z-}} = z\,dm$$

and, at constant T and P,

$$dG' = \overline{G}_{A^{z+}}dm_{A^{z+}} + \overline{G}_{A^{z+}}dm_{A^{z+}} + \overline{G}_{H_2O}dn \tag{ii}$$
$$= (a\overline{G}_{A^{z+}} + y\overline{G}_{A^{z+}})dm + \overline{G}_{H_2O}dn$$

By definition

$$\overline{G}_{A_aY_y} = \left(\frac{\partial G'}{\partial m}\right)_{T,P,n}$$

$$\overline{G}_A^{z+} = \left(\frac{\partial G'}{\partial m_{A^{z+}}}\right)_{T,P,n,m_{Y^{z-}}}$$

and

$$\overline{G}_{Y^{z-}} = \left(\frac{\partial G'}{\partial m_{Y^{z-}}}\right)_{T,P,n,m_{A^{z+}}}$$

As m can be varied at constant n, $\overline{G}_{A_aY_y}$ can be determined experimentally. However, as $m_{A^{z+}}$ and $m_{Y^{z-}}$ cannot be varied independently, neither $\overline{G}_{A^{z+}}$ nor $\overline{G}_{Y^{z-}}$ can be measured. Combination of Eqs. (i) and (ii) gives

$$\overline{G}_{A_aY_y} = a\overline{G}_A^{z+} + y\overline{G}_Y^{z-} \tag{iii}$$

which shows that, although neither $\overline{G}_A^{z+}$ nor $\overline{G}_Y^{z-}$ can be measured, the combination given by Eq. (iii) can be measured.

If the component $A_a Y_y$ occurs in the unit molality standard state, Eq. (iii) is written as

$$G^{\circ}_{A_aY_y} = aG^{\circ}_{A^{z+}} + yG^{\circ}_{Y^{z-}} \tag{iv}$$

and subtraction of Eq. (iv) from Eq. (iii), noting that

$$\overline{G}_i - G_i^{\circ} = RT \ln a_i$$

gives, on rearrangement,

$$a_{A_aY_y} = a^a_{A^{z+}} + a^y_{Y^{z-}} \tag{v}$$

Thus, again, although neither $a_{A^{z+}}$ nor $a_{Y^{z-}}$ can be measured experimentally, which necessarily means that neither $a_{A^{z+}}$ nor $a_{Y^{z-}}$ has any physical significance, the product given by Eq. (v) can be measured and does have a physical significance. Eq. (v) can be written as

$$a_{A_aY_y} = (\gamma_{A^{z+}} m_{A^{z+}})^a(\gamma_{Y^{z-}} m_{Y^{z-}})^y \tag{vi}$$
$$= \gamma^a_{A^{z+}} \gamma^y_{A^{z-}} m^a_{A^{z+}} m^y_{Y^{z-}}$$

The mean ionic molality, $m_{\pm}$, is defined as

$$m_{\pm} = (m^a_{A^{z+}} m^y_{Y^{z-}})^{\frac{1}{(a+y)}} \tag{vii}$$

and the mean ion activity coefficient, $\gamma_{\pm}$, is defined as

$$\gamma_{\pm} = (\gamma^a_{A^{z+}} \gamma^y_{Y^{z-}})^{\frac{1}{(a+y)}} \tag{viii}$$

Thus substitution of Eqs. (vii) and (viii) into Eq. (vi) gives

$$a_{A_aY_y} = (\gamma_{\pm}m_{\pm})^{a+y} \tag{ix}$$

Consider an m molar solution of NaCl. As $|z^+|$ and $|z^-| = 1$ and $a = y = 1$, Eq. (ix) gives

$$a_{NaCl_{(m)}} = (\gamma_{\pm}m_{\pm})^2$$

and Eq. (vii) gives

$$m_{\pm} = (mm)^{1/2}$$

Thus

$$a_{NaCl_{(m)}} = (\gamma_\pm m_{NaCl})^2$$

and Henrian behavior follows

$$a_{NaCl_{(m)}} = m_{NaCl}^2$$

In an m molal solution of $CaCl_2$, as $|z^+| = 2, |z^-| = 1, a = 1,$ and $y = 2$,

$$a_{CaCl_{2(m)}} = (\gamma_\pm m_\pm)^3$$

$$m_\pm = [m(2m)^2]^{1/3}$$

and thus

$$a_{CaCl_{2(m)}} = 4(\gamma_\pm m_{CaCl_2})^3$$

Similarly, in an m molal solution of $Fe_2(SO_4)_3$

$$a_{Fe_2(SO_4)_{(m)}} = 36(\gamma_\pm m_{Fe_2(SO_4)_3})^5$$

15.9 THE GIBBS FREE ENERGY OF FORMATION OF IONS AND STANDARD REDUCTION POTENTIALS

Consider the cell

$$Pt, H_{2(g)} | HCl_{(aqueous)} | Hg_2Cl_{2(s)} | Hg_{(l)}$$

set up as shown in Fig. 15.11. The half-cell reaction at the anode is

$$\frac{1}{2} H_2 = H^+ + e^-$$

and the calomel half-cell reaction at the cathode is

$$\frac{1}{2} Hg_2Cl_2 + e^- = Hg + Cl^-$$

The overall reaction is thus

$$\frac{1}{2} H_{2(g)} + \frac{1}{2} Hg_2Cl_{2(s)} = Hg_{(l)} + HCl_{(m)} \tag{i}$$

and, with HG and Hg_2Cl_2 occurring at unit activity and $p_{H_2} = 1$ atm, the EMF of the cell is

$$\mathscr{E} = \mathscr{E}° - \frac{RT}{\mathfrak{f}} \ln a_{HCl_{(m)}}$$

$$= \mathscr{E}° - \frac{RT}{\mathfrak{f}} \ln(\gamma_\pm m_{HCl})^2$$

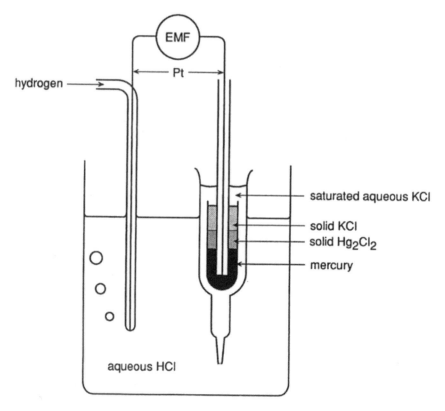

Figure 15.11 The calomel cell, $Pt, H_{2(g)}|HCl_{(aq)}|Hg_2Cl_{2(s)}|Hg_{(l)}$.

This expression can be rearranged as

$$\mathscr{E} + \frac{2RT}{\mathfrak{f}} \ln m_{HCl} = \mathscr{E}° - \frac{2RT}{\mathfrak{f}} \ln \gamma_{\pm} \qquad \text{(ii)}$$

in which the measurable quantities occur on the left-hand side. Extrapolation of the term $\mathscr{E} + (RT/\mathfrak{f}) \ln m_{HCl}$ to infinite dilution, where $\gamma_{\pm} \rightarrow 1$, allows calculation of $\mathscr{E}°$ and hence, from Eq. (ii), calculation of the variation of $\gamma_{\pm}$ with m_{HCl}. The value of $\mathscr{E}°$ at 298 K has been determined as 0.26796 volts, and the variation of $a_{HCl_{(m)}}$ with m_{HCl}^2 is as shown in Fig. 15.12. As $\mathscr{E}° = 0.26796$ volts,

$$\Delta G°_{(i),\,298\,K} = -\mathfrak{f}\mathscr{E}° = -96{,}487 \times 0.26796 = 25{,}855 \text{ J}$$

For

$$Hg_{(l)} + \frac{1}{2}Cl_{2(g)} = \frac{1}{2}Hg_2Cl_{2(s)} \qquad \text{(iii)}$$

$$\Delta G°_{(iii),298\,K} = -105{,}320 \text{ J}$$

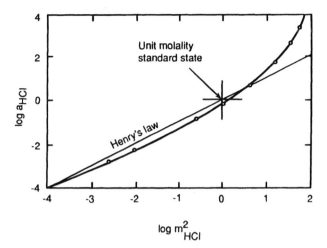

Figure 15.12 The activity of HCl in aqueous solution.

and thus, for

$$\frac{1}{2}H_{2(g)} + \frac{1}{2}Cl_{2(g)} = HCl_{(m)}$$

$$\Delta G_{(iv)}^{\circ} = \Delta G_{(i)}^{\circ} + \Delta G_{(iii)}^{\circ} \qquad (iv)$$
$$= -25,855 - 105,320 \text{ J}$$
$$= -131,175 \text{ J}$$

Thus the standard Gibbs free energy of formation of HCl in aqueous solution at unit activity from H_2 gas and Cl_2 gas, each at 1 atm pressure, is $-131,175$ J at 298 K. For

$$\frac{1}{2}H_{2(g)} + \frac{1}{2}Cl_{2(g)} = HCl_{(g)} \qquad (v)$$

$\Delta G_{(v),298 \text{ K}}^{\circ} = -94,540$ J, and hence, for the change of state

$$HCl_{(g)} = HCl_{(m)}$$

$$\Delta G_{298 \text{ K}}^{\circ} = -131,175 + 94,540$$
$$= -36,635 \text{ J}$$

$$= -8.3144 \times 298 \ln \frac{a_{HCl_{(m)}}}{p_{HCl}}$$

Thus, at 298 K, an aqueous solution of HCl at unit activity exerts a partial vapor pressure of HCl of 3.79×10^{-7} atm.

From Eq. (iv), the standard EMF of the cell

$$Pt,H_{2(g)} | HCl_{(m)} | Cl_{2(g)},Pt$$

is

$$\mathscr{E}^\circ = \frac{-\Delta G^\circ_{(iv)}}{\mathcal{f}} = \frac{131{,}175}{96{,}487} = 1.3595 \text{ volts}$$

and the cell reaction

$$\frac{1}{2} H_{2(g)} + \frac{1}{2} Cl_{2(g)} = HCl_{(m)}$$

or

$$\frac{1}{2} H_{2(g)} + \frac{1}{2} Cl_{2(g)} = H^+_{(m)} + Cl^-_{(m)} \tag{vi}$$

is the sum of the half-cell reactions

$$\frac{1}{2} H_{2(g)} = H^+ + e^- \tag{vii}$$

and

$$\frac{1}{2} Cl_2 + e^- = Cl^- \tag{viii}$$

It is now convenient to introduce the concept of a half-cell, or single electrode potential, the sum of which, in any cell, equals the EMF of the cell. This concept is useful in spite of the fact that it is impossible to construct, and hence to measure the potential of, a cell with a single electrode. In order to give meaning to the concept it is necessary to choose a particular single standard electrode and arbitrarily assign it a potential of zero. In aqueous solutions this standard single electrode is the Standard Hydrogen Electrode, in which hydrogen gas at 1 atm pressure, in contact with a platinum wire, is bubbled through an aqueous solution containing hydrogen ions at unit activity. In the Standard Hydrogen Electrode, hydrogen gas in its standard state is arbitrarily placed at the same potential as hydrogen ions in their standard state in aqueous solution. Consider the cell

$$Pt, H_{2(g, \, 1 \, atm)} \, | \, H^+_{(m)} \, Cl^-_{(m)} \, | \, Cl_{2(g, \, 1 \, atm)}, Pt$$

The variation of the potential with distance from the hydrogen gas/electrolyte interface to the chlorine gas/electrolyte interface is shown schematically in Fig. 15.13. Although the value of $\phi^{\circ,e}$ is not known, the difference between $\phi^{\circ,Cl}$ and $\phi^{\circ,H}$ is experimentally measured as 1.3595 volts. Thus if the absolute potential $\phi^{\circ,H}$ is arbitrarily assigned the value zero, the standard electrode potential at which the reduction of $\frac{1}{2}Cl_2$ to Cl^- occurs at the cathode is 1.3595 volts. Thus with the standard oxidation (and hence, reduction) potential of hydrogen being zero, then, algebraically, from Fig. 15.13, the standard EMF of the cell, $\mathscr{E}^\circ$ is obtained as the standard reduction potential of cathode minus the standard reduction potential at the anode, i.e.,

$$\mathscr{E} = \mathscr{E}^{\circ,Cl} - \mathscr{E}^{\circ,H}$$

$$= 1.3595 \text{ volts}$$

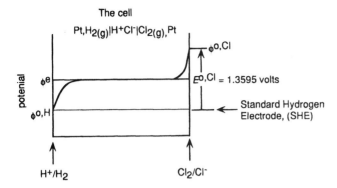

Figure 15.13 The relationship of the standard reduction potential of Cl to the standard hydrogen electrode and the absolute potentials in the cell $Pt,H_{2(g,\ 1\ atm)}$ $|H^+_{(m)}Cl^-_{(m)}|Cl_{2(g,\ 1\ atm)}$, Pt.

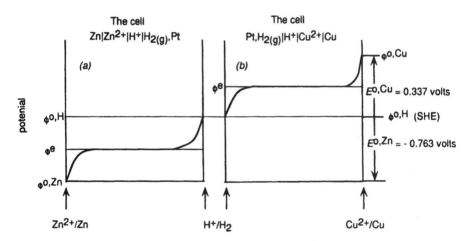

Figure 15.14 The relationship of the standard reduction potentials of Zn and Cu to the standard hydrogen electrode and the absolute potentials in the cells $Zn|Zn^{2+}|H^+|H_{2(g)}$,Pt, and $Pt,H_{2(g)}|H^+|Cu^{2+}|Cu$.

Consider, again, the Daniell cell with pure Zn and Cu as electrodes and Zn^{2+} and Cu^{2+} ions at unity molality. Fig. 15.14a shows a schematic variation of the potential through the electrolyte in the sub-cell which has the cell reaction

$$Zn_{(s)} + 2H^+_{(m)} = Zn^{2+}_{(m)} + H_{2(g)}$$

and Fig. 15.14b shows the corresponding variation through the sub-cell which has the cell reaction

$$H_{2(g)} + Cu^{2+}_{(m)} = 2H^+_{(m)} + Cu_{(s)}$$

With reference to the Standard Hydrogen Electrode, SHE, as shown, the standard reduction potential for Cu^{2+} is given by $\phi^{o,Cu} - \phi^{o,H}$ which is experimentally measured as 0.337 volts. The standard reduction potential of copper is thus 0.337 volts. Also, with respect to the SHE, the standard oxidation potential for Zn is measured experimentally as $\phi^{o,H} - \phi^{o,Zn} = 0.763$. Thus the standard reduction potential of Zn, $\mathscr{E}^{o,Zn}$, is -0.763 volts, and the standard EMF of the Daniell cell is the standard reduction potential of Cu minus the standard reduction potential of Zn, i.e.,

$$\mathscr{E}^o = 0.337 - (-0.763) = 1.100 \text{ volts}$$

The systematic list of standard reduction potentials, part of which is presented as Table 15.1, is called the Electrochemical Series.

Solubility Products

From Table 15.1, the standard reduction potential for sodium, $\mathscr{E}^{o,Na}$, is -2.714 volts, and the standard reduction potential for chlorine, $\mathscr{E}^{o,Cl}$, is 1.3595 volts. Thus, for the equilibrium

$$Na = Na^+ + e^-$$

$$\Delta G^o = -\mathfrak{f}\mathscr{E}^{o,Na} = -96,487 \times 2.714 = -261,870 \text{ J}$$

and for

$$\frac{1}{2}Cl_2 + e^- = Cl^-$$

$$\Delta G^o = -\mathfrak{f}\mathscr{E}^{o,Cl} = -96,487 \times 1.3595 = -131,170 \text{ J}$$

Summing gives

$$\Delta G^o_{(i)} = -261,870 - 131,170 = -393,040 \text{ J}$$

for the reaction

$$Na_{(s)} + \frac{1}{2}Cl_{2(g)} = Na^+_{(m)} + Cl^-_{(m)} \tag{i}$$

For the reaction

$$Na_{(s)} + \frac{1}{2}Cl_{2(g)} = NaCl_{(s)} \tag{ii}$$

$\Delta G^o_{(ii),298 \text{ K}} = -385,310 \text{ J}$, and combination of Eqs. (i) and (ii) gives

$$NaCl_{(s)} = Na^+_{(m)} + Cl^- \tag{iii}$$

Table 15.1 Standard electrode potentials at 298 K, 1 atm (standard state is 1 molal)

Electrode reaction	$\mathscr{E}^{o,x}$, volts
Acid solutions	
$F_2 + 2e^- = 2F^-$	2.65
$S_2O_8^{2-} + 2e^- = 2SO_4^{2-}$	1.98
$Co^{3+} + e^- = Co^{2+}$	1.82
$Ce^{4+} + e^- = Ce^{3+}$	1.61
$\frac{1}{2}Cl_2 + e^- = Cl^-$	1.3595
$Cr_2O_7^{2-} + 14H^+ + e^- = 2Cr^{3+} + 7H_2O$	1.33
$MnO_2 + 4H^+ + 2e^- = Mn^{2+} + 2H_2O$	1.23
$Br_2(l) + 2e^- = 2Br^-$	1.0652
$2Hg^{2+} + 2e^- = Hg_2^{2+}$	0.92
$Hg^{2+} + 2e^- = Hg$	0.854
$Ag^+ + e^- = Ag$	0.7991
$Fe^{3+} + e^- = Fe^{2+}$	0.771
$I_2 + 2e^- = 2I^-$	0.5355
$Fe(CN)_6^{3-} + e^- = Fe(CN)_6^{4-}$	0.36
$Cu^{2+} + 2e^- = Cu$	0.337
$S_4O_6^{2-} + 2e^- = 2S_2O_3^{2-}$	0.17
$Cu^{2+} + e^- = Cu^+$	0.153
$Sn^{4+} + 2e^- = Sn^{2+}$	0.15
$S + 2H^+ + e^- = H_2S$	0.141
$2H^+ + e^- = H_2$	0.000
$Fe^{3+} + 3e^- = Fe$	−0.036
$Pb^{2+} + 2e^- = Pb$	−0.126
$Sn^{2+} + 2e^- = Sn$	−0.136
$Cd^{2+} + 2e^- = Cd$	−0.403
$Cr^{3+} + e^- = Cr^{2+}$	−0.41
$Fe^{2+} + 2e^- = Fe$	−0.440
$Zn^{2+} + 2e^- = Zn$	−0.763
$Al^{3+} + 3e^- = Al$	−1.66
$\frac{1}{2}H_2 + e^- = H^-$	−2.25
$Mg^{2+} + 2e^- = Mg$	−2.37
$Na^+ + e^- = Na$	−2.714
$Ca^{2+} + 2e^- = Ca$	−2.87
$Ba^{2+} + 2e^- = Ba$	−2.90
$Cs^+ + e^- = Cs$	−2.923
$K^+ + e^- = K$	−2.925
$Li^+ + e^- = Li$	−3.045
Basic solutions	
$O_3 + H_2O + 2e^- = O_2 + 2OH^-$	1.24
$Fe(OH)_3 + e^- = OH^- + Fe(OH)_2$	−0.56
$Ni(OH)_2 + 2e^- = Ni + 2OH^-$	−0.72
$2H_2O + 2e^- = H_2 + 2OH^-$	−0.828
$SO_4^{2-} + H_2O + 2e^- = 2OH^- + SO_3^{2-}$	−0.93
$CNO^- + H_2O + 2e^- = 2OH^- + CN^-$	−0.97
$ZnO_2^{2-} + 2H_2O + 2e^- = Zn + 4OH^-$	−1.216
$Cr(OH)_3 + 3e^- = Cr + 3OH^-$	−1.3
$Ca(OH)_2 + 2e^- = Ca + 2OH^+$	−3.03

for which

$$\Delta G^{\circ}_{(iii)} = \Delta G^{\circ}_{(i)} - \Delta G^{\circ}_{(ii)}$$
$$= -393{,}040 + 385{,}310$$
$$= -8.3144 \times 298 \ln\frac{(\gamma_{\pm}m_{NaCl})^2}{a_{NaCl}}$$

In Eq. (iii), the standard state on the left-hand side is pure solid NaCl, and the standard state on the right-hand side is the unit molality standard state. Saturation of the aqueous solution occurs when NaCl has dissolved to the extent that the activity of NaCl in the solution, with respect to solid NaCl as the standard state, is unity. In this state

$$(\gamma_{\pm}\, m_{NaCl})^2 = 22.6$$

or

$$\gamma_{\pm}m_{NaCl} = 4.76$$

Thus, if the ions in solution are behaving ideally, the saturated aqueous solution of NaCl at 298 K is 4.76 molal. When the activity of NaCl in the solution is unity with respect to solid NaCl as the standard state, the term $(\gamma_{\pm}m_{NaCl})^2$ is called the *solubility product*, K_{sp}. Thus, generally, for the salt A_aY_y,

$$K_{sp} = (\gamma_{\pm}m_{\pm})^{a+y} = \exp\frac{-\Delta G^{\circ}_{298\,K}}{298\,R}$$

where $\Delta G^{\circ}_{298\,K}$ is the standard Gibbs free energy change for the change of state

$$A_aY_{a(Raoultian)} = aA^{z+}_{(m)} + yY^{z-}_{(m)}$$

Example 1   Calculate the solubility of AgBr in water at 298 K. From Table 15.1, $\mathscr{E}^{\circ,Ag} = 0.7991$ volts and $\mathscr{E}^{\circ,Br} = 1.0652$ volts. Therefore, for the reaction

$$Ag_{(s)} + \frac{1}{2}Br_{2(l)} = Ag^{+}_{(m)} + Br^{-}_{(m)}$$

$$\Delta G^{\circ}_{298\,K} = -\mathfrak{f}(-0.7991 + 1.0652) = 25{,}775\ J$$

For

$$Ag_{(s)} + \frac{1}{2}Br_{2(l)} = AgBr_{(s)}$$

$$\Delta G^{\circ}_{298\,K} = -95{,}670\ J$$

and hence for

$$AgBr_{(s)} = Ag^{+}_{(m)} + Br^{-}_{(m)}$$

$$\Delta G^{\circ}_{298\,K} = 25{,}675 + 95{,}670$$
$$= 121{,}345\ J$$
$$= -8.3144 \times 298 \ln K_{sp}$$

Thus $K_{sp} = (\gamma_{\pm}m_{AgBr})^2 = 5.4 \times 10^{-22}$ or $m_{AgBr} = 2.3 \times 10^{-11}$ which indicates that AgBr is virtually insoluble in water.

Example 2 Calculate the molalities of H^+ and OH^- in water at 298 K. In Table 15.1 the standard reduction potential for the half-cell reaction

$$H_2O_{(l)} + e^- = \frac{1}{2}H_{2(g)} + OH^-_{(m)}$$

is -0.828 volts, and the standard reduction potential for the reaction

$$H_{(m)} + e^- = \frac{1}{2}H_{2(g)}$$

is zero. Summing gives

$$H_2O_{(l)} = H^+_{(m)} + OH^-_{(m)}$$

for which

$$
\begin{aligned}
\Delta G^\circ_{298\,K} &= -f\mathscr{E}^\circ = -96487 \times (-0.828) \\
&= 79{,}900\ \text{J} \\
&= -8.3144 \times 298 \ln \gamma_{\pm}m_{H+}m_{OH-}
\end{aligned}
$$

Thus, presuming that $\gamma_{\pm} = 1$,

$$m_{H^+}m_{OH^-} = 0.97 \times 10^{-5}$$

or, from the stoichiometry of the dissociation, as $m_{H^+} = m_{OH^-}$,

$$m_{H^+} = m_{OH^-} = 10^{-7}$$

At a molality of 10^{-7}, the assumption that $\gamma_{\pm} = 1$ is reasonable.

The Influence of Acidity

The single electrode potential for the half-cell reaction

$$H^+_{(m)} + e^- = \frac{1}{2}H_{2(g)}$$

is

$$\mathscr{E}^H = \mathscr{E}^{o,H} - \frac{RT}{f}\ln\frac{p_{H_2}^{1/2}}{m_{H^+}}$$

$\mathscr{E}^{o,H} = 0$ and thus

$$\mathscr{E}^H = -\frac{RT}{f}\ln\frac{p_{H_2}^{1/2}}{m_{H^+}} \tag{i}$$

i.e., for a fixed pressure of hydrogen gas, $\mathscr{E}^H$ is a linear function of the logarithm of the molality of the hydrogen ions. The concentration of hydrogen ions in an aqueous solution determines the acidity of the solution, and, conventionally, acidity is quantified by the definition of pH, as

$$pH = -\log[H^+] \tag{ii}$$

$[H^+]$ is the *molarity* of the hydrogen ions, i.e., the number of moles of H^+ per liter of solution. Substitution of Eq. (ii) into Eq. (i) would require that either the half-cell reduction potential be determined with reference to the unit molarity standard state, or that pH be defined as $-\log(m_H^+)$. As, in dilute solutions, molality and molarity are virtually identical, this theoretical difficulty is of no practical significance. In the following discussions, the unit molarity standard state will be used for ions in solution. This standard state is defined as

$$a_A^{z+} \to [A^{z+}] \quad \text{as} \quad [A^{z+}] \to 0$$

Furthermore, in the following discussions it will be assumed that all ions in aqueous solutions behave ideally in which case

$$a_A^{z+} = [A^{z+}] \sim m_A^{z+}$$

With this understanding, Eq. (i) becomes

$$\mathscr{E}^H = -\frac{RT}{\mathfrak{f}} \ln \frac{p_{H_2}^{1/2}}{[H^+]}$$

which, with Eq. (ii), becomes

$$\mathscr{E}^H = -\frac{2.303 \times 8.3144 \times 298}{2 \times 96,487} \log p_{H_2} + \frac{2.303 \times 8.3144 \times 298}{96,487} \log [H^+]$$

$$= -0.0591(pH) - 0.0296 \log p_{H_2} \tag{15.7}$$

or, with $p_{H_2} = 1$ atm, the reduction potential of hydrogen varies with pH as

$$\mathscr{E}^H = -0.0591(pH)$$

In Ex. 2 of the previous section, it was found that the molality (and hence the molarity) of H^+ in H_2O at 298 K is 10^{-7}. Thus the pH of H_2O at 298 K is 7, and the reduction potential of hydrogen in water is $-0.0591 \times 7 = -0.414$ volts. The hydrogen electrode has its standard reduction potential of 0 at pH = 0, i.e., at $[H^+] = 1$.

15.10 POURBAIX DIAGRAMS

Pourbaix diagrams,* or "potential-pH" diagrams, are graphical representations of thermodynamic and electrochemical equilibria occurring in aqueous systems. They

---

*M. Pourbaix, "Atlas of Electrochemical Equilibria in Aqueous Solutions," National Association of Corrosion Engineers, Houston, Tx, 1974.

are thus the electrochemical analogues of the chemical phase stability diagrams discussed in Sec. 13.4. Consider the cell

$$Pt, H_{2(g)} | H_2O | O_{2(g)}, Pt$$

The galvanic cell reaction

$$H_{2(g)} + \frac{1}{2}O_{2(g)} = H_2O_{(l)} \tag{i}$$

is the sum of the half-cell reactions

$$H_2 = 2H^+ + 2e^- \tag{ii}$$

and

$$\frac{1}{2}O_2 + 2H^+ + 2e^- = H_2O \tag{iii}$$

Thus

$$\Delta G_{(i)}^\circ = \Delta G_{(ii)}^\circ + \Delta G_{(iii)}^\circ$$

$\Delta G_{(i)298\ K}^\circ = -237,190$ J and by convention, $\Delta G_{(ii)298\ K}^\circ = 0$. Thus the standard reduction potential for the reaction given by Eq. (iii) is

$$\mathscr{E}_{(iii)}^\circ = \frac{-\Delta G_{(iii)}^\circ}{2\mathfrak{f}}$$

$$= \frac{237,190}{2\mathfrak{f}}$$

$$= 1.229 \text{ volts}$$

With all of the reactants and products occurring in their standard states, electrochemical equilibrium is established when the backing EMF applied to the cell is 1.229 volts. If the backing EMF is less than this value, the cell is a current-producing galvanic cell and the cell reaction occurring is

$$H_{2(g)} + \frac{1}{2}O_{2(g)} \rightarrow H_2O_{(l)}$$

with the oxidation $2H_2 \rightarrow 2H^+ + 2e^-$ occurring at the anode and the reduction reaction $\frac{1}{2}O_2 + 2H^+ + 2e^- \rightarrow H_2O$ occurring at the cathode. However, if the backing EMF is greater than 1.229 volts, the cell is a current-consuming electrolysis cell and the cell reaction occurring is

$$H_2O_{(l)} \rightarrow H_{2(g)} + \frac{1}{2}O_{2(g)}$$

with the oxidation reaction $H_2O \rightarrow \frac{1}{2}O_2 + 2H^+ + 2e^-$ occurring at the anode and the reduction reaction $2H^+ + 2e^- \rightarrow H_2$ occurring at the cathode. Thus, with hydrogen gas and hydrogen ions at unit activity, electrochemical equilibrium is established at the hydrogen electrode when the electrode is at zero potential. If the potential of the electrode is increased to a value above zero the anodic reaction $H_2 \rightarrow 2H^+ + 2e^-$ occurs, and if the potential of the electrode is decreased to a value less than zero, the cathodic reaction $2H^+ + 2e^- \rightarrow H_2$ occurs. Similarly, with oxygen at unit pressure and hydrogen ions at unit activity, electrochemical equilibrium at the oxygen electrode is established when the electrode potential is 1.229 volts. If the potential of the electrode is increased above 1.229 volts the anodic reaction $H_2 \rightarrow \frac{1}{2} O_2 + 2H^+ + 2e^-$ occurs, and if the potential of the electrode is decreased below 1.229 volts the cathodic reaction $\frac{1}{2} O_2 + 2H^+ + 2e^- \rightarrow H_2O$ occurs. Thus, generally, if the potential, $\mathscr{E}$, of an electrode at equilibrium with the electrolyte is increased, an anodic oxidation reaction occurs, and if it is decreased, a cathodic reduction reaction occurs.

As both half-cell reactions (i) and (ii) involve hydrogen ions, the half-cell potentials are functions of the pH of the aqueous solution. At the hydrogen electrode, at 298 K, the electrochemical equilibrium

$$H^+ + e^- = \frac{1}{2} H_2 \qquad (a)$$

is established at

$$\mathscr{E}_{(a)} = -\frac{RT}{\mathnormal{f}} \ln \frac{p_{H_2}^{1/2}}{[H^+]}$$

$$= -0.0591(\text{pH}) - 0.0298 \log p_{H_2}$$

which, with $p_{H_2} = 1$ atm, is drawn as line (a) in Fig. 15.15. At the oxygen electrode, at 298 K, the electrochemical equilibrium

$$\frac{1}{2} O_2 + 2H^+ + 2e^- = H_2O \qquad (b)$$

is established at

$$\mathscr{E}_{(b)} = 1.299 - \frac{RT}{2\mathnormal{f}} \ln \frac{1}{[H^+]^2 \, p_{O_2}^{1/2}}$$

$$= 1.229 - 0.0591 \, \text{pH} + 0.0148 \log p_{O_2}$$

which, with $p_{O_2} = 1$ atm, is drawn as line (b) in Fig. 15.15. Lines (a) and (b) in Fig. 15.15 define the *domain of thermodynamic stability of water* in aqueous solutions under a pressure of 1 atm of H_2 and 1 atm of O_2. Below line (a) the equilibrium pressure of hydrogen gas is greater than 1 atm, and thus hydrogen is cathodically evolved from an aqueous solution, the potential of which is moved below line (a) when the pressure of hydrogen at the electrode is 1 atm. Similarly, above line (b) the equilibrium pressure of oxygen gas is greater than 1 atm, and thus oxygen gas is anodically evolved from an aqueous solution at an electrode, the potential of which is moved above line (b) when the oxygen pressure at the electrode is 1 atm. Water is thermodynamically stable between lines (a) and (b) with oxygen and hydrogen pressures of 1 atm.

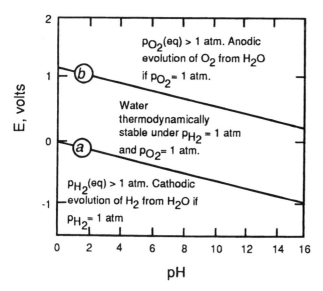

Figure 15.15   The domain of thermodynamic stability of water.

The Pourbaix Diagram for Aluminum

The species participating in the various chemical and electrochemical equilibria are the solids Al and Al_2O_3 and the ions Al^{3+} and AlO_2^-, and the pertinent standard Gibbs free energies of formation are

Reaction	$\Delta G_{298\ K}^{\circ}$, J
Al	0
$\frac{1}{2} H_{2(g)} = H_{(m)}^+ + e^-$	0
$2Al_{(s)} + 1.5O_{2(g)} = Al_2O_{3(s)}$	$-1{,}608{,}900$
$Al_{(s)} = Al_{(m)}^{3+} + 3e^-$	$-481{,}200$
$Al_{(s)} + O_{2(g)} + e^- = AlO_{2(m)}^-$	$-839{,}800$
$H_{2(g)} + \frac{1}{2} O_{2(g)} = H_2O_{(l)}$	$-237{,}200$

Consider first the equilibrium which occurs between the ions in solution. As reduction potentials are being used, the equations for the electrochemical equilibria are set up with the higher oxidation state on the left and the lower oxidation state on the right, i.e., with the balancing electronic charge on the left-hand side of the equation.

The equilibrium between the two dissolved substances.

1. *The equilibrium between Al^{3+} and AlO_2^-:*

$$Al^{3+} \longleftrightarrow AlO_2^-$$

The procedure for deriving the expression for the equilibrium is as follows:
(1) Balance the oxygen with H_2O, i.e.,

$$Al^{3+} + 2H_2O \longleftrightarrow AlO_2^-$$

(2) Balance the hydrogen with H^+, i.e.,

$$Al^{3+} + 2H_2O \longleftrightarrow AlO_2^- + 4H^+$$

(3) If necessary, balance the charge with e^-
This step is not necessary with Al^{3+} and AlO_2^-, as the equilibrium is not electrochemical. The desired expression is thus

$$Al^{3+} + 2H_2O = AlO_2^- + 4H^+ \tag{i}$$

for which

$$\Delta G_{(i)}^\circ = (-839,800) - (2 \times -237,200) - (-481,200)$$
$$= 115,800 \text{ J}$$
$$= -8.3144 \times 298 \times 2.303 \log \frac{[H^+]^4[AlO_2^-]}{[Al^{3+}]}$$

Thus

$$-20.29 = 4 \log [H^+] + \log \frac{[AlO_2^-]}{[Al^{3+}]}$$

or

$$\log \frac{[Al^{3+}]}{[AlO_2^-]} = 20.29 - 4pH$$

Thus $[Al^{3+}] = [AlO_2^-]$ at pH = 5.07 and pH = 5.07 is drawn as line 1 in Fig. 15.16a. At values of pH greater than 5.07 $[AlO_2^-] > [Al^{3+}]$, and at values of pH less than 5.07 $[Al^{3+}] > [AlO_2^-]$.

The equilibrium between the two solids.

2. *The equilibrium between Al and Al_2O_3:* With the higher oxidation state on the left and the lower oxidation state on the right, the procedure for deriving the expression for the equilibrium begins with

$$Al_2O_3 \longleftrightarrow 2Al$$

Then, as before, balance the oxygen with H_2O, i.e.,

$$Al_2O_3 \longleftrightarrow 2Al + 3H_2O$$

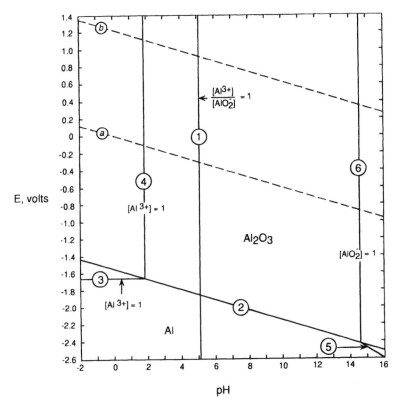

Figure 15.16 (a) Construction of the Pourbaix diagram for Al.

balance the hydrogen with H^+, i.e.,

$$Al_2O_3 + 6H^+ \longleftrightarrow 2Al + 3H_2O$$

balance the charge with e^-, i.e.,

$$Al_2O_3 + 6H^+ + 6e^- \longleftrightarrow 3H_2O + 2Al$$

The required equilibrium is thus

$$Al_2O_3 + 6H^+ + 6e^- = 3H_2O + 2Al \tag{ii}$$

for which

$$\Delta G^\circ_{(ii)} = (3 \times -237,200) - (-1,608,900)$$
$$= 897,300 \text{ J}$$
$$= -z\mathcal{F}\mathscr{E}^\circ_{(ii)} = -6 \times 96,487 \, \mathscr{E}^\circ_{(ii)}$$

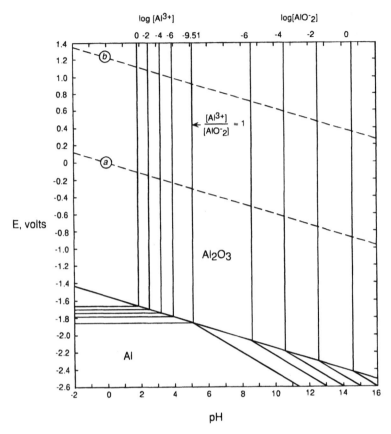

Figure 15.16 (*b*) The Pourbaix diagram for Al.

Thus $\mathscr{E}^{\circ}_{(ii)} = -1.55$ volts and, with $a_{Al} = a_{Al_2O_3} = 1$

$$\mathscr{E}_{(ii)} = \mathscr{E}^{\circ}_{(ii)} - \frac{RT}{z\mathfrak{f}} \ln \frac{1}{[H^+]^6}$$

$$= -1.55 + \frac{8.3144 \times 298 \times 2.303 \times 6}{6 \times 96{,}487} \log [H^+]$$

$$= -1.55 - 0.0591 \text{ pH volts}$$

This is drawn as line (2) in Fig. 15.16*a* and is the line along which Al is in equilibrium with Al_2O_3. Al is the stable solid phase in states below line 2 (at lower values of pH and at more negative electrode potentials), and Al_2O_3 is the stable solid in states above the line.

One solid in equilibrium with a dissolved substance.

3. *The equilibrium between Al and Al^{3+}:* The equilibrium is

$$Al^{3+} + 3e^- = Al \qquad \text{(iii)}$$

for which

$$\Delta G^\circ_{(iii)} = 481{,}200 \text{ J}$$

Thus

$$\mathcal{E}^\circ_{(iii)} = \frac{-481{,}200}{3 \times 96{,}487} = -1.66 \text{ volts}$$

Then, with $a_{Al} = 1$,

$$\mathcal{E}_{(iii)} = -1.66 - \frac{8.3144 \times 298 \times 2.303}{3 \times 96{,}487} \log \frac{1}{[Al^{3+}]}$$

The variation of the concentration of Al^{3+} ions in equilibrium with Al is thus

$$\mathcal{E}_{(iii)} = -1.66 + 1.971 \times 10^{-2} \log [Al^{3+}]$$

With $[Al^{3+}] = 1$ mole per liter, this equation is drawn as line 3 in Fig. 15.16a. The concentration of Al^{3+} in equilibrium is independent of pH and decreases with increasingly negative values of $\mathcal{E}$.

4. *The equilibrium between Al_2O_3 and Al^{3+}:* The sequence in the derivation of the equilibrium equation is

$$Al_2O_3 \leftarrow \rightarrow 2Al^{3+}$$
$$Al_2O_3 \leftarrow \rightarrow 2Al^{3+} + 3H_2O$$
$$6H^+ + Al_2O_3 \leftarrow \rightarrow 2Al^{3+} + 3H_2O$$

Balancing the charge with e^- is not necessary, and the required expression is thus

$$6H^+ + Al_2O_3 = 2Al^{3+} + 3H_2O \qquad \text{(iv)}$$

for which

$$\Delta G^\circ_{(iv)} = (3 \times -237{,}200) + (2 \times -481{,}200) - (-1{,}608{,}900)$$
$$= -65{,}100 \text{ J}$$

which, with $a_{Al_2O_3} = 1$, gives

$$-65{,}100 = -8.3144 \times 298 \times 2.303 \log \frac{[Al^{3+}]^2}{[H^+]^6}$$

or

$$\log [Al^{3+}] = 5.70 - 3pH$$

which is the variation of the concentration of Al^{3+} ions in equilibrium with Al_2O_3. This equilibrium is independent of the electrode potential, and line 4 in Fig. 15.16a, at pH = 1.9, represents a concentration of Al^{3+} of 1 mole per liter in equilibrium with Al_2O_3.

5. *The equilibrium between Al and AlO_2^-:*

$$AlO_2^- \longleftrightarrow Al$$
$$AlO_2^- \longleftrightarrow Al + 2H_2O$$
$$AlO_2^- + 4H^+ \longleftrightarrow Al + 2H_2O$$
$$AlO_2^- + 4H^+ + 3e^- \longleftrightarrow Al + 2H_2O$$

The equilibrium is thus

$$AlO_2^- + 4H^+ + 3e^- = Al + 2H_2O \tag{v}$$

for which

$$\Delta G_{(v)}^\circ = (2 \times -237,200) + 839,800$$
$$= 365,400 \text{ J}$$

Thus

$$\mathscr{E}_{(v)}^\circ = \frac{-365,400}{3 \times 96,487} = -1.26 \text{ volts}$$

With $a_{Al} = 1$,

$$\mathscr{E}_{(v)} = -1.26 - \frac{8.3144 \times 298 \times 2.303}{3 \times 96,487} \log \frac{1}{[H^+]^4[AlO_2^-]}$$

which gives the variation of the concentration of AlO_2^- ions in equilibrium with Al as

$$\mathscr{E}_{(v)} = -1.26 - 0.0789 \text{ pH} + 0.0198 \log [AlO_2^-]$$

This equation, with $[AlO_2^-] = 1$ mole per liter, is drawn as line 5 in Fig. 15.16a. The equilibrium is dependent on both pH and $\mathscr{E}$.

6. *The equilibrium between Al_2O_3 and AlO_2^-:*

$$Al_2O_3 \longleftrightarrow 2AlO_2^-$$
$$H_2O + Al_2O_3 \longleftrightarrow 2AlO_2^-$$
$$H_2O + Al_2O_3 \longleftrightarrow 2AlO_2^- + 2H^+$$

The equilibrium is thus

$$H_2O + Al_2O_3 = 2AlO_2^- + 2H^+ \tag{vi}$$

for which

$$\Delta G_{(vi)}^\circ = 166,500 \text{ J}$$

This gives

$$\log[AlO_2^-] = pH - 14.59$$

as the variation of the concentration of AlO_2^- in equilibrium with Al_2O_3. This equation, with $[AlO_2^-] = 1$ mole per liter is drawn as line 6 in Fig. 15.16a at pH = 14.59.

The Pourbaix diagram for Al, containing iso-concentrations lines for Al^{3+} and AlO_2^- in the fields of stability of Al and Al_2O_3, is shown in Fig. 15.16b. At pH = 5.07 in the Al_2O_3 field of stability $[Al^{3+}] = [AlO_2^-] = 3.10 \times 10^{-10}$ moles per liter. However, in the Al stability field, the 3×10^{-10} moles of Al^{3+} per liter iso-concentration line occurs at $\mathscr{E} = -1.85$ volts, and the 3×10^{-10} moles of AlO_2^- per liter iso-concentration lines lies at $\mathscr{E} = -1.45 - 0.0789$ pH. The domain of thermodynamic stability of water at hydrogen and oxygen pressures of 1 atm is defined by the lines a and b, and the position of line a relative to line 2 in Fig. 15.16a illustrates why Al metal cannot be produced by electrolysis of aqueous solutions. With a hydrogen pressure of 1 atm, attempts to decrease the potential of an electrode to a value less than that given by line a cause the cathodic evolution of hydrogen gas from the electrolyte.

The Solubility of Alumina in Aqueous Solutions

Alumina dissolves in aqueous solutions to form Al^{3+} ions according to

$$Al_2O_3 + 6H^+ = 2Al^{3+} + 3H_2O$$

and to form AlO_2^- ions according to

$$Al_2O_3 + H_2O + 2AlO_2^- + 2H^+$$

The solubility, as Al^{3+}, which varies with pH as

$$\log[Al^{3+}] = 5.70 - 3pH$$

and as AlO_2^-, which varies with pH as

$$\log[AlO_2^-] = pH - 14.59$$

is shown in Fig. 15.17. The solubility of Al_2O_3 as AlO_2^- decreases by 10 orders of magnitude when the pH is decreased from 15 to 5, and this behavior is made use of in the Bayer process for the separation of Al_2O_3 from bauxite. Bauxite, which is the principal aluminum ore, is a mixture of aluminum monohydrate and aluminum trihydrate containing up to 60% Al_2O_3 with Fe_2O_3 as the main impurity. The ore is digested in a strong caustic solution at a temperature between 150° and 250°C and a pressure high enough to suppress boiling. The alumina dissolves as AlO_2^-, and the insoluble residue of Fe_2O_3 (red mud) is removed from the solution by filtration. The pH of the solution is then increased by adding water, and seed crystals of $Al(OH)_3$ are added. The decreasing solubility causes the precipitation of $Al(OH)_3$ by the hydrolysis reaction

$$NaAlO_2 + 2H_2O = NaOH + Al(OH)_3$$

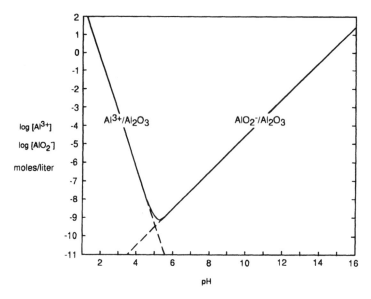

Figure 15.17 The solubility of Al_2O_3 in aqueous solution.

The strong caustic solution is regenerated by removal of water by boiling, and the precipitated hydrate is then calcined at $1200°-1350°C$ to form $\alpha-Al_2O_3$.

15.11 SUMMARY

1. A system which is capable of performing electrical work as the result of the occurrence of a chemical reaction is called a galvanic cell, and the overall chemical reaction is represented by an equation called the cell reaction. If, in such a reaction, dn gram ions of valence z are transported through a voltage difference $\Delta\phi$ maintained between the electrodes of the cell, then

$$\delta w' = z\mathfrak{f}\Delta\phi dn$$

where $\mathfrak{f}$ is Faraday's constant ($= 96,487$ coulombs/mole). If the transportation is conducted reversibly, in which case the electric potential difference between the electrodes of the cell is called the electromotive force or EMF, $\mathscr{E}$, of the cell, then

$$\delta w'_{max} = z\mathfrak{f}\mathscr{E}dn = -dG'$$

and, for the transportation of 1 mole of ions,

$$\Delta G = -z\mathfrak{f}\mathscr{E}$$

2. If the reaction

$$aA + bB = cC + dD$$

is conducted in an electrochemical cell, the EMF of the cell is given by

$$\mathscr{E} = \mathscr{E}^\circ - \frac{RT}{z\mathfrak{f}} \ln \frac{a_C^c a_D^d}{a_A^a a_B^b}$$

where the standard EMF of the cell, $\mathscr{E}^\circ$, is given by

$$\mathscr{E}^\circ = -\frac{\Delta G^\circ}{z\mathfrak{f}}$$

3. For any cell reaction at constant temperature and pressure

$$\Delta G = -z\mathfrak{f}\mathscr{E}$$

Thus, the molar entropy change for the cell reaction is

$$\Delta S = z\mathfrak{f}\left(\frac{\partial \mathscr{E}}{\partial T}\right)_P$$

and the molar enthalpy change for the cell reaction is

$$\Delta H = -z\mathfrak{f}\mathscr{E} + z\mathfrak{f}T\left(\frac{\partial \mathscr{E}}{\partial T}\right)_P$$

4. The composition of an aqueous solution is usually expressed in terms of the molality, m, or the molarity, M, where molality is the number of moles of solute present per 1000 grams of water and molarity is the number of moles of solute present in 1 liter of solution. The unit molality activity scale (which is analogous with the 1 weight percent activity scale) is defined as

$$a_{i(m)} \to m_i \quad \text{as} \quad m_i \to 0$$

where $a_{i(m)}$ is the activity of the solute with respect the unit molality standard state, and the unit molality standard state is located on the Henry's law line at $m_i = 1$. Deviation from ideality is accommodated by an activity coefficient defined as

$$\gamma_{i(m)} = \frac{a_{i(m)}}{m_i}$$

5. The mean ionic molality, $m_\pm$, of the electrolyte, or salt, $A_a Y_y$, when dissolved in water is defined by

$$m_\pm = (m_{A^{z+}}^a \, m_{Y^{z-}}^y)^{\frac{1}{a+y}}$$

and the mean ion activity coefficient, $\gamma_\pm$, is defined as

$$\gamma_\pm = (\gamma_{A^{z+}}^a \, \gamma_{Y^{z-}}^y)^{\frac{1}{a+y}}$$

The activity of the dissolved salt is thus

$$a_{A_aY_y} = (\gamma_{\pm}m_{\pm})^{a+y}$$

6. Any cell reaction is the sum of two half-cell reactions, and the Standard Hydrogen Electrode, at which the half-cell reaction

$$H^+_{(aqueous,\ m=1)} + e^- = \frac{1}{2}H_{2(g,P=1\ atm)}$$

occurs is assigned a potential of zero. This facilitates the assignment of standard reduction half-cell potentials to all other half-cell reactions. The Electrochemical Series is produced by listing the half-cell reactions in decreasing order of the magnitudes of their standard reduction half-cell potentials.

7. The influence of electrode potential and pH on the phase equilibria and solubilities in aqueous electrochemical systems can be presented graphically by Pourbaix diagrams.

15.12 NUMERICAL EXAMPLES

Example 1

Fayalite, $2FeO\cdot SiO_2$, is the only iron silicate compound formed by reaction of FeO with SiO_2 at a total pressure of 1 atm, and the standard Gibbs free energy change for the reaction

$$2FeO_{(s)} + SiO_{2(s)} = 2FeO\cdot SiO_{2(s)}$$

is $-11,070$ J at 1200 K. Calculate the EMF of the cell

$$Fe\,|\,SiO_2\,|\,2FeO\cdot SiO_2\,|\,CaO-ZrO_2\,|\,FeO\,|\,Fe$$

at 1200 K. This is an oxygen concentration cell in which the cell reaction can be written as

$$O_2(\text{higher pressure at the cathode}) \rightarrow O_2(\text{lower pressure at the anode}) \qquad (i)$$

for which the EMF is

$$\mathscr{E} = -\frac{RT}{4\mathfrak{f}}\ln\frac{p_{O_2}(\text{at the anode})}{p_{O_2}(\text{at the cathode})} \qquad (ii)$$

The oxygen pressure at the electrodes are fixed by the chemical equilibrium

$$Fe + \frac{1}{2}O_2 = FeO \qquad (iii)$$

for which

$$K_{(iii)} = \frac{a_{FeO}}{a_{Fe}p_{O_2}^{1/2}}$$

At the cathode the activity of FeO with respect to Fe-saturated pure FeO, is unity and, at the anode, the activity of FeO is that occurring in $2FeO \cdot SiO_2$ saturated with Fe and SiO_2. As

$$K_{(iii)} = \frac{a_{FeO(cathode)}}{a_{Fe(cathode)} P_{O_2(cathode)}^{1/2}} = \frac{a_{FeO(anode)}}{a_{Fe(anode)} P_{O_2(anode)}}$$

and

$$a_{Fe(anode)} = a_{Fe(cathode)} = a_{FeO(cathode)} = 1$$

$$\frac{P_{O_2(anode)}}{P_{O_2(cathode)}} = a_{FeO(anode)}^2$$

and, hence, in Eq. (ii),

$$\mathscr{E} = -\frac{RT}{4\mathfrak{f}} \ln a_{FeO(anode)}^2$$

For $2FeO + SiO_2 = 2FeO \cdot SiO_2$,

$$\Delta G_{1200\,K}^\circ = -11,070 \text{ J}$$

$$= -8.3144 \times 1200 \ln \frac{a_{2FeO \cdot SiO_2}}{a_{FeO}^2 a_{SiO_2}}$$

Thus, at the anode, with $a_{2FeO \cdot SiO_2} = a_{SiO_2} = 1$,

$$a_{FeO(anode)} = 0.574$$

and thus

$$\mathscr{E} = -\frac{8.3144 \times 1200}{4 \times 96,487} \ln (0.574)^2$$

$$= 0.0287 \text{ volts}$$

Alternatively, the anode half-cell reaction can be written as

$$O^{2-} = \frac{1}{2} O_{2(eq.Fe/\underline{FeO})} + 2e^-$$

and the cathode half-cell reaction can be written as

$$\frac{1}{2} O_{2(eq.Fe/FeO)} + 2e^- = O^{2-}$$

or, at the anode,

$$2Fe + 2O^{2-} + SiO_2 = 2FeO \cdot SiO_2 + 4e^-$$

and, at the cathode,

$$2FeO + 4e^- = 2Fe + 2O^{2-}$$

summation of which gives the cell reaction as

$$2FeO + SiO_2 = 2FeO \cdot SiO_2$$

The Gibbs free energy change for the cell reaction is

$$\Delta G° = -z\mathfrak{f}\mathscr{E}° = -11{,}070 \text{ J}$$

and thus

$$\mathscr{E} = \frac{-\Delta G°}{4\mathfrak{f}} = \frac{11{,}070}{4 \times 96{,}487} = 0.0287 \text{ volts}$$

Example 2

A waste liquor consists of a 0.5 molal solution of $CaCl_2$ in water. Calculate the minimum work, per mole of $CaCl_2$, required to separate the liquor into anhydrous $CaCl_2$ and pure water at atmospheric temperature and pressure. The mean ionic activity coefficient of 0.5 molal $CaCl_2$ is 0.448. The minimum work is that required when the separation is conducted reversibly, i.e., $w = -\Delta G$ for the process. 0.5 molal $CaCl_2$ comprises 0.5 moles of $CaCl_2$ and 1000 grams of H_2O or 0.5 moles of $CaCl_2$ and $1000/18 = 55.55$ moles of water. Thus 1 mole of $CaCl_2$ exists in 111.1 moles of water, and the mole fraction of water is $111.1/112.1 = 0.991$.

1. Move 1 mole of dissolved $CaCl_2$ from a concentration of 0.5 molar to the 1 molal standard state.

$$\Delta G_{(1)} = RT \ln \frac{a_{CaCl_2 \,(m=1)}}{a_{CaCl_2 \,(m=0.5)}}$$
$$= -RT \ln 4 \,(\gamma_{\pm} \, m_{CaCl_2})^3$$
$$= -8.3144 \times 298 \ln [4 \times (0.448 \times 0.5)^3] = 7686 \text{ J}$$

2. Transfer the 1 mole of Ca^{2+} ions from the 1 molal standard state to solid Ca at 298 K. From Table 15.1, $\mathscr{E}°^{,Ca} = -2.87$ volts. Therefore, for the reaction

$$Ca^{2+}_{(m)} + 2e^- = Ca_{(s)}$$
$$\Delta G_{(2)} = -(2 \times 96{,}487 \times -2.87) = 553{,}835 \text{ J}$$

3. Transfer the 2 moles of Cl^{2-} ions from the 1 molal standard state to Cl_2 gas at 1 atm pressure at 298 K. From Table 15.1, $\mathscr{E}°^{,Cl} = 1.3595$ volts. Therefore for the reaction

$$2Cl^-_{(m)} = Cl_{2(g)} + 2e^-$$
$$\Delta G_{(3)} = -(2 \times 96{,}487 \times -1.3595) = 262{,}348 \text{ J}$$

4. Transfer 111.1 moles of H_2O from a mole fraction of 0.991 to a mole fraction of 1.0. Assuming Raoultian behavior,

$$\Delta G_{(4)} = -n_{H_2O}RT \ln X_{H_2O}$$
$$= -111.1 \times 8.3144 \times 298 \ln 0.991 = 2486 \text{ J}$$

5. Allow the 1 mole of solid Ca to react with the mole of gaseous Cl_2 to form 1 mole of solid $CaCl_2$ at 298 K. For the reaction.

$$Ca_{(s)} + Cl_{2(g)} = CaCl_{2(s)}$$

$$\Delta G_{(5)} = \Delta G^\circ_{298\ K} = -752{,}100\ J$$

Thus, the change in the Gibbs free energy for the separation process is

$$\Delta G_{(1)} + \Delta G_{(2)} + \Delta G_{(3)} + \Delta G_{(4)} + \Delta G_{(5)} = 74{,}255\ J$$

which is the minimum amount of work required per mole of $CaCl_2$ separated.

PROBLEMS

15.1 The EMF of the galvanic cell

$$Pb_{(s)}\,|\,PbCl_{2(s)}\,|\,HCl_{(aqueous)}\,|\,AgCl_{(s)}\,|\,Ag_{(s)}$$

where all of the components are present as pure solids in contact with an HCl electrolyte, is 0.490 volts at 25°C and, at that temperature, the temperature coefficient of the EMF is -1.84×10^{-4} volts/degree. Write the cell reaction and calculate the Gibbs free energy change and the change in enthropy for the cell reaction at 298 K.

15.2 At 298 K, the EMF of the cell

$$Pb\,|\,PbCl_2\,|\,Hg_2Cl_2\,|\,Hg$$

is +0.5357 volts and the temperature coefficient of the EMF is 1.45×10^{-4} volts/degree. Calculate (*a*) the maximum work available from the cell at 298 K per mole of Pb reacted, (*b*) the change in entropy for the cell reaction, and (*c*) the heat absorbed by the cell at 298 K per mole of Pb reacted when the cell is operating reversibly.

The Hg electrode in the cell is replaced by an Hg–X alloy in which $X_{Hg} = 0.3$ and where X is inert. The EMF of the cell at 298 K is found to increase by 0.0089 volts. Calculate (*d*) the activity of Hg in the alloy at 298 K.

15.3 The solid-state electrochemical cell

$$(Pt),O_2(\text{gas at } p_{O_2})\,|\,CaO\text{–}ZrO_2\,|\,Fe\,|\,FeO,(Pt)$$

is built to measure the partial pressure of oxygen in gases. Write an equation relating the oxygen pressure and temperature of the gas to the EMF to the cell.

15.4 The EMF of the cell

$$Ag_{(s)}\,|\,AgCl_{(s)}\,|\,Cl_{2(g,\ 1\ atm)},Pt$$

is found to be

$$\mathscr{E}(\text{volts}) = 0.977 + 57 \times 10^{-4}(350 - t) - 4.8 \times 10^{-7}(350 - t)^2$$

in the temperature range $t = 100°\text{–}450°C$. Calculate the value of Δc_p for the cell reaction.

15.5 A galvanic cell is set up with electrodes of solid aluminum and solid aluminum-zinc alloy and an electrolyte of molten $AlCl_3$–$NaCl$. When the mole fraction of Al in the alloy electrode is 0.38, the EMF of the cell is 7.43 millivolts at 380°C, and the temperature coefficient of the EMF is 2.9×10^{-5} volts/degree. Calculate (a) the activity of Al in the alloy, (b) the partial molar Gibbs free energy of mixing of Al in the alloy, and (c) the partial molar enthalpy of mixing of Al in the alloy.

15.6 By measuring the EMFs of cells of the type

$$Ni_{(s)} | NiO_{(s)} | CaO\text{–}ZrO_2 | Cu_{(l)} \text{ containing dissolved oxygen}$$

it has been established that e_O^O in liquid copper at 1363 K is -0.16 and that the standard Gibbs free energy change for

$$\frac{1}{2} O_{2(g)} = [O]_{1 \text{ wt\% in Cu}}$$

is $\Delta G° = -74{,}105 + 10.76T$ J. If the EMF of such a cell is 0.222 volts at 1363 K, calculate (a) the activity of oxygen in the liquid copper cathode with respect to a standard state of oxygen gas at 1 atm pressure, (b) the activity of Cu_2O in the cathode metal with respect to Cu-saturated pure solid Cu_2O, (c) the weight percentage of oxygen dissolved in the copper cathode, (d) the maximum solubility of oxygen in liquid copper at 1363 K.

15.7 Calculate the conditions under which an aqueous solution of $[Pb^{2+}] = 1$ mole/liter is in equilibrium with metallic Pb and solid PbO at 298 K. Is any other lead ion present in significant concentration in this solution?

Species	$\Delta G°_{298 \text{ K}}$ J
$PbO_{(s)}$	$-189{,}300$
$Pb^{2+}_{(m)}$	$-24{,}310$
$Pb^{4+}_{(m)}$	$+302{,}500$
$HPbO^-_{2(m)}$	$-339{,}000$
$PbO^{2-}_{3(m)}$	$-277{,}570$
$PbO^{4-}_{4(m)}$	$-28{,}210$
$H_2O_{(l)}$	$-237{,}190$

15.8 Aluminum can be produced by electrolysis of Al_2O_3 dissolved in molten cryolite, $3NaF·AlF_3$. If inert electrodes are used in an electrolysis cell and the cryolite is saturated with Al_2O_3 at 1000°C, what is the decomposition voltage of the Al_2O_3? The Hall-Heroult process for electrolysis of Al_2O_3 uses graphite as the anode material, and the gas which is evolved at the anode is essentially pure CO_2 at 1 atm pressure. Calculate the decomposition voltage of Al_2O_3 in an Al_2O_3-saturated $3NaF·AlF_3$ electrolyte at 1000°C in the Hall-Heroult cell.

15.9 At 298 K the solubility of Cl_2 in H_2O, under a partial pressure of $Cl_2 = 1$ atm, is 0.0618 molal. Calculate the standard Gibbs free energy of formation of an aqueous solution of chlorine and calculate the change in the Gibbs free energy to form a solution of chlorine which is 0.01 molal. Ideal solution behavior can be assumed.

Appendix A

SELECTED THERMODYNAMIC AND THERMOCHEMICAL DATA

TABLE A-1: THE STANDARD GIBBS FREE ENERGY CHANGES FOR SEVERAL REACTIONS

This table lists the standard Gibbs free energy changes for selected reactions in the form

$$\Delta G_T^\circ = A + BT \text{ J}$$

or

$$\Delta G_T^\circ = A + BT \ln T + CT \text{ J}$$

and lists the range of temperature in which the expression is valid.

Example—For the oxidation of solid copper to form solid cuprous oxide according to

$$2Cu_{(s)} + \frac{1}{2}O_{2(g)} = Cu_2O_{(s)}$$

$$\Delta G^\circ = -162,200 + 69.24\,T \text{ J}$$

in the range 298–1356 K. Thus, at the melting temperature of Cu, 1356 K,

$$\Delta G_{1356\,K}^\circ = -162,200 + 69.24 \times 1356$$
$$= -68,310 \text{ J}$$

Table A-1 The standard Gibbs free energy changes for several reactions (standard states are noted by subscript)

Reaction	$\Delta G°$, joules	Range, K
$2Ag_{(s)} + \frac{1}{2}O_{2(g)} = Ag_2O_{(s)}$	$-30{,}540 + 66.11\,T$	298–463
$Al_{(l)} = [Al]_{(1\ wt\%\ in\ Fe)}$	$-43{,}100 - 32.26\,T$	
$2Al_{(l)} + 1.5O_{2(g)} = Al_2O_{3(s)}$	$-1{,}687{,}200 + 326.8\,T$	993–2327
$C_{(s)} + \frac{1}{2}O_{2(g)} = CO_{(g)}$	$-111{,}700 - 87.65\,T$	298–2000
$C_{(s)} + O_{2(g)} = CO_{2(g)}$	$-394{,}100 - 0.84\,T$	298–2000
$C_{(s)} + \frac{1}{2}O_{2(g)} + \frac{1}{2}S_{2(g)} = COS_{(g)}$	$-202{,}800 - 9.96\,T$	773–2000
$C_{(gr)} + 2H_{2(g)} = CH_{4(g)}$	$-91{,}040 + 110.7\,T$	773–2000
$C_{(gr)} = [C]_{(1\ wt\%\ in\ Fe)}$	$22{,}600 - 42.26\,T$	
$CaO_{(s)} + CO_{2(g)} = CaCO_{3(s)}$	$-168{,}400 + 144\,T$	449–1150
$2CaO_{(s)} + SiO_{2(s)} = 2CaO \cdot SiO_{2(s)}$	$-118{,}800 - 11.30\,T$	298–2400
$CoO_{(s)} + SO_{3(g)} = CoSO_{4(s)}$	$-227{,}860 + 165.3\,T$	298–1230
$2Cr_{(s)} + 1.5O_{2(g)} = Cr_2O_{3(s)}$	$-1{,}110{,}100 + 247.3\,T$	298–1793
$2Cu_{(s)} + \frac{1}{2}O_{2(g)} = Cu_2O_{(s)}$	$-162{,}200 + 69.24\,T$	298–1356
$2Cu_{(l)} + \frac{1}{2}O_{2(g)} = Cu_2O_{(s)}$	$-188{,}300 + 88.48\,T$	1356–1509
$2Cu_{(s)} + \frac{1}{2}S_{2(g)} = Cu_2S_{(s)}$	$-131{,}800 + 30.79\,T$	708–1356
$3Fe_{(\alpha)} + C_{(gr)} = Fe_3C_{(s)}$	$29{,}040 - 28.03\,T$	298–1000
$3Fe_{(\gamma)} + C_{(gr)} = Fe_3C_{(s)}$	$11{,}234 - 11.00\,T$	1000–1137
$Fe_{(s)} + \frac{1}{2}O_{2(g)} = FeO_{(s)}$	$-263{,}700 + 64.35\,T$	298–1644
$Fe_{(l)} + \frac{1}{2}O_{2(g)} = FeO_{(l)}$	$-256{,}000 + 53.68\,T$	1808–2000
$3Fe_{(s)} + 2O_{2(g)} = Fe_3O_{4(s)}$	$-1{,}102{,}200 + 307.4\,T$	298–1808
$Fe_{(s)} + \frac{1}{2}S_{2(g)} = FeS_{(s)}$	$-150{,}200 + 52.55\,T$	412–1179
$H_{2(g)} + Cl_{2(g)} = 2HCl_{(g)}$	$-188{,}200 - 12.80\,T$	298–2000
$H_{2(g)} + I_{2(g)} = 2HI_{(g)}$	$-8370 - 17.65\,T$	298–2000
$H_{2(g)} + \frac{1}{2}O_{2(g)} = H_2O_{(g)}$	$-247{,}500 + 55.85\,T$	298–2000
$Hg_{(v)} + \frac{1}{2}O_{2(g)} = HgO_{(s)}$	$-152{,}200 + 207.2\,T$	
$Li_{(g)} + \frac{1}{2}Br_{2(g)} = LiBr_{(g)}$	$-333{,}900 + 42.09\,T$	1289–2000
$Mg_{(l)} + Cl_{2(g)} = MgCl_{2(l)}$	$-603{,}200 + 121.43\,T$	987–1368
$Mg_{(g)} + \frac{1}{2}O_{2(g)} = MgO_{(s)}$	$-729{,}600 + 204\,T$	1363–2200
$2MgO_{(s)} + SiO_{2(s)} = Mg_2SiO_{4(s)}$	$-67{,}200 + 4.31\,T$	298–2171
$MgO_{(s)} + CO_{2(g)} = MgCO_{3(s)}$	$-117{,}600 + 170\,T$	298–1000
$MgO_{(s)} + Al_2O_{3(s)} = MgO \cdot Al_2O_{3(s)}$	$-35{,}560 - 2.09\,T$	298–1698
$Mn_{(s)} + \frac{1}{2}O_{2(g)} = MnO_{(s)}$	$-388{,}900 + 76.32\,T$	298–1517
$N_{2(g)} + 3H_{2(g)} = 2NH_{3(g)}$	$-87{,}030 + 25.8\,T\ln T + 31.7\,T$	298–2000
$2Ni_{(s)} + O_{2(g)} = 2NiO_{(s)}$	$-471{,}200 + 172\,T$	298–1726
$2Ni_{(l)} + O_{2(g)} = 2NiO_{(s)}$	$-506{,}180 + 192.2\,T$	1726–2200
$\frac{1}{2}O_{2(g)} = [O]_{(1\ wt\%\ in\ Fe)}$	$-111{,}070 - 5.87\,T$	
$Pb_{(l)} + \frac{1}{2}O_{2(g)} = PbO_{(s)}$	$-208{,}700 + 91.75\,T$	600–1158
$Pb_{(l)} + \frac{1}{2}O_{2(g)} = PbO_{(l)}$	$-181{,}200 + 68.03\,T$	1158–1808
$Pb_{(l)} + \frac{1}{2}S_{2(g)} = PbS_{(s)}$	$-163{,}200 + 88.03\,T$	600–1386
$PbO_{(s)} + SO_{2(g)} + \frac{1}{2}O_{2(g)} = PbSO_{4(s)}$	$-401{,}200 + 261.5\,T$	298–1158
$PCl_{3(g)} + Cl_{2(g)} = PCl_{5(g)}$	$-95{,}600 - 7.94\,T\ln T + 235.2\,T$	298–1000

Table A-1 Continued

Reaction	$\Delta G°$, joules	Range, K
$\frac{1}{2}S_{2(g)} + O_{2(g)} = SO_{2(g)}$	$-361{,}700 + 76.68\,T$	718–2000
$Si_{(s)} + O_{2(g)} = SiO_{2(s)}$	$-907{,}100 + 175\,T$	298–1685
$3Si_{(s)} + 2N_{2(g)} = Si_3N_{4(s)}$	$-723{,}800 + 315.1\,T$	298–1685
$Sn_{(l)} + Cl_{2(g)} = SnCl_{2(l)}$	$-333{,}000 + 118.4\,T$	520–925
$SO_{2(g)} + \frac{1}{2}O_{2(g)} = SO_{3(g)}$	$-94{,}600 + 89.37\,T$	298–2000
$U_{(l)} + C_{(gr)} = UC_{(s)}$	$-102{,}900 + 5.02\,T$	1408–2500
$2U_{(l)} + 3C_{(gr)} = U_2C_{3(s)}$	$-236{,}800 + 25.1\,T$	1408–2500
$U_{(l)} + 2C_{(gr)} = UC_{2(s)}$	$-115{,}900 + 10.9\,T$	1408–2500
$V_{(s)} + \frac{1}{2}O_{2(g)} = VO_{(s)}$	$-424{,}700 + 80.04\,T$	298–2000
$Zn_{(v)} + \frac{1}{2}O_{2(g)} = ZnO_{(s)}$	$-460{,}200 + 198\,T$	1243–1973

For the oxidation of liquid copper to form solid cuprous oxide according to

$$2Cu_{(l)} + \frac{1}{2}O_{2(g)} = Cu_2O_{(s)}$$

$$\Delta G_T^\circ = -188{,}300 + 88.48T\ \text{J}$$

in the range 1356–1509 K. Thus, at 1356 K,

$$\Delta G_{1356\,K}^\circ = -188{,}300 + 88.48 \times 1356\ \text{J}$$
$$= -68{,}310\ \text{J}$$

TABLE A-2: THE CONSTANT-PRESSURE MOLAR HEAT CAPACITIES OF VARIOUS SUBSTANCES

The constant-pressure molar heat capacities are presented as

$$c_p = a + bT + cT^{-2}\ \text{J/K}$$

or as

$$c_p = a + bT + cT^{-2} + dT^2\ \text{J/K}$$

and the table includes the ranges of temperature in which the expressions are valid.
    Example—For Ag in the range 298–1234 K

$$c_p = 21.30 + 8.54 \times 10^{-3}T + 1.51 \times 10^5 T^{-2}\ \text{J/K}$$

and for graphite in the range 298–1100 K

$$c_p = 0.11 + 38.94 \times 10^{-3}T - 1.48 \times 10^5 T^{-2} - 17.38 \times 10^{-6}T^2\ \text{J/K}$$

Table A-2 The constant-pressure molar heat capacities of various substances ($c_p = a + bT + cT^{-2}$ J/mole·K)

Substance	a	$b \times 10^3$	$c \times 10^{-5}$	Range, K	Remarks
Ag	21.30	8.54	1.51	298–1234 (T_m)	
$Ag_{(l)}$	30.50	—	—	1234–1600	
$Al_{(s)}$	20.67	12.38	—	298–937(T_m)	
$Al_{(l)}$	31.76	—	—	934–1600	
Al_2O_3	106.6	17.78	−28.53	298−2325(T_m)	
Ba(α)	−473.2	1587.0	128.2	298–648	
$Ba_{(\beta)}$	−5.69	80.33	—	648–1003	
BaO	53.30	4.35	−8.30	298–2286(T_m)	
$BaTiO_3$	121.46	8.54	−19.16	298–1800	
$C_{(graphite)}$	0.11	38.94	−1.48	298–1100	$-17.38 \times 10^{-6}T^2$
$C_{(graphite)}$	24.43	0.44	−31.63	1100–4000	
$C_{(diamond)}$	9.12	13.22	−6.19	298–1200	
CO	28.41	4.10	−0.46	298–2500	
CO_2	44.14	9.04	−8.54	298–2500	
$Ca_{(\alpha)}$	25.37	−7.26	—	298–716	$23.72 \times 10^{-6}T^2$
$Ca_{(\beta)}$	−0.36	41.25	—	716–1115	
CaO	49.62	4.51	−6.95	298–1177	
$CaTiO_3$	127.49	5.69	−27.99	298–1530	
$Cr_{(s)}$	24.43	9.87	−3.68	298–2130(T_m)	
Cr_2O_3	119.37	9.30	−15.65	298–1800	
$Cu_{(s)}$	22.64	6.28	—	298–1356(T_m)	
$Fe_{(\alpha/\delta)}$	37.12	6.17	—	298–1183/1664–1809	
$Fe_{(\gamma)}$	24.47	8.45	—	1187–1664	
$Fe_{(l)}$	41.8	—	—	1809–1873	
$H_2O_{(g)}$	30.00	10.71	0.33	298–2500	
$O_{2(g)}$	29.96	4.18	−1.67	298–3000	
$2MgO \cdot 2Al_2O_3 \cdot 5SiO_2$	626.34	91.21	−200.83	298–1738(T_m)	
N_2	27.87	4.27	—	298–2500	
Si_3N_4	70.54	98.74	—	298–900	
$SiO_{2(\alpha\text{-quartz})}$	43.89	1.00	−6.02	298–847	
Ti	22.09	10.46	—	298–1155	
$TiO_{2(rutile)}$	75.19	1.17	−18.20	298–1800	
$Zr_{(\alpha)}$	21.97	11.63	—	298–1136	
$Zr_{(\beta)}$	23.22	4.64	—	1136–2128	
$ZrO_{2(\alpha)}$	69.62	7.53	−14.06	298–1478	
$ZrO_{2(\beta)}$	74.48	—	—	1478–2950(T_m)	

TABLE A-3: THE STANDARD MOLAR HEATS OF FORMATION AND MOLAR ENTROPIES OF VARIOUS SUBSTANCES AT 298 K

Example—For the reaction

$$2Al_{(s)} + \frac{3}{2}O_{2(g)} = Al_2O_{3(s)}$$

$$\Delta H^\circ_{298\ K} = -1,675,700\ J$$

Table A-3 The standard molar heats of formation and molar entropies of various substances at 298 K

Substance	ΔH°_{298}, J	S°_{298}, J/K
Al_2O_3	$-1,675,700$	50.9
Ba	—	62.4
BaO	$-548,100$	72.1
$BaTiO_3$	$-1,653,100$	107.9
$C_{(graphite)}$	—	5.73
$C_{(diamond)}$	1,900	2.43
CH_4	$-74,800$	186.3
CO	$-110,500$	197.5
CO_2	$-393,500$	213.7
Ca	—	41.6
CaO	$-634,900$	38.1
$CaTiO_3$	$-1,660,600$	93.7
$3CaO{\cdot}Al_2O_3{\cdot}3SiO_2$	$-6,646,300$	241.4
$CaO{\cdot}Al_2O_3{\cdot}SiO_2$	$-3,293,200$	144.8
$CaO{\cdot}Al_3O_3{\cdot}2SiO_2$	$-4,223,700$	202.5
$2CaO{\cdot}Al_2O_3{\cdot}SiO_2$	$-3,989,400$	198.3
Cr_2O_3	$-1,134,700$	81.2
$H_2O_{(g)}$	$-241,800$	232.9
N_2	—	191.5
O_2	—	205.1
$SiO_{2,(\alpha\text{-quartz})}$	$-910,900$	41.5
Si_3N_4	$-744,800$	113.0
Ti	—	30.7
TiO	$-543,000$	34.7
Ti_2O_3	$-1,521,000$	77.2
Ti_3O_5	$-2,459,000$	129.4
TiO_2	$-944,000$	50.6
Zr	—	39.0
ZrO_2	$-1,100,800$	50.4

which is thus the standard molar heat of formation of Al_2O_3 at 298 K. The molar entropy of Al_2O_3 at 298 K is 50.9 J/K. By convention, the standard molar enthalpies of elements in their standard states at 298 K are assigned the value of zero.

TABLE A-4: THE SATURATED VAPOR PRESSURES OF VARIOUS SUBSTANCES

The saturated (equilibrium) vapor pressures of substances, in the stated ranges of temperatures, are presented in the form

$$\ln p \text{ (atm)} = -\frac{A}{T} + B \ln T + C$$

Example—The saturated vapor pressure exerted by liquid CaF_2 in the range of temperature $1691-2783$ K is given by

$$\ln p \text{ (atm)} = -\frac{50,200}{T} - 4.525 \ln T + 53.96$$

Thus, at its normal boiling temperature of 2783 K, the saturated vapor pressure of liquid CaF_2 is

$$\ln p \text{ (atm)} = -\frac{50,200}{2783} - 4.525 \ln 2783 + 53.96$$

$$= 0$$

i.e., at the normal boiling temperature, the saturated vapor pressure is 1 atm.

Table A-4 The saturated vapor pressures of various substances $\left[\ln p \text{ (atm)} = -\frac{A}{T} + B \ln T + C\right]$

Substance	A	B	C	Range, K
$CaF_{2(\alpha)}$	54,350	−4.525	56.57	298–1430
$CaF_{2(\beta)}$	53,780	−4.525	56.08	1430–1691 (T_m)
$CaF_{2(l)}$	50,200	−4.525	53.96	1691–2783 (T_b)
$Fe_{(l)}$	45,390	−1.27	23.93	1809 (T_m)–3330 (T_b)
$Hg_{(l)}$	7611	−0.795	17.168	298–630 (T_b)
$Mn_{(l)}$	33,440	−3.02	37.68	1517 (T_m)–2348 (T_b)
$SiCl_{4(l)}$	3620	—	10.96	273–333 (T_b)
$Zn_{(l)}$	15,250	−1.255	21.79	693 (T_m)–1177 (T_b)

TABLE A-5: MOLAR HEATS OF MELTING AND TRANSFORMATION

Example—At the melting temperature of 1234 K the enthalpy change for

$$Ag_{(s)} \rightarrow Ag_{(l)}$$

is 11,090 J. Thus at 1234 K the molar heat of melting of Ag is 11,090 J. The change in molar entropy due to melting at 1234 K is thus

$$\frac{\Delta H_m}{T_m} = \frac{11,090}{1234} = 8.987 \text{ J/K}$$

Table A-5 Molar heats of melting and transformation

Substance	Trans	ΔH_{trans}, J	T_{trans}, K
Ag	$s \rightarrow l$	11,090	1234
Al	$s \rightarrow l$	10,700	934
Al_2O_3	$s \rightarrow l$	107,500	2324
Au	$s \rightarrow l$	12,600	1338
Ba	$\alpha \rightarrow \beta$	630	648
Ba	$\beta \rightarrow l$	7650	1003
Cu	$s \rightarrow l$	12,970	1356
Ca	$\alpha \rightarrow \beta$	900	716
CaF_2	$s \rightarrow l$	31,200	1691
Fe	$\alpha \rightarrow \gamma$	670	1187
Fe	$\gamma \rightarrow \delta$	840	1664
Fe	$\delta \rightarrow l$	13,770	1809
H_2O	$s \rightarrow l$	6008	273
$K_2O \cdot B_2O_3$	$s \rightarrow l$	62,800	1220
MgF_2	$s \rightarrow l$	58,160	1563
$Na_2O \cdot B_2O_3$	$s \rightarrow l$	67,000	1240
Pb	$s \rightarrow l$	4810	600
PbO	$s \rightarrow l$	27,480	1158
Si	$s \rightarrow l$	50,200	1658
V	$s \rightarrow l$	22,840	2193
Zr	$\alpha \rightarrow \beta$	3900	1136
ZrO_2	$\alpha \rightarrow \beta$	5900	1478

At 1187 K the enthalphy change for the transformation

$$Fe_{(\alpha)} \rightarrow Fe_{(\gamma)}$$

is 670 J. The corresponding change in the molar entropy at 1187 K is thus

$$\Delta S_{trans} = \frac{\Delta H_{trans}}{T_{trans}} = \frac{670}{1187} = 0.56 \text{ J/K}$$

Appendix B

EXACT DIFFERENTIAL EQUATIONS

Consider an initial state defined by the parameters x, y, z and a final state at an infinitesimal distance $x + dx$, $y + dy$, $z + dz$. Movement from the initial to the final state causes a change in the volume of the system dV given by

$$dV = V(x + dx, y + dy, z + dz) - V(x, y, z) \tag{B.1}$$

which corresponds to

$$dV = \left(\frac{\partial V}{\partial x}\right)_{yz} dx + \left(\frac{\partial V}{\partial y}\right)_{xz} dy + \left(\frac{\partial V}{\partial z}\right)_{xy} dz \tag{B.2}$$

Each function in parentheses in Eq. (B.2) is a partial derivative of the function $V(x,y,z)$ with respect to one of the variables, that is, the derivative of V with respect to one variable at constant values of the other two variables. The differential dV, which is the sum of the partial derivatives, is called an *exact differential*.

Consider the function

$$V(x,y,z) = x^2y^3 + xz$$

Its partial differentials are

$$\left(\frac{\partial V}{\partial x}\right)_{yx} = 2xy^3 + z; \qquad \left(\frac{\partial V}{\partial y}\right)_{xz} = 3x^2y^2; \qquad \left(\frac{\partial V}{\partial z}\right)_{xy} = x$$

and the exact differential is

$$dV = (2xy^3 + z)dx + 3x^2y^2dy + xdz \tag{B.3}$$

The exact differential given by Eq. (B.2) has the properties

$$\left(\frac{\partial^2 V}{\partial x \partial y}\right) = \left(\frac{\partial^2 V}{\partial y \partial x}\right); \quad \left(\frac{\partial^2 V}{\partial y \partial z}\right) = \left(\frac{\partial^2 V}{\partial z \partial y}\right); \quad \left(\frac{\partial^2 V}{\partial z \partial x}\right) = \left(\frac{\partial^2 V}{\partial x \partial z}\right) \quad \text{(B.4)}$$

Conversely, the differential

$$dV = Xdx + Ydy + Zdz \quad \text{(B.5)}$$

is exact if there exists a function $V(x,y,z)$ for which

$$X = \left(\frac{\partial V}{\partial x}\right)_{yz}; \quad Y = \left(\frac{\partial V}{\partial y}\right)_{xz}; \quad Z = \left(\frac{\partial V}{\partial z}\right)_{xy} \quad \text{(B.6)}$$

and thus, from Eq. (B.4), it follows that

$$\frac{\partial X}{\partial y} = \frac{\partial Y}{\partial x}; \quad \frac{\partial Y}{\partial z} = \frac{\partial Z}{\partial y}; \quad \frac{\partial Z}{\partial x} = \frac{\partial X}{\partial z} \quad \text{(B.7)}$$

The relationships given by Eq. (B.7) are a necessary and sufficient condition for Eq. (B.5) to be an exact differential. The application of Eq. (B.7) shows Eq. (B.3) to be an exact differential as follows.

$$\frac{\partial(2xy^3 + z)}{\partial y} = 6xy^2; \quad \frac{\partial(3x^2y^2)}{\partial x} = 6xy^2$$

$$\frac{\partial(3x^2y^2)}{\partial z} = 0; \quad \frac{\partial(x)}{\partial y} = 0$$

$$\frac{\partial(x)}{\partial x} = 1; \quad \frac{\partial(2xy^3 + z)}{\partial z} = 1$$

THE GENERATION OF AUXILIARY FUNCTIONS AS LEGENDRE TRANSFORMATIONS

The variation of the dependent variable y with the independent variable x can be represented as the locus of points satisfying the relation $y = y(x)$, as shown in Fig. C.1, or as the envelope of a family of tangent lines shown in Fig. C.2. In Figure C.1 every point in the plane is described by two numbers x and y and every line in the plane in Fig. C.2 can be described by two numbers m and ψ, where m is the slope of the line and ψ is its intercept with the y-axis. Then, just as the relation $y = y(x)$ selects a subset of all possible points (x,y), a relation $\psi = \psi(m)$ selects a subset of all possible lines (m,ψ). Knowledge of the intercept ψ of the tangent lines as a function of the slopes m allows the construction of the family of tangent lines and thus the curve for which they are the envelope. Thus the relationship

$$\psi = \psi(m) \tag{C.1}$$

is equivalent to the relation $y = y(x)$, and in Eq. (C.1) m is the independent variable. The computation of the relation $\psi = \psi(m)$ from the known relation $y = y(x)$ is known as a Legendre transformation. Fig. C.3 shows a tangent line of slope m and going through the point x,y. If the intercept is ψ then

$$m = \frac{y - \psi}{x - 0} \tag{C.2}$$

or

$$\psi = y - mx \tag{C.3}$$

Differentiation of the known equation $y = y(x)$ gives $m = m(x)$ and elimination of x and y gives the desired relation ψ and m. The function ψ is known as the Legendre transform of y.

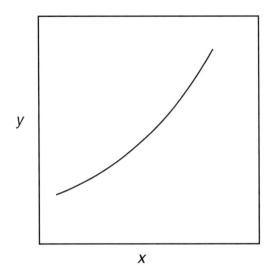

Figure C.1 The locus of points satisfying the relation $y = y(x)$.

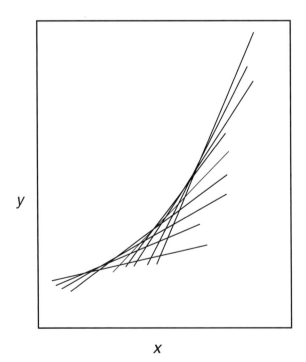

Figure C.2 The family of tangent lines $\psi = \psi(m)$.

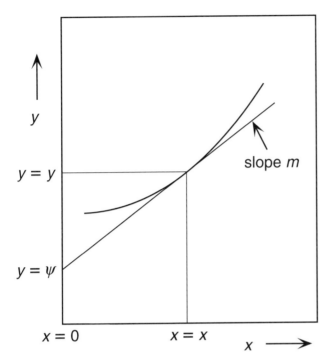

Figure C.3 Geometry procedure for obtaining ψ, the Legendre transform of y.

Eq. (3.12) gives the variation of the dependent variable U on the independent variables S and V for a closed system of fixed composition which is undergoing a process involving a change of volume against the external pressure as the only form of work performed on, or by, the system.

$$dU = TdS - PdV \qquad (3.12)$$

A schematic variation of U with V at constant S is shown in Fig. C.4. Using the geometry procedure outlined in Figure C.3 gives

$$m = \frac{U - \psi}{V - 0}$$

which rearranges to give

$$\psi = U - mV \qquad (C.4)$$

From Eq. (3.12) the slope of the line, m, is $(dU/dV)_s = -P$ and thus Eq. (C.4) becomes

$$\psi = U + PV$$

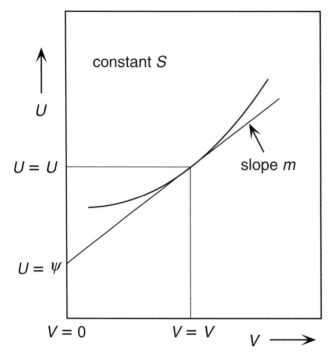

Figure C.4 Geometry procedure for obtaining ψ, the Legendre transform of U at constant S.

which is called the enthalpy H. The enthalpy H is thus a Legendre transform of the internal energy U, given as

$$H = U + PV \tag{C.5}$$

Differentiation of Eq. (C.5) gives

$$dH = dU + PdV + VdP$$

which in combination with Eq. (3.12) gives

$$dH = TdS + VdP \tag{C.6}$$

which is an expression for the variation of H, as the dependent variable, with the independent variables S and P. Comparison of Eq. (C.6) with

$$dH = \left(\frac{\partial H}{\partial S}\right)_P dS + \left(\frac{\partial H}{\partial P}\right)_S dP$$

gives the thermodynamic definitions of T and V as

$$T = \left(\frac{\partial H}{\partial S}\right)_P$$

and

$$V = \left(\frac{\partial H}{\partial P}\right)_S$$

Fig. C.5 shows a schematic variation of U with S at constant V and the geometry procedure gives

$$m = \frac{U - \psi}{S - 0}$$

or

$$\psi = U - mS \qquad (C.7)$$

From Eq. (3.12) the slope of the line, m, is $(dU/dS)_V = T$ and thus Eq. (C.7) becomes

$$\psi = U - TS \qquad (C.8)$$

which is called the Helmotz free energy A. The Helmholtz free energy is thus another Legendre transform of the internal energy U, given as

$$A = U - TS \qquad (C.9)$$

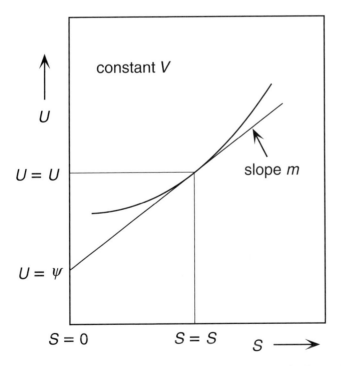

Figure C.5 Geometry procedure for obtaining ψ, the Legendre transform of U at constant V.

Differentiation of Eq. (C.9) gives

$$dA = dU - TdS - SdT$$

which, in combination with Eq. (3.12) gives

$$dA = -SdT - PdV \tag{C.10}$$

which is an expression for the variation of A, as the dependent variable, with the independent variables T and V. Comparison of Eq. (C.10) with

$$dA = \left(\frac{\partial A}{\partial T}\right)_V dT + \left(\frac{\partial A}{\partial V}\right)_T dV$$

gives the thermodynamic definition of S and P as

$$S = -\left(\frac{\partial A}{\partial T}\right)_V$$

and

$$P = -\left(\frac{\partial A}{\partial V}\right)_T$$

The thermodynamic variable which is dependent on T and P is obtained as a Legendre transform of A in which V in Eq. (C.10) is replaced by P. The variation of A with V at constant T gives

$$m = \left(\frac{A - \psi}{V - 0}\right)$$

or

$$\psi = A - mV \tag{C.11}$$

From Eq. (C.10), $m = (dA/dV)_T = -P$ and, thus, Eq. (C.11) becomes

$$\psi = A + PV = U - TS + PV = H - TS$$

which is called the Gibbs free energy G; that is,

$$G = H - TS \tag{C.12}$$

Differentiation of Eq. (C.12) gives

$$dG = dH - TdS - SdT$$

which, in combination with Eq. (C.6) gives

$$dG = -SdT + VdP \tag{C.13}$$

which is an expression for the variation of G, the dependent variable, on T and P as the independent variables. Comparison of Eq. (C.13) with

$$dG = \left(\frac{\partial G}{\partial T}\right)_P dT + \left(\frac{\partial G}{\partial P}\right)_T dP$$

gives the thermodynamic definitions of S and V as

$$S = -\left(\frac{\partial G}{\partial T}\right)_P$$

and

$$V = \left(\frac{\partial G}{\partial P}\right)_T$$

Nomenclature

LIST OF SYMBOLS

a	van der Waals constant
a_i	the activity of species i with reference to a specified standard state
A	Helmholtz free energy (or work function)
b	van der Waals constant
C	the number of components
C	heat capacity
c_p	constant pressure molar heat capacity
c_v	constant volume molar heat capacity
$\mathscr{E}$	electromotive force
$\mathscr{E}^{o,A}$	standard reduction potential of the species A
e_j^i	the interaction parameter of i on j
F	the number of degrees of freedom of an equilibrium
$\mathfrak{f}$	Faraday's constant
f	fugacity
f_i	the Henrian activity coefficient of the species i
$f_{i(wt\%)}$	the activity coefficient of the species i with respect to the 1 weight percent standard state
f_j^i	the interaction coefficient of i on j
G	Gibbs free energy
H	enthalpy
h_i	the Henrian activity of the species i
$h_{i(wt\%)}$	the activity of the species i with respect to the 1 weight percent standard state

599

K	the equilibrium constant
k	Boltzmann's constant
m	mass
n	the number of moles
n_i	the number of moles of the species i
N_O	Avogadro's number
P	pressure
P	partition function
P	the number of phases occurring in a system
p_i	the partial pressure of the species i
p_i^o	the saturated vapor pressure of the species i
q	heat
R	the Gas constant
S	entropy
T	temperature
T_m	melting temperature
T_b	boiling temperature
U	internal energy
V	volume
w	work
X_i	the mole fraction of the species i
Z	the compressibility factor
α	coefficient of thermal expansion
α	the regular solution constant
β	coefficient of isothermal compressibility
γ	ratio of c_p to c_v
γ_i	the activity coefficient of the species i
γ_i^o	the Henry's law constant
ε_i	the energy of the ith energy level
ε_j^i	the interaction parameter of i on j
μ_i	the chemical potential of the species i
(s)	solid
(l)	liquid
(g)	gas

NOTATION FOR EXTENSIVE THERMODYNAMIC PROPERTIES
(Exemplified by G, the Gibbs Free Energy)

G'	the Gibbs free energy of the system containing n moles
G	the Gibbs free energy per mole of the system
ΔG	the change in G due to a specified change in the state of the system
ΔG^M	the integral molar Gibbs free energy change due to mixing of the components to form a solution
$\Delta G^{M,\mathrm{id}}$	the integral molar Gibbs free energy change due to mixing of the components to form an ideal solution

G_i	the molar Gibbs free energy of the species i
G_i^o	the molar Gibbs free energy of the species i in its designated standard state
$\bar{G}_i$	the partial molar Gibbs free energy of i in some specified solution
ΔG_i^M	$= \bar{G}_i - G_i^o$, the partial molar Gibbs free energy of mixing of i
G^{xs}	$= \Delta G^M - \Delta G^{M,id}$, the integral excess molar Gibbs free energy of a solution
$\bar{G}_i^{xs}$	the partial molar excess Gibbs free energy of mixing of i
ΔG_m	the molar Gibbs free energy of melting
ΔG_b	the molar Gibbs free energy of boiling

VALUES OF SELECTED PHYSICAL CONSTANTS

Absolute temperature of the ice point (0°C)	= 273.15 K
Absolute temperature of the triple point of H_2O (by definition)	= 273.16000 K
Faraday's constant	f = 96,487 coulomb/mole
Avogadro's number	$N_o = 6.0232 \times 10^{23}$/gram·mole
Boltzmann's constant	$k = 1.38054 \times 10^{-23}$ joules/degree
Atmosphere	1 atm = 1.01325 bar
	= 101.325 kPa
	= 760 mm Hg
Gas constant	R = 8.3144 joules/degree·mole
	= 82.06 cm^3·atm/degree·mole

Answers

Chapter Two

2.1 (1a) 22.5 liters, (2a) $w = 9244$ J, (3a) $q = 9244$ J, (4a) $\Delta U = 0$, (5a) $\Delta H = 0$, (1b) 19.13 liters, (2b) $w = 5130$ J, (3b) $q = 0$, (4b) $\Delta U = -5130$ J, (5b) $\Delta H = -8549$ J

2.2 (a) $w = 2270$ J, $q = 5675$ J; (b) $w = 0$, $q = 6809$ J; (c) $w = -3278$ J, $q = -13{,}492$ J

2.3 $V = 1.52$ liters, $w = 8.7$ J

2.4 $T = 1620$ K

2.5 (a) $P = 1$ atm, V $= 30.61$ liters, $T = 373$ K; (b) $\Delta U = 2168$ J, $\Delta H = 3000$ J; (c) $c_v = 21.7$ J/mole·K, $c_p = 30$ J/mole·K

2.6 (a) $+123.4$ kJ, (b) -22.5 kJ, (c) 0. Total work $= 100.9$ kJ done by the system.

2.7 $P = 0.3$ atm

2.8 7958 J, $T_{max} = 1116$ K, $T_{min} = 255$ K

Chapter Three

3.1 (a) 5.76 J/K, (b) 0 J/K, (c) -8.65 J/K

3.2

    a $\Delta U = \Delta H = q = w = 0$, $\Delta S = 9.13$ J/K
    b $\Delta U = q = 1247$ J, $\Delta H = 2079$ J, $w = 0$, $\Delta S = 3.59$ J/K
    c $\Delta U = \Delta H = 0$, $q = w = 3654$ J, $\Delta S = 9.13$ J/K
    d $\Delta U = -1247$ J, $\Delta H = q = -2079$ J, $w = -831$ J, $\Delta S = 5.98$ J/K

    totals: $\Delta U = \Delta H = 0$, $w = q = 2322$ J, $\Delta S = 15.88$ J/K

3.3 $T_1 = 300$ K, $T_2 = 600$ K. The isothermal expansion is conducted at 300 K.

3.4 $\Delta H = 42750$ J, $\Delta S = 59.7$ J/K

3.5 The final temperature is 323.32 K, which is greater than 323 K because the heat capacity increases with increasing temperature. Thus the decrease in temperature caused by withdrawing heat q from hot copper is less than the increase in temperature caused by adding heat q to cold copper. The quantity of heat transferred is 1233 J and $\Delta S_{irr} = 0.6$ J/K.

3.6 $T_f = (T_1^{C_1} T_2^{C_2})^{\frac{1}{C_1 + C_2}}$, $w = q_2 - q_1 = [-C_2(T_f - T_2)] - [C_1(T_f - T_1)]$

Chapter Four

4.1 $R \ln 4$, $R \ln 8$, 0, $R \ln (32/27)$

4.2 The total number of distinguishable complexions is

$$\frac{(2n)!}{n!n!}$$

and, with n being a multiple of 4, the number of complexions in the most probable distribution is

$$\left[\frac{n!}{(0.5n)!(0.5n)!}\right]\left[\frac{n!}{(0.5n)!(0.5n)!}\right] \qquad \text{(ii)}$$

The ratio of (ii) to (i) decreases with increasing n.

4.3 $\Delta S_{conf} = 1.02$ J/K

4.4 65.0 grams

Chapter Five

5.1

$$dU = TdS - PdV$$

$$\therefore dS = \frac{dU}{T} + \frac{P}{T}dV$$

$$\therefore \left(\frac{\partial S}{\partial V}\right)_P = \frac{1}{T}\left(\frac{\partial U}{\partial V}\right)_P + \frac{P}{T}$$

$$U = H = PV$$

$$\therefore \left(\frac{\partial U}{\partial V}\right)_P = \left(\frac{\partial H}{\partial V}\right)_P - P$$

$$\therefore \left(\frac{\partial S}{\partial V}\right)_P = \frac{1}{T}\left(\frac{\partial H}{\partial V}\right)_P$$

$$dH = \left(\frac{\partial H}{\partial T}\right)_P dT + \left(\frac{\partial H}{\partial P}\right)_T dP$$

$$\therefore \left(\frac{\partial H}{\partial V}\right)_P = \left(\frac{\partial H}{\partial T}\right)_P \left(\frac{\partial T}{\partial V}\right)_P = \frac{c_p}{V\alpha}$$

$$\therefore \left(\frac{\partial S}{\partial V}\right)_P = \frac{c_p}{T\alpha V}$$

5.2

$$dU = TdS - PdV$$

$$\therefore dS = \frac{dU}{T} + \frac{P}{T}dV$$

$$\therefore \left(\frac{\partial S}{\partial P}\right)_V = \frac{1}{T}\left(\frac{\partial U}{\partial P}\right)_V$$

$$dU = \left(\frac{\partial U}{\partial T}\right)_V dT + \left(\frac{\partial U}{\partial V}\right)_T dV$$

$$\therefore \left(\frac{\partial U}{\partial P}\right)_V = \left(\frac{\partial U}{\partial T}\right)_V \left(\frac{\partial T}{\partial P}\right)_V$$

$$\left(\frac{\partial T}{\partial P}\right)_V = \frac{-\left(\frac{\partial V}{\partial P}\right)_T}{\left(\frac{\partial V}{\partial T}\right)_P} = \frac{\beta}{\alpha} \quad \text{and} \quad \left(\frac{\partial U}{\partial T}\right)_V = c_v$$

$$\therefore \left(\frac{\partial U}{\partial P}\right)_V = \frac{c_v\beta}{\alpha} \quad \text{and} \quad \left(\frac{\partial S}{\partial P}\right)_V = \frac{c_v\beta}{T\alpha}$$

$$c_p - c_v = \frac{VT\alpha^2}{\beta} \quad \therefore c_v = c_p - \frac{VT\alpha^2}{\beta}$$

$$\therefore \left(\frac{\partial S}{\partial P}\right)_V = \frac{c_p\beta}{T\alpha} - V\alpha$$

5.3

$$dA = \left(\frac{\partial A}{\partial T}\right)_V dT + \left(\frac{\partial A}{\partial V}\right)_T dV$$

$$\therefore \left(\frac{\partial A}{\partial P}\right)_V = \left(\frac{\partial A}{\partial T}\right)_V \left(\frac{\partial T}{\partial P}\right)_V$$

$$\left(\frac{\partial A}{\partial T}\right)_V = -S \quad \text{and} \quad \left(\frac{\partial T}{\partial P}\right)_V = \frac{\beta}{\alpha}$$

$$\therefore \left(\frac{\partial A}{\partial P}\right)_V = \frac{-S\beta}{\alpha}$$

5.4

$$dA = \left(\frac{\partial A}{\partial T}\right)_P dT + \left(\frac{\partial A}{\partial P}\right)_T dP$$

$$\therefore \left(\frac{\partial A}{\partial V}\right)_P = \left(\frac{\partial A}{\partial T}\right)_P \left(\frac{\partial T}{\partial V}\right)_P = \left(\frac{\partial A}{\partial T}\right)_P \frac{1}{\alpha V}$$

$$dA = -SdT - PdV$$

$$\therefore \left(\frac{\partial A}{\partial T}\right)_P = -S - P\left(\frac{\partial V}{\partial T}\right)_P$$

$$\therefore \left(\frac{\partial A}{\partial V}\right)_P = \left(-S - P\alpha V\right)\frac{1}{\alpha V}$$

$$= -\left(\frac{S}{\alpha V} + P\right)$$

5.5

$$dH = \left(\frac{\partial H}{\partial T}\right)_V dT = \left(\frac{\partial H}{\partial V}\right)_T dV$$

$$\therefore \left(\frac{\partial H}{\partial S}\right)_V = \left(\frac{\partial H}{\partial T}\right)_V \left(\frac{\partial T}{\partial S}\right)_V$$

$$TdS + \delta q_v = c_v dT \therefore \left(\frac{\partial T}{\partial S}\right)_V = \frac{T}{c_v}$$

$$H = U + PV$$

$$\therefore \left(\frac{\partial H}{\partial T}\right)_V = \left(\frac{\partial U}{\partial T}\right)_V + V\left(\frac{\partial P}{\partial T}\right)_V$$

$$= c_v + \frac{V\alpha}{\beta}$$

$$\therefore \left(\frac{\partial H}{\partial S}\right)_V = \frac{T}{c_v}\left[c_v + \frac{V\alpha}{\beta}\right] = T\left[1 + \frac{V\alpha}{c_v\beta}\right]$$

5.6

$$dH = TdS + VdP$$

$$\therefore \left(\frac{\partial H}{\partial V}\right)_S = V\left(\frac{\partial P}{\partial V}\right)_S$$

$$\left(\frac{\partial P}{\partial V}\right)_S = -\frac{\left(\frac{\partial S}{\partial V}\right)_P}{\left(\frac{\partial S}{\partial P}\right)_V}$$

From problem 5.1:

$$\left(\frac{\partial S}{\partial V}\right)_P = \frac{c_p}{TV\alpha}$$

From problem 5.2:

$$\left(\frac{\partial S}{\partial P}\right)_V = \frac{c_v\beta}{T\alpha}$$

$$\therefore \left(\frac{\partial P}{\partial V}\right)_S = -\frac{c_p}{Vc_v\beta}$$

$$\therefore \left(\frac{\partial H}{\partial V}\right)_S = -\frac{c_p}{c_v\beta}$$

5.7

$$c_p = \left(\frac{\partial H}{\partial T}\right)_P \therefore \left(\frac{\partial c_p}{\partial P}\right)_T = \left[\frac{\partial}{\partial P}\left(\frac{\partial H}{\partial T}\right)_P\right]_T = \left[\frac{\partial}{\partial T}\left(\frac{\partial H}{\partial P}\right)_T\right]_P$$

$$\left(\frac{\partial H}{\partial P}\right)_T = T\left(\frac{\partial S}{\partial P}\right)_T + V \quad \text{and} \quad \left(\frac{\partial S}{\partial P}\right)_T = -\left(\frac{\partial V}{\partial T}\right)_P$$

$$\therefore \left(\frac{\partial c_p}{\partial P}\right)_T = \frac{\partial}{\partial T}\left[-T\left(\frac{\partial V}{\partial T}\right)_P + V\right]_P$$

$$= \frac{\partial}{\partial T}\left[-T\alpha V + V\right]_P$$

$$= -\alpha V - TV\frac{d\alpha}{dT} - T\alpha^2 V + \alpha V$$

$$= -TV\left(\alpha^2 + \frac{d\alpha}{dT}\right)$$

5.8

$$\left(\frac{\partial S}{\partial T}\right)_P = \frac{c_p}{T} \quad \text{and} \quad \left(\frac{\partial S}{\partial P}\right)_T = -\left(\frac{\partial V}{\partial T}\right)_P = -\alpha V$$

$$\therefore \left(\frac{\partial T}{\partial P}\right)_S = -\frac{\left(\frac{\partial S}{\partial P}\right)_T}{\left(\frac{\partial S}{\partial T}\right)_P} = \frac{\alpha VT}{c_p}$$

5.9

$$\left(\frac{\partial P}{\partial V}\right)_S = -\frac{\left(\frac{\partial S}{\partial V}\right)_P}{\left(\frac{\partial S}{\partial P}\right)_V}$$

$$dS = \left(\frac{\partial S}{\partial T}\right)_V dT + \left(\frac{\partial S}{\partial V}\right)_T dV$$

$$\therefore \left(\frac{\partial S}{\partial P}\right)_V = \left(\frac{\partial S}{\partial T}\right)_V\left(\frac{\partial T}{\partial P}\right)_V = -\frac{c_v}{T}\frac{\left(\frac{\partial V}{\partial P}\right)_T}{\left(\frac{\partial V}{\partial T}\right)_P} = \frac{c_v}{T}\times\frac{\beta V}{\alpha V} = \frac{c_v\beta}{T\alpha}$$

$$dS = \left(\frac{\partial S}{\partial T}\right)_P dT + \left(\frac{\partial S}{\partial P}\right)_T dP$$

$$\therefore \left(\frac{\partial S}{\partial V}\right)_P = \left(\frac{\partial S}{\partial T}\right)_P\left(\frac{\partial T}{\partial V}\right)_P = \frac{c_p}{T}\frac{1}{\alpha V}$$

$$\therefore \left(\frac{\partial P}{\partial V}\right)_S = -\frac{c_p}{c_v V\beta}$$

5.10

$$dG = -SdT + VdP$$

$$\therefore \left(\frac{\partial G}{\partial P}\right)_T = V \quad \text{and} \quad \left(\frac{\partial^2 G}{\partial P^2}\right)_T = \left(\frac{\partial V}{\partial P}\right)_T$$

$$dA = -SdT - PdV$$

$$\therefore \left(\frac{\partial A}{\partial V}\right)_T = -P \quad \text{and} \quad \left(\frac{\partial^2 A}{\partial V^2}\right)_T = -\left(\frac{\partial P}{\partial V}\right)_T = -\frac{1}{\left(\frac{\partial^2 G}{\partial P^2}\right)_T}$$

5.11 The process is adiabatic. Therefore $q = 0$ and work done by the gas, $w = P_2V_2 - P_1V_1$. Thus $\Delta H = q - w + (P_2V_2 - P_1V_1) = 0$:

$$\mu_{J-T} = \left(\frac{\partial T}{\partial P}\right)_H$$

but

$$\left(\frac{\partial T}{\partial P}\right)_H = -\frac{\left(\frac{\partial H}{\partial P}\right)_T}{\left(\frac{\partial H}{\partial T}\right)_P} = -\frac{1}{c_p}\left(\frac{\partial H}{\partial P}\right)_T$$

and

$$\left(\frac{\partial H}{\partial P}\right)_T = T\left(\frac{\partial S}{\partial P}\right)_T + V = -T\left(\frac{\partial V}{\partial T}\right)_P + V = -T\alpha V + V$$

$$\therefore \mu_{J-T} = -\frac{1}{c_p}(-\alpha TV + V) = \frac{V}{c_p}(\alpha T - 1)$$

For an ideal gas

$$\alpha = \frac{1}{T}, \quad \text{so} \quad \alpha T = 1 \quad \text{and} \quad \mu_{J-T} = 0$$

5.12

    a 1 $\Delta U = \Delta H = 0$, $\Delta S = R \ln 4$, $\Delta A = \Delta G = -T \Delta S = -RT \ln 4$
        2 $\Delta U = \Delta H = 0$, $\Delta S = R \ln 8$, $\Delta A = \Delta G = -RT \ln 8$
        3 $\Delta U = \Delta H = \Delta S = \Delta A = \Delta H = 0$
        4 $\Delta U = \Delta H = 0$, $\Delta S = R \ln(32/27)$, $\Delta A = \Delta G = -RT \ln(32/27)$
    b $\Delta U = \Delta H = 0$, $\Delta S = R \ln(V_2/V_1) = R \ln 2$, $\Delta A = \Delta G = -RT \ln 2$
    c $\Delta U = c_v(T_2 - T_1)$, $\Delta H = c_p(T_2 - T_1)$, $\Delta S = 0$, $\Delta A = \Delta U - S(T_2 - T_1)$,
       $\Delta G = \Delta H - S(T_2 - T_1)$
    d $\Delta U = c_v(T_2 - T_1)$, $\Delta H = c_p(T_2 - T_1)$, $\Delta S = c_p \ln(T_2/T_1)$,
       $\Delta A = \Delta U - (T_2 - T_1)S_1 - T_2\Delta S$, $\Delta G = \Delta H - (T_2 - T_1)S_1 - T_2\Delta S$
    e $\Delta U = c_v(T_2 - T_1)$, $\Delta H = c_p(T_2 - T_1)$, $\Delta S = c_v \ln(T_2/T_1)$,
       $\Delta A = \Delta U - (T_2 - T_1)S_1 - T_2\Delta S$, $\Delta G = \Delta H - (T_2 - T_1)S_1 - T_2\Delta S$

Chapter Six

6.1 $\Delta H_{1600} = -1.086 \times 10^6$ J, $\Delta S_{1600} = -178.5$ J/K

6.2 $H_{\text{diamond, 1000}} - H_{\text{graphite, 1000}} = 1037$ J. Thus the oxidation of diamond at
        1000 K is 1037 J/mole more exothermic than the oxidation of graphite.

6.3 $\Delta H_{1000} = -80,500$ J, $\Delta S_{1000} = 6.6$ J/K

6.4   Increasing the pressure to 1000 atm increases the molar enthalpy by 707 J.
        This increase in molar enthalpy is achieved by increasing the temperature of
        Cu from 298 to 332 K at a pressure of 1 atm.

6.5   (a) $\Delta H = -435,000$ J, $\Delta S = -94.75$ J/K; (b) $\Delta H = -355,000$ J, $\Delta S = -75.35$ J/K; (c) $\Delta H = -373,000$ J, $\Delta S = -80.15$ J/K

6.6   15.1 kg

6.7   (a) 4745 K, (b) 2330 K

6.8 $\Delta G_{800} = -1.817 \times 10^6$ J. If it is assumed that Δc_p for the reaction is zero,
        $\Delta G_{800} = \Delta H_{298} - 800\Delta S$ is calculated as -1.811×10^6 J, which involves a
        0.3% error.

6.9 $a = 3$, $b = c = 2$, $\Delta H_{298} = 99,700$ J, $\Delta S_{298} = 125.8$ J/K, $\Delta G_{298} = 62,210$ J

6.10   1675 kJ

Chapter Seven

7.1 (a) The triple point for α-β-vapor is $T = 1163$ K, $p = 2.52 \times 10^{-10}$ atm, and the triple point for β-liquid-vapor is $T = 1689$ K, $p = 8.35 \times 10^{-5}$ atm. (b) $T_b = 2776$ K. (c) $\Delta H_{(\alpha \rightarrow \beta)} = 4739$ J, $\Delta H_m = 29{,}770$ J.

7.2 $p_{Hg,373 \text{ K}} = 3.55 \times 10^{-4}$ atm

7.3 Condensation begins at 328 K; at 280 K 82.5% of the $SiCl_4$ has condensed

7.4 Eq. (I) gives the vapor pressure of solid zinc.

7.5 $\Delta H_{b,Fe,3330 \text{ K}} = 342$ kJ

7.6 $p_{CO_2,298 \text{ K},(l)} = 73.3$ atm. The triple point pressure is 5.14 atm, and, as the 1 atm isobar does not pass through the liquid phase field, liquid CO_2 is not stable at atmospheric pressure.

7.7 $P = 2822$ atm

7.8 The slopes of the lines at the triple point are obtained from $dP/dT = \Delta S / \Delta V$

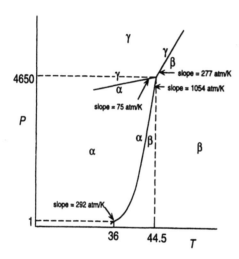

7.9 $T_b = 523$ K

Chapter Eight

8.1 The van der Waals equation containing the reduced variables is

$$\left(P_R + \frac{3}{V_R^2} \right)(3V_R - 1) = 8T_R$$

$Z_{cr} = 0.375$; $(\partial U / \partial V)_T = a/V^2$

8.2 $n_A/n_B = 1$, $P = 1.414$ atm

8.3 The tank contains 565 moles of van der Waals oxygen and 511 moles of ideal gas oxygen. As the gas is purchased by the tank-load, the same price purchases more moles of a van der Waals gas than it does an ideal gas.

8.4 $w = -1384$ J

8.5 (a) $b = 0.0567$ 1/mole, $a = 6.771^2 \cdot$atm/mole^2; (b) 0.170 1/mole; (c) P (van der Waals) $= 65.5$ atm, P (ideal gas) $= 82.1$ atm

8.6 With the virial equation $w = -301$ kJ, with the van der Waals equation $w = -309$ kJ, with the ideal gas law $w = -272$ kJ.

8.7 (a) $f = 688$ atm, (b) $P = 1083$ atm, (c) $\Delta G = 16{,}190$ J with an nonideal contribution of 790 J

Chapter Nine

9.1 $\Delta H = 117{,}400$ J, $\Delta S = 59.63$ J/K

9.2 $\gamma_{Mn} = 1.08$

9.3 (a) The average value of Ω is 4396 ± 6 J which indicates that, with respect to the behavior of G^{xs}, the solution is regular. (b) $\overline{G}_{Fe}^{xs} = 1583$ J and $\overline{G}_{Mn}^{xs} = 703$ J. (c) $\Delta G^M = -9.370$ J. (d) $p_{Mn} = 0.0118$ atm and $p_{Fe} = 3.68 \times 10^{-5}$ atm.

9.4 73,380 J

9.5 $\Omega = -4578$ J, $a_{Sn} = 0.418$

9.6 See Figs. 9.12, 9.13, 9.15, and 9.17

9.7 See Figs. 9.10, 9.11, 9.14, and 9.16

9.8 The temperature is increased by 2.37 degrees (K).

9.10 $\ln \gamma_{Cd} = 0.425 \, X_{Zn}^2 + 0.30 \, X_{Zn}^3$, $a_{Cd} = 0.577$

9.11 $a_{Au} = 0.695$, $a_{Ni} = 0.85$

Chapter Ten

10.1 $T = 1317$ K, $X_{CaF_2} = 0.53$

10.2 (a) $-11{,}140$ J, (b) zero

10.3 (a) 2418 K, (b) $X_{Al_2O_3} = 0.62$, (c) 2444 K, (d) $X_{Al_2O_3} = 0.38$

10.4 -814 J

10.5 $\Omega_l = 38{,}096$ J, $T_{cr} = 2291$ K

10.6 (a) $\Delta H_{m,Ge}^\circ$ from liquidus $= 21{,}527$ J, (b) $\Delta H_{m,Ge}^\circ$ from solidus $= 33{,}111$ J

10.7 The maximum solubility of CaO in MgO is $X_{CaO} = 0.066$, and the maximum solubility of MgO in CaO is $X_{MgO} = 0.15$.

Chapter Eleven

11.1 $X_{CO_2} = X_{H_2} = 0.182$, $X_{H_2O} = 0.0677$, $X_{CO} = 0.568$

11.2 43,800 J

11.3 $CO_2/H_2 = 1.276$

11.4 1771 K

11.5 $P_T = 0.192$ atm, $T = 792$ K

11.6 (a) $p_N = 5.94 \times 10^{-6}$ atm, (b) $P_T = 3.18 \times 10^{-9}$ atm

11.7 13.3 atm, $\Delta H_{573\,K}^\circ = -50{,}900$ J, $\Delta S_{573\,K}^\circ = -110.7$ J/K

11.8 $PCl_5/PCl_3 = 0.371$

11.9 At $P_T = 1$ atm, $p_{H_2} = 1.05 \times 10^{-8}$ atm, $p_{O_2} = 0.0756$ atm. At $P_T = 10$ atm, $P_{H_2} = 3.31 \times 10^{-8}$ atm, $p_{O_2} = 0.756$ atm.

11.10 $X_{H_2} = X_{I_2} = 0.165$, $X_{HI} = 0.669$, $T = 906$ K

Chapter Twelve

12.1 $T = 565$ K

12.2 $T_{m,Ni} = 1731$ K, $\Delta H^{\circ}_{m,Ni} = 17,490$ J, $\Delta S^{\circ}_{m,Ni} = 10.1$ J/K

12.3 (a) $T = 462$ K, (b) $T = 421$ K

12.4 $p_{H_2O} = 1.32 \times 10^{-3}$ atm, and the oxidation reaction is exothermic

12.5 Equilibrium would produce a gas containing 11.4% HCl, 46.6% H_2, and 42% Ar. Therefore equilibrium is not attained.

12.6 The FeO disappears

12.7 $p_{Mg} = 2.42 \times 10^{-2}$ atm

12.8 (a) $T = 1173$ K, (b) $p_{CO_2} = 0.055$ atm, (c) $p_{CO_2} = 1.23$ atm

12.9 $P = 1$ atm ($p_{SO_3} = 7.99 \times 10^{-2}$ atm, $p_{SO_2} = 0.612$ atm, $p_{O_2} = 0.306$ atm)

12.10 99.1% of the sulfur is removed, and p_{S_2} in the effluent gas is 6.3×10^{-11} atm

12.11 $\Delta G^{\circ} = 282,000 - 123T$ J

12.12 0.76 moles of CH_4 are consumed per mole of Fe produced

12.13 Eq. (i) for solid Mg, Eq. (ii) for gaseous Mg, Eq. (iii) for liquid Mg, $T_{m,Mg} = 930$ K, $T_{b,Mg} = 1372$ K

12.14 54.92 g of Zn are oxidized to form ZnO and 29.78 g of Zn are evaporated, which leaves 115.3 g of metallic Zn in the crucible

12.15 4.76 moles of $CaCO_3$ are decomposed per mole of CH_4 burned

12.16 $X_{Hg} = 0.0152$, $X_{O_2} = 0.0071$

12.17 $P_T = 1.651$ atm, $p_{CO} = 1.009$ atm, $p_{CO_2} = 0.642$ atm

Chapter Thirteen

13.1 $a_{Cu} = 0.159$

13.2 $a_{Mg} = 6.4 \times 10^{-4}$

13.3 $a_{PbO} = 0.5$

13.4 $X_{Cu} = 0.018$. Increasing T decreases the extent to which Cu is removed

13.5 $a_C = 0.5$, $p_{H_2} = 0.92$ atm

13.6 $a_{FeO} = 9.9 \times 10^{-5}$

13.7 (a) $p_{H_2}/p_{CO_2} = 2.15$, (b) $a_C = 0.194$, (c) $P_T = 5.16$ atm, (d) the total pressure does not influence p_{O_2}

13.8 With $C = 3$ and $P = 3$, $F = 2$, which are used by specifying $T = 1000$ K and $[X_{Mn}] = 0.001$. (X_{FeO}) $= 1.22 \times 10^{-3}$, $p_{O_2} = 2.33 \times 10^{-27}$ atm.

13.9 For $2A + B = A_2B$, $\Delta G^{\circ}_{1273\text{ K}} = -24,370$ J; for $A + 2B = A_2B$, $\Delta G^{\circ}_{1273\text{ K}} = -23,190$ J

13.10 (a) 10^{-3}, (b) 8.07×10^{-4}, (c) 7.14×10^{-3}, (d) 0.65

13.11 $\Delta G^{\circ} = -567,500$ J

13.12 $p_{O_2} = 5.17 \times 10^{-10}$ atm; $h_{Al(1\text{ wt\% in Fe})} = 7.2 \times 10^{-6}$; $C = 3$, $P = 4$, therefore $F = 1$ which is fixed by specifying $T = 1600°C$

13.13 $T_{max} = 2211$ K, $T_{min} = 1515$ K

13.14 $p_{Mg} = 0.053$ atm

13.15 $a_{MgO}(min) = 0.027$

13.16 $p_{CO} = 0.739$ atm, $p_{CO_2} = 0.0117$ atm, $p_{Zn} = 0.763$ atm

13.17 $a_{Al_2O_3} = 0.129$

13.18 (a) wt% Al = 0.00042, wt% = O = 0.0039; (b) wt% Al = 0.00054, wt% O = 0.0035

13.19 $T_{max} = 1108$ K

13.20 $P = 6917$ atm

13.21 Wustite $0.904 < p_{CO} < 3.196$ atm, cementite $7.43 < p_{CO} < 8.14$ atm

13.22 $T_{max} = 1026$ K

13.23 $a_{Zn} = 0.154$

13.24 [wt% O] = 1.9, $k(1273$ K$) = 2.0$

13.25 The formation of a manganese silicate melt as the deoxidation product decreases the activity of SiO_2 to a value less than unity and thus shifts the equilibrium [Si] + 2[O] = (SiO_2) to the right. For any given value of [wt% Si], the extent of deoxidation is maximized when the product of deoxidation is an MnO-saturated silicate melt in which a_{SiO_2} has its minimum value of 0.02.

Chapter Fifteen

15.1 $Pb_{(s)} + 2AgCl_{(s)} = 2Ag_{(s)} + PbCl_{2(s)}$, $\Delta G° = -94,560$ J, $\Delta S° = -35.5$ J/K

15.2 (a) $-103,400$ J, (b) 27.98 J/K, (c) 8338 J/mole of Pb, (d) $a_{Hg} = 0.71$

15.3 $\ln p_{O_2}$ (atm) $= \dfrac{46,620\mathscr{E}}{T} + \dfrac{63,440}{T} - 15.48$

15.4 $\Delta c_p = -0.093$ J/K

15.5 (a) $a_{Al} = 0.673$, (b) $\Delta \overline{G}^M_{Al} = -2150$ J, (c) $\Delta \overline{H}^M_{Al} = 3329$ J

15.6 (a) 1.62×10^{-6}, (b) 0.5, (c) 0.266 wt%, (d) 0.602 wt%

15.7 pH = 6.33, $\mathscr{E} = -0.126$ volts, $[Pb^{4+}] = 3.1 \times 10^{-62}$ moles/liter, $[HPbO_2^-] = 10^{-9}$ moles/liter, $[PbO_3^{2-}] = 2.5 \times 10^{-47}$ moles/liter, $[PbO_4^{4-}] = 6 \times 10^{-221}$ moles/liter

15.8 2.20 volts, 1.17 volts

15.9 $+6897$ J, -4513 J

Index